Emmanouil Spyrou

Powder Coatings

Chemistry and Technology

3rd Revised Edition

Emmanouil Spyrou: Powder Coatings: Chemistry and Technology

ISBN: 978-3-86630-884-8

Cover: Evonik Industries AG, Marl/Germany

Emmanouil Spyrou, 3rd Revised Edition
Based on
Pieter G. de Lange's, 2nd Edition, Vincentz Network, 2004, and
Tosko A. Misev's, 1st Edition, John Wiley and Sons, 1991
Powder Coatings: Chemistry and Technology
Hanover: Vincentz Network, 2012
European Coatings Tech Files
ISBN 3-86630-884-1
ISBN 978-3-86630-884-8

Please ask for our book catalogue
Vincentz Network, Plathnerstr. 4c, 30175 Hanover, Germany
T (202) 684-6630, F (202) 380-9129
E-mail: books@american-coatings.com, www.american-coatings.com

Layout: Vincentz Network, Hanover, Germany

ISBN 3-86630-884-1
ISBN 978-3-86630-884-8

European Coatings Tech Files

Emmanouil Spyrou

Powder Coatings

Chemistry and Technology

3rd Revised Edition

Emmanouil Spyrou: Powder Coatings: Chemistry and Technology

ISBN: 978-3-86630-884-8

Preface

"If I have seen a little further, it is by standing on the shoulders of giants." (Isaac Newton, 1676) The excellent original book "Powder Coatings, Chemistry and Technology" written by Tosko Aleksandar Misev and later updated by Peter Gillis de Lange has served for my generation as a sort of bible on its sector. Therefore, it has been a great honor while at the same time a tremendous challenge to write a third edition of this book. Due to this respect the original text is maintained where possible, but altered and extended if recent scientific findings made it necessary.

The Stone Age did not end due to a lack of stones. New technologies replace the old ones; novel developments change the look of the world, sometimes at an incredibly fast speed. Just to name a few recent incidents since the last print of the book: The rise of China, severe worldwide financial crises, the changed way of communicating (e.g. smart phones, social media) and radical governmental regulations (e.g. REACH). Of course, these developments had influences on the world of powder coatings, too. Consolidation, relocation and customization are some of the effects. Replacement of toxic ingredients (e.g. TGIC), use of energy saving curing methods and the extension of powder coatings into new application fields are the interconnected technological changes that are covered in this book.

This book is directed to anybody who is involved in developing, producing, testing and marketing of powder coatings, raw materials or application equipment.

What is new compared to the second edition? More than 5,000 recent articles and patents concerning powder coatings have been evaluated and 250 of those have been referenced in this book to ensure that it illustrates the current state-of-the-art. Highlighted core terms and a significant extended index should help finding the desired topic in a shorter time. A list of powder coating related web addresses will enable the reader to locate additional relevant information at the push of a button. Product and company names have been updated as much as possible. Plus more than 30 new photos, diagrams and drawings complete this revised and updated third edition.

I would like to thank Werner Grenda for valuable discussions, Dr. Corey King for corrections of the manuscript and Dr. Michael Ringel for his contribution regarding REACH. Many thanks to the two dozens companies for the excellent additional photos they provided for the 3rd edition of this book.

Emmanouil Spyrou

Marl, Germany, Mai 2012

Contents

1 Introduction

1.1 Historical background

Two thousand five hundred years ago the great Greek philosopher Thales of Miletus (624 to 556 BC), who was dubbed as the "father of science", was the first to discover that amber stone when rubbed attract other objects. The Greek word for amber, ελεκτρον (electron), is the origin of electrostatic forces, which are used nowadays for almost 90% of all powder coating applications.

The appearance of powder coatings is often associated with the ecological and energy related events of the late 1960's and early 1970's. The famous **Rule 66** which was brought in by The Town Council of Los Angeles in 1966 was the first legislative act regulating the environmental aspects of the coatings. Later on similar regulations were introduced in most of the industrially developed countries.

Although the history of powder coatings has been strongly influenced by environmental aspects, first developments in the field began in the 1940's with a simple flame spray application process. Early in 1950's powdered PVC was successfully applied by *Gemmer* in a fluidized bed process on a preheated metal surface [1]. A patent application for *Gemmer's* invention was filed in Germany in 1953 and the patent was issued in 1955. Very soon the fluidized bed technique for application of thermoplastic powders including polyethylene and nylon powder coatings was well established in the USA.

In the late 1950's the first thermosetting powder coatings appeared on the market, mainly as a result of the research work done by Shell Chemicals. The target was development of superior protective ("functional") organic coatings for the company's own underground natural gas and oil pipelines. The first systems were relatively simple physical **dry blends** of epoxy resins, hardeners and pigments dispersed by ball milling techniques. Due to a considerable degree of heterogeneity, the application results were rather inconsistent.

The hot melt mixing methods of the present day for production of powder coatings were preceded by a technique that employed liquid epoxy resins and hardeners. The homogeneous liquid binder/crosslinker blend was prereacted until partially cured ("B stage") solid material was obtained, which was finely ground in the next step. The completely cured "C stage" was obtained by stoving the "B stage" powders at high temperatures. A drawback of this technique was the lack of reproducibility and difficult control of the process [2].

Hot melt compounding on a heated twin roller mill or in a heated **Z-blade mixer** was already a step forward in the development of thermosetting powder coatings, but the immense cleaning problems, created by the fast(er) curing powder coatings, have almost completely excluded the Z-blade mixer and of course the twin roller mill from the machines (extruders) used to produce contemporary powder coatings. However, Z-blade mixers are still used for batch-wise production of thermoplastic powder coatings, where chemical reactivity does not play a role.

Extrusion methods for production of thermosetting powder coatings, which are in current use, were developed in the Shell Chemical Laboratories in England and The Netherlands in

Emmanouil Spyrou: Powder Coatings: Chemistry and Technology

ISBN: 978-3-86630-884-8

the period 1962 to 1964. In 1962 the first decorative epoxy/dicyandiamide (DICY) powder coatings, produced by Wagemakers (now DuPont) in Breda, and soon followed by Libert in Ghent and Van Couwenberghe in Le Havre, appeared on the European market. Also in 1962 SAMES in France developed the first equipment for **electrostatic spraying** of organic powder coatings. This made a considerable contribution to the success of decorative thermosetting powder paint, since for the first time coatings with an acceptable "thin" layer thickness could be applied.

One serious drawback of epoxy/DICY thermosetting powder coatings is their sensitivity to attack by UV light. Exposed to sun they chalk and deteriorate rapidly. This, followed by poor yellowing resistance restricted their use mainly to protective coatings or decorative coatings for interior use where resistance towards yellowing was not a prime target.

Attempts to overcome these problems led to polyester/melamine systems which were introduced in 1970 by Scado BV and UCB in The Netherlands and Belgium [3, 4]. Almost at the same time *Hüneke* reported that the gloss retention and yellowing resistance of powder coatings based on blends of epoxy and polyester resins is considerably improved compared to pure epoxy powders [5].

The real breakthrough in the area was made in 1971 by Scado BV in The Netherlands. It was discovered that coatings with exceptionally good decorative properties can be obtained by hot melt compounding of carboxyl functional polyester resins and epoxy compounds in the form of bisphenol A resins (polyester/epoxy hybrids) or triglycidyl isocyanurate (TGIC systems) [6]. Shortly after, in 1972 a **weathering resistant** powder coating system produced by VP-Landshut in Germany was used for protection of aluminum extrusions and claddings in outdoor architecture in Switzerland [7].

At the same time in 1971 Bayer AG and BASF AG in Germany each offered thermosetting acrylic powder coatings to the market. Although later on attempts to commercialize the acrylic systems failed in Europe and the USA, they were widely accepted in Japan for outdoor use.

In the 1980's polyurethane powder coatings established a solid position in the American market and in Japan with a marginal market share in Europe.

Other developments in the area include a wider acceptance of PVDF powder coatings [8] mainly used for monumental architectural objects and ethylene chlorotrifluoroethylene (ECTFE) powders for protective purposes [9].

The development of new powder coating systems was quickly followed by improvements in the production and application equipment. Although the melt extrusion method, "borrowed" in the1960's from the plastic industry, remained almost the only process used for powder production, contemporary plants are changing substantially and very often use a continuous concept of powder paint production.

Developments in the application equipment led to the introduction of so-called frictional **tribo guns**. The frictional method of electrical charging of powder coatings has reduced considerably the *Faraday* cage effect, improving the penetration of powder coatings into recessed areas during spraying. These developments were followed by corresponding efforts to overcome the inefficient tribo chargeability of polyester (including hybrid) powder coatings, which at this moment dominate the market [10–13].

Reclaiming of powder during application is an advantage that leads to almost **100 % utilization** of the material. This creates, however, serious problems with respect to the color change in the spraying booths. The producers of application equipment have made considerable efforts to speed up the color change. This resulted in the development of systems that allow color change up to 5 minutes which is quite close to the liquid systems [14]. But modern equipment suppliers have their doubts and say, that the shortest possible time for color changing is 10 to 15 minutes.

The **automotive market** is one of the most demanding concerning the protective and decorative characteristics of the coating. The lack of good flow was an inborn weakness of powder systems and the orange peel effect was one of the major concerns. Other problems encountered when powder paints were used as automotive body coatings were inconsistent transfer efficiency, color change time, color contamination, difficulties to reach areas around the doors, and under the hood and trunk areas. Together with relatively high curing temperatures these were the main reasons for powder coatings not being successful competitors to the wet paints for car body finishing. Therefore, for a long time all applications have been limited to underbody and interior trim components and later on to exterior trim parts and steel and aluminum car wheels and the car manufacturers remained reluctant to use powder as a full-body coating.

However, Honda in Japan began in the 1980's with the application of a powder coating **primer-surfacer** on the car bodies, followed by Fiat in Italy. The car manufacturers in the USA were more careful, regarding the speed of introduction of full-body primer-surfacers. General Motors began in 1982 with a powder primer-surfacer on small pick-up trucks and built up experience and know-how, which was implemented in the early 1990's to passenger cars. *Miller and Kerr* [15] summarize the qualities and advantages provided by powder primer-surfacers as follows: overall surface smoothness (improving the appearance of the succeeding layers), no VOC emission, chip and corrosion resistance for the entire vehicle body, almost 100 % utilization and no waste disposal problems.

Especially the combination of the high utilization rate and the elimination of paint sludge, collected in the spraying booths, make powder primer-surfacers very interesting paints from an economical point of view, and can be considered as one of the major driving forces for a "revolution" in the automotive industry. The following examples may illustrate it: In the period 1991 to 1995 General Motors and Chrysler introduced full-body powder primer-surfacers in more than 10 production sites for passenger cars. Some years later Eurostar Works in Austria (a joint-venture of Chrysler and Steyer-Daimler-Puch) introduced powder primer-surfacers as full-body anti-chip coatings for their passenger cars [16]. These developments in the 1990's secured and consolidated the position of the powder primer-surfacer at the automotive producers in the USA. This is confirmed by the fact, that in 2000 more than 10,000 car bodies were coated daily with powder primer-surfacers at both GM and Daimler-Chrysler [17]. This trend is now followed in Europe.

The automotive clear **topcoat** was considered for years as a too difficult field for powder coatings to enter in. But the ecological and economical advantages offered by powder coatings, and, last but not least, the proven success with the powder primer-surfacer, initiated extensive research work at both coating producers and car manufacturers. General Motors, Ford and Chrysler have formed the so-called Low-Emission Paint Consortium (LEPC) to jointly develop new technologies for application of powder clear topcoats on automotive bodies [18, 19]. BMW, Volvo, Audi and Renault are or were among the European car manufacturers, experimenting with powder clear topcoats over metallic base coats.

BMW is the first car manufacturer in the world to use powder clear topcoat in standard production. Up to the end of 2000 powder clear topcoats has been commercially utilized in BMW's German plant for over 500,000 cars [20]. Since 2007 the main supplier of powder clear coats, PPG and BASF, delivered 800 t to BMW each year.

1.2 Market situation and powder economics

The different powder coating systems that have been developed in the last 40 years employ most of the polymers used in the conventional solventborne coatings. At the same time the continuing market growth considerably lowered production costs. Combined with sharp price increases of solvents after the "oil shock" in the 1970's this resulted in powder coating prices comparable with those of the wet coatings. Therefore powder coatings today are recognized not only as environmentally friendly systems, but also as materials that can compete successfully in price with the solvent and waterborne coatings.

Next to environmental and price aspects, quality is an even more important element, especially in high-tech areas of use, where powder coatings are commonly used. In this respect, powder coatings satisfy many of the most stringent requirements by the end users, failing only in cases where an extremely good flow of the coating is expected. Therefore the answer to the question *"why powder?"* contains the popular "**Four** E's" introduced 1986 by *Bocchi*[21]:

- Ecology
- Excellence of finish
- Economy
- Energy

Even a fifth E could be added for efficiency: no need get rid of a solvent to cure. Comparing the relative importance of these factors, it can be said, that at the moment, the high quality and the economic aspects perhaps contribute more to the acceptance of powder coatings by industrial users, than the regulatory compliances alone.

The market shares of powder coatings compared to other types of coating materials are not yet large. However, in a relatively mature field like the paint industry, powder coatings are among the few still enjoying a considerable growth in volume per year. According to available marketing data the world-wide market for powder coatings was 290 kt in 1990, 900 kt in 2001 [22], and about **2000 kt in 2010** [23] as shown in *Figure 1.1*, representing a total value

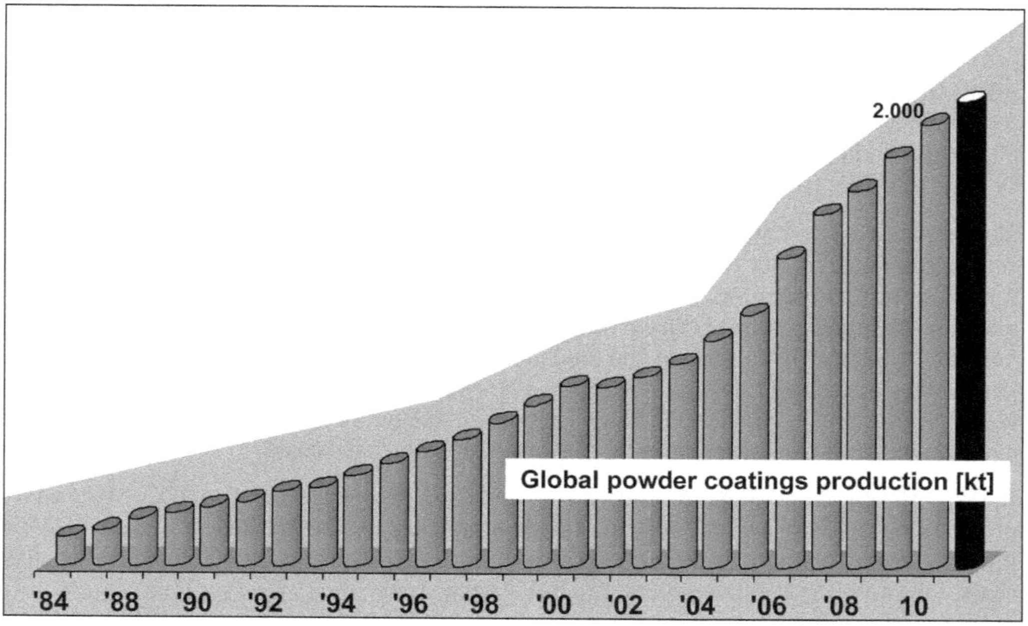

Figure 1.1: Global powder coating production since 1984 [23] *source: Evonik Industries*

of roughly 8 billion € ($ 10 billion). The annual growth rate, which showed double digits (more than 10 %) up to 1991, has been decreased in recent years to 4.5 to 10 % (varying by region), due to an economic recession in the nineties and a serious financial crisis in 2009.

Western Europe, where the powder coatings were invented, shows the lowest annual growth rate in the last 10 years, but the highest market share in industrial coatings (18 %), when compared to North America (15 %) and Japan (15 %) [24]. Inefficient production sides are recently rationalized and eliminated in Western Europe and reopened in lower-cost emerging regions like **China**. *Table 1.1* based on data of SRI Consulting [24] gives an indication of the production of powder coatings in different geographical areas. China accounts for 700 kt in 2007, but it is believed that only half of it meets western quality (and price!) standards. The Chinese impetus is not restricted on production volumes. 2004 only 12 % of the patent applications in the field of powder coatings were filed in Chinese, 33 % in English and 30 % in Japanese. This picture is reversed in 2011: Now China is the formal innovation leader with 37 % patent applications, followed by English (27 %) and Japanese (18 %). It has to be mentioned, though, that similar to the powder coatings not all of the Chinese patents meet the high western standard, yet.

The various types of powder coatings commonly used in different **geographical locations** differ considerably. For example the three major market areas in the world, Europe, USA and Asia, have developed completely different systems for exterior use. In Europe, systems based upon carboxyl functional polyester resins/ß-hydroxy alkylamides (HAA) - which have replaced TGIC almost totally in the last couple of years - are the most popular, in the USA polyurethanes with aliphatic polyisocyanates as crosslinkers have dominated for many years, but ever since 1998 they are competing with the polyester/TGIC/HAA/PT910 (TGIC alternatives) types, while in Japan powder coatings with glycidyl functional acrylic resins as binders and dibasic acids as crosslinkers and polyurethanes are market leaders. TGIC systems are not at all used in Japan [25]. For coatings for indoor applications the so-called hybrids, based on carboxylated polyesters and epoxy resins, are the most popular, both in Europe and the USA. *Table 1.2* presents average figures for the **market shares** of five types of powder coatings [23]. The different regional consumption of these competitive powder coating systems are accounted to historical reasons, differences in prices and availability of resins, toxicological issues, governmental regulations and an unequal intensity of sunlight which the coatings are exposed to [24].

More than **1000 powder coating producers** are active world-wide, most of them serving a limited geographical area. The three largest are responsible for 30 % of the global market if China is excluded. These three multinationals are AkzoNobel, DuPont and Rohm and Haas. In China there are about 600 powder coating formulators, but the majority of them are very small.

Table 1.1: The world market for powder coatings by region for 2000 and 2007 [24]

	2000		2007	
	Kilo tons	%	Kilo tons	%
Western Europe	335	41	377	23
North America	175	22	189	11
China	96	12	700	43
Rest of Asia/Pacific	119	15	183	11
Rest of World	85	10	196	12
Total	**810**	**100**	**1645**	**100**

Table 1.2: Market shares (in %) of different types of powder coatings [23]

	W-Europe	N-America	Asia	ROW	World
Epoxy	6	13	13	9	10
Epoxy-polyester (hybrid)	52	32	52	51	50
PES-TGIC + Derivatives	11	29	26	24	22
PES-ß-Hydroxyalkylamide	28	4	7	15	13
Polyurethane	2	17	2	1	4
Acrylic	1	5	0	0	1

The **metal furniture** industry was one of the first areas invaded by powder coatings, and it is steel that is the important substrate. Hybrids are the main type used for this purpose, while for outdoor furniture (garden furniture for example) polyester/TGIC (and its alternatives) or polyurethanes are the common systems.

The industry of **domestic appliances** is an important consumer of powder coatings. Appliance applications in total represent about 14 % of the market for powder coatings world-wide [26]. More than half of this market is located in Asia Pacific, since the appliance production itself has been shifted largely in this region. In Europe hybrids are the main type used since a mild orange peel effect is acceptable. In the U.S. a smooth finish is preferred and therefore polyurethanes are much more common. Typical products of this industry that are coated with powder coatings are food freezers and refrigerators, water heaters, toasters, washing machine lids, domestic cookers, electric and gas heaters, etc.

Architectural application is still the most important market for powder coatings, holding a market share of 36 %. Aluminum plays a special role in this field. Powder coatings show a penetration of roughly 30 %, whereas anodizing accounts for 55 %. In 2007 45 % of the powder coated architectural aluminum has been located in Europe, 34 % in Asia Pacific and only 3 % in North America. In U.S. coil coating panels are used predominantly, due to cost reasons [24].

The **automotive market** accounts for one third of all powder coating applications. The major part of this is claimed by primer-surfacer, a coating to protect the electrocoat primer and to provide a smooth surface for the basecoat. Primer-surfacer resins are usually made by acrylic/styrene/methyl methacrylate mixtures and cured by dodecanoic diacid (DDDA). But ß-hydroxyalkylamide (HAA) based systems and polyurethanes are used, too, due to superior chip resistance. The system, that has been chosen by all car producers, for automotive topcoats is a blend of an epoxy functional acrylic copolymer containing glycidyl methacrylate (GMA) as co-monomer, and DDDA or its (poly)anhydride as crosslinker. Very positive experiences are reported with such acrylic clear topcoats [27]. In 2007 almost half of the powder world market for automotive has been located in Europe, one third in North America [24].

The market growth of **exterior powder coatings** has continued to expand due to the wide acceptance of polyester/HAA systems in Europe and polyurethanes in the USA. Exterior coatings held a market share of 9 % of all powder coatings in 2008 [26].

Another market for powder coatings is in corrosion resistant types for e.g. **pipes and reinforcement bars** (rebars) holding a share o about 6 %. One third of this amount was produced in China, one fourth in the rest of Asia Pacific in 2007 [24]. This is the area where pure ("fusion-bonded") epoxies dominate. In Europe the rebar market has been largely ignored, but its importance from a technical point of view has been recognized and this is an area where a considerable growth can be expected.

When powder coatings economics is compared with the other VOC (volatile organic compound) compliance coating options, one has to take into account all of the expenses connected with the coating process. On-line operating advantages reported for powder coatings compared with other environmental friendly systems are an application utilization of the material from 95 % up to 99 %; 30 % less energy consumption (than with conventional low-solids enamels), combined with a reduction of labor expenses of 40 to 50 %; reduction of waste material accounts for almost 90 % and there are approximately four to six times fewer rejects due to surface defects [28].

Figures with respect to the **economy of the powder** coatings process have been published by the PCI [28] in 1999. The major cost factor for powder coating formulations is raw materials which make up for two thirds of the total costs. In powder coating applications of medium sized job shops the powder coating itself attributes only for 20 % of total costs, the main factor is labor (40 %) [24].

Recent developments in application equipment that make the clean - out of the line easier and the change of the color faster and the trend in developing powders with lower curing temperatures and higher reactivity, which allows higher line speeds thus saving energy, make the economics concerning powder coatings even more attractive. The "Four E's" - excellence of finish, ecology, economy and energy - are attributes the powder coatings future can certainly rely upon.

1.3 REACH

No chemical legislation ever before has influenced the research, the production and the sales of coating raw materials as fundamentally as REACH - The chemicals legislation of the European Community. Therefore a short description of REACH will follow in this chapter.

REACH (Regulation EC (No) 1907/2006) is the chemicals legislation of the European Community (EC) which came into force in 2007. REACH consists of three main legislative processes: registration, evaluation and authorization of chemicals.

Registration process

All substances which are produced within the EC in quantities of at least 1 metric ton per year have to be registered by the EC manufacturer. Furthermore all substances which have to be imported in quantities of at least 1 metric ton have to be registered by the EC importer. These substances are either imported as pure substances, as substances in mixtures or as monomers covalently bound in imported polymers. A registration is also needed for certain substances contained in imported articles. For registration a Technical Dossier has to be submitted to the European Chemicals Agency (ECHA). The Technical Dossier contains information regarding substance identity, uses of the substance and information about physicochemical, toxicological and eco-toxicological properties. For this purpose available studies should be used, either own studies or studies that have to be shared between the registrants after receipt of a Letter of Access from the study owner. If no study or other alternative data (e.g. read across, qualitative or quantitative structure-activity relationships) are available, physicochemical, toxicological and eco-toxicological studies have to be conducted. With regard to animal welfare, multiple testing of a substance has to be avoided, i.e. studies with vertebrates shall be conducted only once. The registration costs can rise up to several hundred thousands of Euros depending on the substance properties, the annual quantity of the substance, data availability and number of registrants.

Furthermore a Chemical Safety Report has to be submitted to ECHA for substances registered in quantities of at least 10 metric tons per year and per registrant. The Chemical Safety Report consists of an exposure assessment and a risk assessment, which has to be conducted for hazardous substances based on their physicochemical, toxicological and eco-toxicological properties.

A simplified registration is allowed for intermediates which are produced and used under strictly controlled conditions.

Transition periods are granted, if the substance is pre-registered by the potential registrant. Within the transition periods the potential registrant can produce or import the pre-registered substances without registration. The transition period for substances produced or imported in quantities of at least 1000 metric tons per year as well as for substances with carcinogenic, mutagenic or reprotoxic properties (≥1 ton/year) and for substances that are dangerous to the environment (R 50/53 substances ≥100 tons/year) expired in 2010. The transition period for substances produced or imported in quantities of at least 100 metric tons per year ends in 2013 and for substances produced or imported in quantities of at least 1 metric ton per year the transition period ends in 2018.

There are substances that are exempted from REACH registration. Whereas article 2 of the REACH regulation lists substances that are regulated by other legislations, annexes IV and V of the regulation include substances that are exempted from registration (e.g. natural substance such as glucose, carbon dioxide, nitrogen, crude oil).

Polymers are exempted from registration as well but the monomers that are used to manufacture the polymer have to be registered either by the producer of the monomer, by the EC importer of the monomer or by the EC importer of the polymer. Since powder coating raw materials are often polymers, **powder coating have a natural advantage** over other environmentally compliant coatings systems, e.g. UV-curing formulations.

Substances produced or imported for product and process orientated research and development (PPORD) are exempted from registration for at least five years. However the manufacturer or importer has to notify the substance to ECHA.

Evaluation process

The evaluation process will be conducted by ECHA and the national competent authorities of the Member States. Within the dossier evaluation process, ECHA checks the registration documents for compliance and evaluates the testing proposals that are included in the Technical Dossier. The substance information and the chemical safety assessment are evaluated by ECHA in cooperation with the competent authorities of the different Member States.

Authorization process

The authorization process aims to control risks from Substances of Very High Concern (SVHC) and to replace the substances by suitable alternatives substances or to develop alternative technologies. Therefore, the uses of these substances have to be authorized by the European Commission.

The following categories of substances are considered to be Substances of Very High Concern:

- carcinogenic, mutagenic and reprotoxic substances
- persistent, bioaccumulative and toxic substances (PBT substances)
- very persistent, very bioaccumulative substances (vPvB substances)
- substances of equivalent concern (e.g. endocrine disruptors)

Substances of the mentioned categories will be listed in Annex XIV. If a substance is listed in this annex, it can be used only, if an authorization dossier was submitted and approved by the European Commission.

More detailed information about REACH and current activities of ECHA is available at the ECHA homepage (http://echa.europa.eu).

1.4 References

[1] *Gemmer, E.*, Industrie-Anzeiger, 75, 1095 (1953)
[2] *Harris, S.T.*, The Technology of Powder Coatings, Portcullis Press Ltd., Redhill, England, 1976
[3] Scado BV, US Pat. 3.624.232, 1970
[4] UCB SA, DE Pat. 2.352.467, 1972
[5] *Hüneke von H.*, Metalloberfläche, 24, (9), 315, 1970
[6] Unilever N.V., DE Pat. 2.163.962, 1971
[7] *Bodnar, E.*, Product Finishing, August 1988, p.22
[8] *Meiyden, B. van der*, in Thermoset Powder Coatings, Ed. John Ward, FMJ International Publications Ltd, Redhill, 1989, p.54
[9] *Maguire, G.*, Product Finishing, October 1988, p.66
[10] *Misev, A.T., Binda, P.H.G., Hardeman, G.*, to Stamicarbon BV, NL Pat. 8.800.640, 1988
[11] *Misev, A.T., Binda, P.H.G.*, to Stamicarbon B.V., NL Pat 8.802.913, 1988
[12] *Binda, P.H.G., Misev, A.T.*, to Stamicarbon B.V., NL Pat. 8.802.748, 1988
[13] Hoechst AG, EP 315 082, 1987; Hoechst AG, EP 315 083, 1987; Hoechst EP 315 084, 1987
[14] *Floyd, R.*, Polymers Paint Colour Journal, 179, (4236), 270, (1989)
[15] *Miller and Kerr*, Proceed. Powder Coating '92 (1992) p. 79
[16] FARBE&LACK, 102, 8, (1996) p. 16
[17] *Harris, S.T.*, Focus on Powder Coatings, Nov. 2000, p. 3
[18] *Cole, G.*, Industr. Paint & Powder, 69, (12), 24 (1993)
[19] *Bailey, J.*, Industr. Paint & Powder, 70, (7), 12 (1994)
[20] Modern Paint and Coatings, 91, (2), 13 (Feb. 2001)
[21] *Bocchi, G.J.*, Modern Paints and Coatings, 76, 44, (1986)
[22] *Harris, S.T.*, Focus on Powder Coatings, Oct. 2002, p. 8
[23] *Grenda, W., Spyrou, E.*, Congress Papers European Coatings Congress (ECS), Nuremberg, 113, 2011
[24] *Linak, E., Kishi, A., Yang, V.*, Thermosetting Powder Coatings, Specialty Chemicals SRI CONSULTING, 2008
[25] *Rankl, F-J.*, (CEPE), Proc. Pulverlack Kongress, Hamburg, Sep.'96, p. 6, 1996
[26] *Moran, B.*, Market research report, Powder Coatings: Materials, Coatings, Technologies, CHM051A, BCC research, 2009
[27] *Schmidt, H. and Fink, D.*, Surface Coatings International, 2, p. 66, 1996
[28] Powder Coating - The Complete Finisher's Handbook, Edited by the PCI, p. 301-304, 2nd Ed. 1999

2 Thermoplastic powder coatings

The first powder coatings produced were based on thermoplastic polymers which melt at the application temperature, and solidify upon cooling. Several factors such as relatively simple methods of manufacturing and application, no involvement of complicated curing mechanisms, raw materials that in many cases belong to commodity polymers, acceptable properties for many different applications etc. contributed to the popularity of these coatings in the market very soon after their appearance in the beginning of the 1950's. At the same time, however, weaknesses such as high temperature of fusion, low pigmentation level, poor solvent resistance and bad adhesion on metal surfaces necessitating the use of a primer can be listed. These problems inherent to the thermoplastic powder coatings were successfully overcome later on by the thermosetting powders which very quickly took the largest part (90 %) of the market.

Despite the disadvantages, thermoplastic powder coatings can offer some distinguished properties. Some of them possess excellent solvent resistance (polyolefins), outstanding weathering resistance (polyvinylidene fluoride), exceptional wear resistance (polyamides), a relatively good price/performance ratio (polyvinyl chloride) or high aesthetic appearance (polyesters). These properties combined with the simplicity of the system, created a considerable market share for thermoplastic powder coatings.

Thermoplastic coatings are offered in a variety of performance classes (*Figure 2.1*). Besides the engineering and commodity polymers, which will be discussed in the following chapters, there are high performance polymers, e.g. PEEK (polyether ether ketones).

PEEK is a fully aromatic, semi-crystalline thermoplastic polymer with a maximum crystallinity of 48 %. It is a member of the class of polyaryl ether ketones (PAEK) high performance polymers. PEEK has a glass transition point of 143 °C and a melting range of ca. 340 °C. Characteristics of the polymer include a very high heat resistance and therefore service temperature, high rigidity, low water absorption, high hardness, good strength, low sliding friction, excellent chemical and hydrolysis resistance, low flammability, very low emission of smoke and toxic fumes during burning and its good electrical characteristics. PEEK offers one of the highest resistances against radiation among polymers.

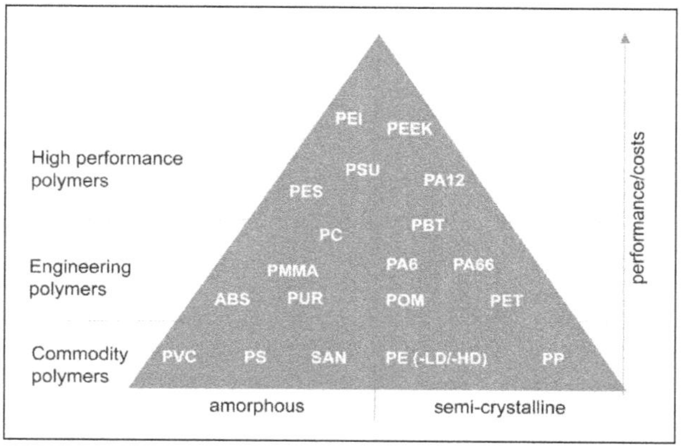

Figure 2.1: Performance/cost ratio of thermoplastics

source: Evonik Industries

Due to its high price PEEK is normally only used in applications where special properties are needed. Examples include valves, cable insulation, bearings, pump parts and sealings. Evonik Industries offers industrial grade PEEK under the trade name "Vestakeep".

ISBN: 978-3-86630-884-8

2.1 *Vinyl powder coatings*

Two binders are used for the manufacture of the so-called vinyl powder coatings; polyvinyl chloride (PVC) and polyvinylidene fluoride (PVDF). On the basis of their polymer nature both powder coatings can be included in the same group, although they differ considerably in their performance. While PVC powder coatings are predominantly intended for indoor application because of their limited outdoor durability, PVDF powder coatings are among the best coating systems with respect to their weathering resistance.

Vinyl polymers belong to a group of resins having a vinyl radical as the basic structural unit. Polyvinyl chloride and copolymers of vinyl chloride are the most significant members of this group being among the first thermoplastics to be applied by powder techniques.

2.1.1 *PVC powder coatings*

Polyvinyl chloride powder coatings were introduced on the market in the time when the thermosetting powder coatings were in very early stages of development. PVC based coatings offered many advantages over the other thermoplastic materials available as binders for coating production. These coatings have very good resistance to many solvents, which is a rather poor characteristic of the thermoplasts, combined with resistance towards water and acids. They have excellent impact resistance, salt spray resistance, food staining resistance, and good dielectric strength for electrical applications.

Polyvinyl chloride $(-CH_2-CHCl-)_n$ is one of the **cheapest polymers** produced by the industry on a large scale. Its basic properties include chemical and corrosion resistance, good physical strength and good electrical insulation. Polyvinyl chloride (PVC) is by nature a brittle polymer, but the flexibility of the material can be easily adjusted by using an appropriate amount of a suitable plasticizer.

Polymerization of vinyl chloride into PVC homopolymer or its co-polymerization with different co-monomers is carried out by a free radical mechanism. Most PVC resins are produced by emulsion or suspension polymerization of vinyl chloride in an aqueous system containing an emulsifying agent or suspension stabilizer. However, bulk and solution polymerization processes are also carried out on an industrial scale.

Emulsion polymerization of vinyl chloride can be performed in both a batch and continuous way. The reaction temperature is maintained between 40 and 50 °C by cooling the reactor in order to remove the heat developing during the polymerization. The reaction medium is de-ionized water containing enough proper surfactant to obtain a stable emulsion. The initiators used are peroxides soluble in water, such as hydrogen peroxide or different persulphates. Since the monomer itself is a gas at room temperature, the polymerization is performed under pressure in autoclaves. The pressure in the reactor falls as the polymerization proceeds. After reaching conversion of ca. 90 %, the content of the reactor is discharged and the unpolymerized vinyl chloride is recovered.

Suspension polymerization of vinyl chloride is an important process in the commercial manufacture of PVC. In principle it is a batch process, although attempts have been made to develop a continuous technique for suspension polymerization of PVC. The polymerization is carried out by first charging to the reactor the required amount of de-ionized water and adjusting the pH depending on the suspending agent used and then the dispersing agent and the initiator. The monomer is charged after sealing the reactor and evacuating the oxygen. The polymerization reaction is carried out under pressure at 40 to 60 °C, controlling the temperature by appropriate cooling of the reaction mass. Although a great amount of research has been done in this area, it is interesting to note that the compositions of the reaction mass do not differ substantially from those used in the very early days of development of suspension PVC. The differences are mainly limited to the choice of the suspending agent, which is in most cases polyvinyl alcohol obtained by saponification of polyvinyl acetate, gelatin, methyl cellulose and copolymers of vinyl acetate with maleic anhydride. The initiator is water insoluble peroxide, such as lauryl peroxide, or azobisisobutyronitrile.

The type of suspending agent plays a very important role in obtaining primary particles with high porosity. Gelatin normally produces glassy spherical particles which have poor plasticizer absorption characteristics, whereas polyvinyl alcohol gives particles of a porous nature which readily absorb plasticizers to give dry powder blends. Thus, a patent of Air Products and Chemicals, Inc. [1] discloses a process for suspension polymerization of vinyl chloride giving a polymer specially suitable for production of powder coatings. The use of an excessive amount of secondary suspending agent causes two effects necessary for the critical powder coating application. Firstly, it reduces the size of the primary particles within the polyvinyl chloride grain, thus raising the surface to volume ratio and allowing plastication of the primary particles by the plasticizer. Secondly, very high porosity is gained, allowing complete and uniform plasticization of the resin grain in its entirety.

Solution polymerization of vinyl chloride is almost exclusively used for manufacturing copolymers containing vinyl acetate. The co-monomers are dissolved in a suitable solvent such as cyclohexane or n-butane, and the polymerization is carried out at 40 to 60 °C catalyzed by a free radical initiator which is soluble in the reaction mass. The copolymer begins to precipitate after a certain molecular mass is reached, which depends on the co-monomers' ratio, polymerization temperature and type of solvent. The last step of the process includes filtering of the final product and then washing to remove the residual diluent and any traces of organic peroxide which would have a detrimental effect on the heat stability.

The **bulk polymerization** of vinyl chloride has been developed by Pechiney-St. Gobain, and plays an important role in the commercial production of PVC. Although bulk polymerization is associated with a homogeneous system, this is not the case with the bulk polymerization of vinyl chloride; namely, at very early stage of the reaction the polymer formed precipitates from the reaction medium in a form of insoluble material dispersed in the monomer. The system is therefore heterogeneous for a significant portion of the whole conversion. Since the polymer precipitates from the monomer, and there are no solvents present in the system, the concentration of the monomer available for polymerization remain constant with time. Therefore, at constant temperature the average molecular weight of the polymer obtained is apparently independent of the monomer conversion.

Bulk polymerization of vinyl chloride is performed in variants of the original Pechiney two-stage process. Vinyl chloride containing initiator is charged in a reactor and the polymerization is carried out at 50 to 60 °C. Already at 1 % conversion the polymer begins to precipitate. After reaching conversion of about 10 % the entire content of the reactor is discharged in an autoclave where a further quantity of vinyl chloride is added. When the conversion is at about 20 % the material is in the form of a wet powder; at 40 % conversion the residual monomer is completely absorbed by the polymer and the mass takes on the form of a dry powder. The design of the reactor permits further polymerization which usually proceeds up to 90 % conversion. The residual monomer is then removed under vacuum and the polymer is discharged and sieved. Polymer particles with sizes larger than desired undergo additional grinding and sieving.

Polyvinyl chloride is an **amorphous polymer** containing very small amounts of imperfect crystallites. The average molecular weights for the most commercial PVC resins are in the range 50,000 to 120,000. Better physical properties are obtained by PVC having a high molecular weight, but on the other hand, for applications in powder coatings, PVC resins with low molecular weight are more suitable because of lower melt viscosity and lower softening temperatures. Although different glass transition temperatures ranging from 68 to 105 °C are reported, depending on the method of polymerization and especially the temperature at which the polymerization is carried out, the normal commercial grades of PVC have T_g of between 80 and 85 °C.

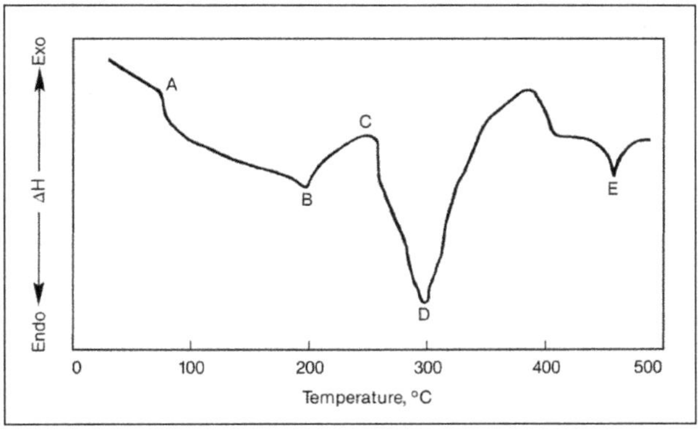

Figure 2.2: Differential thermal analysis curve of unplasticized PVC [3] *source: Federation of Societies for Coatings Technology*

Figure 2.2 represents the **differential thermal analysis** curve of polyvinyl chloride. Five major regions of different thermal behavior can be noticed above room temperature. *Matlack* and *Metzger* [2] and *Smith* and *Koleske* [3] assign the glass transition temperature to the change of the slope at point A. The endothermic effect between 165 and 210 °C is a result of the melting of the crystallites present in the polymer. At 250 °C they define the point C as a region of oxidative attack, while at 300 °C (point D) the dehydrohalogenation takes place with evolving gaseous hydrogen chloride. Finally, at about 450 °C (point E) the endotherm is attributed to the occurrence of depolymerization.

Polyvinyl chloride is soluble in a wide range of organic solvents. The best solvents for PVC are tetrahydrofuran and cyclohexanone, but it is also easily solubilized by other ketones, halogenated hydrocarbons and nitrated aromatic solvents. However, from the standpoint of a powder coating producer it is important that PVC polymers are not soluble in alcohols and aliphatic hydrocarbons, and powder coatings having PVC as a binder possess considerable **solvent resistance** with respect to the most common solvents used in the industry.

The performance characteristics of the vinyl powder coatings have been compared by *Christensen* and are summarized in *Table 2.1* [4]. Comparisons of this type are always discussible. In this case, the outdoor durability of the vinyl polymers should be reconsidered, keeping in mind new developments particularly those of polyester thermosetting coatings. The table, however, gives a good indication of the characteristics of the vinyl powders with respect to the other thermoplastic materials.

The **main problems** of PVC powder coatings are related to the stability of the binder during processing, and also its brittle nature. These weak points were successfully overcome

Table 2.1: Performance characteristics of various powder coating materials [4]

	Vinyl	Polyester	Epoxy	Nylon	Polyethylene
Exterior durability	+++	++	0	+	0
Salt spray resistance	+++	+	++	++	+
Impact resistance	+++	+	+	+++	0
Flexibility	+++	0	0	++	+++
Food staining	+++	+++	+++	++	+++
Dielectric strength	++	+++	+++	+	+++

+++ = Excellent, ++ = Very good, + = Good, 0 = Fair

due to the enormous amount of work that was done on stabilization and plasticization of polyvinyl chloride. The result was development of a number of heat stabilizers, plasticizers and lubricants permitting problem-free hot compounding of the vinyl resins.

Because it is a rigid material not having sufficient flexibility polyvinyl chloride as such can not be used for powder coating purposes. In fact, only in a few applications can PVC be used without the addition of **plasticizers**. A wide range of high boiling materials has been developed for plasticization of PVC, such as esters of phthalic acid, phosphoric acid and aliphatic diacids. Liquid polymers or oligomers with relatively low glass transition temperatures are also used for the same purpose.

Phthalate esters are the cheapest plasticizers for PVC, proving to be very satisfactory over all properties. Di-2-ethylhexyl phthalate and diisooctyl phthalate have been the most widely used, offering a good price/performance ratio. But recently diisononyl phthalate gained more and more attention and market share due to a lower migration rate. It is also characterized with better oxidation stability and low-temperature flexibility.

Phosphate esters are used to obtain PVC products with **self-extinguishing properties**. Trixylyl phosphate is the most widely used material among the great variety of phosphate esters. The same self-extinguishing effect is obtained by the use of chlorinated paraffins. They in general cannot completely replace the so-called primary plasticizers of phthalate or phosphate type, but they are quite competitive for price reasons. The common chlorinated paraffins used for plasticizing PVC have 42 to 56 % chlorine content and a chain length of C_{15} to C_{25}.

Linear polyesters produced by esterification of dibasic acids with dibasic alcohols are plasticizers that provide good solvent resistance. They are especially suitable for PVC coatings in contact with hydrocarbon fuels or other organic solvents. However, polyvinyl chloride plasticized with polyesters exhibits relatively poor low-temperature flexibility.

The addition of the plasticizer to PVC improves the flexibility and the impact resistance of the polymer, but at the same time reduces the tensile strength, modulus and hardness of the plasticized material. The desired balance between **hardness** and **flexibility** can be achieved by careful choice of the type and amount of plasticizer. Although many efforts have been made to predict the behavior of the plasticized system by systematic approaches to this problem, the desired set of properties is usually achieved by making a series of formulations.

The curve representing **modulus-temperature behavior** of externally plasticized PVC with dioctyl phthalate (DOP) exhibits a typical drop of the modulus at the glass transition temperature at 80 to 90 °C with the rubbery plateau region afterwards in the non-plasticized PVC *(Figure 2.3)* [3]. A further increase of the temperature leads to melting of the crystallites, resulting in an additional drop of the modulus. The addition of DOP shifts the rubbery plateau to lower temperatures, but at the same time decreases the modulus, in this way affecting the hardness of the film.

The type and amount of added plasticizer must be determined in a way that will result not only in a good compromise between flexibility, impact resistance and hardness, but also in a way that will not affect the powder storage stability. Due to a concern over plasticizer leaching into food the market share of PVC has decreased, in particular in the western hemisphere.

Polyvinyl chloride that exhibits good mechanical properties has a molecular weight above 40,000. In pigmented systems, due to such a high molecular weight, there is a considerably more pronounced increase of the **melt viscosity** compared to the thermosetting binders. The melt viscosity of PVC pigmented with TiO_2 shows an asymptotic increase at concentrations of the pigment between 15 and 25 % (by weight). This is almost four times lower than the concentration of the alkyd resins [3].

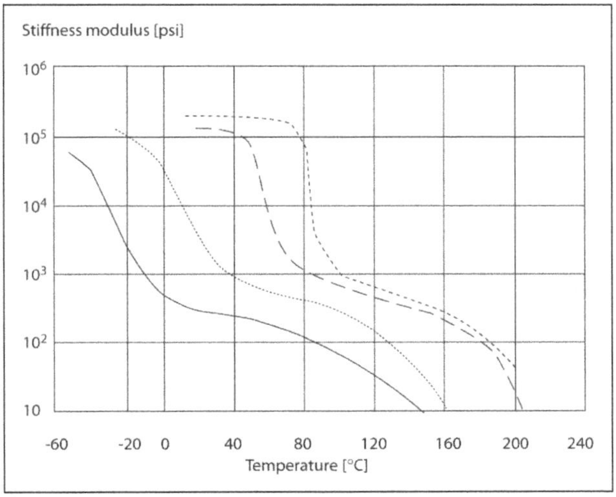

Figure 2.3: Modulus temperature behavior of externally plasticized PVC[3] *source: Federation of Societies for Coatings Technology*

High melt viscosity of the pigmented PVC powder coatings is one of the reasons for their application in thick films in order to obtain good flow. Very often the low price of PVC powder coatings is compensated by the thick film application, losing in this way the advantage of having cheap and good quality protection. *Smith* and *Koleske* have suggested, that for film thickness below 50 μm, the melt viscosity during cure should be in the range between 50 and 500 Pa · s [3]. This corresponds to a weight averaged molecular mass of between 20,000 and 40,000 (Mn = 10,000 to 20,000), which is 2 to 4 times lower in value than the commonly used molecular masses.

PVC powder coating with a proper formulation should always contain a suitable **stabilizer**. An enormous amount of work has been done to elucidate the mechanism of thermal degradation of polyvinyl chloride. It is beyond the scope of this book to go into detail about the degradation phenomena of PVC. The overall reaction of decomposition involving hydrogen chloride will be presented simply as an illustration to help explain the role of the stabilizers:

Equation 2.1

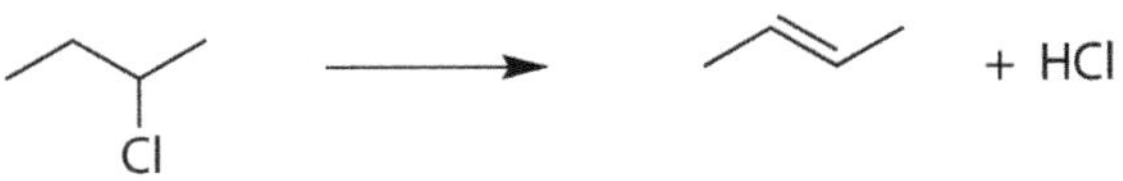

The common property of all materials used for stabilization of polyvinyl chloride is the ability to react with hydrogen chloride. It seems rather strange that the stabilizer reacts with the product of degradation, i.e. it is active after the degradation has happened. Therefore, it can be assumed that the stabilizer decreases the rate of degradation, lowering the concentration of hydrogen chloride that may act as a catalyst for the degradation process. This can be accepted for some simple stabilizers such as the metal stearates. The experiments with labeled stabilizers of a metal soap nature or dialkyl tin dicarboxylates showed, however, that the soap or carboxylate moieties are retained to a greater extent than the metal or the alkyl groups. This indicates that the stabilizer is involved in some coordinate complex with the polymer chain, and that its action is not limited to a simple scavenging action of the evolved hydrogen chloride only.

A wide variety of materials has been developed for thermal stabilization of PVC. Such materials include mixed salts of calcium and zinc with rather inferior stabilizing properties, barium-cadmium soaps, lead sulfates, tin mercaptides, dibutyltin derivatives, di(n-octyl) tin maleate, amino crotonic esters, epoxy compounds, etc., though lead- and cadmium compounds are now forbidden in many countries for their toxicity.

A typical vinyl powder coating formulation containing a plasticizer and stabilizer is given in *Table 2.2* [5].

Table 2.2: Vinyl powder coating formulation [5]

source: Communication Channels Inc.

Ingredient	Parts by weight
Bakelite vinyl resin QYNJ (1)	100.3
Bakelite vinyl resin QYJV (1)	9.3
Flexol plasticizer EP-8 (1)	6.0
Flexol plasticizer 10-10 (1)	36.0
Mark C (2)	2.0
Thermolite 31, PVC stabilizer (3)	6.0
TiO_2-rutile	20.0

(1) Union Carbide Corp.; (2) Argus Chemical Corp.; (3) M&T Chemicals, Inc.

Additives known as **lubricants** are often used in the powder coating formulation. They influence both the behavior during processing and the properties of the finished product. The primary requirements of the lubricants are to improve the flow properties of the fused compound, and to help the movement of the material through the extruder during the production of the powder coating by lubricating the interface between the PVC and the metal surface.

Two grades of lubricants are usually used for this purpose. The internal types which are compatible with PVC reduce the melt viscosity of the compound, improve the flow properties and reduce the friction in the processing equipment. The external lubricants which are not compatible with PVC act in this case only as a lubricating layer on the surface of the PVC where it comes into contact with the metal surface of the processing equipment.

Representatives of internal lubricants are long-chain fatty acids, calcium stearate, alkylated fatty acids and long-chain alkyl amines. External lubricants include fatty acid esters, high molecular weight alcohols, synthetic waxes, and low-molecular weight polyethylene.

PVC powder coatings applied by the **fluidized bed** method have been the most widely used materials for coating wire goods in thick layers. In this application they offer a good compromise between price and performance. They are characterized by uniformity of the coating, no sagging and good edge coverage. The usual thickness of the PVC powder coatings applied by the fluidized bed method is between 200 and 400 µm. To achieve good uniformity of the coating in thinner layers, the electrostatic spray technique is required.

Typical products coated with PVC powders are appliances and dishwasher racks. In the latter case, PVC coated materials can withstand the normal testing conditions of total vapor and water immersion at 80 to 100 °C without blistering. Detergent resistance and resistance to food stains are excellent. This is combined with the important characteristic of allowing the water to run off from the coated products. PVC powder coatings provide another important characteristic to dish racks by acting as sound dampers, thus preventing the development of noise during filling or discharging of the racks. Chemical and detergent resistance makes them suitable for protection of dishwasher tubes.

The non-toxic nature of the PVC powder coatings allows them to be used for refrigerator racks, freezers and shelves. The stain resistance and low odor of PVC powder coatings are appreciated characteristics in coatings for shelves made of wire, expanded metal or sheets. For the same reasons they are used as metal furniture coatings.

Good dielectric properties of PVC powder coatings are among the pre-requisites they fulfill for application in the electric and electronic industries. The soft touch feeling makes them suitable for covering various types of handles, automotive seat belt anchors, bus bars, tools etc.

Outdoor durability is not a strong point of PVC powder coatings. The main reason is the sensitivity of the binder to the ultraviolet spectrum below 350 nm. The development of effective UV stabilizers has opened up new possibilities for outdoor use of PVC powders.

A large market for **exterior use** of PVC powders developed in USA includes the electrical transformer, whose manufacturers give a 20 year guarantee for their products coated with PVC powder [6]. The successful development of thermosetting powder coatings specially designed for outdoor application is certainly a big obstacle for PVC powders to compete in the market for outdoor coatings.

2.1.2 Fluor based powder coatings

The best representative of vinyl polymers exhibiting **outdoor durability** is polyvinylidene fluoride (PVDF). Powder coatings based on PVDF have the highest resistance to degradation on outdoor exposure, very good abrasion resistance, excellent chemical resistance, very low surface friction, ice release properties, and very low fluid absorption.

PVDF is produced by free radical polymerization of vinylidene fluoride, using peroxides as initiators, or coordination catalysts of the *Ziegler-Natta* type. Different patents describe various methods for polymerization of vinylidene fluoride including emulsion, suspension or solution polymerization [7-12]. The structure in the repeating unit of polyvinylidene fluoride is $-CH_2-CF_2-$. Hydrogen and fluorine atoms are spatially symmetrical, which optimizes cross-bonding forces between polymer molecules. Polyvinylidene fluoride is a **crystalline polymer** with a melting point between 158 and 197 °C. It exists in two different crystalline forms: the so-called alpha form has a *trans-gauche-trans-gauche* conformation, while the other beta form possesses a planar zigzag conformation [13]. The polymorphism of PVDF is the reason why the melting point, which is in a considerably broad temperature range, is not well defined. The relatively high melting point allows PVDF to be used permanently in a relatively broad temperature range from -40 up to +150 °C which corresponds with its glass transition temperature and the lower border of the melting range.

Polyvinylidene fluoride is characterized by a high mechanical and impact strength and very good abrasion resistance combined with excellent flexibility and hardness. Even better flexibility and elongation is achieved by copolymerizing it with hexafluoropropylene [14]. Carbon steel coated with this copolymer could be a cost-effective alternative to stainless steel.

PVDF homopolymer is resistant to attacks by most corrosive chemicals such as acids, alkalis, strong oxidizing agents, etc. It is also not soluble in the common solvents used in the coating industry. Some highly polar solvents can only soften the surface of the polyvinylidene films temporarily. The only chemicals that can damage the surface of the PVDF films are fumic sulfuric acid and strong primary amines. PVDF complies with the requirements of the Food and Drug Administration (FDA) for materials that can be used in the food processing industry and allowed to come in direct contact with food.

PVDF powder coatings have been recognized as materials with **exceptional properties** such as low friction and abrasion, oil and water repellency, extremely good outdoor durability, excellent flexibility, corrosion and chalking resistance, resistance to chemicals and aggressive industrial atmospheres rich in SO_2. They are very easy to maintain because of extremely low dirt pick-up properties. These unique properties are due to the small bond polarization of the fluorine-carbon bond, which is responsible for the low surface energy of the PVDF used as a binder in these coatings. The very high bond energy of the carbon-fluorine bond (477 kJ/mol) provides exceptional weathering resistance [15, 16]. PVDF can be used as a sole binder in the manufacture of powder coatings, especially when high weathering resistance is required, although this is not a common case in practice. The main reasons include the high melt viscosity resulting in the formation of pinholes during thin film applications, rather poor adhesion to metal substrates, and the relatively **high price**. 3M filed a patent application [17] describing the addition of hydroxyl or amine functionalized

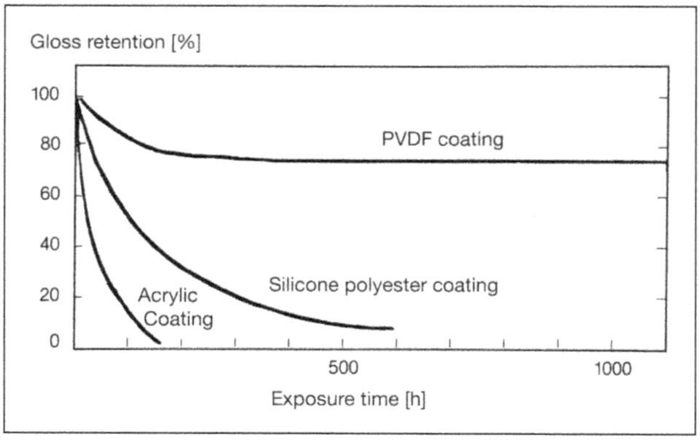

Figure 2.4: Gloss retention of PVDF coating on exposure in a Dew Cycle Weather-O-Meter [15] *source: Elsevier Sequoia SA*

aromatic compounds like fluorphenol or 2-naphtol to fluoropolymers to enhance adhesion to metals and ceramics. Daikin Industries claimed [18] that adhesion as good as on chromium phosphate primers is achieved by the combination of amido and/or imido containing fluoro resins.

In order to improve the melt flow properties, adhesion and aesthetic appearance of the powder coatings, acrylic resins are usually blended with PVDF. It is common that the binder for PVDF powder coatings contains **30 % acrylic resins**. Higher amounts of acrylic resin decrease the weathering resistance of the coatings, although it still remains superior to other manufactured organic coating materials known so far.

Alternating copolymers of tetrafluoroethylene and ethylene are claimed [19] to have excellent mechanical strength and processability making them suitable for chemical plant equipment, reactors, vessels and pipes.

The results of a study of the weathering resistance of PVDF coatings containing 25 to 40 % acrylic resin baked at 300 °C are presented in *Figure 2.4* [15]. Similar results with respect to accelerated weathering in QUV apparatus are obtained by PVDF powder coatings containing 30 % acrylic resin, baked for 10 minutes at 230 °C *(Figure 2.5)* [20, 21].

PVDF powder coatings have low gloss in the range of 30 ± 5 % (at 60°). This is certainly a restriction for a whole range of application of PVDF powder coatings for decorative purposes.

The process of production of PVDF powder coatings does not differ from that of the other powder coatings. It consists of extrusion of the resins and pigment premix in a single or twin screw extruder followed by granulation and drying of the granulate. The next step is cryogenic grinding and sieving in order to obtain powder particles below 50 µm [20]. An alternative method has been described in a Japanese patent [22]. Fluor containing polymers are dissolved in isophorone at 200 °C, spray-dried to spherical particles, applied electrostatically on stainless steel and heated at 300 °C to give a pinhole free coating.

The very **low surface energy** of PVDF, which provides low dirt pick-up properties to the coating, at the same time, causes difficulties concerning adhesion to the substrate. This, in general a weak point of thermoplastic powder coatings, is even more emphasized in the case of PVDF powders. As already mentioned, the blends with acrylic resins improve adhesion, but even in this case, direct application of the PVDF powder coatings to a metal substrate is not recommended. For good adhesion, PVDF powder coatings are applied over an epoxy stoving primer [20, 21]. Systems based on polyurethane primer with a PVDF top coating are also reported [23]. A Chinese publication [24] recommends an addition of fluoroethylenepropylene/polyphenylene sulphide mixtures for enhanced adhesion on metals.

The problem of **bad adhesion** can be overcome in the way described in the US patent from 1974 [25] where a two-coat system based on polyvinylidene fluoride is used. The first primer coating is produced by physical mixing of polyvinylidene fluoride of a particle size from a 60 to 200 mesh and silica sand of a particle size from a 150 to 325 mesh. The adhesion of the coating was tested by exposure to steam at 100 °C. The time before the occurrence of blisters, taken as the relevant parameter for estimation of the adhesion properties of the system, ranged from 7 hours in the case of the primer containing no silica sand to more than 480 hours for the one having a PVDF/silica ratio of 100/40 or more. Similar results were obtained with a primer in which graphite of the same particle size was used instead of silica sand [25].

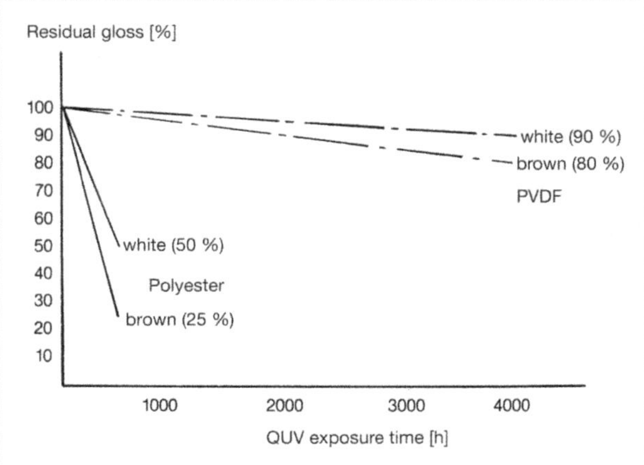

Figure 2.5: Accelerated weathering (QUV) of PVDF white and brown coatings compared with polyester powder coatings [21]
source: Technology Communicative Markt BV

Recently Arkema published a patent [26] in which maleic anhydride was grafted on PVDF by means of γ-radiation. These modified PVDF resins show a significant improvement in adhesion on untreated metal substrates like aluminum.

Although for the coil coatings based on PVDF as a binder a 20 year guarantee is normally given [15, 23, 27], for PVDF powder coatings containing 30 % acrylic resins, a 10 year guarantee for a maximum of 50 % loss of the initial gloss has been granted [20].

PVDF powder coatings are mainly used as **architectural coatings** for monumental-type structures. Architectural panels for roofing and wall claddings, and aluminum extrusion window frames are the main surfaces on which PVDF powder coatings have been applied.

2.2 *Polyolefinic based powder coatings*

The discovery of the new coordination catalysts for polymerization of olefins in the first halve of the 1950's resulted in the development of a new class of polyolefins. This invention has not only made a big commercial impact, but has also contributed very much to the development of polymer science.

By the use of coordination catalysts, or the so-called *Ziegler-Natta* catalysts after the names of the inventors, polypropylene and low pressure polyethylene are produced on an industrial scale, being two of the **most important thermoplastic polymers** of this century. They are the main representatives of the polyolefins used in the manufacture of powder coatings. Being non polar, high molecular weight crystalline polymers, with carbon-carbon bonds in the polymer backbone, they possess a unique balance of toughness, chemical and solvent resistance. It is obvious that such materials could be of great inter-

Table 2.3: Typical properties of polyethylene obtained by the three basic processes [30]
source: John Wiley & Sons, Inc.

	High pressure process	Medium pressure process	Ziegler process
Crystallinity	65 %	95 %	85 %
Relative rigidity	1	4	3
Softening temperature	104 °C	127 °C	124 °C
Tensile strength [bar]	138	380	241
Elongation [%]	500	20	100
Relative impact strength	10	3	4
Density [g/cm³]	0.92	0.96	0.95

est when used as binders for protective coatings. However, their insolubility in the solvents commonly used by the paint industry restricted their application to the powder coatings only. In fact, one of the first powder coatings which were used in a fluidized bed application at the beginning of the 1950's, was based on powdered polyethylene [28].

The empirical formula of polyethylene $-(CH_2-CH_2)_n-$ is rather simple and does not indicate the many implicit structural possibilities that can exist, offering products with different properties and uses. The molecular weight of polyethylene and molecular weight distribution, the amount and length of side chains that depend on the manufacturing conditions, together with different degree of crystallinity are parameters that influence to a great extent the properties of the polymer. The crystallinity of polyethylene in particular, directly reflects the degree of packing of polymer chains and thus the density of the material and is an important parameter distinguishing several classes of commercial products in the market. According to the **ASTM classification**, commercial grades of polyethylenes are grouped into four types based upon densities regardless of the molecular weight or molecular weight distribution:

Type	**Density** [g/cm³]
I	0.910 to 0.925
II	0.926 to 0.940
III	0.941 to 0.959
IV	0.960 and higher

When the density or crystallinity of polyethylene is reduced, then the impact resistance, cold flow, tackiness (blocking), clarity and permeability increase. On the other hand, a reduction in density produces a decrease in stiffness, tensile strength, hardness, abrasion resistance, brittleness temperature, softening point, fusion temperature, chemical resistance and shrinkage.

Both, low and high density grades polyethylene are successfully used for the manufacture of powder coatings. The high density grade is suitable for industrial application, while the low density is used for domestic applications [29]. The first process with a commercial potential of polymerizing ethylene under high pressure and temperature was discovered in the laboratories of Imperial Chemical Industries, Ltd. in 1933. Polyethylene produced in this way (temperatures of 100 to 300 °C and pressures above 1000 bar) has a high molecular weight, broad molecular weight distribution, a high degree of branching, moderate crystallinity and low density.

By the discovery of *Ziegler* in 1953 polyethylene was produced in an astonishingly simple process at atmospheric pressure and almost room temperature by using alkyl aluminum compounds combined with titanium tetrachloride as catalyst. This was an outstanding event not only from the technological point of view, but also because it offered polymer that was highly oriented with a high degree of crystallinity, almost without side chains, and properties that differed very much from those obtained by the high pressure process.

Another method of production of polyethylene was developed in USA about the time of Ziegler's discovery, and employed medium pressure and temperature during polymerization in the presence of metal oxides as catalysts. These processing conditions lie in between high pressure polymerization and the process of Ziegler.

Table 2.3 presents the **basic differences** in the properties of polyethylene polymers obtained by these three different methods [30]. Other important differences from the standpoint of the powder coating chemist are presented in *Table 2.4* [30].

The other representative of polyolefins that is used as a binder for powder coatings, polypropylene, can be produced by a wide variety of catalysts, including cationic, anionic and

Table 2.4: Important differences between different polyethylene types with respect to their application in powder coating[30] *source: John Wiley & Sons, Inc.*

	Low density	Medium density	High density
Resistance to acids	Resistant	Very resistant	Very resistant
Oxidizing acids	Attacked	Slowly attacked	Slowly attacked
Resistance to alkalis	Resistant	Very resistant	Very resistant
Resistance to organic solvents	Resistant below 60 °C	Resistant below 60 °C	Resistant below 80 °C
Clarity	Translucent	Translucent	Opaque
Crystalline melting point [°C]	108 to 126	126 to 135	126 to 136
Heat resist. (contin. use [°C])	82 to 100	104 to 121	121
Density [g/cm³]	0.910 to 0.925	0.926 to 0.940	0.941 to 0.965
Elongation [%]	90 to 800	50 to 600	15 to 100

free radicals. The technique for polymerizing polyethylene under high pressure has been unsuccessful for polypropylene. The explanation is that polypropylene radicals tend to chain transfer to C_3H_6 giving the unreactive allyl radical. Therefore, the free radical polymerization of propylene has almost no commercial importance.

The polymerization process catalyzed by cationic catalysts of *Friedel-Crafts* type results in amorphous polypropylene of low molecular weight. There is a typical drawback of the polypropylene obtained in this way, which restrict the application of the polymer in the powder coating field; namely, the polymer chains obtained by cationic polymerization are without structural order. As a consequence, crystalline polymers can not be made by this route.

The production of polypropylene with properties useful for most modern applications became possible with the discovery of the *Ziegler-Natta* catalysts. The polymerization of propylene by the anionic coordination mechanism produces a stereo regular crystalline polymer with a melting point high enough from a practical stand point.

Three types of polypropylene can be distinguished with respect to the stereo regularity of the chains. Atactic polypropylene without a regular structure and obtained by cationic polymerization is an amorphous, soft and tacky material with little or no physical strength. On the other hand, isotactic and syndiotactic polypropylenes are obtained by anionic polymerization. These structures are schematically presented below:

Equation 2.2

The ordered structures of the syndiotactic and isotactic polypropylene allow a great degree of crystallinity resulting in a polymer with high physical strength, solvent resistance and chemical resistance.

The anionic polymerization of propylene mainly gives the isotactic type of polymer. This is one of the **lightest commercial plastics** with a density of only 0.9 g/cm³. The melting point is in a range from 165 to 170 °C for the commercial grades, while the melting point of 100 % isotactic polypropylene is 183 °C.

Syndiotactic polypropylene is not produced on a commercial scale. Its melting point is about 20 °C lower than that of the isotactic type, and it is more soluble in ether and hydrocarbon solvents.

The same catalysts used for production of high density polyethylene at atmospheric pressure are suitable for polymerization of polypropylene. The catalytic system usually includes titanium chloride, an aluminum compound, preferably trialkyl aluminum or aluminum chloride, and optionally a *Lewis* base or other modifier.

Polypropylene has a number of desirable properties that make it a versatile material for production of powder coatings. It possesses good surface hardness and is scratch and abrasion resistant. It is essentially unaffected by most chemicals and possesses excellent solvent resistance. It is somewhat more brittle than polyethylene at normal temperature, which is caused by the relatively high glass transition temperature ranging between 25 and 35 °C depending on the degree of crystallinity. The brittleness of the commercial grades and impact resistance are markedly improved by copolymerization with other olefins. Considerable amounts of polypropylene are sold as copolymer containing 2 to 5 % of ethylene. The resulting polymer has increased flexibility, toughness, impact resistance and clarity, and somewhat lower melting point.

The fact that the fragility of polyethylene and polypropylene is increased by cooling with liquid nitrogen or by soaking with alcohol before grinding is used as the basis for several techniques to make fine powders. Also some polymerization processes yield polyethylene directly as fine powder. High-pressure polyethylene, however, is produced as a solid resin and must be ground to produce powder.

Because of the carbon-carbon bonds in the molecular chains, polyethylene and polypropylene powder coatings exhibit very good chemical resistance. Due to the nonpolar paraffinic nature of the binder they have a reputation for coatings with excellent solvent resistance. Therefore, a very **important application** field for polyethylene and polypropylene powder coatings is the protection of chemical containers, pipes and pipelines for keeping and transporting different chemicals and solvents.

Being inert materials, polyolefins have **poor adhesion to metal** or other substrates. Therefore, before polyethylene and polypropylene powder coatings can be used successfully, the surface to be coated must first be primed or adhesion promoters must be added to the powder coating to improve adhesion. A family of adhesion promoting polymers, called "Hercoprime". has been developed, which, when blended with polyolefins, and specially with polypropylene, give a one-coat system with good adhesion [6, 31]. These polymers are produced in the form of small particles whose size is of a range typical for powder coatings. Their density is similar to that of polypropylene, and they can be used by simple tumble blending with polypropylene powder coatings. For pigmented systems, a 15 % addition of these adhesion promoting resins provides good adhesion for most of the substrates. Clear propylene coatings need 5 to 10 % of primer promoter to exhibit satisfactory adhesion properties [32]. Different modifications of polyolefins have been tried in order to improve the adhesive properties. In some cases these are simple blend of polyethylene or polypropylene with polymers containing acid groups as adhesion promoters. A Dainippon patent [33] describes a modified polyolefin composition prepared from 60 to 97 part of polyolefin, and 3 to 40 parts of copolymer based on 20 to 60 % acid monomers, 40 to 80 % styrene and 0 to 40 % alkylmethacrylates. It is claimed that the powder coating possesses excellent adhesiveness, coatability and color stability. In examples of another patent referring to improved adhesion of polyolefin powder coatings [34], 0.05 to 40 parts of a polypropylene modified product containing 15 to 30 % by weight grafted maleic anhydride and 99.5 to 60 parts of polyolefins are used as a binder composition for polyolefin powders. Excellent adhesion of the coating is combined with excellent mechanical properties.

At the PCE 2002 Congress in Nuremberg, Plascoat [35] also came with a thermoplastic coating powder, based on an "alloy" of acid modified polyolefins. They claim their powder has excellent adhesion to the metal substrate, without the need of a primer. The material also provides good abrasion and impact resistance and is resistant to stress cracking and adverse weather conditions.

A powder composition with improved adhesion properties is obtained by physical mixing of polyethylene powder with powder of copolymer comprising 60 to 95 % of ethylene and 5 to 40 % of glycidyl methacrylate [36]. The coating is applied in a fluidized bed on a preheated surface at 300 °C followed by postcuring at 200 °C for 20 minutes. It is claimed that the adhesion properties of this coating are much improved in comparison with pure polyethylene powder. A Chinese research group filed a patent application [37] describing the beneficial effect of silane modified nano-silica for the adhesion and corrosion resistance of polyethylene coated steel pipes.

Ethylene vinyl alcohol copolymers are described [38] to be highly resistant against hydrocarbons and organic solvents. In addition they show outstanding oxygen barrier properties which make them suitable for chemical packaging, fuel tanks and protective clothing.

Polyethylene powder coatings are even more problematic with respect to adhesion properties than polypropylene powders. Priming of the surface is usually necessary prior to application of the polyethylene powder coating, since the adhesion promoting resins do not perform as well as in case of polypropylene. A patent assigned to Hoechst describes a two-step method for heavy duty protection of pipelines for the transport of oil [39, 40]. In the first step, a bisphenol A epoxy resin with amino hardener is sprayed over preheated metal pipe at 270 °C in a 100 μm thick layer. The still soft, highly viscous layer is coated with a powder coating consisting of 75 % polyethylene and 25 % copolymer of vinyl acetate-ethylene-acrylic acid-alkyl acrylate in a weight ratio of 3:76:1:20. The total thickness is very high (above 2 mm) providing extremely good mechanical protection. Pipes can be transported without damage over rocky ground, they are heat stable and they resist desert conditions.

Three-layer polyolefin pipe coatings have gained widespread **international acceptance**. The first layer is thermosetting epoxy powder or solventless liquid coating with a thickness between 60 and 100 μm. The intermediate polymer layer with a thickness between 250 and 400 μm is a polyethylene powder coating and is applied over the epoxy primer, followed by the third layer as a topcoat which is applied by extrusion with a coating thickness between 1.5 and 3.0 mm. These systems are well established in Europe, the Middle East and Asia, offering excellent corrosion resistance and interfacial properties of epoxy coatings, combined with excellent chemical, solvent and mechanical resistance of the thick polyethylene coatings. The Hajira-Bijaipur-Jagdispoor 2800 km long pipeline system in India is an example of one of the major operations to be coated with three-layer polyethylene coatings [41].

A one-step application procedure refers to a coating system that combines the excellent **anticorrosive properties** of zinc dust, adhesion of the epoxy powder coating and good solvent, chemical and weather resistance of the polyethylene powder coating [42]. An epoxy powder coating comprising bisphenol A epoxy resin as the main binder, dicyandiamide as the hardener, barium sulfate and carbon black as the filler and pigment, catalytic amounts of amine accelerator and flow additive with particle size of 75 to 150 μm, is dry blended with clear polyethylene powder and zinc dust in the ratio of 70:30:5. Powder coating prepared in this way has been applied on steel panels using an electrostatic spray gun in a one-step operation. After curing for 3 minutes at 150 °C and raising the temperature at a linear rate to 215 °C for 10 minutes, a flat finish is obtained with a clear polyethylene top coating and a zinc-rich bottom layer covered with an epoxy intermediate thick layer. The main proposed application is to coat the interior of underground oil or gas pipes. The zinc layer produces a galvanized finish on the interior of the pipe, while the epoxy part overlaying the zinc serves to protect the zinc from abrasion as well as providing an integral coating with high corrosion resistance. The polyethylene top coating is an electrically non-conducting layer, and thus prevents or minimizes electrolytic corrosion, providing at the same time excellent solvent resistance. The same coating technique is proposed for applications where the polyethylene top coating contributes in improving **exterior durability**, for example in coating automobile wheel rims.

Kodokian and fellow researchers [43] at DuPont describe a novel approach of reacting a thermoplastic ethylene copolymer with a solid bisphenol A epoxy, which provides the possibility of a new and tough corrosion resistant powder coating system, that is not readily available from pure thermoset powders. They use an ethylene copolymer with (meth)acrylic acid for reaction with the epoxy resin and catalyze it with 0.5 to 1 % TBPBr (tetrabromophosphonium bromide). They suggest end uses for these modified thermoplastic systems as automotive and industrial metal coatings and even as coatings for heat sensitive substrates.

Powder coatings based on polyolefin polymers have a **low aesthetic appearance**, and, therefore, are used mainly for protective rather than decorative purposes. For pigmentation, dry blending techniques with high shear mixers can be successfully used. This is not of course a suitable technique for highly decorative coatings, but in general, for protective polyolefin powder coatings, it seems to be a method that gives good compromise between processing costs and coating performances.

In the case of polypropylene powder coatings, high gloss can be obtained by water-quenching immediately after fusion. Because of the rapid cooling, the spherulite size of the crystalline polypropylene is smaller, resulting in a film with less surface roughness and imperfections combined with higher flexibility. Coatings with better aesthetic appearance can be obtained by using higher molecular weight polyolefins, since the crystallization rate considerably decreases due to the reduced polymer chain mobility in the melt.

Physical aging of high molecular weight polypropylene has in principle a less emphasized effect on decreasing impact strength. High molecular weight polypropylene powder coatings retain their initial impact strength over a six-month period after quenching, while lower molecular weight grades exhibit slight decreases with time.

Polyolefinic powder coatings are resistant to a variety of hostile environments. **Solvent resistance** at room temperature is excellent against all common solvents used in industry. However, at elevated temperatures, non-polar solvents, such as hydrocarbons and aromatics, will swell or in extreme cases dissolve the coating. Resistance against water, salt solutions, detergents, acids and bases is excellent. Only very strong oxidizing agents, such as nitric acid, will attack the coatings at room temperature.

Although polyolefinic powder coatings are not recommended for outdoor use where decorative purposes are requested, stabilized versions of polypropylene can be used for the manufacture of powder coatings with reasonable gloss retention when exposed to weathering conditions. The main use of polyethylene and polypropylene powders, however, remains in the field of **functional coatings** with excellent protective properties. Applications for these functional coatings include appliance parts, such as dishwasher racks, refrigerator racks, washer tubes, metal containers, drums, pipes, electrical applications and industrial equipment components [44]. Polyolefinic powder coatings comply with FDA regulations, and can be used for covering surfaces which come in direct contact with food products.

2.3 Nylon based powder coatings

The early work on polyamides resulting in production of the first industrial polymers known under the name nylon was done by *Carothers* in the laboratories of DuPont. A period of 10 years of intensive work starting from 1929, in which the basic patents disclosing the process of polycondensation of diacids and diamines or self-condensation of amino acids were issued [45-48], was necessary to build up the base for commercialization of polyamides. In the same period a process of preparation of polyamides from lactams was protected by I.G. Farbenindustrie AG in Germany [49]. Although the experimental work on polyamides began with amino acids capable of self-condensation to form linear polymers, the first polymer produced on an industrial scale was the so-called nylon-6,6, prepared from adipic acid and hexamethylene diamine; production started in 1939 in DuPont's plant for manufacturing nylon products [50]. The success of nylon textile yarns initiated enormous research work on the development of new types of polyamides and the extension of application outside the textile industry. This resulted in the development of a very **large family of polyamides**, which were used as molding powders, extruded films and tubes, surface coatings, adhesives, printing inks, etc. As usually happens, the commercial name nylon became a synonym for the whole group of polymers, and it is used interchangeably with the name polyamide, a practice widely followed in this book.

There are three general routes used to produce polyamides. The first commercial polyamide, the so-called nylon-6,6 , was produced by condensation of adipic acid and hexamethylene diamine. This is done in a two-step procedure. In the first step a salt between the diacid and diamine is formed, usually in an aqueous medium. In the second step of the process the salt is dehydrated by heating. The elimination of water as a by-product results in the formation of an amide group connecting the monomers in a repeating unit:

Equation 2.3

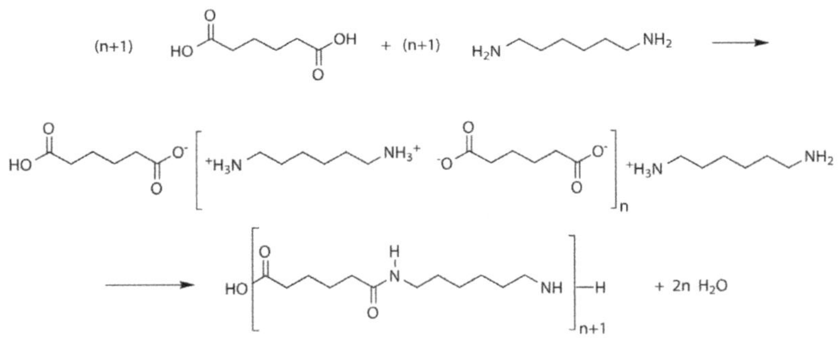

The formation of the salt and its purification is an essential step in the process providing a stoichiometric ratio between the reaction partners, necessary to avoid terminating the polymer chain by having an excess of either reagent present. The polymerization reaction is carried out at relatively high temperatures between 260 and 290 °C, without a catalyst or in the presence of acid compounds which exhibit the catalytic effect.

There is no principle difference between the process of **polycondensation** of diamines and diacids and the method of producing polyamides starting from amino acids:

Equation 2.4

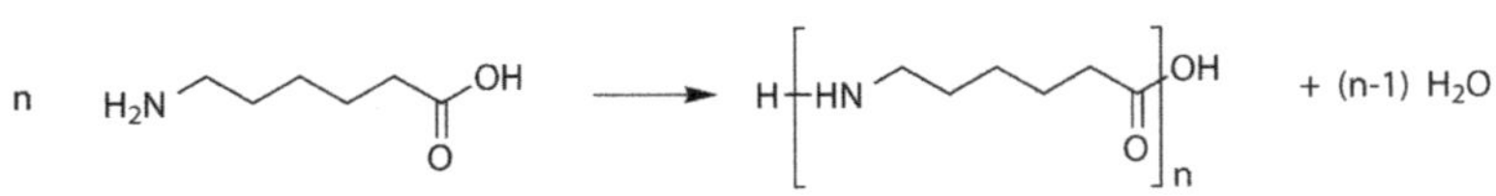

In both cases it is important that the combination of diamines and diacids or the amino acid itself is not capable of forming a five- or six-membered ring as a result of the condensation reaction. Since these rings are thermodynamically preferred structures, the condensation will lead to the formation of cyclic materials rather than polymers. Therefore, amino acids suitable for polymerization into polyamides should contain more than 6 carbon atoms.

A very important commercial route for preparing polyamides is the thermal **polymerization of lactams**. The ring opening may be affected by initiation with water leading to hydrolysis of the lactam to the corresponding amino acid, or by direct addition of amino acid as the polymerization promoter:

Equation 2.5

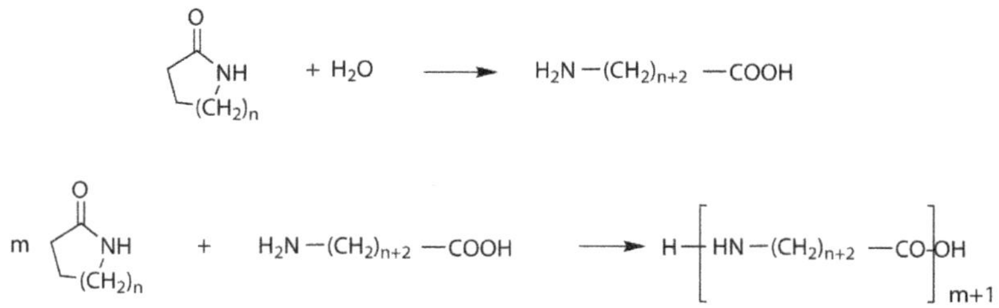

A worldwide accepted **nomenclature for polyamides** was established very soon after their introduction in the market. When the polymer is obtained by condensation of diamines and diacids, the word polyamide (or generic name nylon) is followed by two numbers. The first represents the number of carbon atoms in the diamine, and the second the number of carbon atoms in the diacid. For example, nylon-6,6 is the name of the polyamide produced by condensation of hexamethylene diamine and adipic acid. When the product is obtained by self condensation of amino acids or corresponding lactams, then the number which follows the name polyamide (or nylon) is equal to the number of carbon atoms in the amino acid in question. For example, nylon-6 is the name of the polymer obtained by condensation of 6-aminohexanoic acid or its lactam.

The considerable number of different raw materials available for synthesis of polyamides makes it possible to produce polymers which cover wide variety of structures having different physical properties. Because of the structure regularity, most of the commercial types of polyamides are **crystalline materials** with relatively well-defined melting points. Compared to the crystalline aliphatic polyesters, melting temperatures of polyamides are much higher. This is caused by the high polarity of the amide groups and the existence of interchain hydrogen bonds. As expected, polyamides melt at increasingly higher temperatures as the concentration of the amide linkages increases. As in the case of crystalline polyesters, polyamides with an even number of CH_2-groups between the amide polar linkages, have higher melting points compared to those with an odd numbers of CH_2-groups. *Figure*

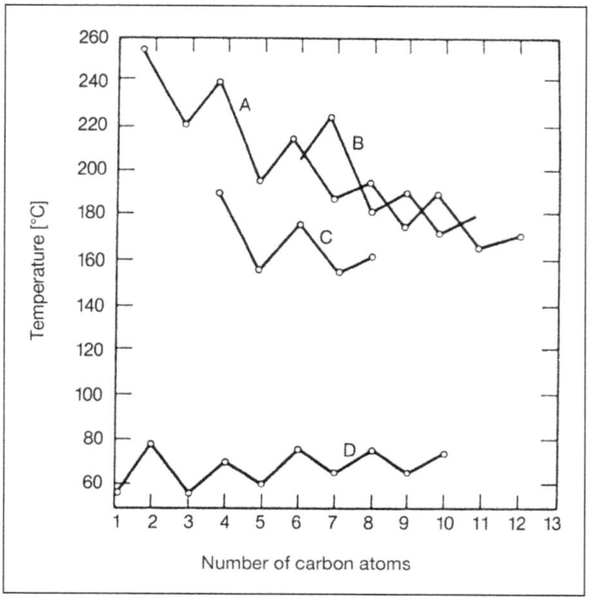

Figure 2.6: Effect of frequency of polar linkages on melt temperature for (A) polyamides, (B) poly(ω-amino acids), (C) polyurethanes, (D) polyesters [30, p.522]

source: John Wiley & Sons, Inc)

2.6 gives a comparison between the melting temperatures of different polymers with different numbers of carbon atoms in the repeating unit.

Normal baking temperatures for powder coatings are below 200 °C. Therefore, those polyamides having low temperatures of melting are given preference over the others, because of energy saving, measures. Nylon-11 has a relatively low melting point of 185 °C. Together with nylon-12 which has a melting point of 178 °C they are practically the only polyamides that have found wide applications as binders for powder coatings. Despite their wide availability and comparatively low price, nylon-6, nylon-6,6 and nylon-6,10 with melting points of 215, 250 and 210 °C respectively, have not been significantly accepted by powder coatings producers.

Nylon-11 is a condensation product of 11-aminoundecanoic acid. It was introduced to the market in 1955 by the company Aquitaine-Organico in France under the name "Rilsan". Nylon-12 was pioneered during the early 1960's by Chemische Werke Huels AG (now Evonik Industries) in Germany under the brand "Vestamid" (nylon-12 granules) and "Vestosint" (nylon-12 powders) and can be defined as polylauryl lactam.

The monomer for synthesis of nylon-11 is prepared from castor oil by its hydrolysis to ricinoleic acid followed by oxidation to undecylenic acid, hydrobromation to 11-bromoundecanoic acid and its amination to 11-amino-undecanoic acid:

Equation 2.6

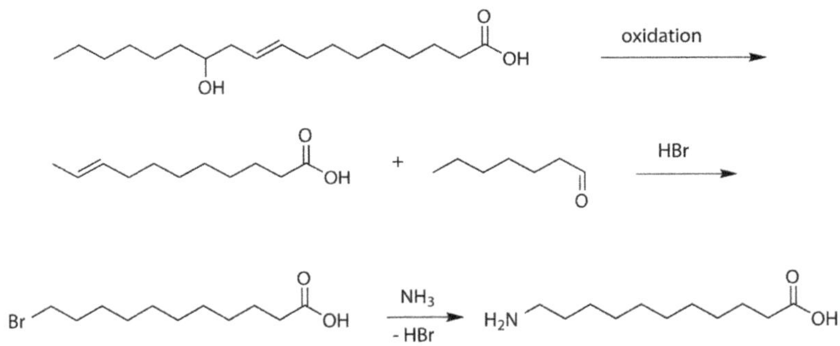

Melt polycondensation of 11-aminoundecanoic acid is carried out at 215 °C under nitrogen in order to suppress the occurrence of side reactions.

The production of nylon-12 is based on self-condensation of lauryl lactam according to the reaction scheme already given. Lauryl lactam itself is made in a multi step synthesis starting with butadiene *(Equation 2.7)*.

Equation 2.7

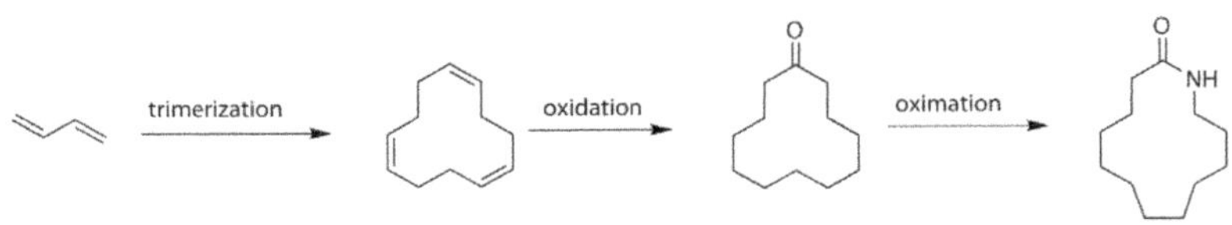

Both, nylon-11 and nylon-12 are practically insoluble in the common organic solvents. However, they are soluble or easily attacked by phenols, formic acid, mineral acids and similar compounds even at room temperature. At higher temperatures they are soluble in mixtures of alcohols and halogenated hydrocarbons, nitro alcohols and calcium chloride-methanol mixtures.

When exposed to light, nylon-11 and nylon-12 like all other polyamides have a tendency to lose tensile strength. This is especially the case in the presence of air. The **deterioration of mechanical properties** is usually accompanied by discoloration of the polymer. Light, oxidative and thermal degradation of polyamides have been the subject of many investigations. The most probable mechanism involves the free-radical chain reaction which is peroxide catalyzed. It is believed that oxidation sites exist at the methylene group adjacent to the carbonyl or the amide group [50, p.41]. *Gaumet* et al. [51] identified carboxylic acids as the major critical photo degradation product by means of FTIR. By using the Xenotest 1200 device according to the ISO 4892 standard they were able to simulate natural outdoor aging of powder coatings. The peroxide-catalyzed chain degradative mechanism is supported by the fact that small amounts of antioxidants drastically reduce thermal degradation of polyamides. Different types of **stabilizers** have been developed in order to increase the service life of the polyamides. Copper- and manganese salts and organic antioxidants such as N,N'-di-ß-naphtyl-p-phenylenediamine have been reported as effective stabilizers for polyamides [52].

Nylon-11 and nylon-12 are unaffected by water at room temperature, or even by boiling water. Stability against the action of alkalis is also very good. However, the stability in **acidic media** is not a strong point of polyamides in general.

The first nylon powder coatings were introduced in Europe using nylon-11 and nylon-12 as a binder system [6, p.182]. Several unique properties of nylon based powder coatings give them an advantage over other powder coatings. They are characterized by very good toughness, high abrasion resistance combined with superior impact resistance even at low temperature. Their low coefficient of friction and unusual wearing properties make them **excellent interfacial coatings** to reduce metal-to-metal wear or noise. Good thermal insulation is another distinguishing feature of nylon powders. Additionally, most of these properties are retained over a wide range of working temperatures. Nylon powder coatings have exceptionally good **resistance** against most of the solvents commonly used in practice. This is to be expected and emanates from the presence of the amide groups characterized by the formation of strong hydrogen bonds. The resistance to organic acids, inorganic salts and alkalis in low concentration is also good.

The possibility of **stabilization** of nylon makes this material attractive not only for indoor functional coatings, but also for **outdoor use**, where the weather resistance of the coating is combined with good chemical and humidity resistance, high impact and abrasion resistance, and high wearing properties. Although weathering data for nylon powder coatings are not widely available, there are indications that this could also be a strong point in connection with this system. *Blackmore* [53] has reported excellent results with coatings based on nylon, having a service life time of more than 10 years. Similar results are obtained with external cladding panels coated with nylon based powders [54]. According to *McAndrew* [55] cavitation corrosion of nylon is much less observed than in other coatings, including epoxies. Even stainless steel doesn't meet the high performance of nylon.

Problems which could be encountered with nylon powder coatings derive from their **easy contamination** by other materials such as PVC, epoxies, polyethylene etc. Even the epoxy booth-sprayed cyclones ruined nylon powder coatings reclaimed through them as reported by *Blackmore* [53]. However, modern nylon coatings have suitable anti-crater additives incorporated into them and, do not suffer from these problems [56].

Like most thermoplastic powder coatings, the use of a primer is recommended in order to improve adhesion to the substrate. Due to the semi-crystalline character of nylon-11 and nylon-12 the coating films show shrinkage of ca. 1 %, which increases the adhesion properties especially on wire articles. In general, suitable primers for nylon powder coatings are solventbased coatings used in conventional systems.

Atofina [57] markets "Primgreen", a **waterbased primer**, which promotes the adhesion of nylon powder coatings to metal parts. The primer contains less than 20 % of volatile organic compounds and, therefore, helps minimizing the VOC emissions during handling and application. "Primgreen" can be applied by spraying or dipping.

In some cases where the end user has not given high demands with respect to adhesion, a single coating procedure applying the nylon powder coating to an unprimed surface can be used.

A possible **improvement of adhesion** properties of nylon powder coatings is suggested by *Wertz* [58] by melt blending minor amounts of a reactive epoxy resin and polyvinyl acetal into the coating composition, while omitting an epoxy curing agent from the formulation. It could be expected that during the melting of the coating the epoxy compound would not react with the hydrogen belonging to the amide group of the polyamide. Surprisingly it does not affect the mechanical properties, but it does provide an improvement in the adhesion. Epoxy resins with epoxy equivalent weights of between 200 and 2000 are recommended for this purpose, in an amount from 2 to 15 parts by weight per 100 parts of nylon, with inclusion of 1 to 10 parts by weight of polyvinyl acetal. The formulation given in *Table 2.5* [58] illustrates an example of such a powder coating based on nylon-11 as the main binder.

Guichenuy et al. [59] reported that addition of aminopropyl triethoxysilan improves the adhesion of polyamides on a steel surface significantly. ToF-SIMS revealed covalent Fe-O-Si bonds, similar to the deposition of a silane layer prior to coating application. Aminosilane and polyamide do not react with each other and thus are forming interpenetrating networks [60].

Polyamide powder particles are usually **precipitated out of solution** of inert solvents. According a recent Japanese patent [61] it is possible to use fine polyamide powder as seeds to come to spherical particles with a narrow particle size distribution.

The **fluidized bed** method is the most commonly used application technique for nylon powder coatings. In this way, thick coatings in the order of 200 to 700 μm can be applied in a one-step coating procedure. Depending on the velocity of the conveyor chain on which the articles are hung and the mass of the article to be coated, the fusion temperatures of nylon powder coatings are between 200 and 230 °C [56]. The **temperature most recommended is 220** °C [6, p.204]. For thin film application, electrostatic spraying can be successfully used with either positive or negative charge. However, it has been noticed, that the positive charge deposits greater quantities of powder in a given time. Normal spraying equipment for the other powder coatings, operating at between 30 and 70 kV, is also suitable for nylon powders. The film thickness obtained by electrostatic spraying is usually between 100 and 150 μm. Recently *McAndrew* [62] reported that flame spraying could be an alternative application method for polyamides.

The powder grain shape and the particle size distribution influence the fluidizing behavior, flowability, spraying and transportation behavior of a nylon powder extremely. Precipitated nylon powders (e.g. "Vestosint", see *Figure 2.7*) have a potato-look **grain shape** which shows due to reduced electrostatic charging better processing properties than ground nylon powders. The more nearly spherical the powder grain, the better is the fluidizing behavior of the sintering powder and hence the better the flow into critical regions such as the crossover points of the wires. Moreover, the fluidized powder should contain only a small percentage of fine particles and no coarse particles, in order to avoid coating defects and excessive dust in the powder basin.

Table 2.5: Nylon powder coating with improved adhesion [58]

Material	Parts by weight
Nylon-11 (1)	100
Epoxy resin (2)	7
Polyvinyl butyral (3)	3
Calcium carbonate	20
TiO_2	8

(1) Rilsan BMNO – (Atofina); (2) Epikote 1001 – (Momentive); (3) Mowital B30H – (Eastman)

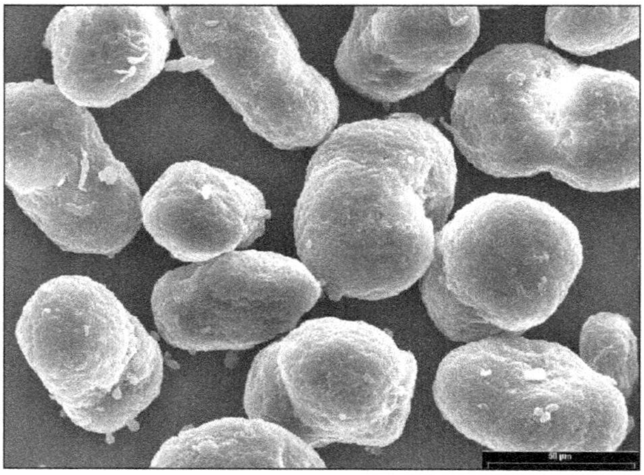

Figure 2.7: The grain shape of polyamide 12 ("Vestosint")
source: Evonik Industries

An advantage of nylon powder coatings concerning the application properties is the **high line speed**, that, until recently (see Chapter 6), could not be achieved with thermosetting powders. As thermoplastic materials, nylon powder coatings need only melting and fusion into continuous film without additional crosslinking. Moreover, the immediate cooling of the coated surface by immersing in water or by spraying with cold water without damaging the surface (because of low dimensional changes with temperature), also reduces the residence time of the article on the conveyor chain. The quenched cooling in water gives a high gloss finish, due to the minimized crystallization, while the slower air cooling produces matt finishes.

It has been found that infra-red ovens are not suitable for fusion of nylon powder coatings. They can be used in cases where the coated articles are flat sheets with an uniform thin film thickness. However, the high thermal insulation coefficient of the polymer and the sharp melting point are two factors that prevent the successful use of infra-red curing ovens when three-dimensional parts have to be covered with a thick nylon layer. In such cases **conventional recirculating air ovens** of the tunnel or box design with air temperature between 200 and 230 °C are the best choice [56].

Due to their specific properties, nylon powder coatings have found broad application fields. Their resistance to salt water and inertness to fungus, bacteria and barnacles, makes them very suitable coatings for articles that will be immersed or in touch with sea water during their service lives.

Nylon powder coatings are **non-toxic** and odor and taste free. This, combined with the fact that they are not attacked by fungus, and do not facilitate bacterial growth, makes them very attractive candidates for medical equipment, coating machine parts and piping used in the food processing industry, or any other place where food comes in direct contact with the coated surface. Powder coatings based on nylon have approval in all industrial countries for use with potable water and foodstuffs [56].

Another strong point of the use of nylon powder coatings is their excellent impact resistance, which is retained over a broad temperature range from -40 to 80 °C.

The low coefficient of friction, excellent abrasion resistance and low dirt pick-up, promote their **application** for coating dishwasher baskets, shopping karts, luggage trolleys, metal furniture, safety devices, sport equipment, farm machinery, etc.

Valve stems and seats, pump housings, degreasing baskets and trays, pump housing for domestic washing machines and large pipes for pipelines, are items that are also coated with nylon powder coatings, having the advantage of extremely good solvent resistance and resistance to weak alkalis and detergents.

McAndrew et al. [63] report that polyamide powder coatings are in particular suitable for bare brass, bronze, ductile cast iron, carbon steel, and stainless steel. Polyamides are claimed to be superior to epoxy coatings on these metals.

An interesting application of nylon powder coatings is in the coverage of handles of all types. Not only abrasion and scratch resistance are important elements in encouraging the use of nylon powders for this purpose, but their low thermal conductivity gives the handle a warm feeling, which makes these materials very interesting for coating of tool handles, door handles, steering wheels, etc.

High performance polyamides are usually based on special components. Polyphthalamides are aromatic, semi-crystalline, thermoplastic copolyamides produced by polycondensation. They are characterized by melting points above 300 °C. The glass transition temperature is often about 120 °C, but by selection of special monomers it can be adjusted in the range from 80 to 150 °C. Typical raw materials for polyphthalamides are terephthalic acid mixed with other diacids, like isophthalic acid or adipic acid, in combination with diamines. The most important diamines are hexamethylene and decamethylene diamine. The first mentioned case is abbreviated PA6T and the last is named PA10T. Polyphthalamides are especially suitable for parts subjected to high temperature. PA6T exhibits higher humidity resistance than polyamide 66 and a better dimension stability under humid conditions. PA10T is partially natural resourced and thus bio-based. Evonik Industries offers PA6T and PA10T in shape of granules and powder under the trade name "Vestamid HT plus".

2.4 Polyester powder coatings

Powder coatings based on thermoplastic polyesters as binders are prepared from linear high molecular weight polymers produced by **polycondensation** of dibasic acid and diols.

There are two main methods by which high molecular weight polyesters can be produced. A direct esterification of diacids with diols results in the formation of polyester and water as a by-product:

Equation 2.8

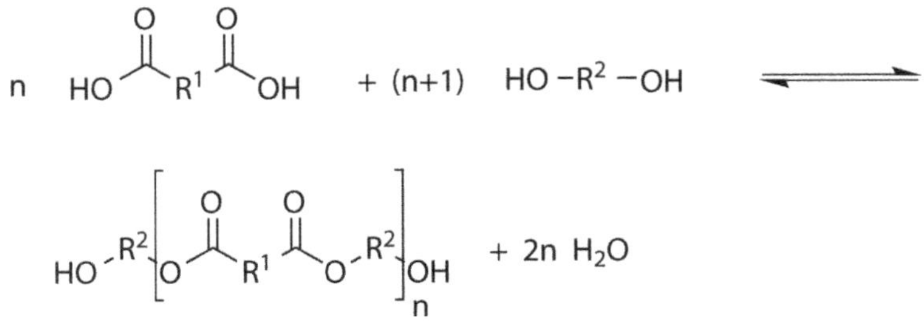

The other method is transesterification of diesters of diacids and diols under vacuum at high temperature, leading to a formation of polyester and corresponding diol as a by-product:

Equation 2.9

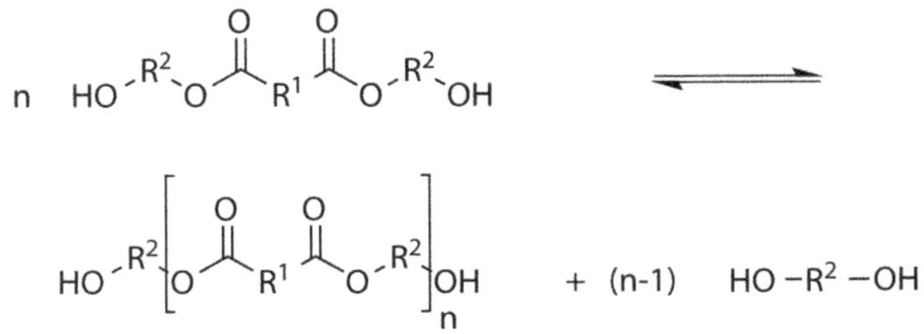

Direct esterification is a less suitable production method because it is very difficult to keep the correct mole ratio between the reactants in order to reach desired molecular mass. On the other hand, the process of esterification is reversible; in order to obtain a high degree of polycondensation, the by-product (water in this case) has to be permanently removed from the reactor so that the equilibrium state is avoided. *Schulz*[64] has derived an equation giving the dependence between the degree of polymerization P, the equilibrium constant K and the mole concentration of the reaction water C_W in the reactor:

$$P = (K / C_W)^{1/2}$$

In order to reach a high degree of polymerization the concentration of water should be as low as possible. This can be obtained by means of a vacuum. However, together with the reaction water a certain amount of glycol is removed, too. This makes it even more difficult to keep a well-defined mole ratio between the diacids and diols. This problem can be successfully solved by the so-called **azeotropic cooking technique**. Here recycling of an inert solvent is used to help the removal of water. Since the resins for making powder coatings must be solvent-free, an additional step in removing the solvent at the end of the process is necessary. It is very difficult to perform this step in the same reactor under vacuum, due to the high molecular weight of the resin and its high melt viscosity. The resulting heavy foaming during the stripping off of the solvent under vacuum can not be prevented. Therefore only a spray drying technique can be used for this purpose, which makes the process expensive.

Transesterification under high vacuum is a much more convenient way of producing high molecular weight polyesters. During the process, the resulting diol is permanently stripped off from the reactor by means of a vacuum, thus approaching the correct ratio for obtaining the desired molecular weight. The molecular weight of the polyester is dependent only on the efficiency of the glycol removal. Using the equation of *Schulz* in the case of polyethyleneglycol terephthalate at transesterification at 280 °C when the equilibrium constant has a value of 4.9, *Griehl* and *Förster*[65] have calculated that to obtain a molecular mass of 15,000 the concentration of ethylene glycol should be lower than 0.75 mmol/l. In order to keep the concentration of the glycols at this low level, a vacuum in the range of 0.7 to 1.3 mbar has to be used at relatively high transesterification temperatures. Technically such problems have been successfully solved so that at present this is the process most commonly used for production of high molecular weight polyester resins on an industrial scale.

The film forming process of thermoplastic polyester coatings does not involve any additional increase in molecular weight. Therefore, to obtain a coating having good mechanical properties it is necessary to start with a polyester resin with a relatively high molecular weight. For such applications typical molecular weights are higher than 15,000.

Another important characteristic of polyester resins for thermoplastic powder coatings is their ability to crystallize. A certain **degree of crystallinity** is necessary to assure a high enough **melting point**, which at the last instance determines the stoving temperature range of the coating during its application. For example, the melting point of the crystalline polyethylene terephthalate is 256 °C. However, polyethylene isophthalate with a much lower degree of crystallinity has a melting point of 108 °C, while the fully amorphous polyethylene orthophthalate melts at a temperature of only 63 °C [66]. By the correct choice of diacids and diols and their combinations, polyesters with melting points varying over a rather wide range can be synthesized. *Table 2.6*, compiled on the basis of a series of published results by *Korshak* and coworkers [67-74], gives an indication of the melting points of different polyester resins based on aliphatic and aromatic diacids. It is quite remarkable that polyester resins obtained from diacids with an odd number of carbon atoms have lower melting points, not only from their neighbors with an even number of carbon atoms, but also compared to all polyesters, that are based on diacids with even number of carbon atoms. The same tendency, although not so emphasized, can be noticed with the glycols. It can also be seen that aromatic diacids with a carboxyl group in the para position impart very high melting points to polyester resins. It is interesting that mixed polyesters based on aliphatic and aromatic diacids or different glycols have melting points which are not only in between the melting points of the pure polyesters, but for certain combination the melting points are even lower than the lowest melting point of the pure polyester, showing a minimum for a certain fixed ratio [74]. This phenomenon is often used to produce polyester resins with a rather high degree of crystallinity and a relatively low melting point.

Thermoplastic polyester resins for powder coatings differ in the **raw materials** used for their preparation and their ratio in the polyester formulation. Known processes include a mixture of polyesters made out of neopentyl glycol and terephthalic acid, copolyesters of cyclohexyl dimethanol, terephthalic and isophthalic acid in combination with acetylcellulose, where high gloss of the coating is obtained, and copolyesters of ethylene glycol and terephthalic acid containing 8 to 20 % of isophthalic acid, which exhibit good adhesion to metal substrate and good bending processability [75]. The main objection that could be addressed to these polyester resins is the rather inferior impact resistance and high fusion temperature in the neighborhood of 250 °C; a fluidized bed application would require a preheated surface up to 350 °C [76]. An interesting approach involves copolyester derived from a dicarboxylic acid composition comprising 40 to 60 mol-% of terephthalic acid and at least 70 mol-% of 1,4-butane diol with a melting point in the range between 100 and 150 °C determined by differential thermal analysis [77]. In the two-step production process, the blend of dimethyl terephthalate, dimethyl isophthalate and if necessary dimethyl adipate is transesterified in the first step with 1,4-butane diol using zinc acetate as the ester exchange catalyst at a temperature of 150 to 220 °C. In the second step, with addition of antimony trioxide as the polymerization catalyst, the reaction mixture is heated to 270 °C and subjected to glycol elimination under a vacuum of 0.7 mbar until a specified molecular mass is obtained.

Thermoplastic polyester powder coatings are based in most cases on **high molecular weight** crystalline or semi crystalline linear polyester resins, alone or in combination with other binders, in order to achieve desirable properties of the fused films. Problems which have to be overcome, when thermoplastic polyester resins are used as binders, are poor **adhesion** to the substrate, crack formation upon rapid cooling or bending, and poor luster in spite of the fact that the polyester resins offer reasonable weathering resistance. Addition of cellulose esters such as cellulose acetate butyrate or propionate improves the gloss of the coating to a considerable extent.

Thermoplastic polyester powder coatings should have a certain degree of crystallization when cooled down to room temperature, after the fusion of powder particles has been completed. The melting and crystallization of thermoplastic polyester resins is a reversible process. The crystalline regions act as crosslinking points eliminating the necessity of chemical crosslinking and the thermal curing step. The **degree of crystallization** is

Table 2.6: Melting points [°C] of polyester resins made from different pairs of glycols and diacids [67-74]

Glycol	Diacid								
	1	2	3	4	5	6	7	8	9
Ethylene glycol	159	-22	102	-19	47	25	70	256	330
Trimethylene glycol	66	-25	43	35	36	41	49	217	246
Tetramethylene glycol	103	-24	113	36	58	38	64	222	255
Pentamethylene glycol	49	-26	32	22	37	39	53	134	160
Hexamethylene glycol	70	-48	52	28	55	52	65	148	195

1 = oxalic acid; 2 = malonic acid; 3 = succinic acid; 4 = glutamic acid; 5 = adipic acid; 6 = pimelic acid; 7 = sebacic acid; 8 = terephthalic acid, 9 = p,p'-diphenyl dicarboxylic acid

important to obtain surface hardness and a reasonably high service temperature of the coating. On the other hand, too high a crystallinity reduces adhesion to the substrate due to an increase in the cohesion energy during crystallization.

Good adhesion to the substrate can be obtained when the diacids used to synthesize the polyester resin are a combination of isophthalic and terephthalic acid in a certain optimal ratio that adjusts the crystallinity of the system. It is claimed that polyethylene isophthalate copolyesters containing 8 to 20 mol-% of an isophthalic acid component with intrinsic viscosity of 0.7 to 1.0 give coatings with good adhesion and excellent surface hardness, and corrosion and weathering resistance [75]. However, it seems that although this recipe can reduce some disadvantages, it still cannot provide good overall **properties**, particularly impact resistance, thermal shock resistance and a sufficiently low fusion temperature. The improvement of adhesion and impact resistance by incorporation of isophthalic acid is limited by the very slow crystallization rate and low melting point of the polyester. The latter decreases with increasing the isophthalic content to values that are not acceptable for practical reasons. On the other hand, a large amount of terephthalic acid would increase the rate of crystallization enormously, lowering the impact and shock resistance of the formed film.

An interesting approach in solving this problem is described by Toray Industries, Inc. [77] where a ratio between the isophthalic and terephthalic acid is kept quite high (40 to 60) in order to improve the adhesion and the impact resistance of the coating. The reduced rate of crystallization is compensated for by using nucleating agents. In this way enough crystallinity can be obtained during the film forming process after the fusion of the powder particles. Useful **nucleating agents** are said to be talc, aluminum stearate, barium stearate, kaolin or clay. A typical powder coating composition according to the examples of the patent [77] consists of polyester resin, titanium dioxide, 0.1 to 1.5 % (by weight based on the resin content) of nucleating agent and flow additives.

Thermoplastic polyester powder coatings have found an application in the processing of **food cans** where they are used for covering the welding seams. The enormously high speeds at which cans are being manufactured allow only a few seconds to melt down the powder. Thermosetting coatings cannot complete the curing process within such a short time. High molecular weight polyesters with a molecular weight of 25,000 and a fusion temperature of 175 °C have been developed for this purpose [78].

Thermoplastic polyester powder coatings are characterized by very high **decorative properties** giving a porcelain-like finish. The surface of the coating has a high gloss and excellent flow. Adhesion to metal substrates, in contrast to other thermoplastic powder coatings, is excellent, so previously priming the substrate with an adhesion promoter is not necessary. The abrasion resistance, scratch resistance and hardness of the coatings are excellent, combined at the same time with good flexibility and impact resistance. They possess excellent electrical properties, considerable chemical resistance, long-term durability and an unlimited shelf-life [32]. They do not undergo chemical curing during film formation, and therefore are very suitable for **high-speed running lines**. The melting temperature of the powder coating can be easily adjusted by the proper choice of polyester resin, which makes them suitable for a variety of stoving schedules.

Several weak points, however, are inherent in the thermoplastic powder coatings. Due to the high molecular mass, acceptance of the pigments and fillers, and dispersion efficiency during processing are very low. The **low pigmentation level** has as a consequence low hiding power of the coating. From a practical point of view this is not always a problem, since the powder coatings are in principle applied in somewhat thicker layers, but it has of course a negative influence on the final price.

Low grinding efficiency during the fine grinding step after extrusion is another typical problem related to thermoplastic polyester powder coatings. The grinding of the coating is hampered by the high modulus of the polyester resins, thus decreasing the capacity of the plant. Often, to keep plant capacity at a certain level, **cryogenic grinding** has to be performed; this prevents the powder particles sticking as a result of heat development during the grinding operation.

Polyester powder coatings have the **lowest solvent resistance** among thermoplastic coatings. This is easy to understand if the structure of the binder is kept in mind. Fortunately, since these coatings are predominantly used for decorative purposes, this weakness is not very important from a practical point of view.

Relatively low hydrolytic **stability** of the ester group restricts the application of these coatings mainly to indoor use. In the new types of polyesters, glycols with bulky substituents in the α position with respect to the hydroxyl group are used. In these cases the ester groups formed during the process of polyesterification are considerably sterically hindered. Although the hydrolytic stability of these polyesters is considerably improved, outdoor use should be recommended only after careful weathering tests. Nevertheless *Duarte* et al. [79] reported that polyethylene terephthalate (PET) powder was tested for protection against corrosion in automobile fuel tanks, which are usually based on lead-tin-coated steel. The corrosion resistance of the coating against gasoline, diesel oil and alcohol was remarkable. In a Japanese publication [80] a PET homopolymer powder coating was described, which was combined with a sacrificial zinc undercoating to show an outstanding **corrosion resistance** against salt-water spray (6000 h) and seaside weathering (5 years).

Because of their excellent aesthetic properties, thermoplastic polyester powder coatings are used for metal furniture, guardrails, inside architectural elements and panels, shelves, aluminum extrusions, tools and machine components. Good dielectric properties make them suitable for **applications** including transformers, housing for electrical equipment and instruments. Although the weathering resistance of some polyester resins is satisfactory, thermoplastic powder coatings still cannot compete with the thermosetting counterparts with respect to exterior durability. They are therefore used mainly for inside applications, or on articles which are not permanently exposed to weathering conditions.

2.5 References

[1] Air Products and Chemicals, Inc., EP 0.264.861 (1987)
[2] *Matlack, J.D., Metzger, A.P.*, Polymer Letters, 4, 875 (1966)
[3] *Smith, O.W., Koleske, J.W.*, J. Paint. Techn., 45, (582), 60 (1973)
[4] *Christensen, J.K.*, Polyester Powder Coatings, Paper FC72-941, SME Powder Coating Conference, Cincinati, Ohio, 3/1972
[5] *Hagan, J.W.*, Paint and Varnish Prod., 4, 45 (1972)
[6] *Miller, P.E., Taft, D. D.*, Powder Coating, Society of Manufacturing Engineers, Dearborn, Michigan (1974) p.165
[7] Pennwalt Corp., US Pat. 3.193.539 (1965)
[8] Pennwalt Corp., US Pat. 3.245.971 (1966)
[9] Kureha Chem. Comp., FR Pat. 1.419.741 (1965)
[10] Daikin Kogyo Co. Ltd., GB Pat. 1.178.227 (1967)
[11] E.I. DuPont de Nemours & Co., Inc., DE Pat. 1.806.426 (1969)
[12] Kali-Chemie AG, GB Pat. 1.057.088 (1967)
[13] *Miyazawa, T., Ideguchi, Y.*, J. Polymer. Sci., B3, 541 (1965)
[14] *Kosar, W. P.; Morris, S.*, Journal of Coatings Technology and Research, 4(1), 51-58, 2007
[15] *Munekata, S.*, Progress in Organic Coatings, 16, 113 (1988)
[16] *Sheppard, W.A.*, Organic Fluorine Chemistry, Benjamin, New York (1969), p.1
[17] 3M Innovative Properties Company, WO 2008048884, WO 2008048888, 2008
[18] Daikin Industries, EP 2258777, 2009

[19] *Yamamoto, S.; Spencer, P.J.L.*, Polymers in Aggressive and Corrosive Environments, Conference Proceedings, Cologne, Germany, Sept. 8-9, 203-214, 2004
[20] "Sigma PVDF-Powder Coating", Brochure of Sigma Coatings B.V., July (1987)
[21] *Sietses, W.*, Aluminium, Nr. 3, p.16, May (1988)
[22] *Iya, T.*, JP 2011225677, 2011
[23] *Naden, V.*, Industrial Finishing, 3, 20 (1988)
[24] *Bian, J.; Wang, W.; Guan, C.; Zhao, Y.*, Journal of University of Science and Technology Beijing, 12(6), 572-576, 2005
[25] Kureha Kagaku Kabushiki Kaisha, US Pat. 3.824.115 (1974)
[26] Arkema, WO 2006042764, 2006
[27] *Moorman, R.*, Industrial Finishing, 3, 20 (1988)
[28] *Crowley. J.D., Teague, G.S., Curtis, L.G., Foulk, R.G., Ball, F.M.*, J. Paint. Techn., 44, (571), 56 (1972)
[29] Anon., Finishing Industries, June (1977) p.20.
[30] Adopted from Encyclopedia of Polymer Science and Technology, Vol. 6, p. 280, Ed. *Bikales, N.M.*, John Wiley & Sons, Inc., New York (1967)
[31] *Ball, F.M., Curtis, G.L.*, J. Powder Coat., 2, (1), 1 (1979)
[32] *Paul, S.*, in Surface Coatings - Science and Technology, John Wiley & Sons, Chichester (1985) p. 671
[33] Dainippon Ink Chem., Jap. Pat. 57.202.335(1981)
[34] Dainipon Ink Chem., Jap. Pat. 58.049.736 (1981)
[35] "Plascoat PPA 571" in Powder Coater's Daily, Preview of PCE 2002 Congress, Nuremberg, Jan. (2002)
[36] Sumitomo Chemical, Can. Pat. 1.142.675 (1980)
[37] *Bi, H.* et al., CN 102277042, 2011
[38] *Houssier, D.*, Polymers in Aggressive and Corrosive Environments, Conference Proceedings, Cologne, Germany, Sept. 8-9, 221-226, 2004
[39] Hoechst AG, EP 0.057.823 (1981)
[40] Hoechst AG, EP 0.100.992 (1982)
[41] *Connely, G., Gaillard, G.*, Proceedings UK Corrosion 87, Pt 26, no. 37, October (1987) p. 377
[42] Beatrice Foods Company, US Pat. 3.860.557(1975)
[43] *Kodokian, G. K., et al.* Proceed. Intern. Waterborne, High-Solids & Powder Coat. Symp. New Orleans, 1-3 Mar. (2000)
[44] *Jezl. J. L.* in Encyclopedia of Polymer Science and Technology, Vol. 11, Ed. Bikales, N. M., John Wiley & Sons (1969) p.611
[45] DuPont, US Pat. 2.071.250, (1936) (patent appl. from 1931)
[46] DuPont, US Pat. 2.071.253 (1937)
[47] DuPont, US Pat. 2.130.523 (1938)
[48] DuPont, US Pat. 2.130.948 (1938)
[49] I.G.Farbenindustrie, DE Pat. 748.253. (1943)
[50] *Floyd, D.E.* in Polyamide Resins, Reinhold Publ. Corp., New York (1966) p. 2
[51] *Gaumet, S.; Delprat, P.; Audenaert, M.*, European Coatings Journal 6, 50-54, 2004
[52] *Todareva, L.G., Mikhailov, N.V., Potemkin, Z.I., Kovaleva, M.V.*, Visokomolekul. Soedin 2, 1728 (1960)
[53] *Blackmore, C.E.*, Product Finishing, August (1988) p. 37
[54] *Westmore, E.R.*, Finishing Industries, February (1979) p. 53
[55] *McAndrew, T.P.; Audenaert, M.; Petersheim, J.*, Conference, American Water Works Association, 15, 1-11, 2005
[56] Anon., Finishing, 13, (6), 39 (1989)
[57] Anon., Kunststof & Rubber, 54, (7), 23 (July 2001)
[58] The Polymer Corporation, US Pat. 4.248.977 (1981)
[59] a) *Guichenuy, M.; Watts, J. F.; Abel, M.-L.; Brown, A. M.; Audenaert, M.; Amouroux, N.*, Surface and Interface Analysis , 36(8), 685-688, 2004
b) *Guichenuy, M.; Watts, J. F.; Abel, M.-L.; Audenaert, M.*, Surface and Interface Analysis, 38(4), 168-171, 2006
[60] *Watts, J. F.; Guichenuy, M.; Abel, M.-L.*, Proceedings of the Annual Meeting of the Adhesion Society, 30th, 84-86, 2007
[61] Shinto Fine Co., JP 2005307096, 2005
[62] *McAndrew, T.P.; Cere, F.*, Proceedings of the International Thermal Spray Conference, Seattle, May15-18, S101/1-5, 2006.
[63] *McAndrew, T.P.; Audenaert, M.; Petersheim, J.; Garcia, D.; Richards, T.*, Polymer Preprints (American Chemical Society, Division of Polymer Chemistry) 45(2), 155-156, 2004
[64] *Schulz, G.V.*, Z. Phys. Chem. A 182, 127 (1938)
[65] *Griehl, W., Förster, F.*, Faserforschung und Textiltechnik, 7, (10), 463, (1956)
[66] *Mleziva, J., Hanzlik, V., Kincl, J., Miklas, Z.*, Poliesteri, Tehnika, Sofia (1969) p. 90
[67] *Korshak, V.V., Vinogradova, S.V., Vasova, E.S.*, Dokl. Akad.Nauk SSSR, 94, 61 (1954)
[68] *Korshak, V.V., Vinogradova, S.V.*, Dokl. Akad. Nauk. SSSR, 89, 1017 (1953)

[69] *Korshak, V.V., Vinogradova, S.V., Frunze, T.M.*, Zurnal Obsh. Him., 27, 1600 (1957)
[70] *Korshak, V.V., Vinogradova, S.V.*, Zurnal Obsh. Him , 26 , 539, 544, (1956)
[71] *Korshak, V.V., Vinogradova, S.V.*, Izv. Akad. Nauk. SSSR, Otd. Him. Nauk (1953) 121; (1954) 1097
[72] *Korshak,V.V.*, Izv. Akad. nauk SSSR, Otd. Him. Nauk (1950) 51
[73] *Korshak, V.V., Vinogradova, S.V., Beljakov, V.M.*, Izv. Akad. Nauk SSSR, Otd. Him. Nauk (1957) 730, 737
[74] *Korshak, V.V., Vinogradova, S.V.*, Heterocepnie Polyerfiri, Izd. Akad. Nauk SSSR, Moskva (1958) p. 275
[75] Jap. Pat. 097.023 1974, cited in EP 0.213.887 (1986)
[76] *Watase H.*, Jap. Pat. 59.041.367 (1982)
[77] Toray Industries Inc., EP 0.213.887 (1986)
[78] *Schmitthenner, M.*, Proc. 14th Intern. Conf. Org. Coat. Science and Technol., Athens p. 293, 1988
[79] *Duarte, L.T.; Paula e Silva, E.M.; Branco, Jose R.T.; Lins, V.F.C.*, Surface and Coatings Technology, 182 (2-3), 261-267, 2004
[80] *Shinozaki, S.; Handa, T.; Watanabe, M.; Asakura, K.; Toyoda, E.; Ichino, T; Kuwaki, N.*, Zairyo to Kankyo, 54(8), 396-398, 2005

3 Thermosetting powder coatings

Thermoplastic powder coatings do not undergo curing during stoving of the coating. The film formation is a result of melting and sintering of the powder particles using surface tension as the main driving force for this process. However, for good mechanical properties, the polymer must have a high **molecular weight**. Since there is no additional increase in the molecular weight of the thermoplastic powder coating, the initial molecular weight of the binder must be high enough to ensure good flexibility, hardness, impact and scratch resistance of the formed film.

In most cases the molecular weight of the polymers exhibiting good mechanical properties is in the range between 20,000 and 200,000 [1]. This is the range of molecular weights of the thermoplasts, which are in use for production of powder coatings. Certain exceptions are the linear polyesters, which in some cases possess good impact resistance and tensile strength at molecular weights higher than 10,000 [2]. The high molecular weight of a polymer binder worsens the processing characteristic of the coating; namely, high molecular weight polymers have very poor wetting properties. Therefore, the amount of pigment and filler that can be successfully incorporated in the coating based on a thermoplastic binder in principle is much lower than in the case of thermosets. This has an influence on the covering power of the coating, on the aesthetic value (especially in the case of high gloss coatings) and also on the price of the coating formulation.

In general, powder coatings based on thermoplastic binders have very poor **adhesion** on metal surfaces. Only coatings which use thermoplastic high molecular polyester resins as binders can be directly applied onto a metal surface without the use of a primer. Using a primer as an adhesion promoter adds to the expense in the coating process and is usually avoided unless special corrosion protection is requested.

The problems inherent to the thermoplasts are successfully overcome by the **thermosetting polymers**. During stoving of the powder coatings based on thermosetting polymers a crosslinking reaction takes place. Therefore, polymers with much lower molecular weights and consequently lower melt viscosities can be used. As a result larger amounts of pigments and fillers can be successfully dispersed and incorporated in the coating. The pigment/resin ratio is limited by the flow properties of the coating during the film formation rather than by the efficiency of the dispersion process.

The crosslinked nature of the cured film makes these coatings exceptionally resistant to solvents. This also improves chemical resistance in general, since the crosslinked film is not very permissive for the attacking agent.

Finally, there are many possibilities to introduce polar groups in the resin backbone, or polar groups are formed due to the curing reaction (like hydroxyl groups in the curing of carboxyl polyesters and epoxy curing agents). These considerably improve the adhesion properties. In general, thermosetting powder coatings do not require primers as intercoats.

Next to the binder thermosetting powders contain crosslinkers involving different curing chemistries in the coating composition. This makes the process of film formation more complex. A broader knowledge and more sophisticated experimental techniques are neces-

Emmanouil Spyrou: Powder Coatings: Chemistry and Technology

ISBN: 978-3-86630-884-8

sary to understand, follow and elucidate the phenomena of film formation. Therefore this Chapter 3 includes sections describing the crosslinking chemistry, the monitoring of the curing process and the types of crosslinkers, before referring to the thermosetting powder coatings themselves.

3.1 Curing reactions used in powder coatings

The crosslinking of powder coatings enables a three-dimensional network to be formed. In this way the low molecular weight binders are transformed during film forming into crosslinked polymers exhibiting desirable physico-chemical and mechanical properties. Unlike many solventbased coatings, which are soft and tacky after evaporation of the solvent and which are converted into harder non-tacky materials upon subsequent crosslinking, powder coatings after fusion into continuous films are hard and non-tacky materials even without curing. However, the flexibility, impact resistance and solvent resistance of the uncured thermosetting powder coatings are far beyond the standard levels typical for industrial coatings. Here the main purpose of the crosslinking is to increase the toughness of the coatings and to make the cured film resistant to the attack of common industrial solvents.

Network formation depends on the average functionality of the system. When the average functionality of the system binder/crosslinker is two or only slightly over two, full conversion of the functional groups will lead to a high molecular linear or branched polymer, but not to a network formation. Films obtained by such a system that depend on the molecular weight after curing, may have good mechanical properties, but certainly will not show good solvent resistance. On the other hand, too high a functionality and excessive crosslinking can lead to the formation of a brittle film exhibiting unsatisfactory postforming properties and low impact resistance.

Curing of powder coatings is a rather complex problem from the theoretical point of view. For complete understanding of the theories describing the **curing process**, mathematical knowledge, which is not very typical for a resin chemist, is often required. The easiest treatment of the problem to understand is that given by Carothers, who was among the first scientists to deal with so-called **non-linear polymerization** [3]. Although his theory suffers from serious errors, especially in choosing the criteria for infinite network formation, it is widely used as a basis for calculating formulations of many different types of polycondensation reins. However, the theory of *Carothers* is not suitable for a satisfactory treatment of the crosslinking phenomena. *Flory* [4] and *Stockmayer* [5] have derived the basic relations between the extent of the reaction and the resulting structure of the polymers undergoing crosslinking or the basic equations describing the non-linear polymerization process. In the 1960's *Gordon* has developed a new treatment of the phenomenon using the theory of stochastic branching processes with cascade substitution to derive equations describing the process of crosslinking [6, 7]. In the 1970's *Macosko* and *Miller*, starting with elementary probability and utilizing the recursive nature of network polymers, developed simpler equations to predict the number average and mass average molecular weight of the polymers in the pre- and post-gel stadium [8, 9]. The theory of *Gordon* has been used to predict the curing behavior of powder coatings [10].

The theoretical treatment of network formation is beyond the scope of this book. The interested reader should consult the enormous literature dealing with crosslinking phenomena and network formation, which is partly cited in the references at the end of this section [11-30].

Various types of curing reactions are employed in thermosetting powder coatings. The crosslinking reactions typical of thermosetting liquid coatings in principle can be used for hardening powder coatings. However, certain differences emanating from the nature of powder coatings exclude many of the crosslinking reactions, typically used in conventional liquid systems. This section will describe the mechanism of crosslinking reactions used in commercial powder coatings today.

3.1.1 Acid/epoxy curing reaction

The reaction between epoxy and acid functional groups is the most **important curing reaction** used in practice for crosslinking thermosetting powder coatings. The so-called hybrid powder coatings, in which carboxyl functional polyesters are cured with bisphenol A epoxy resins and polyester powder coatings where triglycidyl isocyanurate or diglycidyl esters are used as curing agents, are the best representatives of this group of products.

The reaction mechanism between the acid and epoxy groups has been a matter for wide-ranging investigation [31-47]. *Schechter* et al. [31-33] anticipate four possible reactions in the curing of epoxy resins with polybasic acids:

1. **Ring opening** addition of epoxy group to the acid group leading to formation of the corresponding hydroxyl ester:

Equation 3.1

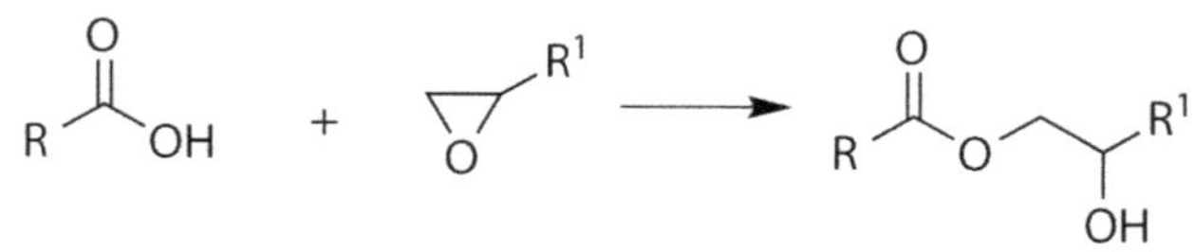

2. **Esterification** between the acid and the hydroxyl group formed in the previous reaction:

Equation 3.2

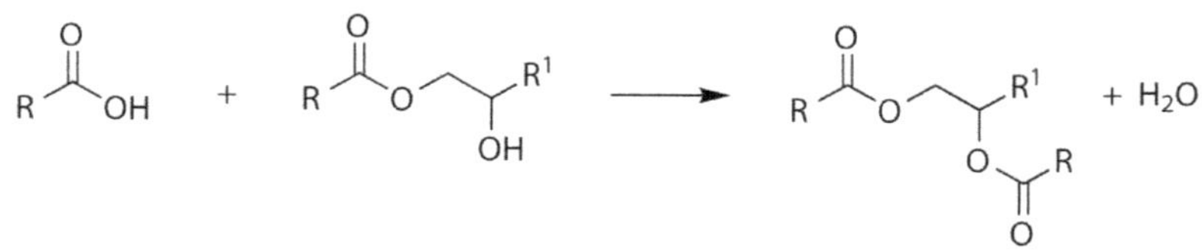

3. **Ring opening** addition of the epoxy group to the hydroxyl group formed in reaction (3.1) resulting in the corresponding ether alcohol:

Equation 3.3

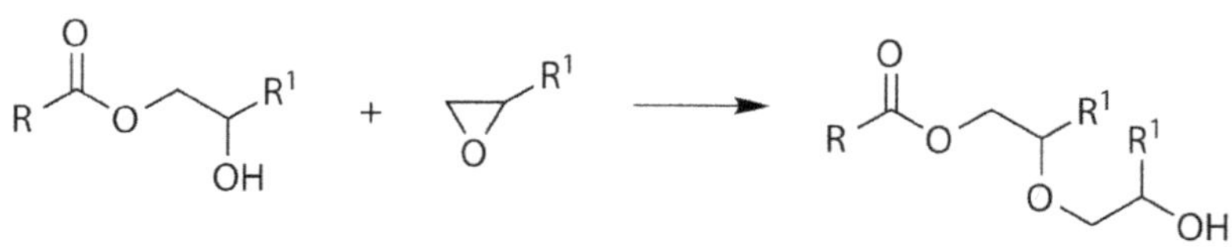

4. **Hydrolysis** of the epoxy ring by water:

Equation 3.4

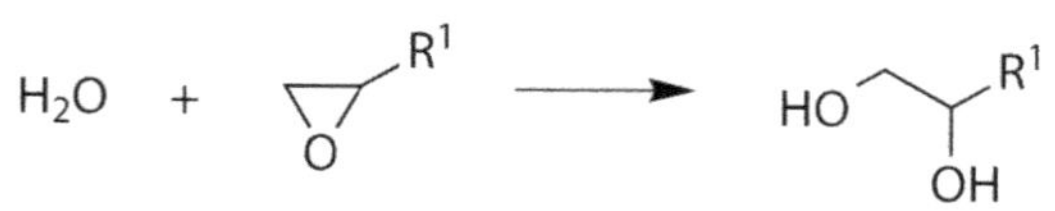

The extent of the above reactions depends on the **catalyst** used in the system. In the absence of catalyst the first three reactions proceed at the same rate to a comparable extent. For a non-catalyzed curing process, *Shechter, Wynstra* and *Kurkji* [32] report that when the curing takes place in a stoichiometric ratio between the reactive groups, the ratio between the reaction products according to reactions (3.1), (3.2) and (3.3) is between 2:1:1 and 1:2:2 respectively.

Reaction (3.3) occurs only when the curing is catalyzed by strong acid catalysts or at higher temperatures. Acidic catalysts do not find a regular use powder coating formulations where the catalyst employed is usually of a basic nature. Higher temperature on the contrary is the standard in powder coatings, therefore, the etherification reaction has to be taken into account [48].

In an excess of epoxy groups, and after the consumption of acid groups, the etherification reaction proceeds according to the scheme (3.3) [32]. Two fundamentally **different reaction mechanisms** are anticipated for the base catalyzed acid-epoxy reaction [45, 48]. When tertiary amines are used as catalysts, it is believed that **decomposition** of the acid-catalyst salt generates active carboxylate anion species, according to the following reaction:

Equation 3.5

$$RCOOH + B \rightleftharpoons \underset{\text{salt}}{RCOO^-HB^+} \overset{K_{eq}}{\rightleftharpoons} RCOO^- + HB^+$$

In the next step of the reaction the carboxylate anion attacks the **epoxy ring**, resulting in the formation of an alkoxide anion and the corresponding ester:

Equation 3.6

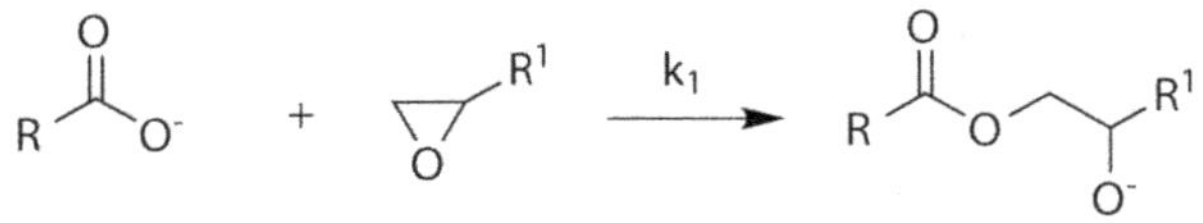

The alkoxide anion formed in reaction (3.6) reacts to generate new carboxylate anions or free base catalyst according to the following reaction schemes:

Equation 3.7

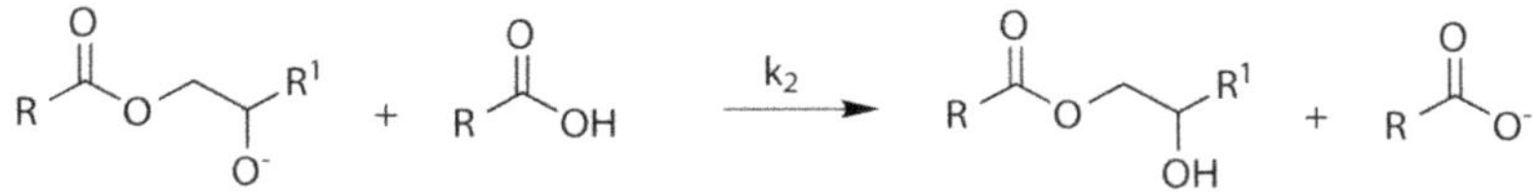

Equation 3.8

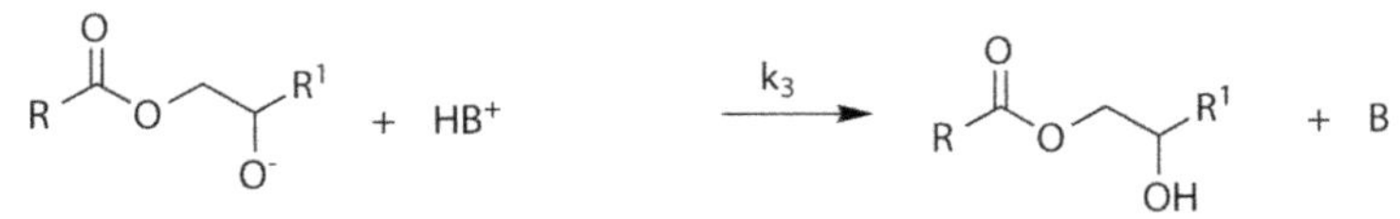

It is supposed that both reactions (3.7) and (3.8) are much faster than reaction (3.6) which will be responsible for the overall **kinetics of the system**. Accordingly, the rate expression for this mechanism will be:

$$S_1 = k_1\,[RCOO^-]\,[epoxy] = k_1K_{eq}\,[RCOOH]\,[epoxy]\,[B]\,/\,[BH^+]$$

The **reaction rate** is obviously controlled by the concentration of the catalyst, as well as by the rate of decomposition of the acid-catalyst salt, or the value of the equilibrium constant K_{eq}. In other words, the reaction rate will also depend on the type of the base used as a catalyst.

A different reaction mechanism is supposed to be valid for the quaternary ammonium salts which are also widely used as catalysts. The first reaction to be considered is the **dissociation** of the salt:

Equation 3.9

$$AB \overset{K_{eq}}{\rightleftharpoons} A^+ + B^-$$

The next steps involve a direct attack of the base on the epoxy generating alkoxide anion:

Equation 3.10

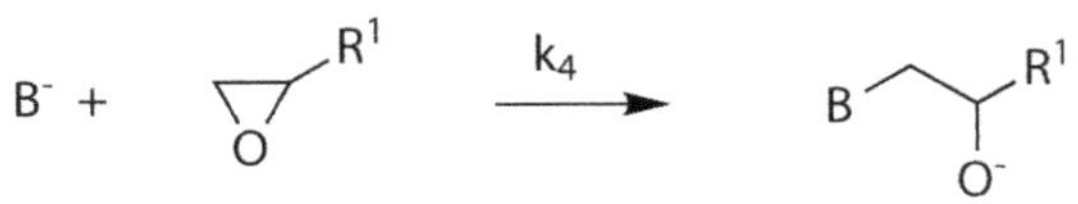

carboxylate anion:

Equation 3.11

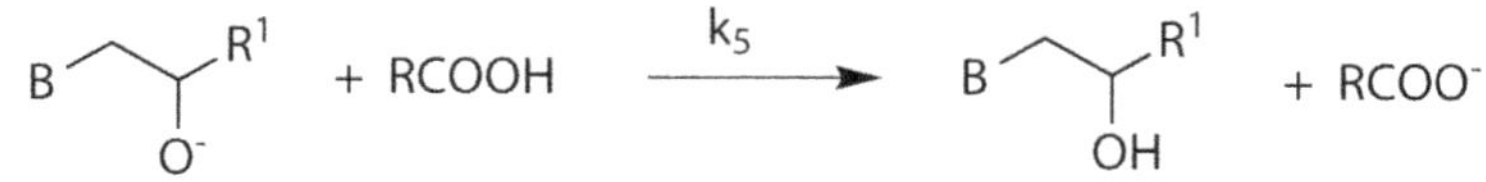

and again alkoxide and carboxylate anions in the next sequences:

Equation 3.12

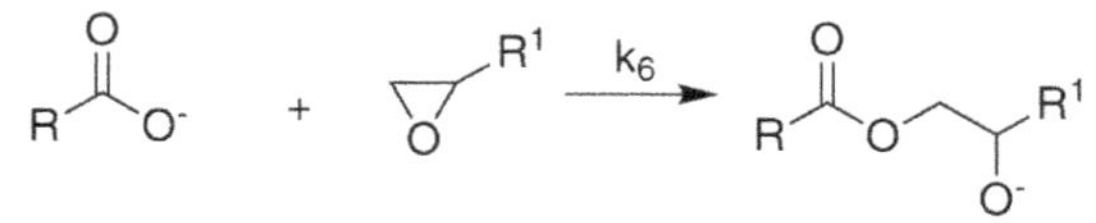

Equation 3.13

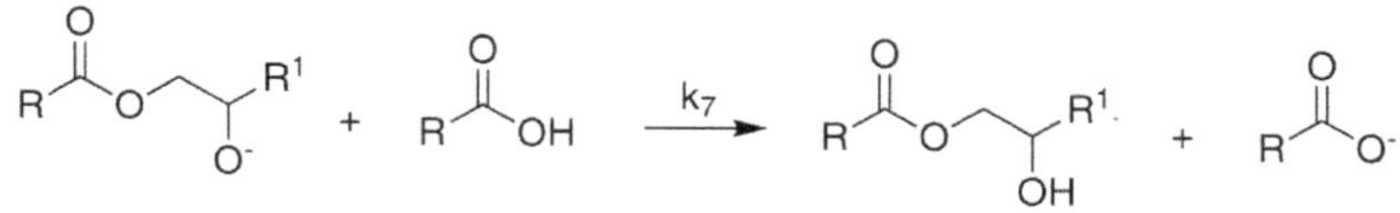

Reactions (3.11) and (3.13) are considered as fast, having no influence on the overall rate of curing. The dissociation step also plays no role, because the result of dissociation, the anion B (reaction 3.9), reacts irreversibly with the epoxy, thus shifting the equilibrium completely to the right. The important reactions for the kinetics of curing are (3.10) and (3.12), involving the following expressions for the reaction rate:

$$S_4 = k_4\ [B]\ [\text{epoxy}]$$

$$S_6 = k_6\ [RCOO^-]\ [\text{epoxy}]$$

Regardless of the values of the constants k_4 and k_6, the concentration of the carboxylate anion will be constant and equal to the original concentration of B because it regenerates in the sequences of the reactions presented above. Therefore, the reaction rate will depend in practice on the **concentration of the catalyst** and the epoxy groups, but not on the type of catalyst. The type of catalyst will determine the ratio between the rate constants k_4 and k_6. If $k_4 < k_6$, a delay should be expected before the reaction runs at its proper rate. The experimental results obtained by *Henig, Jath* and *Mohler*[49] and *van der Linde* and *Belder*[45] support the above described mechanisms.

3.1.2 *Acid anhydride/epoxy curing reaction*

Anhydrides of polybasic acids have lost their importance as curing agents for epoxy powder coatings mainly for toxicological reasons. However, this system offers several advantages compared to curing with polybasic acid, such as **higher crosslinking density** and relative insensitivity with respect to the deviations of the anhydride/epoxy ratio from the stoichiometrical ratio.

Different articles deal with the mechanism and kinetics of the polyaddition reaction of epoxy resins with acid anhydrides. In the case of uncatalyzed system, the following reactions of curing have been proposed by several authors [50-55]:

Equation 3.14

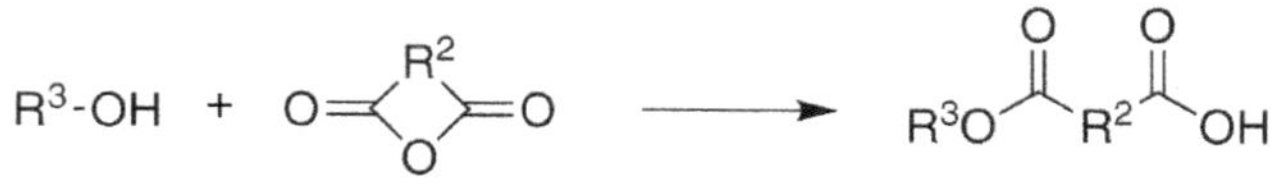

Equation 3.15

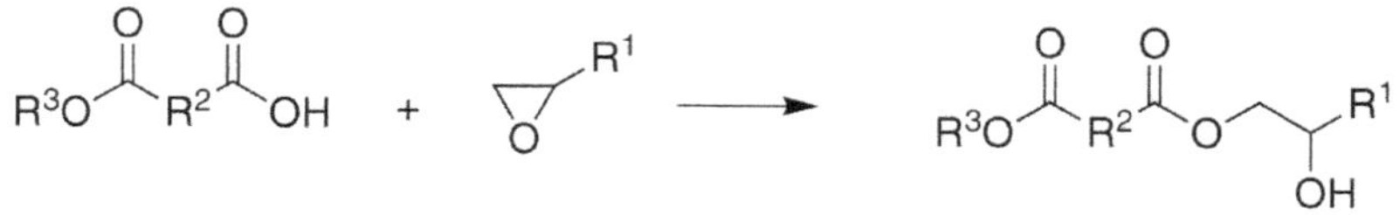

Equation 3.16 

Reaction (3.14) shows that for initiation of the curing a hydroxyl containing compound must be added to the system. In the case of higher homologues of bisphenol A epoxy resins, the hydroxyl groups belonging to the polymer backbone can initiate the reaction themselves. The reaction can be initiated by water as well as by carboxyl containing compounds, since their reaction with the epoxy groups leads to the formation of the hydroxyl group. The other two reactions, (3.15) and (3.16), are in fact the same, as in the case of curing of epoxies with polyacids (reactions 3.1 and 3.3).

A different curing mechanism has been proposed by *Shechter* and *Wynstra* [31] and *Newey* [56] in the case of **base catalyzed curing**. They have found no evidence of the hydroxyl/epoxy reaction leading to the formation of ether linkages. However, *Kaplan* [57] in a study of the effect of water on the epoxy-acid anhydride reactions concluded that dependent on the temperature, type of anhydride and epoxy, and the presence and type of solvent, an etherification reaction (reaction 3.3) proceeds to a certain extent.

The more recent work of *Fedtke* and *Domaratius* [58] supports the findings of *Fischer* [59] that the etherification reaction does not occur essentially when the epoxy-acid anhydride system is catalyzed by tertiary amine. Studying the system of phenyl glycidyl ether, phthalic anhydride and N,N-dimethyl aniline as catalyst, they have proposed an initiation step that involves a reaction of the tertiary amine with the epoxy group, resulting in the **formation of zwitterion** containing quaternary nitrogen atom and an alkoxide anion. This reaction mechanism had been proposed somewhat earlier by *Matejka* et al. [60]:

Equation 3.17 

This zwitterion can react either with the epoxide or with acid anhydride. *Fedtke* and *Domaratius* propose that the reaction with anhydride is much faster, thus practically excluding the formation of ether linkages due to the possible reaction of the zwitterion with epoxy groups:

Equation 3.18 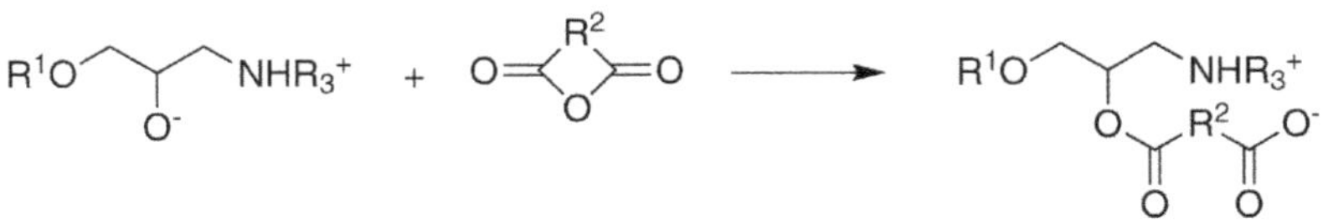

They accept the suggestion of *Tanaka* and *Kakiuchi* [61] and *Luston* and *Manasek* [46] that the product of reaction (3.18) undergoes isomerization to an unsaturated compound:

Equation 3.19 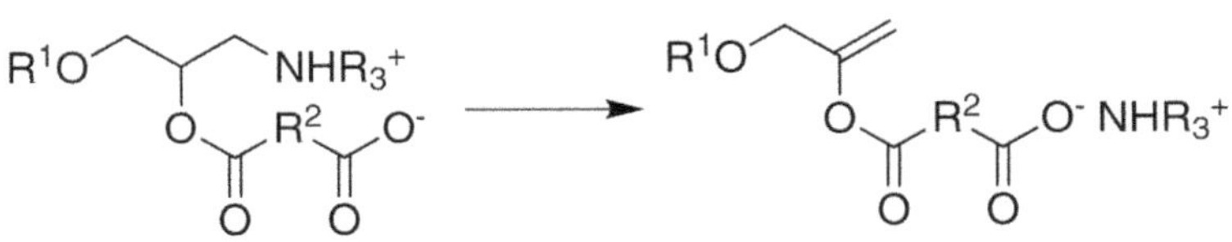

The propagation steps are similar to 3.11–3.15, whereas the termination step occurs by reversible decomposition of the active centers:

Equation 3.20

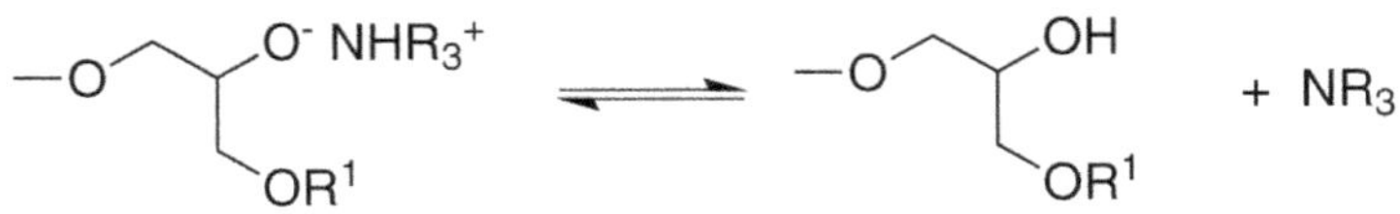

Equation 3.21

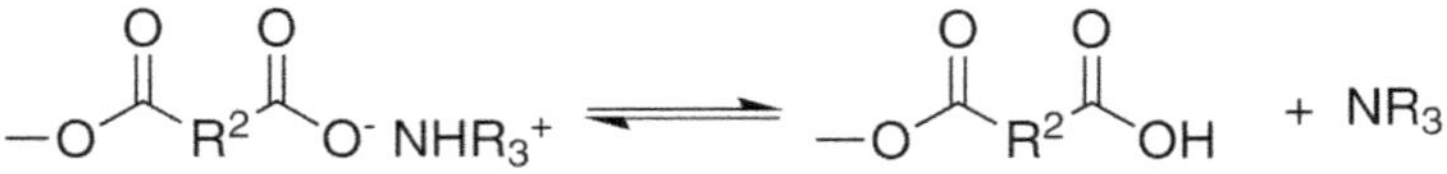

In the presence of proton donating compounds, the initiation step is somewhat different. The active center of catalysis is formed by interaction of the proton donor with amine, which does not exclude the initiation according to reactions (3.17) to (3.19):

Equation 3.22 $$R^4H + NR_3 \longrightarrow R^{4-}\ NHR_3^+$$

The propagation reaction remains the same, as in the case of base catalyzed curing without the presence of a proton donating compound.

The ratio between the two initiation steps (3.17 to 3.19) and (3.22) depends on the acidity of the proton donor. The greater the interaction between the tertiary amine and the proton donating compound, the lower the initiation by the tertiary amine only.

The effect of the proton-donating compounds should be considered, because in technical epoxy resins, a proton donor is always present, either in the form of a secondary hydroxyl group belonging to the higher homologues of bisphenol A epoxy, or in the form of a carboxyl group because of partial hydrolysis of the acid anhydride by humidity.

Fedtke and *Domaratius* have shown that the rate constant for the epoxy/acid anhydride reaction in the presence of benzoic acid as the proton donor is twice as high as in the presence of phenol for a system consisting of phenyl glycidyl ether, phthalic anhydride and N,N-dimethyl benzyl amine. The explanation is the higher acidity of benzoic acid compared to phenol. This is rather contradictory to the **reactivity ratio** suggested by *Tanaka* and *Mika* [48, p.189], which decreases in the following order: phenol > acid > alcohol. On the other hand, acid anhydrides become less reactive in the order maleic > phthalic > tetrahydrophthalic > hexahydrophthalic > methyltetrahydrophthalic [62]. The **proposed mechanism** involving isomerization of epoxy groups to a derivative of allyl alcohol is questionable, since according to some papers [63, 64] this has not been experimentally confirmed.

Other proposed mechanisms not involving the formation of allylic derivatives are those of *Tanaka* and *Kakiuchi* [54, 55], *Fischer* [65] and *Feltzin* [66]. The differences between these mechanisms are rather minor and can easily be incorporated into the mechanism proposed by *Fedtke* and *Domaratius*, avoiding reaction (3.19) which suggests the formation of an allyl derivative.

3.1.3 Epoxy/amino curing reaction

Epoxy compounds react readily with primary and secondary amines giving secondary or tertiary amines as reaction products respectively:

Equation 3.23

Equation 3.24

It can be expected that the resulting tertiary amine can act as a catalyst promoting the reaction between the hydroxyl groups present in the epoxy resin or obtained as a result of the curing process, with the epoxy rings leading to the formation of ether linkages similar to (3.3).

Curing between polyamines and polyepoxides is one of the **most extensively studied reaction** . In the very early work of Shechter, *Wynstra* and *Kurkiy* [32], *Kakurai* and *Noguchi* [67], *Dannenberg* [68] and the recent work of *Gross* et al. [69–71] it was confirmed that the rate of etherification is rather low at room temperature. However, at increased temperatures, as in the case of curing of powder coatings, and especially when the epoxy groups are in excess, the epoxy-hydroxyl curing reaction should be taken into consideration. It has been shown that at higher temperatures, depending on the type of amine, even at a stoichiometric ratio between the amine and epoxy groups, a considerable amount of epoxy groups is used up in reaction with the hydroxyl groups available from the epoxy resin (in the case of higher homologues of bisphenol A epoxy resins) or is borne as the reaction product of curing [72]. The etherification reaction is also catalyzed by tertiary amines, which can be used as accelerators for the curing process. On the other hand, the catalytic effect of the intramolecular tertiary amino group in the reaction products (ß-amino alcohols) is less pronounced due to their rather high steric hindrance. It is suggested [69] that reactions with epoxies via additions of the hydroxyl groups at epoxy/amine reaction products only take place after completing the amino additions.

Gross, Kollek and *Brockmann* [69] have studied the curing of epoxy resins with amines at the model system of 1,2-epoxy-3-(4-methyl-phenoxy)-propane with n-propyl amine as the source of primary amino groups and diethyl triamine as the compound containing primary and secondary amino groups. They have concluded that there is a considerable difference between the **reactivity of the primary and secondary amino groups**. They have also found the same behavior to be valid for the aromatic amines, with the remark that they react 10 times more slowly than their aliphatic counterparts.

Dusek states that the ratio between the rate constants in reactions (3.23) and (3.24) k_2/k_1 varies between 0.5 and 0.15 depending on the structure of the amines. This value is even lower for aromatic amines than for aliphatic ones [73, 74].

Tanaka and *Mika* [48, p.179] relate the difference in the reactivity between the primary and secondary amines to the **steric factors**, and support this finding by the fact that the reactivity of various amines towards phenyl glycidyl ether diminishes in the order: n-butyl amine > diethyl amine > di-n-propyl amine > di-n-butyl amine > diisoamyl amine. The same is valid for the aromatic amines, which have the following order of reactivity: methyl aniline > ethyl aniline > n-propyl aniline > n-butyl aniline.

Gross et al. add another element to the steric factors. According to them, the reactivity of the amines towards epoxy groups is also ruled by their **nucleophilicity**, since the increased steric hindrance of the secondary amino group also includes a decreased nucleophilic character.

The presence of alcohols or acids has a considerable **accelerating effect** on the curing reaction. The catalytic effect of the various alcohols is proportional to their acidity or their pK_a values. The added compounds have decreasing catalytic effect in the order: acids > phenols > water > alcohols > aromatic hydrocarbons > ethers. *Shechter, Wynstra* and *Kurkiy* [32] explain this phenomenon by the formation of a hydrogen bond in the transition state, which accelerates the opening of the epoxy ring:

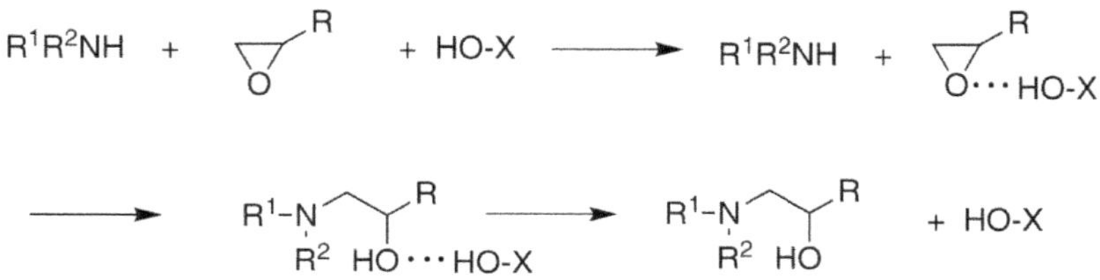

Equation 3.25

A similar mechanism is proposed by *Smith* [75]. Although the hydroxyl groups accelerate the curing reaction, the reactivity of the secondary amino group, which is the product of the reaction between the primary amino group and the epoxy ring (ß-amino alcohol), is lower than the reactivity of the secondary amino groups belonging to amino hardeners containing no hydroxyl groups. *Gross* et al. [69] explain this by the formation of the intramolecular hydrogen bridge bond with the nitrogen atom of the secondary amino group. The nucleophilicity is thereby reduced, as is the affinity to epoxy groups. In the first approximation the following order of reactivity results:

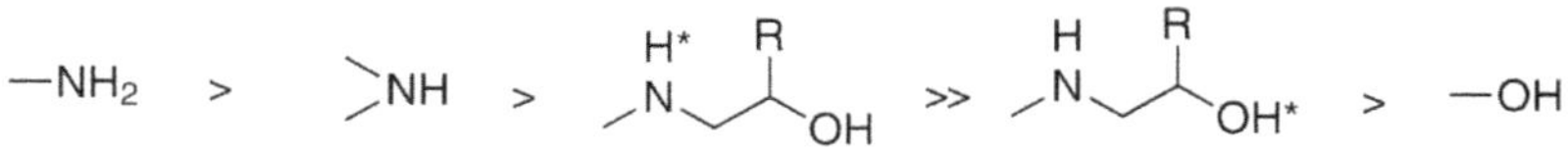

where (*) indicates the reactive site.

It can be assumed that in the case of curing di-functional epoxies with primary diamines, the reaction products at the first stage of the reaction are primarily of a linear character. As the reaction proceeds, the linear high molecular material is crosslinked by the reaction of the formed secondary amino groups with the epoxide.

3.1.4 Polyphenols/epoxy curing reaction

The reaction between the epoxide and aliphatic hydroxyl groups does not proceed at a rate that is suitable for normal curing conditions used in practice with powder coatings, unless it is catalyzed with strong acids such a BF_3. Because of the strong corrosive effect of the catalyst during the service life of the coating, these systems are not used in practice to formulate powder coatings.

However, epoxy groups readily react with phenolic hydroxyl groups when catalyzed with base catalysts. This is a widely used reaction for preparing higher homologues of bisphenol A epoxy resins, starting from low molecular weight bisphenol A/epichlorohydrin precondensates. Curing of the epoxy powder coatings with phenol or bisphenol A type novolacs is catalyzed with tertiary amines or quaternary ammonium salts. The curing mechanism is almost the same as in the case of the carboxy/epoxy curing reactions (3.11 to 3.13). The reaction rate is proportional to the concentration of the catalyst and to the concentration of the epoxy and phenolic groups. The base catalyzed curing reaction between resins containing hydroxyl groups of phenolic type and epoxies proceeds at a proper rate in the temperature range for curing powder coatings.

3.1.5 Polyetherification

Epoxy resins can undergo catalyzed polymerization with a variety of Lewis bases and acids as well as numerous salts and complex catalysts. Despite the fact that the Lewis bases and acids initiate anionic and cationic polymerizations respectively, the resulting network structure is of the same polyether nature. The cationic polymerization of epoxy resins is not commercially used for curing powder coatings and will be discussed later in this book (Chapter 3.4.5).

The **anionic step polyetherification** is a curing reaction used in epoxy powder coatings for obtaining a network structure by base catalyzed self-crosslinking of epoxy resins. The catalysts used in practice are different imidazoles and tertiary amines. Polyetherification can be regarded as a stepwise process starting with an initiation (reaction 3.26), propagation (reaction 3.27), and termination (reaction 3.28).

Equation 3.26

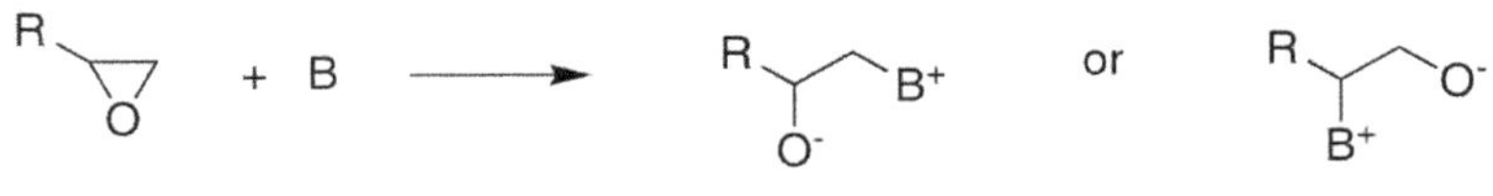

Equation 3.27

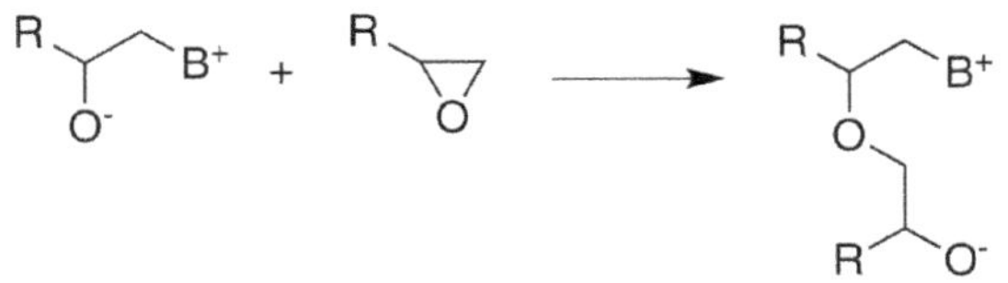

Equation 3.28

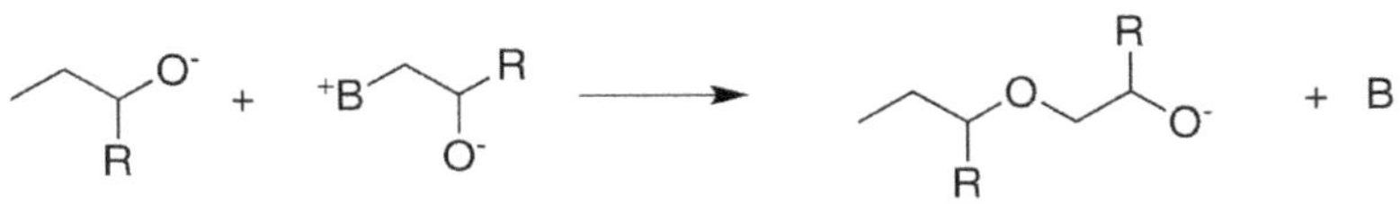

Many excellent articles and monographs dealing the anionic polymerization of epoxy resins explain the mechanism of the reaction [31, 76-90]. The above reaction scheme can be considered as a simplification of the actual mechanism, which is rather complex and, despite all the work done in this field, not yet completely understood.

Next to the previously proposed scheme, it is very probable that the following reaction proposed by *Berger* and *Lohse* leading to the formation of unsaturated alcohols takes place [86]:

Equation 3.29

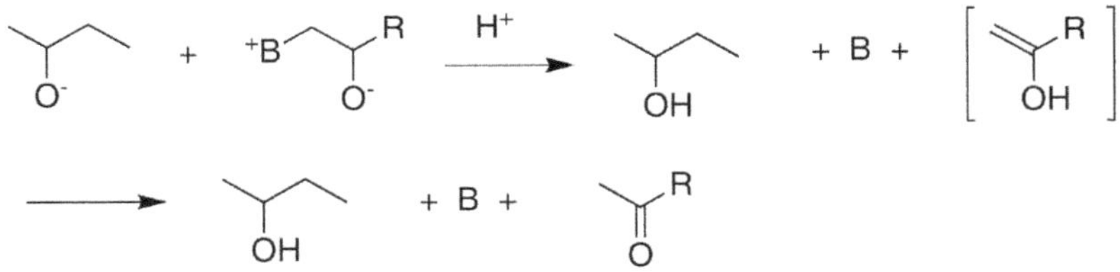

This is very similar to the termination mechanism proposed much earlier by *Narracott* [91] as an explanation of the accumulation of unsaturated compounds during polyetherification catalyzed by tertiary amines (3.19)

The statistical and **kinetic** theoretical approaches to polyetherification [88-90] which have been recently developed do not consider the complications that accompany the process, such as chain transfer, reinitiation and other side reactions. *Dusek* [73], comparing both approaches, has found that the kinetic treatment is correct within the assumption of the proposed mechanism, while the statistical method leads to serious deviations from experimental results. However, the statistical method does approach the results obtained by the kinetic treatment when larger fragments are taken as the building blocks of the polyether structure. Reasonably good results obtained by the kinetic method confirm the overall validity of the proposed model, within the scope of the assumptions involved. *Bressers* and *Goumans* [92] have studied the curing mechanism of epoxies with imidazoles, using DSC, LC and NMR techniques in a model system of phenyl glycidyl ether/1-benzyl-2-methyl imidazole. However, they could not give a full explanation of the multipeaked thermograms, which were also observed by other investigators [83], since the exact elucidation of the initiating species for polyetherification has been hampered by their instability.

3.1.6 Isocyanate/hydroxyl curing reaction

The main feature of the isocyanates is their **reactivity** towards nucleophiles at room and elevated temperatures. The following reactions leading to the formation of urethanes (3.30), substituted ureas (3.31 and 3.32), amides (3.33), allophanates (3.34), biurets (3.35), acylureas (3.36), uretdiones (3.37), isocyanurates (3.38), thiourethanes (3.39) and oxazolidones (3.40) are typical for isocyanates:

Equation 3.30

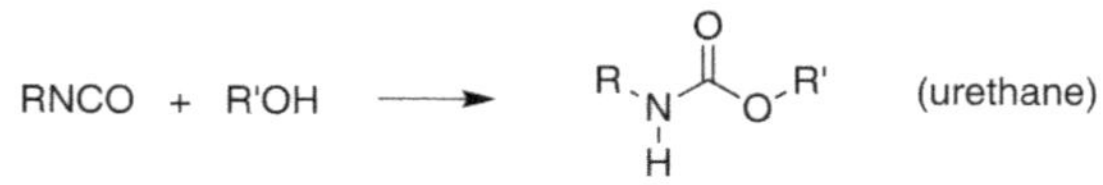

Equation 3.31

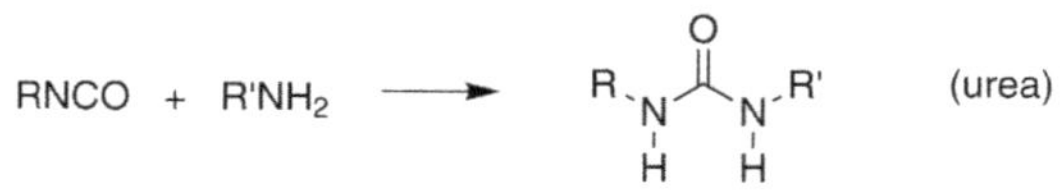

Equation 3.32

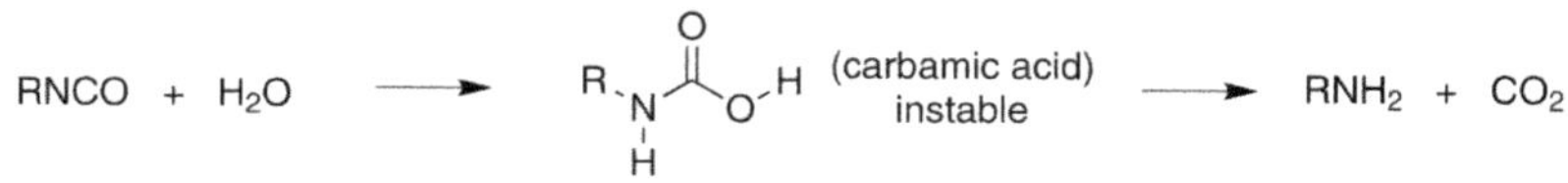

Equation 3.33

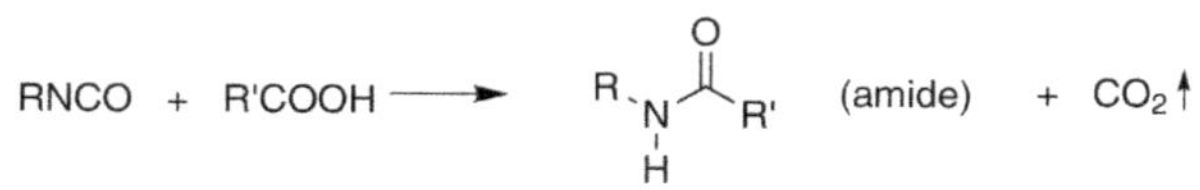

Equation 3.34

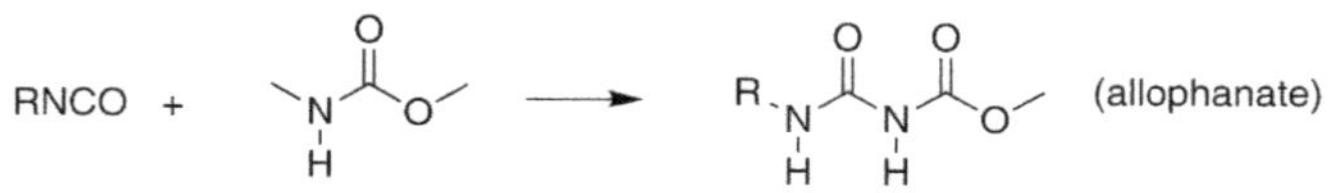

Equation 3.35

Equation 3.36

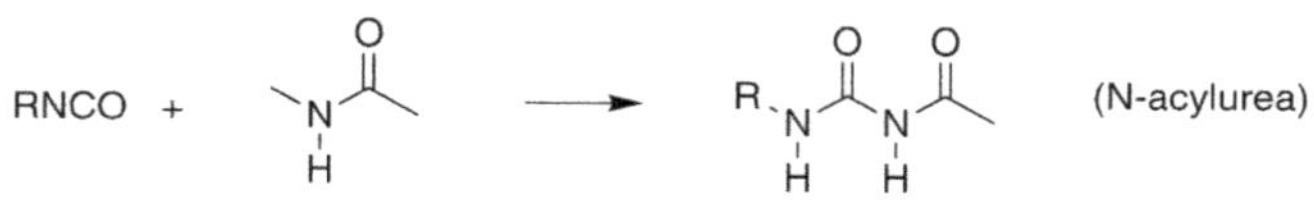

Equation 3.37

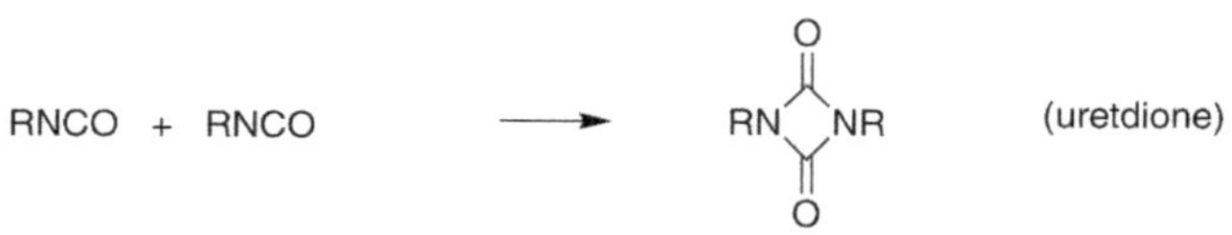

Equation 3.38

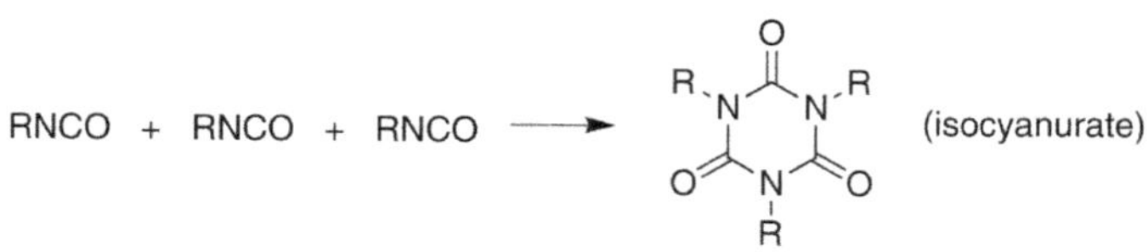

Equation 3.39

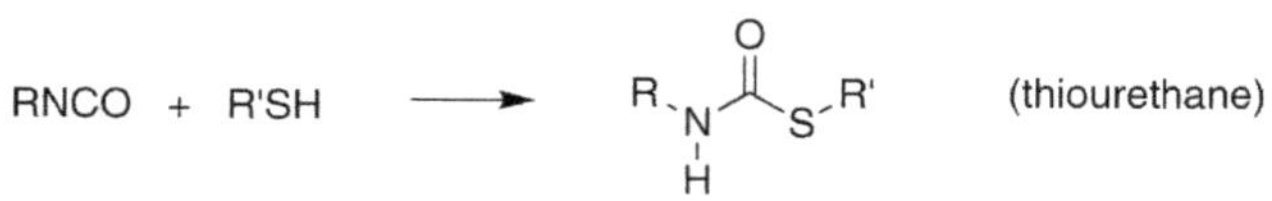

Equation 3.40

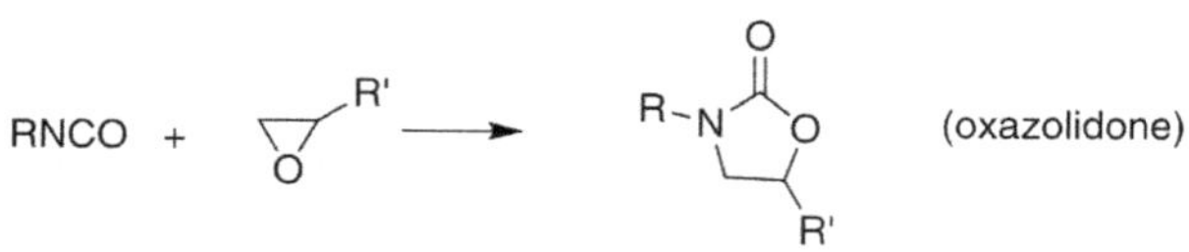

These reactions illustrate the richness of the **isocyanate chemistry** and indicate the large potential for using isocyanates as curing agents for coatings. Various types of coating systems utilize many of the reactions mentioned above for obtaining a crosslinked network at room and elevated temperatures. In the case of powder coatings, only reaction (3.30) between isocyanate and hydroxyl groups is used in practice in the so-called **polyurethane** powder coatings.

Reactions of isocyanates with hydroxyl compounds have been studied extensively by many researchers. Different analytical techniques and models have been used in attempts to determine the kinetic parameters of the curing process or to elucidate the reaction mechanism. The simplest is the titration method using di-n-butyl amine [93, 94], forming urea with the isocyanate within seconds and titration of the excess of amine with HCl-solution. Spectroscopic methods such as IR, UV and NMR have also been used for the same purposes [93-103]. The method combined with the conventional titration method has been used to determine the competing reactions in the system of phenyl isocyanate/n-butanol in the presence of different catalysts [104, 105]. A different approach is the modeling of the structure/property relations during network formation and a comparison of the experimental results with the theoretical predictions based on assumed reaction mechanism [106-110].

Reactions between isocyanates and alcohols proceed remarkably easily at ordinary temperatures with evolution of heat. The reaction rates differ, however, with the type of isocyanate and the co-reactant. Aromatic isocyanates react with alcohols faster than aliphatic isocyanates. The curing reaction between adducts of toluene diisocyanate with different polyols and hydroxyl containing polymers proceeds with very high speed, even at room temperature. In the same conditions, when aliphatic isocyanates are used as the crosslinker the curing proceeds only with the help of a suitable catalyst. Assuming a pseudo first-order reaction, *Squiller* and *Rosthauser* have compared the reaction rate constants of hexamethylene diisocyanate (HDI)/n-butanol and toluene diisocyanate (TDI)/n-butanol reactions. In an uncatalyzed system at 30 °C having a stoichiometrical ratio between the reactants, they have confirmed the well-known fact of different reactivities of the two isocyanate groups in TDI, finding for the rate constants values of $28 \cdot 10^4$ and $7 \cdot 10^4\ s^{-1}$, or an average value of $17 \cdot 10^4\ s^{-1}$. This is almost **a hundred times higher than the rate constant HDI**, for which they found a value of $0.14 \cdot 10^4\ s^{-1}$ [111].

The **kinetics of the curing** between aromatic isocyanates and alcohols has been extensively studied in a model system of phenyl isocyanate and n-butanol by many investigators. The reports agree with a second-order reaction in both catalyzed and uncatalyzed systems, dependent on both phenyl isocyanate and n-butanol [112-121]. Different catalysts are used to speed up the reaction between the isocyanates and alcohols. However, the same **catalysts** produce different activities with aromatic and aliphatic isocyanates. With both aromatic and aliphatic isocyanates the formation of urethane linkage can be promoted by organometallic compounds or metal salts of organic acids. Tin compounds are widely used for this purpose, dibutyltin dilaurate and tin octoate being the predominant representatives among them. Due to toxicological reasons there is a recent trend to switch to tin-free catalysts like bismuth based compounds. Tertiary amines are also used, although they are less effective than the metal compounds. Moreover, many of them catalyze the urethane formation as well as the formation of allophanates and trimerization, leading to cyclization into isocyanurate rings [105, 122]. For many years it has been known that the catalytic activity of tertiary amines increases with the basicity of the amine and decreases with the steric shielding of the amine nitrogen [123-125]. Less data are published with respect to the activity of the organometallic compounds as catalysts in aliphatic isocyanates/alcohol reactions. Based on kinetic studies, *Squiller* and *Rosthauser* [111] have established the following relative order of **catalytic activity** for the reaction of hexamethylene diisocyanate in excess of n-butanol, assuming the reaction to be of pseudo first order:

DMTDC > DBTDLM > DBTL> DBTM > stannous octoate

This comparison is done on the basis of the relative rate constants at 30 °C, when the catalysts are used in the same concentrations on a weight basis. The abbreviations have the following meanings: DMTDC = dimethyltin dichloride; DBTDLM = dibutylin dilauryl mercaptide; DBTDL = dibutylin dilaurate and DBTM = dibutylin maleate.

Britain and *Gemeinhardt* have screened the effectiveness of various compounds showing catalytic effects by determining the **gelation time** of a system consisting of toluene diisocyanate (80:20 ratio of 2.4/2.6 isomer mixture) and polypropylene triol in a stoichiometric ratio [126]. The gelation time was interpreted as a measure of the speed of the isocyanate/hydroxyl reaction. The same weight of compounds to be tested for catalytic activity was used in all tests. *Table 3.1* presents part of the results obtained by screening hundreds of compounds.

The relative reactivities of the two different isocyanate groups of isophorone diisocyanate have been compared in several papers. The secondary isocyanate group is more reactive than the primary isocyanate group.

For reactions under catalyzed mild conditions a reactivity ratio between the two groups of 10:1 has been noticed [127]. *Lorenz, Decker* and *Rose* have found that the selectivity of the reaction decreases with increasing temperature, and have reported that the uncatalyzed

Table 3.1: Catalytic activity of different compounds in system of TDI/polyoxypropylene triol at 70 °C, expressed as a gel time in minute [126]

Compound tested	Gel time [min]	Compound tested	Gel time [min]
Blank	>240	Dibutyltitanium dichloride	13
Bismuth nitrate	1	Lead benzoate	8
Tetrabutyl titanate	8	Ferric chloride	6
Lead oleate	4	Triethylene diamine	4
Ferric acetylacetonate	16	Antimony trichloride	13
Sodium propionate	32	Antimony pentachloride	90
Potassium acetate	10	Cobalt benzoate	32
Lithium acetate	60	Cobalt octoate	12
Tetrabutyltin	>240	Trioctyl aluminium	32
Tributyltin chloride	200	Aluminium oleate	70
Dibutyltin dichloride	109	Zinc naphtenate (14.6 % Zn)	60
Butyltin trichloride	4	Zinc octoate (22 % Zn)	65
Stannic chloride	3	1-Methyl-4-(dimethylaminoethyl)-piperazine	90
Stannous octoate	4	Stannous oleate	8
N-Ethylethylenimine	32	N,N,N',N'-Tetramethylenediamine	60
Stannous tartrate	>240	Dibutyltin di(2-ethylhexoate)	4
Triethylamine	120	N-Ethylmorpholine	180
Dibutyltin dilaurate	8	Dimethylaniline	>240
Dibutyltin diisooctyl maleate	4	2-Methylpyrazine	>240
Dibutyltin oxide	3	Nicotine	240
Titanium tetrachloride	50		

relative reactivities of the two NCO groups at 40 to 110 °C in excess of butanol are in the range of 4.5 to 5.1 to 1 [128] in favor for the secondary isocyanate group. *Ono, Jones* and *Pappas* report that the relative reactivities of the primary and secondary isocyanate group depend on the catalyst used [103]. They have found that when 1,4-diazabicyclo [2.2.2] octane (DABCO) is used as a catalyst, the primary NCO group has a higher reactivity, while in the case of dibutlyltin dilaurate (DBTDL) they have identified the secondary NCO group as the more reactive one. This remarkable inversion of **selectivity** is explained by different catalyzing mechanism by *Spyrou* (carbonyl-O-DBTL interaction in the latter case, carbonyl-C-DABCO interaction in the former case) [214].

The high reactivities of free isocyanates prevent their use in a powder coating composition. The relatively high reactivity towards water will result in homopolymerization of the polyisocyanates through formation of polyureas, and the reaction with the main binder will lead to crosslinking during the storage before application.

The reactivity of the isocyanate group can be drastically reduced by blocking it with compounds that give a stable reaction product at room temperature and undergo deblocking at elevated temperatures, re-forming the parent isocyanate. Another possibility is to use the so-called internally **blocked isocyanates**, which are cyclic condensation products of polyisocyanates undergoing displacement reactions at elevated temperatures with alcohols to form polyurethanes.

A variety of blocking agents have been used in practice. The preparation of the blocked material and deblocking follows the scheme now presented in a simplified way where HBl is a blocking agent:

Equation 3.41

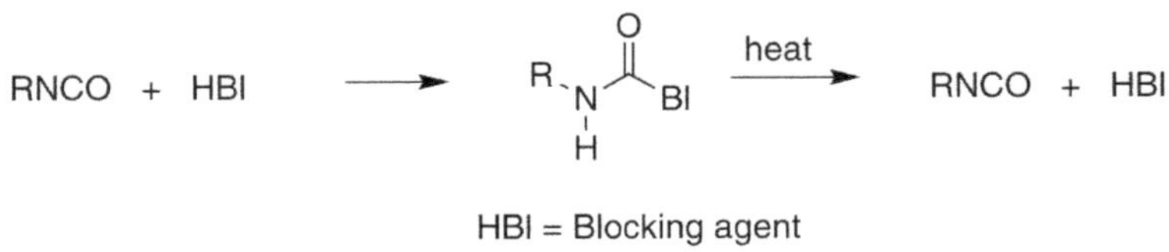

However, the reaction of the blocked isocyanate with a compound containing an active hydrogen atom, such as an alcohol for example, could conceivably involve either a two-step process (elimination-addition) in which the urethane formation is preceded by the unimolecular thermal dissociation of the blocked isocyanate to free isocyanate (reaction 3.42) or a direct bimolecular displacement (addition-elimination) reaction (reaction 3.43):

Equation 3.42

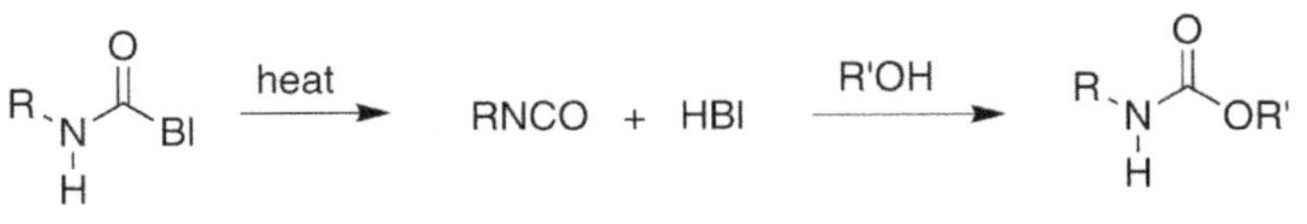

Equation 3.43

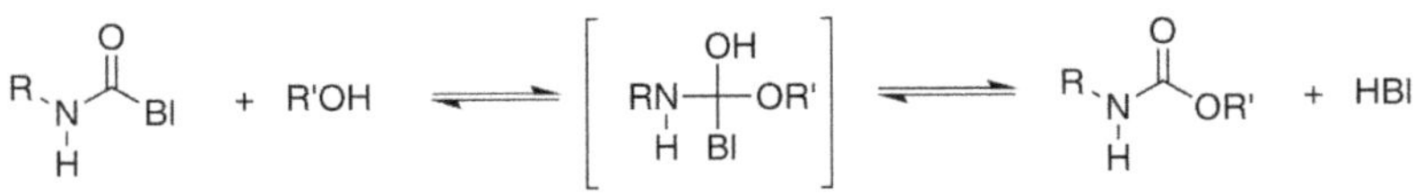

The observation that various ester exchange catalysts facilitate curing by isocyanate adducts can be considered to support the theory of having an interchange mechanism [130]. Evidence of a displacement reaction mechanism has been provided by *Gaylord* and *Sroog* [131] in the case of a reaction between unsubstituted urethane and alcohol. At the same time, comparable reactions involving N-monosubstituted urethanes proceeded through primary dissociation to the free isocyanate. Since blocked isocyanates of the urethane classes are N-substituted, they would be expected to react through an initial dissociation step [132]

(elimination-addition). *Anagnostou* and *Jaul* [133] have studied deblocking reactions of adducts of 4,4'-diphenyl methane diisocyanate (MDI) with caprolactam, methyl ethyl ketoxime and benzotriazole using differential scanning calorimetry (DSC) and infrared (IR) spectroscopy. *Table 3.2* represents the results obtained by performing the deblocking reaction of pure MDI/blocking agent adduct and the same in the presence of polyethylene glycol with a molecular weight of 400, which is able to undergo reaction with the isocyanate group. The start of the deblocking was noticed to be the start of the endotherm at the DSC scan for pure MDI adducts and the start of the exotherm where polyethylene glycol is present. It has been found that **deblocking temperatures** in all cases were lower when hydroxyl containing material is present in the system. The authors suggest that this can be explained by a bimolecular reaction between the primary hydroxyl group on the blocked site or by a solvating effect on the adduct itself. They have also concluded that the addition of catalyst does not alter the data obtained with uncatalyzed systems very much.

Mirgel and *Nachtkamp* [134] suggest that the mechanisms do not exclude each other and that both proceed at a rate dependent on the reaction conditions and types of reactants. According to this mechanism the regeneration of the isocyanate group is not absolutely necessary for crosslinking. This is represented by the alternative mechanism according to which the crosslinking is a result of addition of the polyol to the blocked isocyanate group followed by elimination of the blocking agent. In this **addition-elimination** reaction which is analogous to an ester interchange no isocyanate is involved. The well-known gelation of the system of blocked polyisocyanate/polyamine, even at room temperature [135, 136], can be considered as a support for this mechanism.

All compounds bearing an active hydrogen atom may be considered for use as **blocking agents** for isocyanate groups. In two excellent articles *Wicks* [137, 138] gives a review of different blocked isocyanates either used or having a potential use in the coating industry. The following classification is suggested by *Mirgel* and *Nachtkamp* for a variety of compounds that can be used for **blocking the isocyanate group** [134]:

X-H	Protonic acids
-O-H	Alcohols, phenols
-S-H	Thioalcohols, thiophenols
=N-OH	Oximes, hydroxamic acid esters
>N-H	Amines, amides, imides, lactams
-CO-CH-CO-	β-Dicarbonylic compounds

Next to this, the heterocyclic ring opening reactions can be considered as a route to isocyanate formation, although some of these heterocyclic compounds are not strictly speaking blocked isocyanates since they are not synthesized from isocyanates. They can be more properly classified as isocyanate precursors. The fact that no volatile reaction products are formed during the curing process is certainly an advantage of these compounds.

Table 3.2: Initial and maximum temperatures in DSC scans of blocked MDI and blocked MDI/polyethylene glycol systems [132]

Blocking agent	Blocked MDI		Blocked MDI + polyethylene glycol	
	Initial [°C]	Maximum [°C]	Initial [°C]	Maximum [°C]
ε-Caprolactam	158	162	153	155
Methyl-ethyl ketoxime	120	130 to 139	107	not defined
Benzotriazole	214	222	179	184

The ease with which a blocked isocyanate RNHCOBl will dissociate will depend to a large extent upon the strength of the bond between Bl and the carbonyl group. Different blocking agents used in practice to prevent premature reaction between the isocyanate group and the functional group containing active hydrogen exhibit different deblocking temperatures with the same isocyanate. The **deblocking temperature** also depends on the type of isocyanate. Aromatic isocyanates undergo deblocking at lower temperatures compared to aliphatic isocyanates when blocked with the same blocking agent. Aliphatic isocyanates, for example, blocked with aliphatic alcohols undergo deblocking at about 250 °C, while the same isocyanates blocked with phenolic compounds have a deblocking temperature that is approximately 50 °C lower. On the other hand, the deblocking temperatures of aromatic isocyanates blocked with aliphatic alcohols and phenolic compounds are in the region of 180 and 120 °C, respectively [139].

The dissociation process of the aromatic isocyanates is significantly affected by the substituents on the aromatic ring. Electron-withdrawing groups decrease stability of the substituted isocyanates and consequently lower the dissociation temperature, while electron-releasing groups have an opposite effect [133, 140]. The deblocking temperature of isocyanate adducts is largely determined by the acidity of the blocking agent and the isocyanate, or their pK_a value. An increase in the acidity of the blocking agent or isocyanate usually has the effect of reducing the required curing temperature. Aliphatic alcohols, lactams, and phenols have pK_a values of ca. 18, 17 and 10 respectively, while the pK_a values of the aliphatic and aromatic amines from which the corresponding isocyanates can be derived are 30 and 25 [141]. The **order of the deblocking temperatures** – aliphatic alcohols > lactams > phenols and aliphatic isocyanates > aromatic isocyanates – fits with the order of the pK_a values of the corresponding compounds.

A large number of compounds have the potential to be used as blocking agents, but only a few of them have achieved technical importance in practice. In the case of powder coatings **ε-caprolactam and methyl ethyl ketoxime** are used on a larger commercial scale although many different compounds are disclosed in numerous patents. The structure of these blocking agents and the structure of the blocked isocyanate groups are as follows:

ε-caprolactam methyl ethyl ketoxime

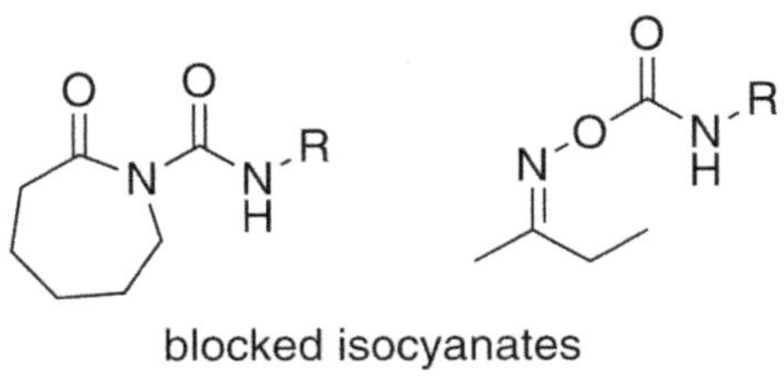

blocked isocyanates

Wicks [137] has investigated the deblocking reaction of caprolactam blocked phenyl isocyanate. He found that the reaction is reversible with an equilibrium dependent on the temperature at which the deblocking is carried out. At 160 °C for example, in a solution of chlorobenzene, the equilibrium is reached within 5 minutes, when about 40 % of the blocked adduct has dissociated. From the activation parameters of the reaction he suggests that a six-membered cyclic transition state is involved:

McLafferty, Figlioti and *Camilleri* [142] have studied the influence of the carboxyl groups normally present in the hydroxyl functional polyesters used for powder coatings on the curing behavior of a system based on caprolactam blocked isophorone diisocyanate oligomer and polyester resin. They have drawn the interesting conclusion that the balance of hydroxyl and carboxyl end groups in urethane powder coatings affects significantly the curing characteristics and performance of the coatings, especially in the case of aliphatic isocyanates.

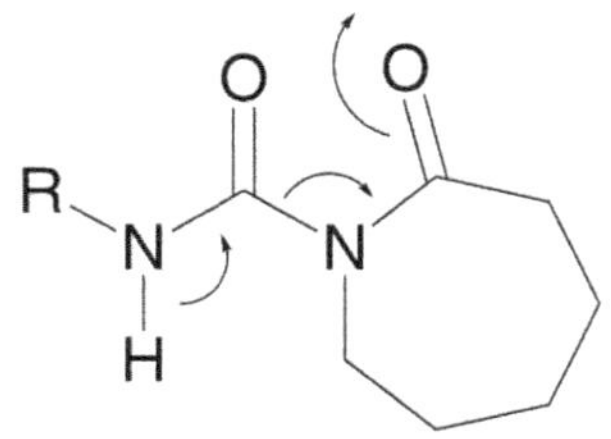

The deblocking temperatures of the oxime blocked isocyanates are lower compared to caprolactam adducts. Methyl ethyl ketoxime blocked MDI undergoes deblocking at a temperature that is 38 °C below the deblocking temperature of MDI/ε-caprolactam (*Table 3.2*).

Carlson et al. [143] have used thermal-mechanical analysis (TMA) and Fourier transform infrared spectrometry (FTIR) for cure **kinetics characterization** of trimer of isophorone diisocyanate blocked with methyl ethyl ketoxime, 2-ethyl-4-methylimidazole, phenol and caprolactam. The results obtained, showing the relatively free NCO-concentration at various temperatures, are depicted in *Figure 3.1*. They have found that in the case of methyl ethyl ketoxime the first hint of deblocking appears at 120 °C, with significant levels above 140 °C. The deblocking (2) reactions for caprolactam (4) and phenol (3) take place above 160 °C. In contrast, 2-ethyl-4-methyl imidazole (1) blocked IPDI trimer undergoes deblocking slightly above room temperature. In all cases they have found first-order kinetics.

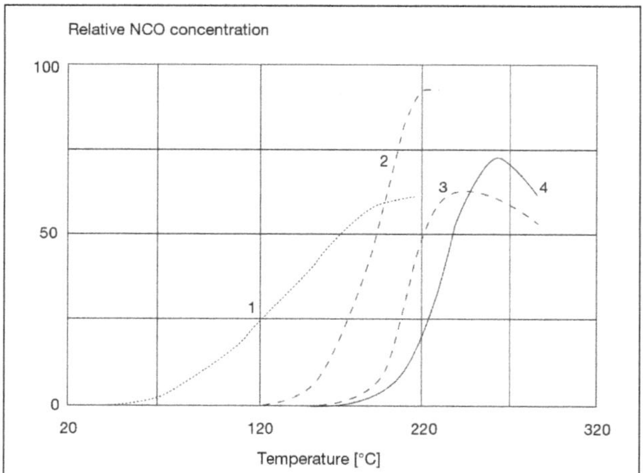

Figure 3.1: Effect of blocking agent on deblocking reaction [143]
1: 2-ethyl-4-methyl imidazole; 2: methyl ethyl ketoxime; 3: phenol; 4: caprolactam

source: Advanced Urethane Sci. Techn. Vol. 9, 1984, p. 9, Figure 4 C

The presence of **typical catalysts** such as 1,4-diaza[2.2.2]bicyclooctane (DABCO), tetrabutylammonium chloride (TBAC) and dibutyltin dilaurate (DBTDL) did not change the reaction order and none catalyzed the deblocking as reported previously by *Wicks* [138]. The results for DABCO are difficult to interpret due to the side reactions. However, all of them (especially DBTDL) have shown a catalytic effect towards the curing reaction.

Interesting **blocking agents** with attractive low deblocking temperatures are described by *Cooray* and *Spencer* [144]. They belong to the group of pyrazole derivatives. Pyrazole blocked aliphatic isocyanates remain stable up to 80 °C, but begin to dissociate at 120 °C at a considerable rate due to cyclic transition state illustrated next to:

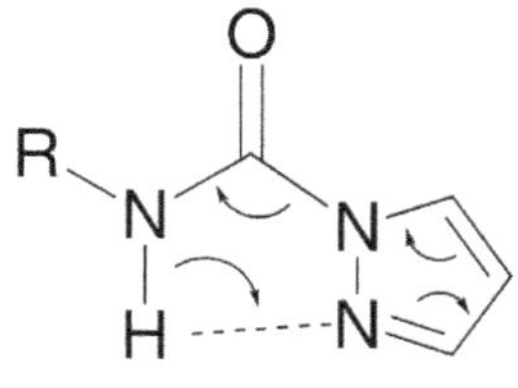

The proton attached to the urea nitrogen atom forms the fifth membered ring. Similarly to the mechanism proposed earlier by *Wicks* [137], *Cooray* and *Spencer* found a ring opening reorganization for the **deblocking mechanism** as the temperature is raised above 100 °C. Below this temperature, proton transfer to the pyrazole is almost eliminated by intermolecular hydrogen bonding between adjacent urea groups in the blocked urethane.

An interesting list of blocking agents that undergo deblocking in a relatively low temperature range has been given by *Guise, Freeland* and *Smith* [145]. As well as the oximes, compounds such as hydroxylamines, hydroxypyridines, hydroxyquinolines and pyrazolinones are considered as suitable blocking agents offering a good compromise between short curing cycles at temperatures down to 100 °C and acceptable storage stability. All of them possess a common structural feature, namely an -OH group (or a group capable of enolization) and a basic nitrogen function.

A drop of the curing temperature down to 15 min at 130 °C was described in a patent [146] using **diethyl malonate** as a blocking agent. The generated bond is an ester linkage and not an urethane bond, though.

Relatively low deblocking and curing temperatures are observed in the case of **acylurethanes**; their use as curing agents for powder coatings has been described by *Brindoepke* [141]. These compounds can be obtained from aliphatic carboxylic acids and carbamates as illustrated below:

Equation 3.44

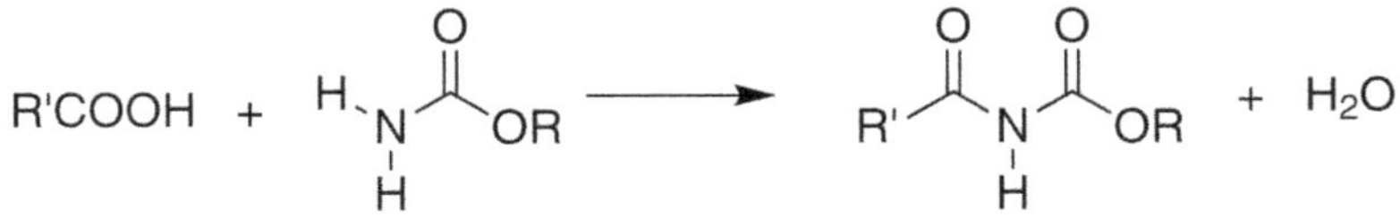

In principle they have the structure of normal blocked isocyanates. The lower deblocking temperature compared to that of the isocyanate blocked with the same blocking agent is due to the reduction of the basicity by the acylation.

The use of **internally blocked isocyanates** in the form of uretdiones (NCO dimers) is becoming more popular because of the obvious reason of having no by-products during the curing. Uretdiones are obtained by dimerization of diisocyanates in the presence of phosphine or pyridine containing catalysts. This is a reversible process which yields monomeric isocyanates under mild heating (*Equation 3.37*).

Uretdione formation is facilitated particularly when the reactivity of the isocyanate groups in the diisocyanate is not equal, as in the case of toluene diisocyanate or isophorone diisocyanate. The reason for this is the selective generation of dimers and not oligomers or polymers, since the latter ones are more difficult to process. Uretdiones are nowadays used as "internally" blocked isocyanate hardeners. See Chapter 3.3.2.2.

Arnold, Nelson and *Verbanc* [147] have followed the ring opening reactions of aromatic isocyanates with amines and found that biuret is the normal product of the reaction. This even at room temperature fast reaction can be used to determine the **content of uretdione** by back titration of the excess of amine. Under forced conditions (T >120 °C), the reaction proceeds to the formation of two moles of substituted urea:

Equation 3.45

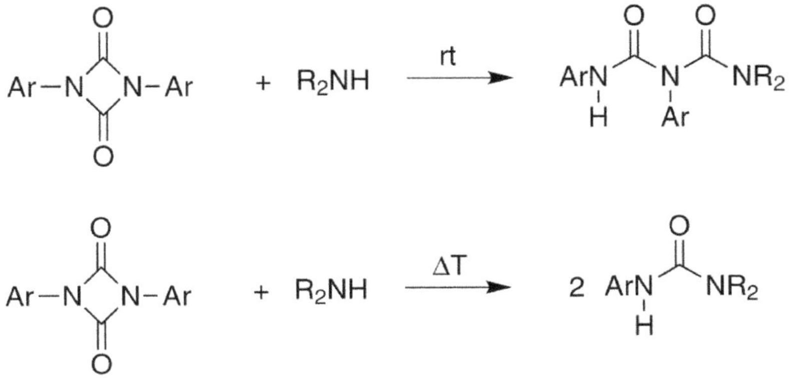

On the other hand, the same reaction with alcohols leads to the formation of the corresponding allophanate [148], if low temperature and/or special catalysts (e.g zincoctoate, zincacetylacetonate) are applied. At higher temperature the uretdione is splitted off and the restored isocyanates form urethanes with the alcohols.

Equation 3.46

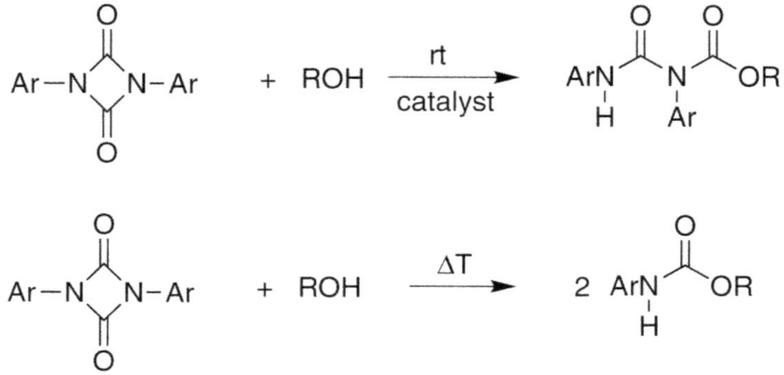

3.1.7 *Curing with amino resins*

The chemistry of amino resins as curing agents for coatings has been a matter of interest and investigation for more than 60 years. The enormous number of articles dealing with this matter has tried to elucidate the curing mechanism of this system which despite its long history seems still to be not well understood.

There is almost no work published with respect to the use of amino resins as crosslinking agents for powder coatings. The published patent applications understandably do not deal with the chemistry of the process to the same extent that would be expected from a scientific paper. Therefore, the material presented below refers to the chemistry of curing with amino resins in solventbased coatings. Disregarding the differences emanating from the absence of solvents, the curing principles and the curing reactions should also be valid to a great extent for powder coatings.

There are four different types of alkylated amino formaldehyde resins that are used in the coating industry: urea, melamine, benzoguanimine and glycoluril-formaldehyde resins. All of them contain the same type of reactive groups that will be discussed in further detail. However, the **reactivity** of these groups differs considerably. Urea-formaldehyde resins are the most reactive crosslinking agents, while benzoguanimine resins are at the bottom of the scale regarding reactivity. Since the coatings containing urea-formaldehyde resins as crosslinkers have low humidity and corrosion resistance, they have never been considered as suitable in application areas that are typical for powder coatings. The chances of the other three gaining wider use in the powder coatings world are rather high. Benzoguanimine resins provide superior chemical resistance properties, humidity, corrosion and detergent resistances, and despite the inferior exterior durability they have a big potential for indoor applications. For many years melamine resins have acquired a solid reputation as crosslinkers for stoving coatings with good **outdoor durability**. The reports on glycoluril-formaldehyde resins indicate that they can be compared with the melamine resins with respect to their exterior durability [149]. The following functional groups can exist in the amino resins [150]:

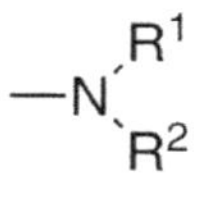

R^1, R^2 = -H, $-CH_2OH$, $-CH_2OR$, $-CH_2OCH_2OR$

Depending on the way of preparation, all of the above mentioned reactive species can be present in the amino resins. It is possible to prepare resins in which almost all functional groups are CH_2OR but resins containing exclusively one of the other groups do not exist. Partially alkylated commercial products contain combinations of the shown functionality.

The network formation during the curing process is highly influenced by the presence of polymeric material in the amino crosslinker since the **self-crosslinking of the triazine rings** during the synthesis of amino resins increases their average functionality. All commercial types of melamine resins contain variable levels of self-condensed triazine rings which are connected by methylene ($-N-CH_2-N-$) or methylene ether groups ($-N-CH_2-O-CH_2-N-$). Keeping in mind the structure of the functional groups given above, a variety of combinations of functional groups attached to two nitrogen atoms bridged by methylene or methylene ether groups are possible. *Blank* [150] describes 12 possible different functional groups in melamine resin bridges. Even the methylene ether bridge can be considered as a functional group, able to undergo a normal condensation reaction with an ether group. This illustrates how complicate the structure of the amino resin is, and how difficult it is to describe exactly the curing process where amino resins are used as crosslinkers.

The methods used for **characterization** of the functional groups in the amino resins and following the curing process which are described in the excellent article by *Christensen* [151] and many other papers, include wet chemical analysis [152-154], elemental analysis [155], analysis of the volatile products produced during curing [150, 156, 157] infra red (IR) spectroscopy [155, 158-163], 13C-NMR spectroscopy [164-167], proton NMR spectroscopy [150, 165, 168], size exclusion chromatography (SEC) [169, 170], high pressure liquid chromatography (HPLC) [154, 165, 170-172], gel permeation chromatography (GPC) [155], swelling of the crosslinked films [173] and dynamic mechanical analysis (DMA) [159, 173-176]. The reaction between polymers containing hydroxyl functional groups and HMMM (hexamethoxymethyl melamine) is transetherification. This reaction proceeds when HMMM is heated with higher alcohol in the presence of acid catalyst followed by liberation of methanol:

Equation 3.47

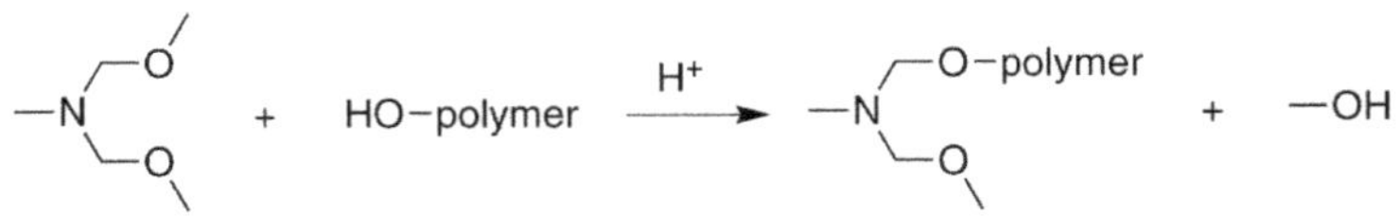

One of the complications that occurs in attempts to quantify the curing process is the self-condensation of the amino resins which proceeds to a different extent depending of the curing conditions and the catalyst used. *Saxon* and *Lestienne* [159] proposed a **self-crosslinking mechanism** in the case **of HMMM** catalyzed by p-toluene sulfonic acid. According to this article, the reaction is initiated by protonation of the ether oxygen of the methoxymethyl group, evolution of methanol, and formation of the carbonium ion. Propagation involves an alternate attack on nitrogen and oxygen. The initial carbonium ion displaces a methoxymethyl group from the nitrogen of a second HMMM molecule, thus forming a methylene crosslink, and displaced methoxymethyl ion by reaction with a third molecule of HMMM liberates methylal and regenerates a triazine carbonium ion to continue the reaction. The analytical data disclosing the composition of the reaction products of HMMM cured at 150 °C in the presence of paratoluene sulphonic acid, being 83 % methylal and 17 % methanol, support the above mechanism.

However, the same authors have found that in crosslinking a hydroxyl functional acrylic resin with HMMM under the same curing conditions, the ratio of methylal to methanol decreases from 83:17 to 5:95, indicating that **crosslinking** of the **acrylic polymer** is about 20-times as rapid as HMMM self-condensation under the selected experimental conditions.

Somewhat different results are obtained by *Koral* and *Petropoulos* [177]. They found that under dry nitrogen the self-condensation of HMMM proceeds by evolution of methylal as the only reaction product.

Wohnsiedler [178] and *Blank* [150] found the formation of the following volatile products: formaldehyde, water, methanol and several ether containing structures. *Table 3.3* presents the results of the analysis of the voatile products during the curing of HMMM with polyols at three different ratios, obtained by Blank.

The experiments with **different catalyst levels** and different temperatures have shown that up to 100 °C the level of catalysis does not play a significant role in the composition of the volatile reaction products. However, at 120 °C, higher amounts of catalyst cause increased rates of self-condensation and demethylolation, leading to considerably higher amounts of methanol in the volatile reaction products.

Different papers dealing with the mechanism of crosslinking with amino resin agree to a large extent with the general scheme presented below [150, 158, 170, 171, 176, 179, 180]:

Equation 3.48 - 51

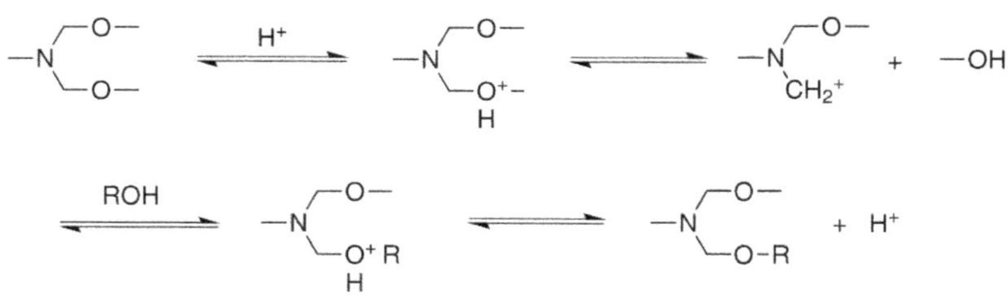

Berge, Gudmundsen and *Ugelstad* [152, 153] postulate that the formation of the carbonium ion (reaction 3.49) is the slowest step in determining the maximum overall reaction rate of curing that can be obtained. This reaction mechanism is valid for a curing reaction catalyzed by strong acids such as p-toluene sulfonic acid, dodecylbenzene sulfonic acid, naphtalene sulfonic acid, etc.

In the case of partially methylated amino resins when weak acid catalysts are used, for example the residual carboxyl groups of the polyester or alkyd resins, or carboxyl groups introduced in acrylic copolymers by acrylic acid, a different curing mechanism has been proposed:

Equation 3.52

Table 3.3: Volatile pruducts developed during curing of HMMM/polyol blend for 20 min/150 °C. Catalyst dodecylbenzene sulfonic acid (0.84 wt%) (Source [150]*)*

Polyol/HMMM ratio:	**75/25**	**62.5/37.5**	**50/50**	**62.5/37.5**	
				dry	**wet**
OH/triazine	4.5/1	2.5/1	1.5/1	2.5/1	
Mol. volatiles/triazine:					
HCHO	0.2	0.5	0.3	0.4	0.8
MeOH	5.0	3.1	2.4	2.6	3.5
Methylal	–	–	0.1	–	–
H_2O	1.0			0.7	0.5

Equation 3.53 - 3.55

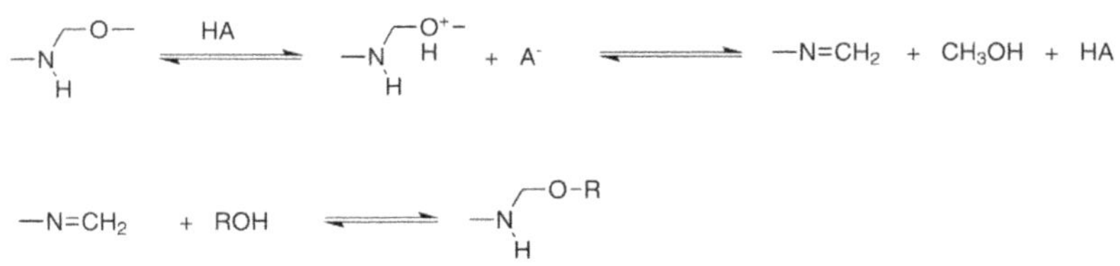

The **rate determining step** in this so called general acid catalysis is the abstraction of the hydrogen atom at the nitrogen to which the methoxymethyl group is attached (reaction 3.54). In the case of HMMM, which lacks this atom, this scheme is obviously not applicable. None of the experiments with HMMM reveal any general acid catalysis [153]. *Yamamoto, Nakamichi* and *Ohe* [155] have determined the change of viscosity in a system of acrylic resin/HMMM catalyzed with p-toluene sulphonic acid as a strong acid catalyst and monodecyl phthalate as a weak acid catalyst. They have used two types of acrylic polymers having the same hydroxyl value, acid value, molecular weight and glass transition temperature, being different only with respect to the nature of the hydroxyl groups. In one case the hydroxyl functionality in the acrylic resin was introduced through 2-hydroxyethylmethacrylate (primary hydroxyl groups) and in the other case through 2-hydroxypropylmethacrylate (secondary hydroxyl groups). The change in relative viscosity during the curing process at 140 °C is shown in *Figure 3.2* and at different curing temperatures in *Figures 3.3*. In the **strong acid catalyst** case with p-toluene sulphonic acid added, the relative viscosity rises rapidly after only four minutes of curing. However, in the case of monodecylphthalate catalyst, both viscosities of primary and secondary hydroxyl systems hardly change with time. In the case of weak acid catalysis, the change in viscosity differs between the primary and secondary hydroxyl acrylic systems at different temperatures, indicating the higher reactivity of the primary hydroxyl groups. Although the curing reaction proceeds at higher temperatures, even at 180 °C, 17 to 20 minutes are necessary to compensate for the viscosity drop with temperature and to reach the initial viscosity.

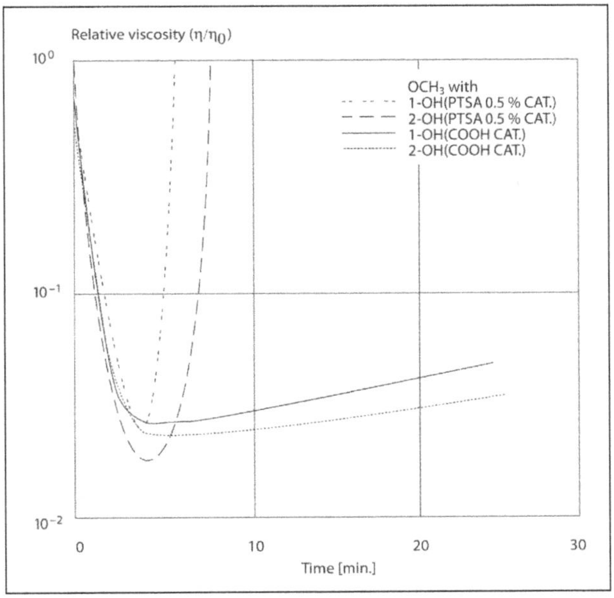

Figure 3.2: Relative viscosity changes as a function of curing time baked at 140 °C for catalytic crosslinking reaction systems, which consist of two types of acrylic polyols (1. primary OH and 2. secondary OH groups) and HMMM. Initial viscosity is designated as η_0 [155]

source: Federation of Societies for Coatings Technology

The influence of different catalysts on the curing rate is quite obvious. Especially in the case of fully methylated amino resins it seems that the use of strong acid catalysts for obtaining acceptable curing time at relatively low curing temperatures is inevitable. However, coatings that have been cured with strong catalysts such as p-toluene sulphonic acid are **sensitive to moisture** that diffuses across the air/film interface and may cause blistering. Water resistance can be improved by substitution of p-toluene sulphonic acid with more hydrophobic acid, such as dinonylnaphthalene disulphonic acid [181].

Considerable work has been done with respect to the **kinetics** of the

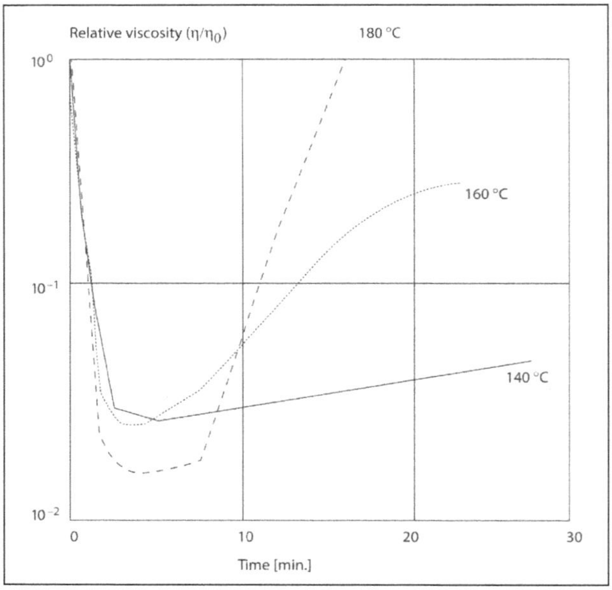

Figure 3.3: Relative viscosity changes as a function of curing time baked at 140, 160 and 180 °C respectively (acrylic polyol with primary hydroxyl groups) catalyzed with monodecyl phthalate [155]

source: Federation of Societies for Coatings Technology

curing process with amino resins, predominantly with HMMM. The already presented reactions which may take place during curing show the complexity of the system, as well as any difficulties that may occur in attempts to make a quantitative description of the curing. Therefore, to a first approximation, many authors describe the cure of fully alkylated melamines by a single reaction. This is the formation of polymer-melamine links via the reaction of alkoxy groups on the melamine with hydroxy groups on the polymer. This approximation can be considered to be valid to a great extent at moderate curing temperatures and normal catalyst levels.

Holmberg [182] concluded that the principle cure reaction between the hydroxyl groups and alkoxymethyl groups is Sn2, the rate being proportional to the concentration of both alkoxymethyl and hydroxyl groups. *Lazzara* has found a good fit of the kinetics data he has obtained to the **second-order kinetics** [157]. The same reaction order has been proposed by *Yamamoto* et al. [155] in the case of weak acid catalyzed systems. However, according to them, the self-condensation reaction, which occurs even when weak acid catalyst is used, does not follow second-order kinetics, but rather that of pseudo-third order (first order for OH and second order for OCH_3), in spite of the non-disappearance of the hydroxyl groups. Although they were not able to provide experimental evidence, they postulate that the residual polar hydroxyl groups play the role of activator on a carboxylic acid catalyst.

In contrast to *Holmberg, Berge, Gudmundsen* and *Ugelstad* [153] and *Blank* [183] have proposed a Sn1 type reaction on the basis that formation of the carbonium ion by the acid catalyst is a rate-determining step. *Blank* has also cautioned against drawing conclusions as to the mechanism from systems that are highly viscous and approaching gelation.

Using the same curing mechanism, which postulates that formation of the carbonium ion is the slowest step that determines the overall kinetics of the curing process, *Bauer* and *Budde* [168] have developed a **kinetic model for crosslinking fully methylated melamines** which accurately predicts the observed dependence of the extent of reaction on bake time and melamine concentration. Using the model and the extent of reaction, a rate constant for the reaction has been determined and found to be proportional to the catalyst concentration and to the square root of the acid strength of the catalyst. From a closer examination of the implications of the accepted curing mechanism they have concluded that the reaction is first order only if the methanol concentration in the film during the curing process is zero, which obviously is not the case in practice.

Alkoxy groups also react with other functional groups, including carboxylic acids which are typically present in the powder coating resins. Very often these groups are incorporated in the polymer on purpose to promote cure. The relative reactivity of the carboxyl groups depends on the level of catalyst and the type of amino resin.

Bauer and *Dickie* found that the hydroxyl groups are more reactive with fully alkylated melamines in the presence of strong acid catalysis, while carboxyl groups are more reactive with partially alkylated melamines [162, 163].

A combination of the kinetic model fitting to a great extent the experimental results and theory of *Macosko* and *Miller* [184] of non-linear polymerization was used by *Bauer* et al. [162, 163, 168, 185, 186] to develop a model describing the network formation during the crosslinking process. The mathematics involved in the derivation of the model is rather complex and beyond the scope of this book. One advantage of this model is that the data required as input can be obtained from practical coatings.

3.1.8 Transesterification

Transesterification as a cure mechanism is well-known. In general, weak catalysts, such as zincoctoate are used to cure activated esters or hydroxy functional compounds. In a few Japanese patents [187] a 2-hydroxyethyl acrylate containing methacrylate-styrene copolymer was cured with a latent strong acid formed by, e.g. diphenylamine and trifluoromethane sulfonate, at 120 °C for 20 min to give a coating with pencil hardness 2H, and good solvent, acid and bending crack resistance.

In 1994 *Craun* [188] introduced the ENCAT (**epoxy nucleophile catalyzed transesterification**) cure system, which is claimed to be suitable for use in the powder coating technology.

In short, ENCAT consists of first preparing a methyl succinate half ester, reacting it with a type 4 epoxy resin to form a full epoxy ester, and further reacting this ester with an additional amount of epoxy resin in the presence of a nucleophile.

But ENCAT uses a strongly basic alkoxide, which is formed during cure by the combination of the nucleophile with the epoxy compound, because the epoxide ring is opened by nucleophilic attack.

Craun tested two nucleophiles, namely 1,4-diazabicyclo (2,2,2) octane (better known as **DABCO**) and tetrabutylphosphonium acid acetate (**TBPA**).

Transesterification is "initiated", rather than catalyzed, by the basic alkoxide, as the latter is consumed in the reaction.

The **rate of transesterification** depends on various factors, namely:

- the **concentration** of the nucleophile and epoxy "co-initiators",
- the **basicity** and **nucleophilicity** of the nucleophile,
- the **structure** of the ester and hydroxyl groups, entering into the reaction.

Craun described the preparation of epoxy esters, capable of self curing to give powder coatings with excellent film properties, and illustrated it with the following example.

Example: Preparation of epoxy esters

An epoxy ester is prepared from 5300 g "Epikote 1004", 494 g succinic anhydride and 173 g methanol. Methanol and succinic anhydride are heated to reflux to form the methyl succinate half ester. Then the solid epoxy resin is added, together with 15 g of benzyl dimethylamine and heated for 2 hours at 120 °C to form the epoxy ester. The formed epoxy ester has an acid value of 0.08 meq acid/g and an epoxy content of 0.06 meq epoxy/g.

Then powder coatings are prepared by dry blending 1 kg epoxy ester with 100 g of a solid epoxy resin (with epoxy equivalent weight of 400); 760 g of TiO_2; 9 g of benzoin and 10 g of a polyacrylate flow agent. Now the two nucleophiles are added (in two separate trials), using 3 % on total resin.

Since the epoxy ester has a very low functionality, additional epoxy resin is added to the formulation.

Film properties, such as mechanical properties, pencil hardness, solvent resistance, are significantly better with TBPA than with DABCO. According to *Craun* this is possibly due to the higher volatility of DABCO at the applied stoving temperature of 150 °C for 20 minutes.

Craun finds an extremely high pencil hardness of 5 H for the TBPA addition and only 1 H for the DABCO cured film.

In studies of transesterification by gas chromatography, the epoxy ester appears to self cure with ENCAT. The terminal methyl succinate half ester transesterifies with hydroxyl groups on the epoxy ester backbone, liberating methanol. These secondary hydroxyl groups are activated by their close proximity to the ester carbonyls of the epoxy ester, to give the excellent film properties, shown by this type of curing.

3.1.9 Radiation curing

Contemporary powder coatings with curing temperatures of 140 to 200 °C are not suitable for substrates that are sensitive to heat (such as wood, MDF, plastics and paper). The attempt to lower the stoving temperature (**low temperature curing**, LTC) by over-catalyzing the system, disturbs the balance of properties in the sequence storage stability-melting-film flow-curing, resulting in a coating system with a rather poor storage stability and/or film forming properties. An obvious solution of this problem is the use of **latent catalysts** that can be activated, after the melting and film formation have been completed. As suggested by *Pappas* and *Hill* [189] a good combination of storage stability and high curing speed can be achieved in the case of first-order curing reactions with large pre-exponential *Arrhenius* parameter. The problem is that most of the thermosetting (powder) coatings are cured by a bimolecular reaction obeying the *Arrhenius* law (see Chapter 4.6). **Polyurethane powder coatings**, using blocked polyisocyanates as crosslinkers, could be an exception, assuming that during the crosslinking process the formation of the urethane linkage by a second order-condensation between the isocyanate and hydroxyl groups is preceded by a first-order deblocking reaction that determines the overall curing kinetics. However, there is enough evidence, that the curing with blocked isocyanates is a combination of first- and second-order reactions, where parallel to the deblocking of the isocyanate group a bimolecular addition of the hydroxyl group to the blocked isocyanate group takes place, too [190-194].

An attempt to use latent catalysts in cationic curing of powder coatings has been published by *Hilbom et al.* [195], but there is no evidence of commercial existence of such a system.

It is, therefore, not surprising, that research efforts have been focused on systems, where the curing can be fully separated from the melting, coalescence and flow in the film formation of powder coatings.

It is self-evident, that the ultraviolet (UV) and electron beam (EB) **radiation curing** techniques offer a unique opportunity for achieving this goal.

Already in 1971 an unsaturated polyester resin combined with dimethallyl phthalate has been described in a patent by BASF [196]. In the work of *Hilbom*, next to the thermally activated cationic polymerization, an initiation by UV light has been reported, too [195]. Two types of polymerization reactions are generally used in UV curable systems, namely: free radical initiated UV cure and cationic UV cure. Both types may be applied in a radiation curable powder coating.

With **radical polymerization** a photo-initiator forms, during UV radiation, free radicals, which initiate the polymerization of the unsaturated binder system. In a simplified form the UV-curing reaction by free radicals can be represented by the following equations:

		hν	
Initiation reaction	R - R	→	2 R$^\bullet$
Starting reaction	R$^\bullet$ + M	→	R - M$^\bullet$
Propagation or growth	R-M$^\bullet$ + nM	→	R - $M^\bullet_{(n+1)}$
Termination reactions	2 R - M$^\bullet$	→	R - M - M - R
	R - M$^\bullet$ + R-M-M$^\bullet$	→	R - M + R - M = M

The first step of the UV-curing mechanism is the splitting of the photoinitiator into two radicals. Then an inactive monomer molecule M is converted into an activated nucleus R - M$^\bullet$, which reacts with another inactive monomer to give an activated dimer. The activity is not lost during this addition, and for this reason the chain keeps on growing by adding new monomers (nM). Until finally, by a termination process, the activity is destroyed and further propagation of the chain is stopped. The termination or cessation process can take place either by collision between two growing chains; or by collision of a growing chain with a radical; or by collision of a growing chain with impurities or with the surface of the substrate.

With cationic polymerization (only used for epoxy resins) the photoinitiators are onium salts, producing cations of strong Lewis or Brònstedt acids upon irradiation. These cations (H^+ protons) attack and open the oxirane (or epoxy) rings of the epoxy resin molecules. After which the cationic polymerization starts and continues as a **homopolymerization of oxirane groups**. Thus the network mainly consists of two different types of structural elements:

- the poly (β-hydroxyether) chains of the solid epoxy resins, and
- the polyethylene-oxide (PEO) chains, formed by the cationic polymerization of the epoxy groups; these PEO chains form the links between the epoxy chains.

The cure mechanism is influenced by addition of aliphatic alcohols with one or more OH groups to the formulation. These alcohols react with the cation and a new chain is initiated by proton transfer.

The cationic polymerization of solid epoxy resins has been reported and patented by Ciba-Geigy [197]. This homopolymerization can be started with thermal latent catalysts (e.g. metal triflates), too, but has not found a broad application, yet, due to the compromise in flexibility and corrosion resistance. *Garcia* et al. published a paper [198] in which they describe the addition of meldrums acid as an “expanding monomer”, leading to much better adhesion and corrosion resistance compared to conventional cationic cured epoxy resins. This way curing cycles of 25 min at 110 °C could be reached.

Comparing radical and cationic UV-curing leads to some typical differences:

Radical polymerization	**Cationic polymerization**
• shrinking	• very little shrinking
• no inhibition by water	• inhibition by water
• inhibition by oxygen	• no inhibition by oxygen
• very fast curing (in seconds)	• longer curing (but also sec.) and dark curing

A paper of *Witting* and *Gohmann* [199] describes free radical as well as cationic radiation curing of powder coatings. For the free radical UV cure they use solid unsaturated polyester resins as binders and (meth)acrylamide derivatives as crosslinkers. That means the curing reaction is a **copolymerization**.

Acrylic resins can also be homopolymerized [200, 201], which means, that in principle no co-reactants or crosslinkers are needed. But for better film flow, higher crosslink density, a greater freedom of formulation and improved film properties, a combination with solid unsaturated polyester resins is sometimes preferred [202, 203]. Also methacrylated polyester resins as sole binders are known [204].

DSM Resins [205] have developed a non-acrylate binder system for radiation curable powder coatings. The solid binder consists of two compounds, a resin and a co-reactant (or crosslinker). The resin is an unsaturated polyester, in which maleic or fumaric acid is incorporated. The co-reactant is a polyurethane, containing vinyl-ether unsaturation. Both materials are solid at room temperature and fully compatible. The glass transition temperature (T_g) of this binder system is sufficiently high to allow for processing and storage at room temperature.

The UV-curing of the binder is based on the free radical initiated **1/1 copolymerization** of the electron-rich vinyl ether groups with the electron-poor maleate or fumarate groups [206]. The mechanism of this maleate/vinylether alternating copolymerization can be illustrated as follows [207]:

1. Formation of donor acceptor complex:

   MA+VE → [MA-VE]

   Homopolymerization of donor acceptor complex:

   PI → R$^\cdot$ + [MA-VE] → R-MA-VE$^\cdot$ + [MA-VE] → R-MA- VE-MA-VE$^\cdot$

2. Copolymerization via cross-propagation mechanism:

   PI → R$^\cdot$ + MA → R-MA$^\cdot$ + VE → R-MA-VE$^\cdot$ + MA → R-MA-VE$^\cdot$

The polymerization can involve a copolymerization via cross-propagation of the two monomers MA and VE and/or homopolymerization of a donor-acceptor complex. Kinetic studies of a maleate-vinyl ether system strongly suggest that the photoinitiated copolymerization occurs via a cross-propagation mechanism [205].

As a result of the 1/1 copolymerization, a minimum of unreacted double bonds will remain after curing, if the system contains a stoichiometric balance of maleate and vinyl ether groups. The radical-initiated polymerization of maleate-vinyl ether systems will not be hampered by high humidity, in contrast to cationic curing, which will be inhibited by atmospheric moisture [209], as well as by water containing pigments, extenders, additives and substrates. Atmospheric oxygen will inhibit free radical-induced polymerizations, but it has been reported, that this effect on a maleate-vinyl ether system is significantly less pronounced than for acrylate resins [208].

One of the advantages, that the chemistry of the MA-VE binder system offers is the fact, that the components do not, under controlled conditions, homopolymerize. Another advantage is that the UV curable powder is thermally stable.

Measurements by DSM Resins [210] have shown that the viscosity of MA-VE powder paint is only 26 Pa s at 100 °C and that no viscosity increase occurs up to 200 °C. This is in contrast to UV curable powder systems, containing (meth)acrylic unsaturations, because they can react under influence of heat alone. The physical storage stability of this system is good at 35 °C for 4 weeks.

The heat sensitive substrates for which UV-curing powders are developed, cannot resist high temperatures, for their distortion temperature is around 100 °C. This means during the melting of the applied paint, the substrate may not absorb too much heat. Preferably, wood should not be heated above 80 °C, in order to avoid deformation and gassing. The **combination of**, respectively, **IR** for melting and formation of a smooth film, **and** - separately - **UV** for curing, seems to be the most efficient technology for the application of the radcure powder paint on wood, MDF and other heat sensitive substrates.

Electron beam curing (EBC) instead of UV-curing is also a possibility for the application of the radcure powder paint on metal, wood, MDF and other heat sensitive substrates. The same powder coating formulations can be used, with the exception of the photoinitiators, which are superfluous due to the much higher energy of the electron beams.

Electron beams have energies varying between 125 keV and 2.5 MeV, while UV radiation has an energy of 3.3 eV (=electron volt) at λ =360 nm. The electron bombardment leads, with unsaturated polyester resins and other unsaturated polymeric materials, to the formation of free radicals, without the aid of photoinitiators, which immediately causes polymerization.

The free radicals can be formed simultaneously in three different ways, namely by ionization, excitation and electron capturing:

1. **Ionization**

   During the bombardment with electrons an electron can be "shot away" from the molecule:

   $$AB \rightarrow AB^{+} + e$$

   Immediately the ionized molecule dissociates into a free radical and a radical ion:

   $$AB^{+} \rightarrow A^{*} + \cdot B^{+}$$

2. **Excitation**

   The radiated energy is not sufficient for ionization, but high enough to cause an "excited" situation in the molecule:

   $$AB \rightarrow AB^{*}$$

   The excited molecule dissociates into free radicals:

   $$AB^{*} \rightarrow A\cdot + \cdot B$$

3. **Electron capture**

   This is the inverted process of sub 1, for it is also possible, that an electron is captured by a molecule, which also means ionization:

   $$AB + e \rightarrow AB^{-}$$

   The generated ion here too decomposes into a free radical and a radical ion:

   $$AB^{-} \rightarrow A\cdot + \cdot B^{-}$$

Polymerization takes place in the same way (by propagation, termination) as has been described for UV radiation. In contrast with UV radiation, EBC is not hindered by hiding pigments, due to the very high energy of the irradiated electrons. The penetration depth is much higher than with UV radiation and is only dependent on the applied energy and the density of the substrate to be coated, in such a way, that this depth is directly proportional to the energy and inversely proportional to the density.

In literature there is not much published about **EB-curing** of powder coatings. *Maetens*[204] has done tests with TiO_2 pigmented powder coatings, based on methacrylated unsaturated polyesters (same formulations as for UV-curing, but without photoinitiators) and applied on steel. After melting and separately curing by EBC at a dose of 5 Mrad, good flexibility as well as boiling water resistance of the coatings could be recorded.

The greatest disadvantage of EBC is the high investment for this technology, although less expensive devices have been introduced into the market, recently. Calculations have shown that EBC only will become economic, when production rates are higher than 3000 m^2 of coated material per hour. Maybe this is the reason, that there is no evidence, at the moment, of commercial existence of EB-curing of powder coatings.

Another **radiation method** for powder coatings was introduced under the title "Curing within Seconds" by *Bär* and *Sedlmeyr*[211] at the PCE 98 Congress in Amsterdam, namely high intensity **Near Infra Red ("NIR") radiation**. NIR radiation is located in the electromagnetic spectrum on the opposite side of the visible light in the wavelength area between 740 and 1200 nm.

With infra red heating the thermal energy is transferred by electromagnetic radiation. The short gap between the wavelength of visible light and the commencement of short wave infra red, thus in the NIR range, is most suitable for rapid energy conversion, as most materials, used in powder coatings, are semitransparent in this wavelength range.

The energy conversion is, according to the radiation laws of *Stefan-Bolzmann* and *Planck*, proportional to the difference of the fourth power of the temperature of the NIR emitter and that of the object. That means, that an increase in the emitter temperature will lead to an extremely increase of the energy density. The radiation intensity increases by the factor T^4. At the same time, there is a relocation of the radiation maximum towards shorter wavelengths, with an increase of temperature (according to *Wien's* relocation law).

It is evident, that at a temperature of 3500 K an emitter (or radiator) reaches a radiation density, which is much higher than that of an emitter with 2000 K in the short-wave IR area. Increasing the emitter temperature from 2000 to 3500 K therefore has an enormous effect on the achieved energy density of this NIR radiation (several orders of magnitude higher than for conventional infra red sources).

High performance halogen lamps with filament temperatures of up to 3500 K are used by *Bär*[211] as the radiation source and the high energy density of the radiation makes it necessary to be controlled by registering the surface temperature by a radiation pyrometer, which is specially adapted to the applied radiation characteristics.

The **working principle** of the NIR technology can be explained as follows.

NIR technology

When applying the normal technique of heating a powder coated metallic substrate by convection or by short/middle wave IR radiation, the surface is heated initially and the heat is then absorbed throughout the powder layer to the substrate by conduction. Much of this heat is then carried off by the substrate (very fast if it is metallic), because it acts as a natural heat conductor and can be seen as a "heat sink". Only when the substrate is completely heated, the boundary layer between substrate and coating has reached the necessary temperature, in order to start the curing process.

In comparison, NIR radiation applies the energy directly and homogeneously throughout the powder film, and in-depth heating is complemented by a degree of energy reflection from the substrate surface. The essential difference is the speed with which the NIR technology spreads the heat uniformly over and through the film, so that, on achieving the required curing temperature, crosslinking occurs at the same rate, at all depths of the coating.

Tests, carried out with both (specially adapted) polyester and hybrid powder coatings, showed that, due to the extremely high radiation density and the depth effect of the NIR technology, a homogeneous heating of the coating up to the curing temperature followed by **curing** can be achieved within **5 to 10 seconds!**

For the complete curing of the powder coating, the temperature at the substrate is held for another few seconds. After cooling down, the coating is then completely cured. In this way a total cure time of much less than one minute is realized, compared to conventional cure times of 15 minutes for the same powders.

The dramatically shortened curing times allow the estimation (according to *Bär*), that the radiation of the NIR technology not only activates a temperature reaction of the powder, but might also accelerate the curing reaction directly.

Bär and *Sedlmeyer*[212] claim to have confirmed this estimation by further investigations and they offer the following explanation:

The material transmission of powders is only given in the spectrum of the visible light and near infrared wave length range, while in the UV as well as in the IR range above 2 μm high energy absorption is given. In addition, powders show absorption bands (e.g. OH) within the near infra red wave length spectrum, which are activated by the energy dense NIR lamps. Therefore the achieved molecular thermal mechanism is initiated with NIR emitters.

The emitted NIR energy is, due to the partial binder transparency, penetrated into the powder coating and absorbed within the complete coating layer. The absorbed energy within the coating is heating up the coating, but also activating the curing directly due to the NIR radiated thermal activation of polar powder modular groups.

NIR-curing process is driven by:

- direct heating,
- an additional curing mechanism, due to activation of polar molecules;
- three-dimensional heating, due to high NIR transparency of the powder and therefore large NIR coating penetration.

Based on these facts, the **NIR-curing process results** in (according to *Bär* et al.):

- an instant anions heating in combination with a higher curing temperature,
- an accelerated curing mechanism, due to the direct activation, and
- a very limited thermal conduction into the substrate (lower substrate temperature) and reduced heat losses.

It should be remarked, that the NIR-curing is not restricted to NIR powders. But, since conventional powders are not specially adapted to the NIR technology and the NIR-curing process is very dependent on the powder behavior, the overall applicability is dependent on the powder formulation. Of course NIR-curing is predestined in particular for flat substrates, as all radiation curing techniques.

3.1.10 References

[1] *Govarkier, V.R.*, Polymer Science, John Wiley & Sons, New York, 1976, p. 98
[2] *Batzer, H., Lang, H.*, Makromol. Chem., 15, 211, 1955
[3] *Carothers, W.H.*, Trans. Faraday Soc., 32, 39, 1936
[4] *Flory, P.J.*, Principles of Polymer Chemistry, Ch. 9, Cornell University Press, Ithaca, New York, 1953

[5] *Stockamyer, W.H.*, J. Appl. Polym. Sci., 9, 69, 1952
[6] *Gordon, M.*, Proc. Roy. Soc. London, A268, 240, 1962
[7] *Gordon, M., and Malcom, G.N.*, Proc. Roy. Soc. London, A295, 29, 1966
[8] *Macosko, C.W., and Miller, D.R.*, Macromolecules, 9, 199, 1976
[9] *Miller, D.R., and Macosko, C.W.*, Macromolecules, 9, 206, 1976
[10] *Tiemersma-Thoone, G.P.J.M., Scholtens, B.J.R., Dusek, K.*, Proceedings of the First International Conference on Industrial and Applied Mathematics, Paris-La Vilette, June 29-July 3, 1987
[11] *Good, I.J.*, Proc. Camb. Phil. Soc., 45, 360, 1948
[12] *Good, I.J.*, Proc. Camb. Phil. Soc., 56, 367, 1960
[13] *Good, I.J.*, Proc. Roy. Soc. London, A272, 54, 1963
[14] *Gordon, M., Scantlebury, G.R.*, Trans. Faraday Soc., 60, 605, 1964
[15] *Gordon, M., Ward, T.C., and Whitney, R.S.*, Polymer Networks, Eds. A.S. Chompff, and S. Newman, Plenum Press, New York, 1970
[16] *Saito, O.*, Polym. Eng. and Sci., 19, (4), 234, 1979
[17] *Ilavsky, M., Dusek, K.*, Polymer, 24, 981, 1983
[18] *Dusek, K., Ilavsky, M., and Matejka, L.*, Polym. Bulletin, 12, 33, 1984
[19] *Ilavsky, M., Bogdanova, L.M., and Dusek, K.*, J. Polym. Sci., Polym. Phys. Ed., 22, 265, 1984
[20] *Dusek, K., Vojta, V.*, Brit. Polym. J., 9, 164, 1977
[21] *Dusek K., Ilavsky, M.*, J. Polym. Sci., Pol. Phys. Ed., 21, 1323, 1983
[22] *Dusek, K.*, Makromol. Chem. Supl., 2, 180, 35, 1979
[23] *Flory, P.J.*, Macromolecules, 15, 99, 1982
[24] *Dusek, K., Ilavsky, M., Lunak, S.*, J. Polym. Sci. Symp., 53, 29, 1975
[25] *Dusek, K., Gordon, M., Ross-Murphy, S.B.*, Macromolecules, 11, 236, 1978
[26] *Dusek, K., Bleha, M., Lunak, S.*, J. Poly. Sci., Polym. Chem. Ed., 15, 2393, 1977
[27] *Charlesworth, J.M.*, J. Polym. Sci., Polym. Phys. Ed., 17, 1557, 1571, 1979
[28] *Topolkaraev, V.A., Zhorina, L.A., Vladimirov, L.V., Berlin, Al., Zelenetskii, A., Prut, E.V.*, Enikolopyan, N.S., Vysokomol. Soed., Sec. A, A21, 1655, 1979
[29] *Allen, G., Holmes, P.A., Walsh, D.J.*, Faraday Disc., 57, 1974
[30] *Allen, G., Booth, C., Price, C.*, Polymer, 8, 397, 1967
[31] *Shechter, L, Wynstra, J.*, Ind. Eng. Chem., 48, 86, 1956
[32] *Shechter, L, Wynstra, J., Kurkjy, R.E.*, Ind. Eng. Chem., 48, 94, 1956
[33] *Shechter, L, Wynstra, J., Kurkjy, R.E.*, Ind. Eng. Chem., 49, 1107, 1957
[34] *Malkemus, J.D., Swan, J.D.*, J. Amer. Oil Chemist' Soc., 34, 342, 1957
[35] *Wrigley, A.N., Smith, D.F., Stirton, A.J.*, J. Amer. Oil Chemists' Soc., 36, 34, 1959
[36] *Kakiuchi, H., Tanaka, Y.*, J. Org. Chem., 3 1, 1559, 1966
[37] *Tanaka, Y., Takeuchi, H.*, Tetrahedron, 24, 6433, 1968
[38] *Tanaka, Y., J.* Org. Chem., 32, 2405, 1967
[39] *Tanaka, Y., Okada, A., Suzuki, M.*, Can. J. Chem., 48, 3258, 1970
[40] *Isaacs, N.S.*, Tetrahedron Letters, 1965, 4549
[41] *Albert, A., Phillips, J.N.*, J. Chem. Soc., 1956, 129
[42] *Brown, H.C., Mihm, X.R.*, J. Amer. Chem. Soc., 77, 1723, 1955
[43] *Brown, H.C., Gintis, D.*, J. Amer. Chem. Soc., 78, 5387, 1956
[44] *Brown, H.C., McDaniel, D.H., Haflinger, O.*, Determination of Organic Structures by Physical Methods, Eds. *F.C. Nochod and W.D. Phillips*, Vol. I, Academic Press, New York, 1962
[45] *Linde van der, R., Belder, E.G.*, VIIth International Conference in Organic Coatings Science and Technology, Athens, July 1981, p. 359
[46] *Luston, J, Manasek, Z., Kulickova, M.*, J. Macromol. Sci. Chem., A12, 995, 1978
[47] *Matejka, L., Lovy, J., Pokorny, S., Bouchal, K, Dusek, K.*, J. Polym., Sci., Polym. Chem. Ed. 21, 2873, 1983
[48] *Tanaka, Y., Mika, T.F.*, in "Epoxy Resins-Chemistry and Technology", Eds. *May, C.A. and Tanaka, Y., Marcel Dekker*, Inc., New York, 1973, p. 179
[49] *Henig, A., Jath, J., Mohler*, H., FARBE&LACK, 86, 313, 1980
[50] *Fisch, W., Hofmann, W.*, J. Polym. Sci., 12, 497, 1954
[51] *Fisch, W., Hofmann, W.*, Makromol. Chem., 44-46, 8, 1961
[52] *Fisch, W., Hofmann, W., Koskikallio*, J., J. Appl. Chem., 6, 429, 1950
[53] *Dearborn, E.C., Fuoss, R.M., White*, A.F., J. Polym. Sci., 16, 201, 1955
[54] *Tanaka, Y., Kakiuchi, H.*, J. Appl. Polym. Sci., 7, 1063, 1963
[55] *Tanaka, Y., Kakiuchi, H.*, J. Polym. Sci., 2A, 3405, 1964
[56] *Newey, H.A.*, Gordon Res. Conf. on Polymers, New London, Connecticat, July, 1955
[57] *Kaplan, S.I.*, Polym. Eng. Sci., 6, 65, 1966
[58] *Fedtke, M., Domaratius, F.*, Polymer Bulletin, 15, 13, 1986
[59] *Fischer, R.F.*, J. Polym. Sci., 44, 155, 1960

[60] *Matejka, L., Lovy, J., Pokorny, S., Bouchal, K, Dusek, K.*, J. Polym., Sci., Polym. Chem. Ed. 21, 2873, 1983
[61] *Tanaka, Y., Kakiuchi, H.*, J. Makromol. Chem., 1, 307, 1966
[62] *Tzirkin, M.Z.*, Soviet Plastics, 7, 17, 1963
[63] *Fedke, M., Tarnow, M.*, Plaste und Kautchuk, 30, 70, 1983
[64] *Kushch, P.P., Lagodzinskaya, G.W., Komarov, B.A., Rozenberg, B.A.*, Visokomol. Soed., B21, 708, 1979
[65] *Fischer, R.F.*, J. Polym. Sci., 44, 155, 1966
[66] *Feltzin, J.*, ACS Meeting, Chicago 1964, Divis. Org. Coat. Plast., Paper No. 40
[67] *Kakurai, T., Noguchi, T.*, J. Soc. Org. Synt. Chem. Japan, 18, 485, 1960
[68] *Dannenberg, H.*, SPE Trans., 3, 78, 1963
[69] *Gross, A., Kollek, H., Brockmann, H.*, Int. J. Adhesion and Adhesives, 8, 4, 225, 1988
[70] *Gross, A., Brockmann, H., Kollek, H.*, Int. J. Adhesion and Adhesives, 7, 1, 33, 1987
[71] *Gross, A., Kollek, H., Schormann, A., Brockmann, H.*, Int., J. Adhesion and Adhesives, 8, 3, 147, 1988
[72] *Kwei, T.K.*, J. Polym. Sci., 1A, 2985, 1963
[73] *Dusek, K.*, Makromol. Chem., Macromol. Symp., 7, 37, 1987
[74] *Dusek, K.*, Adv. Polym. Sci., 78, 1, 1986
[75] *Smith, T.I.*, Polymer, 2, 95, 1961
[76] *Gaylord, N.G.*, Polyethers, Part 1, Interscience, New York, 1963
[77] *Ishii, Y., Sakai, S.*, in Ring Opening Polymerisation, Eds. *K.C. Frisch and S.L. Reegen, Marcel Dekker*, New York, 1969
[78] *Lidarik, M., Stary, S., Mleziva, I.*, Visokomol. Soed., 5, 1748, 1963
[79] *Tanaka, Y., Tomoi, M., Kakiuchi, H.*, J. Macromol. Sci., 1, 471, 1967
[80] *Steiner, E.C., Pelletier, R.R., Trucks, R.O.*, J. Am. Chem. Soc., 86, 4678, 1964
[81] *Farkas, A., Ströhm, P.*, J. Appl. Pol. Sci., 12, 159, 1968
[82] *Dearlove, T.*, J. Appl. Pol. Sci., 14, 1615, 1970
[83] *Barton, J., Scheperd, P.*, Makromol. Chem., 176, 919, 1975
[84] *Ricciardi, F., Romanchik, W., Jouille, M.*, J. Appl. Polym. Sci., 21, 1475, 1983
[85] *Berger, J., Lohse, F.*, J. Appl. Polym. Sci., 30, 531, 1985
[86] *Berger, J., Lohse, F.*, Eur. Polym. J., 21, 435, 1985
[87] *Yamake, T., Fukui, K.*, Bull. Chem. Soc. Japan, 42, 2112, 1969
[88] *Markevich, M.A., Irzhak, V.I.*, Visokomol. Soed., 28, 60, 1986
[89] *Bokare, V.M., Ghandi, K.S.*, J. Polym Sci., Polym. Chem. Ed., 18, 587, 1980
[90] *Dusek, K.*, Polym. Bull., 13, 321, 1985
[91] *Narracott, E.S.*, Brit. Plastics, 26, 120, 1953
[92] *Bressers, H.J.L., Goumans, L.*, Crosslinked Epoxies, Walter Gruyter and Co., Berlin, 1987, p. 233
[93] *David, D.J., Staley, H.B.*, Analytical Chemistry of Polyurethanes, High Polymer Series, XVI, Part III, Wiley Interscience, New York, 1969
[94] *Kresta, J.E., Hsieh, K.H.*, Macromol. Chem., 179, 2779, 1978
[95] *Flynn, K.G., Nenortas, D.R.*, J. Org. Chem., 28, 3527, 1963
[96] *Abbate, F.W., Urlich, H.*, J. Appl. Polym. Sci., 13, 1929, 1969
[97] *Naegele, W., Wendisch, D.*, Org. Magn. Reson., 2, 439, 1970
[98] *Butler, G.B., Corfield, G.C.*, J. Macromol. Sci. Chem., A5, 37, 1971
[99] *Nguyen, A., Marechal. E.*, J. Macromol. Sci. Chem., A16, 881, 1981
[100] *Entelis, S.G., Nesterov, O.V.*, Kinet. Katal., 7, 464, 1966
[101] *Weij van der, F.W.*, J. Polym. Sci. Polym. Chem. Ed., 19, 381, 3063, 1981
[102] *Usmani, A.M.*, J. Coat. Technol., 56, 716, 99, 1984
[103] *Ono, H.K., Jones, F.N., Pappas, S.P.*, J. Polym. Sci Polym. Lett. Ed., 23, 509, 1985
[104] *Wong, S.W., Frisch, K.C.*, J. Polym. Sci., Polym. Chem. Ed., 24, 2867, 1986
[105] *Wong, S.W., Frisch, K.C.*, J. Polym. Sci., Polym. Chem Ed., 24, 2877, 1986
[106] *Dusek, K.*, British Polym. J., 17, 2, 185, 1985
[107] *Shy, L.Y., Eichinger, B.E.*, British Polym. J., 17, 2, 200, 1985
[108] *Bauer, D.R., Dickie, R.A.*, ACS Pol. Mat. Sci. Eng., 52, 550, 1985
[109] *Nelen, P.J.C.*, Proc. XVII Fatipec, Lugano, 1984, p.283
[110] *Wu, D.T.*, ACS Pol. Mat. Sci. Eng., 52, 458, 1985
[111] *Squiller, E.P., Rosthhauser, J.W.*, Proceedings of the ACS Div. of Polymeric Mat.: Sci. and Eng., Preprints, 55, 640, 1986
[112] *Kogon, I.C.*, J. Org. Chem., 24, 438, 1959
[113] *Burkus, J.*, J Org. Chem., 27, 474, 1962
[114] *Okada, H., Iwakura, Y.*, Makromol. Chem., 66, 91, 1963
[115] *Rand, L., Thir, B., Reegen, S.L., Frisch, K.C.*, J. Appl. Polym. Sci., 9, 1787, 1965
[116] *Okuto, H.*, Makromol. Chem., 98, 148, 1966
[117] *Rossmy, G.R., Kollmeier, H.J., Lidy, W., Schator, H., W, M.*, J. Cell. Plast., 13, 26, 1977

[118] *Furukawa, M., Yokoyama, T.,* J. Polym. Sci. Polym. Lett. Ed., 17, 175, 1979
[119] *Bechara, I.S., Mascioli, R.L.,* J. Cell. Plast., 15, 321,1979
[120] *Rossmy, G.R., Kollmeier, H.J., Lidy, W., Schator, H., Wiemann, M.,* J. Cell. Plast., 17, 319, 1981
[121] *Bailey, F.E., Jr., Critchfield, F.E.,* J. Cell. Plast., 17, 333, 1981
[122] *Potter, T.A., Williams, J.L.,* J. Coat. Technol., 59, 749, 63, 1987
[123] *Alzner, B.G., Frisch, K.C.,* Ind. Eng. Chem., 51, 715, 1959
[124] *Farkas, A., Mills, G.A., Erner, W.E., Maerker, J.B.,* Ind. Eng. Chem., 51, 1299, 1959
[125] *Burkus, J.,* J. Org. Chem., 26,779, 1961
[126] *Britain, J.W., Gemeinhardt, P.G.,* J. Appl. Polym. Sci., 4, 207, 1960
[127] Huels Technical Bulletin 22 ME 377-7
[128] *Lorenz, O., Decker, H., Rose, G.,* Angew. Makromol. Chem., 122, 83, 1984
[129] *Spyrou, E.,* Farbe und Lacke, 106, 126-130, 2000
[130] *Jonson, P.C.,* in Advances in Polyurethane Technology, Ed. Buist, *J.M. and Gudheon, H.,* McLaren and Sons Ltd., London, 1968, p. 19
[131] *Gaylord, N.G., Sroog, C.E.,* J. Org. Chem., 18, 1362, 1953
[132] *Griffin, G.R., Willwerth, J.L.,* I &EC Product Research and Dev., 1, 4, 265, 1962
[133] *Anagnostou, T.J., Jaul, E.,* J. Coat. Technol., 53, 673, 35, 1981
[134] *Mirgel, V., Nachtkamp, K.,* Polymers Paint Colour Jornal, 176, 4163, 200, 1986
[135] *Hudson, G.A., Hixenburgh, J.C., Wells, E.R., Saunders, J.H., Hardy, E.E.,* Off. Dig., 32, 213, 1960
[136] *Furuya, Y., Goto, S., Itoko, K., Uraski, I., Morita, A.,* Tetrahedron, 24, 2367, 1968
[137] *Wicks, Z.W., Jr.,* Progress Org. Coat., 9, 3, 1981
[138] *Wicks, Z.W., Jr.,* Progress Org. Coat., 3, 73, 1975
[139] *Myers, R.R., Long, S.S.,* Treatise on Coatings, Book 1, *Marcel Dekker,* New York, 1967, p. 470
[140] *Lateef, A.B., Reeder, J.A., Rand, L.,* J. Org. Chem., 36, 2295, 1971
[141] *Brindoepke, G.,* Proceed. XII Intern. Conf. Org. Coat. Sci. Techn., Athens, 1986, p.45
[142] *McLafferty, J.J., Figlioti, P.A., Camilleri, L.T.,* J. Coat. Technol., 58, 733, 23, 1986
[143] *Carlston, G.M., Neag, C.M., Kuo, C., Provder, T.,* Advanced Urethane, Sci. Techn., 9, 47, 1984
[144] *Cooray, B., Spencer, R.,* Paint and Resin, October 1988, p.18
[145] *Guise, G.B., Freeland, G.N., Smith, G.C.,* J. Appl. Polym. Sci., 23, 353, 1979
[146] a) Bayer Materialscience, EP 1479705, 2004
[146] b) *Faecke, Thomas; Subramanian, Ramesh; Dvorchak, Michael; Feng, Sharon,* Proceedings of the International Waterborne, High-Solids, and Powder Coatings Symposium, 31st 40-52, 2004
[147] *Arnold, R.G., Nelson, J.A., Verbanc, J.J.,* Chem. Rev., 57, 47, 1957
[148] *Singh, P., Boivin, J.L.,* Can. J. Chem., 40, 935, 1962
[149] *Parekh, G.G.,* J. Coat. Technol., 51, 658, 101, 1979
[150] *Blank, W.J.,* J. Coat. Technol., 51, 656, 61, 1979
[151] *Christensen, G.,* Progress in Org. Coat., 8, 211, 1980
[152] *Berge, A., Gudmundsen, S., Ugelstad,* J., Europ. Polym. J., 5, 171, 1969
[153] *Berge, A., Gudmundsen, S., Ugelstad,* J., Europ. Polym. J., 6, 981, 1970
[154] *Chiavarini, M., Del Fanti, N., Bigatto, R.,* Angew. Makromol. Chem., 56, 15, 197
[155] *Yamamoto, T., Nakamichi, T., Ohe, O.,* J. Coat. Technol., 60, 762, 51, 1988
[156] *Kooistra, M.F.,* J. Oil Col. Chem. Assoc., 62, 432, 1979
[157] *Lazzara, M.G.,* J. Coat. Technol., 56, 710, 19, 1984
[158] *Dannenberg, H., Forbes, J.W., Jones, A.C.,* Anal. Chem., 32, 365, 1960
[159] *Saxon, R., Lestienne, F.C.,* J. Appl. Polym. Sci., 8, 475, 1964
[160] *Hornung, K.H., Biethan, U.,* FARBE&LACK, 76, 461, 1970
[161] *Dorffel, J., Biethan, U.,* FARBE&LACK, 82, 1017, 1976
[162] *Bauer, D.R., Dickie, R.A.,* J. Polym. Sci., Polym. Phys., 18,1997, 1980
[163] *Bauer, D.R., Dickie, R.A.,* J. Polym. Sci., Polym. Phys., 18, 2015, 1980
[164] *Tomita, B., Hatono, S.,* J. Polym. Sci., Polym. Chem. Ed., 16, 2509, 1978
[165] *Dijk van, J.H., Brakel van, A.S., Dankelman, W., Groenenboom, C.J.,* XV Fatipec Congress, Book II, p. 326, Amsterdam 1980
[166] *Bauer, D.R., Dickie, R.A., Koenig, J.L.,* Ind. Eng. Chem. Prod. Res. Div., 24, 121, 1985
[167] *Bauer, D.R., Dickie, R.A., Koenig, J.L.,* J. Polym. Sci., Polym. Phys. Ed., 22, 2009, 1984
[168] *Bauer, D.R., Budde, G.F.,* J. Appl. Polym. Sci., 28, 253, 1983
[169] *Alderslay, J.W.,* Brit. Polym. J., 1, 101, 1969
[170] *Anderson, D.G., Netzel, D.A., Tessari, D.J.,* J. Appl. Polym. Sci., 14, 3021, 1970
[171] *Feurer, B., Gourdenne, A.,* Bull. Soc. Chim. Fr., 12, 2845, 1974
[172] *Schindlbauer, H., Anderer, J.,* Angew. Makromol. Chem., 79, 157, 1979
[173] *Collete, J.W., Corcoran, P., Tannenbaum, H.P., Zimmt, W.S.,* J. Appl. Polym. Sci., 32, 4209, 1986
[174] *Hill, L.W., Kozlowski, K.,* J. Coat. Technol., 59, 751, 63, 1987

[175] *Oshikubo, T., Yoshida, T., Tanaka, S.,* Proc. Xth Int. Conf. Org. Coat. Sci.Techn., p. 317, Athens, Greece, 1984
[176] *Nakamichi, T.,* Prog. Org. Coat., 14, 23, 1986
[177] *Koral, J.N., Petropoulos, J.C.,* J. Paint Technol., 38, 501, 600, 1966
[178] *Wohnsiedler, H.P.,* A.C.S., Div. Org. Coatings & Plastics Chem., 2 0/2, p. 52, 1960
[179] *Santer, J.O.,* Prog. Org. Coat., 12, 309, 1984
[180] *Bauer, D.R.,* Prog. Org. Coat., 14, 193, 1986
[181] *Calbo, J.* Coat. Technol., 52, 75, 1980
[182] *Holmberg, K.,* J. Oil Colour Chem. Assoc., 61, 359, 1978
[183] *Blank, W.,* J. Coat. Technol., 54, 26, 1982
[184] *Miller, D.R., Macosko, C.W.,* Macromolecules, 9, 206, 1976
[185] *Bauer, D.R., Budde, G.F.,* Ind. Eng. Chem., Prod. Res. Dev., 20, 674, 1981
[186] *Bauer, D.R., Dickie, R.A.,* J. Coat. Technol., 54, 57, 1982
[187] Dainippon Ink and Chemicals, JP 2004346152, 2004 JP 2005041900, 2005
[188] *Craun, G.P.,* J. Coat Technol. 67, (841), 23 (1995)
[189] *Pappas, S.P. and Hillo, L.W.* J. Coat Technol. 53, (675), 43 (1981)
[190] *Jonson, P.C.* in Advances in Polyurethane Technology, Eds. *J.M. Buist and H. Gudheon* McLare & Sons, Ltd. London 1968, p.19
[191] *Gaylord, N.G. and Sroog, C.E.* J. Org. Chem. 18, 1362 (1953)
[192] *Anagnostou, T.J. and Jaul E.,* J. Coat Technol. 53 (673) 35 (1981)
[193] *Sal'nikova, G.A. Shmidt, Yu, A, and Valodorskaja, Yu., I.* Dokl. Akad. Nauk. SSSR, 181 669 (1968)
[194] *Mirgel, V. and Nachtkamp, K.,* Pol. Paint, Col. J. 176 (4163, 200 (1986)
[195] *Hilborn, J.G. Jönson, S.E. Stödeman, P.M.E. and Skolling, O.S.G.,* Proceed. Int. Waterborne, High Solids and Powder Coatings Symp. New Orleans, Feb. 1991, p. 54
[196] BASF, DE Pat. 2 164 256 (1971)
[197] a) Ciba-Geigy, EPA 667381 (1994)
b) *Finter, J., Frischinger, I., Haug, Th. and Marton, R.,* Proc. 4th Nuremberg Congress April 1997
c) *Reisinger, M.,* Proc. RadTech '99 - Europe, Berlin, Nov. 1999, p. 628
[198] *Garcia, S. J.; Suay, J.,* Progress in Organic Coatings, 57(4), 319-331, 2006
[199] *Wittig, M. and Gohmann, T.* Pol. Paint, Col. J. 184, (4343) 34, (1994)
[200] DSM Resins, WO 93/25596 (1993)
[201] BASF, EPA 650978, 650979 (1994) and EPA 678562 (1995)
[202] *Hardeman, G. and Witte, F.M.,* Proc. "Pulverlack Kongress" Hamburg 1996, p. 41
[203] a) Bayer, EPA 350730 (1990)
b) Hoechst, EPA 585742 (1994)
[203] c) *Fink, D. and Brindöpke,* G., Proc. 3rd Nuremberg Congress, March 1995, Paper 29 and in EuroCoat J. 9/1995, p. 606
[204] a) *Maetens, D.,* Proc. RadTech '98 - USA, Chicago, April 1998, p. 170
b) *Buysens, K.,* Proc. RadTech '99 - Europe, Berlin, Nov. 1999, p. 622 and in Congress Papers PCE 2000, Amsterdam, 19-21 Jan. 2000, p. 253
[205] a) DSM Resins, EPA 636669 (1995)
b) *Witte, F.M., Kieft, G., Stuiver, H.J., Twight, F. and Van der Linde, R.,* Proceed. RadTech '95-Europe, Maastricht, p. 437
[206] *Zahora, E.P.* et al., Modern Paint Coat., Oct. 1994, p. 120
[207] *Udding-Louwrier, S., Witte, F.M. and De Jong, E.S.,* Prod. Finishing, Apr. 1997, p. 26
[208] *Decker, C. and Decker, D.,* Proc. RadTech Conf., Orlando 1994, p. 602
[209] *Cunningham, W.,* Adhesives Age, 39, 24 (1996)
[210] *Witte, F.M., European* Coatings Journal, 3/1996, p. 115
[211] *Bär, K.K.O., Sedlmeyr, M.,* Congr. Pap. PCE 1998, Amsterdam, 20-22 Jan., p. 71, 1998
[212] *Bär, K.K.O., Sedlmeyr, M.,* Congr. Pap. PCE 2000, Amsterdam, 19-21 Jan., p. 321, 2000

3.2 Monitoring the curing process

The characteristics of the cured film of thermosetting powder coatings are dependent on the **degree of curing**. Not only mechanical properties such as hardness, flexibility and impact resistance, but also the appearance of the film surface, flow properties during the film formation, chemical resistance of the cured film, gloss retention and the service life of powder coatings are directly related to the curing process. The curing schedules in practice vary considerably depending on the characteristics of the **stoving line**. Therefore it is of essential importance for the powder coating chemist to be able to follow the curing process by a suitable technique, to interpret the gathered data correctly, and to use them as elements in adjusting the powder coating formulation to the demands of the end user.

A variety of techniques have been developed to evaluate the degree of curing or to follow the curing process at specific time sequences or continuously. These methods and techniques include **infrared (IR) spectroscopy, nuclear magnetic resonance (NMR)**, chemical analysis, dielectric analysis, measurement of certain mechanical properties of the film such as hardness, flexibility and impact resistance, determination of the solvent resistance and degree of swelling of the cured film, thermal analysis such as **differential scanning calorimetry** (DSC) and **thermogravimetry** (TGA) and **thermal mechanical analysis** such as torsional braid analysis (TBA) or **dynamic mechanical analysis** (DMA). According to *Knappe* [42] the **dielectrical analysis** (DEA), is a powerful tool to monitor the hardening process of powder coatings online, based on fact that the mobility of ions is reduced with increasing network formation.

Most of these techniques are suitable for characterization of the curing process in the case of powder coatings. Unfortunately many of them supply either insufficient and not always relevant data for quantitative interpretation, or data that cannot be directly related to the phenomena of film formation. Spectroscopic methods, which are probably the best analytical tools for characterization of the curing mechanism from a chemical point of view, do not say much about **film formation** which is a rather complex physico-chemical process.

In practice, the mechanism of the curing reactions for existing powder coating systems is relatively well known. Therefore, techniques that provide data that can be easily related to the film formation process and used for formulating coatings are of the greatest importance for the powder paint chemist. For this purpose, the thermal and thermal mechanical analytical techniques probably provide the most useful data, although they do not give any direct information about the chemical changes that take place. However, by these methods, one can measure the changes in the fundamental properties of the coating during curing which can be related to the structure of the material.

3.2.1 Differential scanning calorimetry (DSC)

Differential scanning calorimetry (DSC) belongs to a group of thermal analytical methods that are based on detection of the thermal effects in the sample resulting from physical and chemical changes as a function of temperature. The latter is usually programmed at a linear rate. The principle on which this technique operates is based on determining the energy required to keep the sample and the blank reference at the same temperature. The DSC technique could be considered as a follow-up of the **differential thermal analysis** (DTA) which was developed much earlier with the basic principles established by *Roberts-Austen* at the end of the former century. Contrary to DSC in DTA the difference in temperature between the sample and the reference is observed while both are being heated (or cooled) through a pre-selected temperature program, which as in the case of DSC is mostly linear.

There are two types of DSC instruments that have been widely used for calorimetric measurements: the **heat-flux DSC** and the **power-compensational DSC**.

Principle of the heat-flux DSC

The principle scheme of the heat-flux DSC instrument is presented in *Figure 3.4.* Sample holder S and reference R (usually an empty holder) are placed on a disc D which serves for heat transfer to the sample and the reference. Heat supplied by the furnace F is transferred from the disc up to the sample and the reference via the sample pans. In the case of a dynamic run, the furnace temperature T_c increases or decreases at a linear rate. The differential heat flow to the sample and reference is monitored by thermocouple wires which are connected to the underside of the disc. The heat flow dH/dt to the sample is equal to the difference between dQ_s/dt and dQ_r/dt representing the total heat flow to the sample holder plus the sample and reference holder.

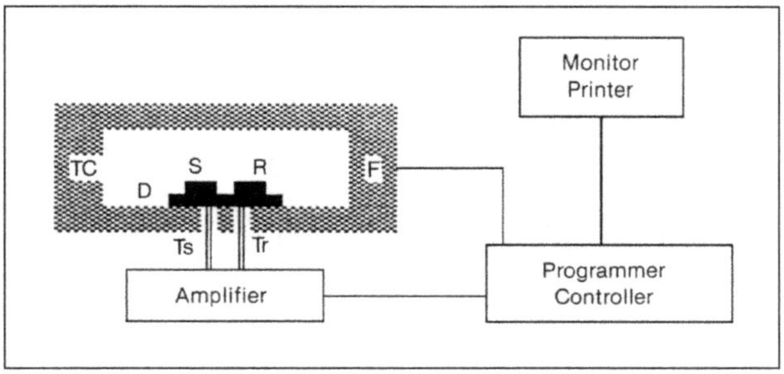

Figure 3.4: Scheme of heat-flow DSC apparatus

For known thermal characteristics of the instrument, the thermal effect of the process in terms of the energy consumed or emitted by the sample can be determined by measuring the temperatures of the sample and reference. The inherent problem emanating from the fact that the thermal characteristics of the instrument are dependent on the actual temperature and change permanently during the dynamic run is solved in different ways depending on the producer of the instrument.

Power-compensational DSC apparatus

A different method is used in power-compensational DSC apparatus. For the powder paint chemist, DSC has a considerable advantage over DTA. This emanates from fundamental differences between the techniques. In DSC acquisition of the calorimetric data resulting from any thermal effects is determined directly from the energy input in order to compensate for the difference in temperature between the sample and the reference. With DTA, the calorimetric information is obtained indirectly from thermometric data. Not only the interpretation of the results obtained from DTA is more complicated, but the reliability of the technique is in many cases doubtful for quantitative interpretation. In other words, since the thermal conductivity of the sample can change during the transition, the proportionality between the temperature differences and the energy changes is in general unknown. In such cases, DTA can provide more or less qualitative results. In contrast, in DSC the peak area is always a true measure of the energy input required to keep the sample and reference temperature equal, independent of the changes in the thermal behavior of the sample or the thermal characteristics of the instrument itself. With proper design the performances of the DTA apparatus can get close to those of DSC. However, the ease of handling and processing the acquired data and their reliability made DSC the most common instrument used for executing thermal analysis in powder coating laboratories.

Next to the **kinetic parameters** of the curing process, which can be obtained by the DSC technique, several additional facts relevant to powder coating can be gathered. These include glass transition temperatures (T_g) of the powder coating and the components used in the coating composition, the crystalline melting point of the polymers (specially in the case of thermoplasts) and the melting point of the crosslinkers, the onset flow temperature (T_{of}) of the powder coating and the onset cure temperature (T_{oc}). Indirectly, from the kinetic parameters, the conversion of the curing reaction with time at a certain temperature can be easily calculated. This gives an indication of the necessary curing time at a given temperature or vice versa, which can easily be related to common practical problems.

Three different methods can be used for obtaining **kinetic information** from dynamic thermal experiments. The easiest method developed by *Borchardt* and *Daniels* [2] which is widely used to examine the cure behavior of powder coatings involves analysis of only one exotherm. A deficiency of this method is its lack of consistent accuracy in measuring kinetic parameters [3]. Much more reliable results concerning the activation energy

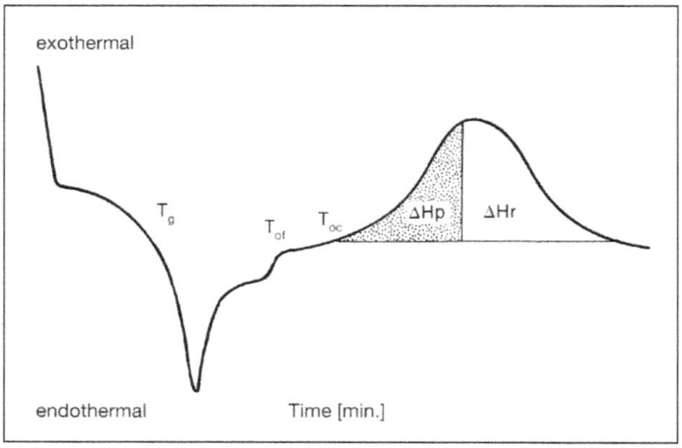

Figure 3.5: Typical DSC curve of a polyester/epoxy powder coating

and frequency factor for different types of reactions can be obtained by the method based on variations in the peak exotherm temperature with heating rate [4]. The last method also uses experiments at different heating rates and analyses the dependence of the temperature to reach a constant conversion on the heating rate [5, 6]. Although the last two methods provide much more reliable results, the method that acquires the kinetic parameters from analysis of only one exotherm seems to be most attractive. In fact its accuracy is quite acceptable in practice for most cases dealing with powder coatings.

Figure 3.5 represents a typical DSC scan of a powder coating in which the carboxyl functional polyester binder is cured with bisphenol A epoxy resin. This graph will be used to explain the methods that use the data obtained from DSC for kinetic analysis of the curing process. The characteristic points of the curve marked with T_g and T_{of} will be discussed later. The exothermic reaction of curing starts at a considerable rate at temperature T_{oc}, a designation for the temperature onset cure. As already stated, the advantage of the DSC technique compared to DTA is in the basic principle of the technique of measuring the power required to prevent the temperature change in the sample compared to the reference due to the thermal effects during the process that is monitored. This means that dH/dt, the time rate of change of the enthalpy of the curing reaction, can be directly followed. This makes the **kinetic analysis** very simple since the thermal energy effects measured directly with the instrument can be correlated to the reaction speed.

In **kinetic calculations** based on data acquired by DSC the rate of change in certain materials accompanied by thermal effects has to be related to the temperature. The heat evolved or absorbed should be proportional to the number of moles reacting, as expressed by:

Equation 3.56 $dH/dt = -(\Delta H/N_0)\, dN/dt$

where N_0 is the initial number of moles of the reactant, $-dN/dt$ is the loss of moles per unit time, dH/dt is the corresponding enthalpy change and ΔH is the enthalpy of the reaction.

The area under the thermal peak in *Figure 3.5* is a measure of the enthalpy of the reaction and it can be assumed that the total area under the peak (ΔH_t) corresponds to the total thermal effect of the process. Since the degree of conversion is proportional to the respective enthalpy change, the relation between the initial number of moles of reactant (N_0), the number of moles at time t, the part of the peak area (H^p) at time t, the residual peak area (ΔH_r) and the total area under the peak is ($\Delta H_t = \Delta H_p + \Delta H_r$). The temperature dependence is assumed to follow the well-known *Arrhenius* equation:

Equation 3.57

$$k = k_0 \exp\left(\frac{-E_a}{RT}\right)$$

Here k_0 is the pre-exponential constant or the so-called frequency factor, E_a is the activation energy of the reaction, R is the gas constant and T is the absolute temperature in Kelvin.

Combining Equations 3.56 and 3.57 gives after proper substitution

Equation 3.58

$$\ln\left(\frac{dH}{dt\Delta H_t}\right) = \ln k_0 \frac{E_a}{RT} + n\ln\left(\frac{\Delta H_r}{\Delta H_t}\right)$$

The three unknowns in Equation 3.58, k_0, E_a and n, can be obtained using three of the many points of the DSC curve. Contemporary DSC instruments perform multilinear regression analysis using ln ($dH/dt\Delta H_t$), 1/T and ln ($\Delta Hr/\Delta H_t$) as variables that have been evaluated from the DSC data. In this way the accuracy of the results is increased since any number of curve points can be evaluated.

Knowing the activation energy, the pre-exponential constant and the reaction order, conversion of the reaction with time can easily be calculated at any given temperature. An example of such applied kinetic calculations obtained on a "Mettler TA3000" DSC system for a curing reaction of polyester/TGIC powder coating is presented in *Figure 3.6.*

The DSC method has been widely used to examine the curing behavior of powder coatings. The interested reader can consult a number of published papers in which the curing kinetics of different powder coating systems, curing mechanism or activity of different catalysts have been studied by the DSC technique [8-17]. Apart from the kinetic data, which can be obtained by analysis of the DSC curve, other relevant data can also be obtained from a single DSC run. The **glass transition temperature** of the powder coating, whose importance will be discussed later in Chapter 4, is the temperature at which there is a 'sudden' increase in the specific heat (C_p) of the powder coating. This is manifested by a shift in the baseline of the DSC curve (*Figure 3.5,* point T_g). The International Confederation of Thermal Analysis (ICTA) proposes an evaluation procedure to be used to determine the glass transition temperature (*Figure 3.7*) [18, 19].

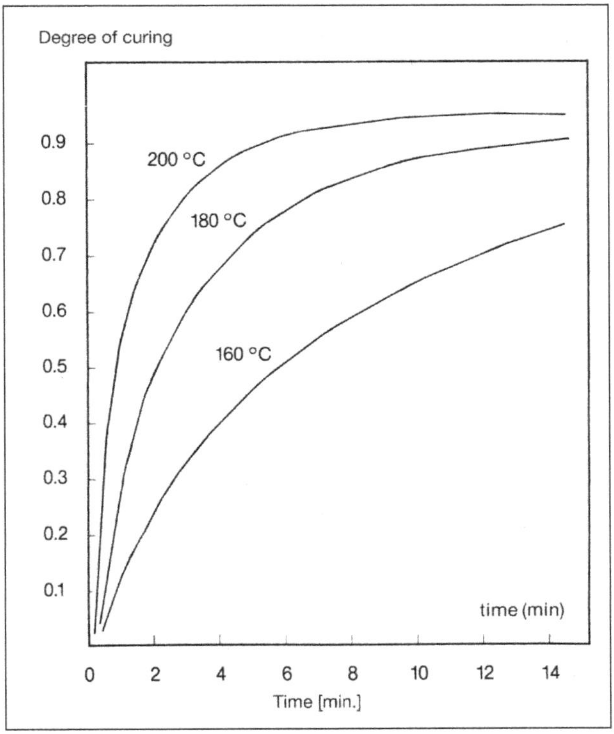

Figure 3.6: Degree of curing as a function of time at 160, 180 and 200 °C for a polyester/TGIC powder coating

According to this procedure three regression lines R_1, R_2 and R_3 are applied to the DSC curve: the regression line before (R_1) and after (R_3) the event and the regression line at the inflection point (R_2). These three lines define the temperature T_{g1} as the first significant deviation from R_1, T_{g2} as the intersection between R_1 and R_2 and T_{g3} as a mid point between the two intersection points. T_{g2} is considered to be the relevant value for the glass transition temperature.

It should be noted that the values for the T_g obtained by DSC are dependent on the heating rate chosen during the experiment. This is not an unexpected phenomenon since, as will be seen later, the glass transition temperature is related to the mobility of the polymer chains. This is

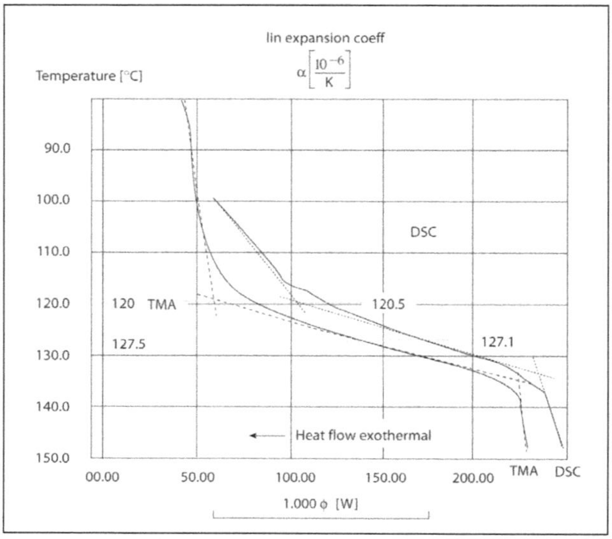

Figure 3.7: Evaluation of the characteristic glass transition temperatures by the ICTA procedure for DSC curves [18]

source: Dr R. Riesen, Mettler Instruments AG/Greifensee/Switzerland

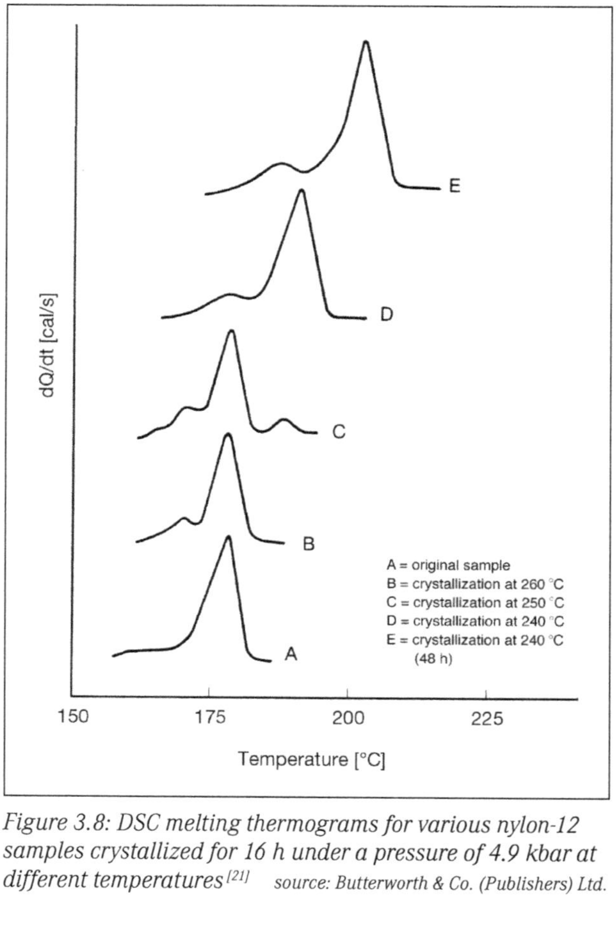

Figure 3.8: DSC melting thermograms for various nylon-12 samples crystallized for 16 h under a pressure of 4.9 kbar at different temperatures [21] *source: Butterworth & Co. (Publishers) Ltd.*

a kinetic dependent process and the rate of the experiment will influence the results obtained. Values of T_g determined by the mid-point method will increase with an increasing heating rate. *Gargallo* and *Russo*, for example, using the DSC technique, have obtained T_g values of 26, 29, 34 and 39 °C for heating rates of 4, 8, 16 and 32 °C/minute respectively for the case of poly(neopentyl) methacrylate [20]. In order to avoid confusion, the T_g values obtained by DSC measurements should always be accompanied by the data concerning the heating rate.

Other useful information that can be obtained by the DSC technique is the temperature at which the powder coating begins to melt. The larger contact surface between the sample and the walls of the sample holder after melting influences the heat transfer characteristics of the system. This is manifested by the shift in the baseline. The temperature at which this shift is observed is called the **temperature onset flow** (point T_{of} in *Figure 3.5*) [14]. Depending on the reactivity of the system, the exothermal peak is observed to start at the so-called **temperature onset cure** (T_{oc} in *Figure 3.5*). The distance between T_{of} and T_{oc} gives an indication of the time available for the powder coating to flow. The size of the plateau between the melting of the powder and the start of the exothermal reaction on the DSC curve cannot be quantitatively correlated to the flow properties of the powder coating. Relations between the factors influencing the flow are rather complex and taking only the time available for flow as a single element is an oversimplification of the problem.

However, it gives an indication of the activity of the catalysts used in the system, and especially an indication about their **latency**, which is very often a desired characteristic of a good catalyst.

The DSC technique can be used to determine the **melting temperature** of the crystalline polymers, which is an important parameter for thermoplastic powder coatings. Since the melting of the crystallites is an endothermal process, it is easily registered by DSC as a sharp endothermal peak with its minimum corresponding to the melting temperature. *Stamhuis* and *Pennings* have investigated the crystallization of nylon-12 [21]. Thermograms of nylon-12 samples crystallized under various conditions are shown in *Figure 3.8*.

3.2.2 Thermogravimetry analysis (TGA)

The DSC technique can be successfully used in the cases of curing where the thermal effect of the curing reaction is not accompanied by the other phenomena characterized by emission or absorption of energy. In principle the cure kinetics of processes involving reactions between pairs of functional groups such as acid/epoxy, acid anhydride/epoxy, acid anhydride/hydroxyl, epoxy/amine and epoxy/phenolic can be successfully followed by DSC. However, in cases where the curing of powder coatings involves **simultaneously several chemical or physical transformations**, each having its own thermal effect, the interpretation of the results obtained by DSC is rather complex and in many cases impossible.

An example of such a curing process is the crosslinking of hydroxyl functional polymers with blocked isocyanates in polyurethane powder coatings. The curing process starts with deblocking of the polyisocyanate. This reaction, as well as evaporation of the blocking agent from the systems, is endothermal. The crosslinking between hydroxyl groups and isocyanate in the next step is an exothermal reaction. Therefore, the DSC curve obtained as a result of these three processes cannot supply data relevant and reliable for kinetic analysis. A similar example is the curing of hydroxyl functional polymers with amino resins, where methanol or other higher alcohol is a by-product of the curing reaction.

In such cases, thermogravimetry is a technique that can be used with partial success for acquiring useful data. Thermogravimetric analysis (TGA) is a rather old thermal analysis technique. Although the first thermobalances had been constructed at the beginning of the last century, the first commercial instrument appeared in 1945, based on the work of *Chevenard, Wache* and *De la Tullaye* [22]. This thermobalance was used for the first time in the polymer field in 1948 by *Duval* [23]. Contemporary instruments with a sensitivity of 0.1 μg and a precision of one part in 100,000 of the mass change were developed in the 1960s by several producers [24]. The TGA instrument is designed to produce a continuous record of weight as a function of temperature. The sample can usually be subjected to a linear temperature change or may be isothermally maintained at a given temperature. *Figure 3.9* presents a schematic diagram of a typical TGA instrument. It consists of the following main parts: furnace (F), furnace temperature programmer, recording balance, a part of which is the sample holder with a temperature sensor (S), and recorder.

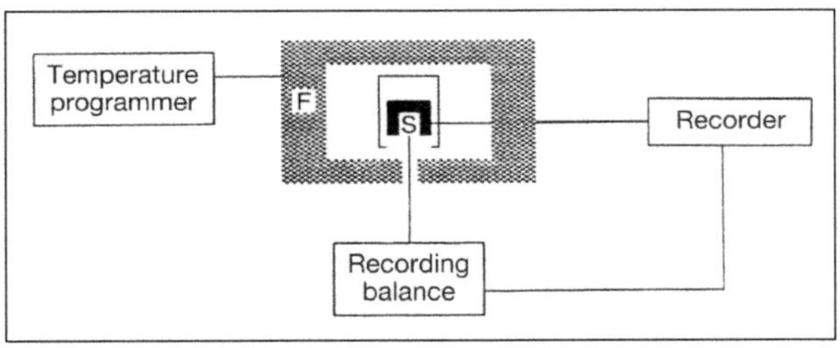

Figure 3.9: Schematic diagram of TGA instrument

The most important component of the TGA instrument is the recording balance. The wide variety of recording balances can be divided into three main groups based on their mode of operation: a) deflection-type instruments, b) null-type instruments and c) those based on changes in resonance frequency [24, p.10].

Deflection balances involve the conversion of balance beam deflections into an electrically recorded signal by means of an appropriate displacement measuring transducer. Four different types of deflection balance are published in the literature [7, p.282].

The automatic **null-type balance** incorporates a sensing element that detects a deviation of the balance beam from its null position. A restoring force of either electrical or mechanical mass loading is then applied to the beam through an appropriate electronic or mechanical linkage, restoring it to the null position. The null-type balance principle is now used in almost all commercially available TGA instruments [24]. They can convert the analog signal output to the derivative of the mass change or digitize it by an analog-to-digital converter for processing by digital computers.

The use of the TGA for evaluation of the kinetic parameters of the curing process of powder coatings makes sense only when the deblocking mechanism is the **slowest step determining the overall curing rate** and when the evaporation of the blocking agent is fast enough and thus does not affect the equilibrium of the curing reaction. The second assumption can be considered as valid for methoxymethyl amino resins, since the evaporation of the methanol which is the by-product of the curing reaction at the curing temperature is rather high.

However, even in such a case, *Bauer* and *Dickie* have shown that although very short, the residence time of the free methanol in the film during the curing process affects the reaction rate in the case of solventbased acrylic coatings cured with HMMM [25]. In the case of polyurethane powder coatings the curing rate should not be affected by the presence of caprolactam above the deblocking temperature, assuming that the curing mechanism is a two-step process in which the urethane formation is preceded by unimolecular thermal dissociation of the blocked isocyanate to free isocyanate. As already discussed, it is quite possible that in the presence of compounds containing active hydrogen, the curing mechanism can also be a direct bimolecular displacement reaction. Therefore, it is difficult to predict how reliable the kinetic parameters can be when evaluated from data acquired by thermogravimetric analysis of the cure of powder coatings.

Although TGA has been widely used in determining the kinetics of polymer systems, it is very difficult to find any paper referring to the use of TGA for following the curing process of powder coatings.

The **method of *Freeman* and *Carrol***[26], in spite of its limited precision [27], is quite convenient for processing the acquired TGA data, since it belongs to the so-called derivative methods.

If w is the weight of the reactive portion of the sample, which means the actual weight at time t minus the final weight of the sample at the end of the experiment, k is the reaction rate constant and F linear heating rate F the following equation suitable for graphical determination of E_a and n is obtained:

Equation 3.59

$$\Delta \ln\left(-\frac{dw}{dt}\right) = -\left(\frac{E_a}{R}\right)\Delta\left(\frac{1}{T}\right) + n\Delta \ln w$$

Freeman and *Carrol* suggest the following form of Equation 3.60:

Equation 3.60

$$\frac{\Delta \log(-dw/dt)}{\Delta \log w} = \left[\frac{(-E_a/R)\Delta(1/T)}{2.303\Delta \log w}\right] + n$$

Plotting the values for $\Delta\log(-dw/dt)/(\Delta\log w)$ versus $\Delta(1/T)/\Delta\log w$), the values of E_a and n can be obtained from the slope and the intercept of the curve.

It has been found convenient to have the logarithm of the experimental rates plotted initially against 1/T; so that the obtained curve may be used to select a set of values for constant increments in 1/T: Keeping 1/T constant a subsequent plot of $\Delta\log(-dw/dt)/\Delta(\log w)$ versus $1/\Delta(\log w)$ leads to a value of *n* as the intercept and E_a as the slope of the curve.

Several other derivative methods include the method of multiple heating rates [28, 29], the method of maximal rates suitable only for determination of the activation energy [7, p.320] and the method which assumes a prior correct analytical form of the kinetic equation for determination of the kinetic parameters [30].

A variety of integral methods have been developed for evaluation of kinetic data obtained in TGA experiments. Most of them are based on the best fit of curve obtained from the assumed analytical form of the equation expressing the kinetic of the process with the experimental TGA trace. The equation usually selected is similar to that used in the DSC experiments:

Equation 3.61 $$f(\alpha) = (1-\alpha)^n$$

Here the meaning of α is the same as before, i.e. the degree of completion of the reaction with respect to the initial sample, and n represents the reaction order.

From the several **integral methods** proposed by different authors [31-35] the one developed by *Coats* and *Redfern* [31] seems to be the most suitable from a practical point of view. This method assumes the validity of Equation 3.61 and takes into account the equation describing the temperature dependence of the degree of conversion in a linear temperature program during the execution of the TGA experiment.

The assumption $T_0 = 0$ can be considered as valid since the reaction rate at low temperature is negligible. The method of *Coats* and *Redfern* takes into account the first three terms of the asymptotic approximation.

Thus the plot of either

Equation 3.62A

$$\log\left[\frac{1-(1-\alpha)^{1-n}}{T^2(1-n)}\right] \mathit{versus}\ \frac{1}{T}\quad \mathit{for}\ n \neq 1$$

or

Equation 3.62B

$$\log\left[-\log\left(\frac{1-\alpha}{T^2}\right)\right] \mathit{versus}\ \frac{1}{T}\quad \mathit{for}\ n = 1$$

should result in a straight line with a slope $-E_a/2.3R$ and intercept reading $\log[(ZR/E_aF)(1-2RT/E_a)]$ or $\log(ZR/E_aF)$ for a correct value of *n* [31].

3.2.3 Thermal and dynamic mechanical analysis (TMA, DMA)

The film formation of thermosetting powder coatings is a complex process. It starts with melting and coalescence of the individual powder particles into a continuous film followed by crosslinking of the low molecular weight binder. This results in a permanent change in the physical-chemical properties of the material, until the final film properties are obtained by completion of the curing reaction to a desired degree.

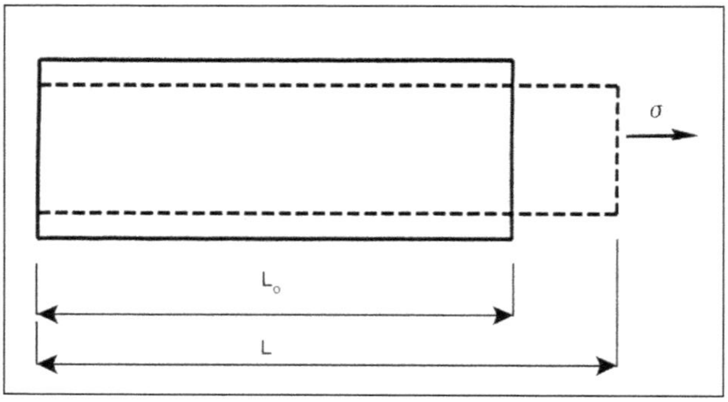

Figure 3.10: Behavior of a perfect elastic body, subjected to stress σ

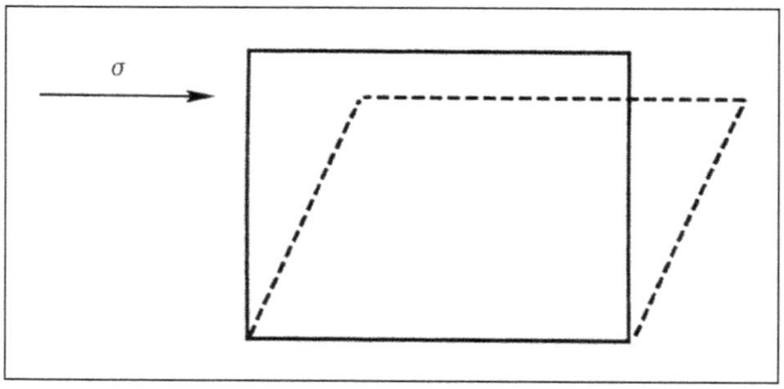

Figure 3.11: Shear type of deformation of an elastic body

For following the degree of conversion of the functional groups involved in the curing reactions DSC and TGA are helpful techniques. As already discussed, by using these techniques, kinetic data relevant for predicting the **curing behavior** of the system can be obtained. Using these kinetic parameters, the degree of conversion of the functional groups during curing at different temperatures can be calculated as a function of time or vice versa. However, DSC and TGA do not provide information with respect to the changes of **mechanical and rheological properties** of the coating during the film forming process. Such information can be gathered by dynamic mechanical analysis (DMA). They can be correlated with the flow behavior and mechanical properties of the powder coating as a function of the resin/crosslinker ratio and resin structure. Combined with the kinetic data obtained by DSC or TGA, mechanical properties can be followed as a function of the **degree of curing**. As well as DSC, the DMA technique can be used to determine the physical transitions in the coating, temperature onset flow and temperature onset cure.

Dynamic mechanical analysis

The dynamic mechanical analysis (DMA) as a technique for characterization of polymeric materials exists for a relatively long time. However, the use of DMA for characterization of cured coatings and for following the progress of curing reactions is not widely described in the literature typical for the coating field. Therefore a short introduction of the theory backing up the DMA seems to be useful before starting with a description of the technique.

According to Hook's law, which assumes perfect elasticity of a material body, the relation between the tensile stress (σ) and strain (ε) is described by

Equation 3.63 $$\sigma = E \cdot \varepsilon$$

where E is defined as Young's modulus, serving as a measure of the stiffness of the material. This means that materials with a higher modulus are more resistant to being stretched. If the sample initial length is L_0 and after being subjected to stress σ the final length is L, then the strain ε is defined as (*Figure 3.10*).

Equation 3.64 $$\varepsilon = (L - L_0) / L_0$$

It is obvious that the stress should be defined in terms of force per unit area.

In another type of deformation the sample can be subjected to a shearing or twisting. In the case of a perfect elastic body, a cube of material fixed with one side to an immobile surface subjected to the action of a force parallel to the surface will take another position, as presented in *Figure 3.11*. As in the case of tensile deformation, the shear stress and shear rate will be connected through the shear modulus G:

Equation 3.65 $\sigma = G \cdot \varepsilon$

The above equation should be compared to the one used to define a perfect liquid or so-called Newtonian liquid:

Equation 3.66
$$f = \eta \left[\frac{ds}{dt} \right]$$

where f and s represent the shear stress and strain respectively, t is the time and η is the viscosity of the liquid as a measure of the resistance of the liquid to flow. The Equation 3.66 is valid for many low molecular materials above their melting point.

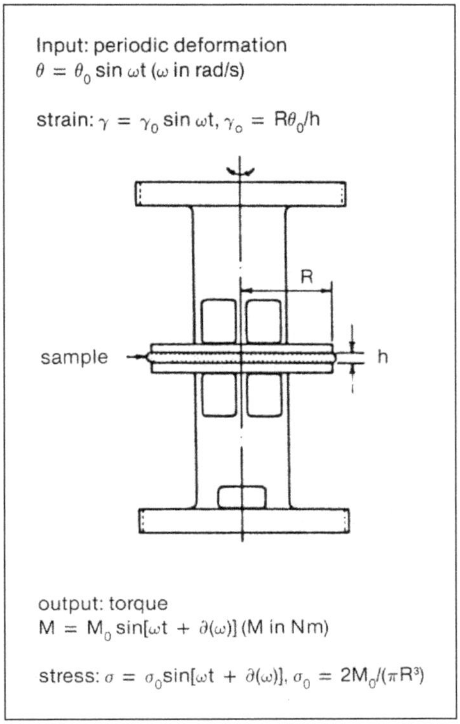

Figure 3.12: Measuring principle of the mechanical spectrometer[36] source: DSM Resins BV

Equation 3.66 defines that, in an ideal Newtonian liquid, the ratio of the force that opposes the flow to the rate of flow is constant. On the other hand, in an ideal Hookean solid (Equation 3.65) the ratio of the force that opposes deformation to the extent of deformation is constant. While neither Equation 3.65 nor 3.66 accurately describes polymer behavior, they represent two important limiting cases.

Powder coatings, as is the case with most of the polymeric materials, neither obey Equation 3.65 describing the behavior of a perfect elastic body under stress nor Equation 3.66 describing the behavior of a perfect liquid under the same conditions. They are defined as **viscoelastic materials** exhibiting some of the properties of solids and some of the properties of liquids. This means that under an applied force they undergo deformation as solid materials, but also undergo flow in the manner of liquids. Depending on the magnitude of these phenomena powder coatings can be identified as **viscoelastic liquids** immediately after melting, when the flow is the usual response to the applied force and as **viscoelastic solids** as the curing progresses, when the response of the cured film to the applied force is a deformation.

The quantities E and G in Equations 3.63 and 3.64 refer to measurements not dependent on time, or in other words to static measurements (the definition of a static measurement here is debatable, since the experiment is performed within a certain time scale anyway). In the case of dynamic mechanical measurements, the sample in question is subjected to cyclic or repetitive motions. In this type of experiment, time as a factor influencing the results should be introduced.

Figure 3.12 represents schematically the principal construction of the DMA instrument.

It is common practice in dynamic mechanical measurements for the applied force to vary sinusoidally with time, by a rate usually specified by the frequency f in cycles per second or $\omega = 2\pi f$ in radians per second. For Hookean solids having the behavior of a perfect elastic body the strain will follow the same pattern. In the case of powder coatings, when such a sinusoidal excitation is applied to the sample, the strain will alternate sinusoidally also.

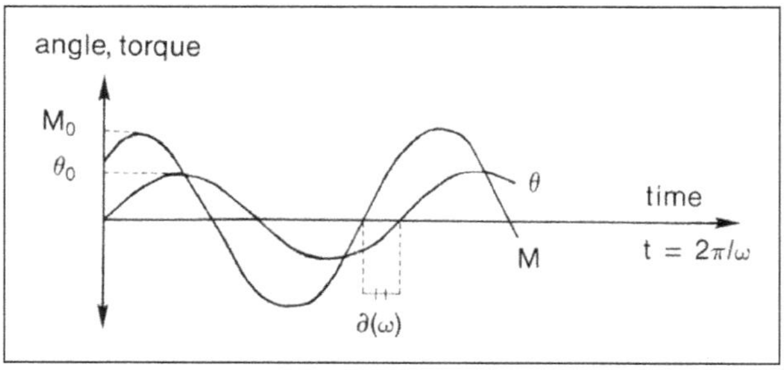

Figure 3.13: Stress and strain as a function of time in sinusoidal excitation of a viscoelastic specimen

However, depending on the temperature of the sample and the degree of curing, the strain will be out of phase with the stress. This phase lag results from the time necessary to make molecular rearrangements and is associated with relaxation phenomena.

In the following equations δ is the phase angle, σ_0 is the peak stress and ε_0 is the peak strain (*Figure 3.13*).

The ratio between the peak stress and peak strain is called the dynamic modulus G_d:

Equation 3.67 $$G_d = \sigma_0/\varepsilon_0$$

Equation 3.68 $$G^* = \sigma/\varepsilon = G' + iG''$$

The relations between the different components of the modulus are summarized in the so-called Argand diagram in *Figure 3.14.*

The in-phase component G' is called the real part of the modulus or the **storage modulus** because it is related to the energy stored in the specimen as a potential energy which is released in the periodic deformation. The out-of-phase component G", also called the imaginary part of the modulus, is associated with the viscous loss of energy, or the energy which is dissipated as heat when the specimen is deformed.

The ratio between the loss and the storage component of the modulus equals the tangent of the phase shift (the Argand diagram, *Figure 3.14*)

Equation 3.69 $$\tan\delta = G''/G'$$

This so-called **loss tangent** δ is a dimensionless term representing the ratio of the energy lost (dissipated as heat) per cycle to the energy stored and hence recovered per cycle. The loss tangent is temperature dependent and is at a maximum near the glass transition temperature. Thus, by an isofrequent scan at different temperatures, the glass transition temperature of the polymers can be located at the maximum of the curve representing tan δ versus temperature. As in the case of determination of T_g by the DSC technique, here again it should be stressed that the T_g determined by DMA is the frequency dependent value [37].

When Equation 3.66 is related to the dynamic mechanical measurements, the first derivation of the strain with time can be replaced by the oscillation frequency. Thus, the determination of the dynamic modulus of the specimen at a certain frequency enables us to obtain the value of the dynamic viscosity η of the system:

Equation 3.70 $$\eta = G_d/\omega$$

According to the theory of rubber elasticity [38, 39] the **degree of crosslinking** above the glass transition temperature can be expressed through the concentration of the elastically active network chains, V_e:

Equation 3.71 $$V_e = G'_{(0)}/(RT)$$

where $G'_{(0)}$ is the value of the storage modulus at zero frequency, R = 8.31 J/(mol K) and T is expressed in Kelvin.

In principle, powder coating can be scanned over a wide frequency range or a wide temperature range in order to study motional transitions and changes in the mechanical and rheological properties during curing. While the frequency scans at constant temperature are suitable for the theoretical study of the relaxation phenomena, the temperature scans at constant frequency provide more relevant data for following the progress of the curing reactions.

Scholtens, van der Linde and *Tiemersma-Thoone* [36] suggest the following experiment design in dynamic mechanical analysis to acquire data that are relevant for the curing process of powder coatings.

A compression molded sample (about 10 minutes at 90 °C for polyester/TGIC systems) is placed between two concentric plates of the instrument. Firstly, an oscillation run, starting at log ω = −0.2 up to +2.0 with steps of log ω = 0.2, is performed at 110 to 120 °C. At this temperature the curing of the powder coating can be neglected. This oscillation run provides information on the melt viscosity of the powder coating.

Figure 3.15 shows the DMA isotherms of unpigmented polyester/epoxy hybrid powder coating catalyzed with different amounts of dimethyl benzyl amine. The phase angle and dynamic modulus are presented as functions of the frequency. A remarkable increase in dynamic modulus and consequently in viscosity can be observed in the system with ten times higher catalyst concentration. This is an indication that some pre-reaction has taken place during the mixing of the components in the extruder

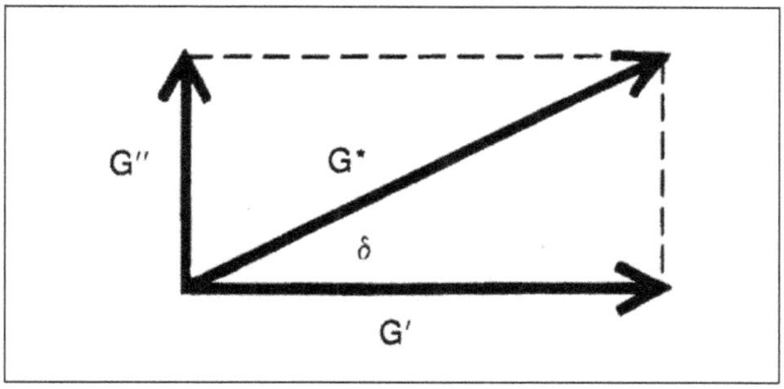

Figure 3.14: Definition of the storage G' and loss G" moduli and tan δ under application of a sinusoidal strain to a viscoelastic specimen

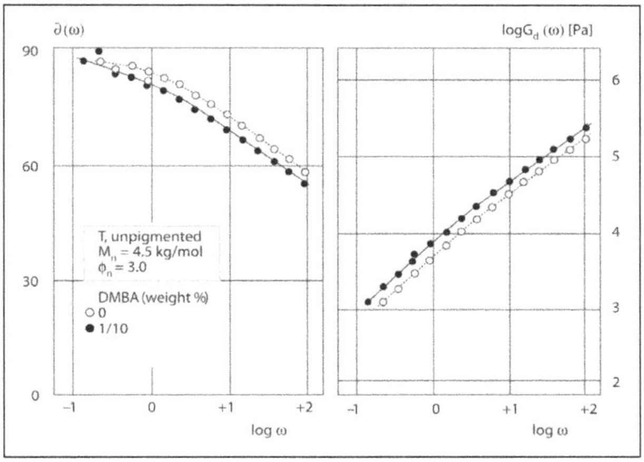

Figure 3.15: DMA isotherm at 110 °C for an unpigmented polyester/TGIC powder coating at four concentrations of DMBA catalyst [36] *source: DSM Resins BV*

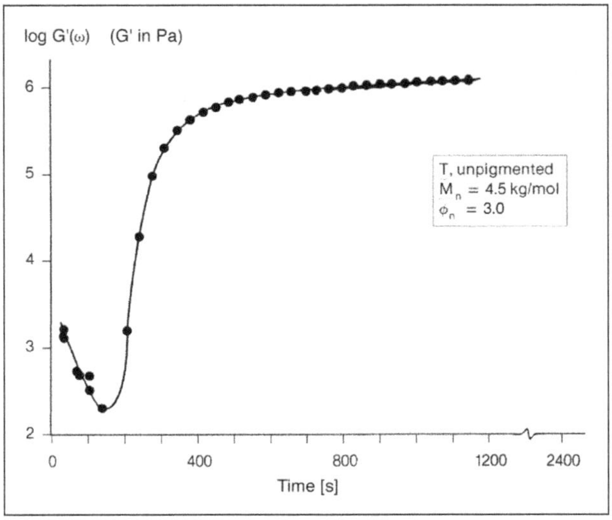

Figure 3.16: Isofrequent runs of polyester/TGIC powder coating at T = 110 to 200 °C at a heating rate of 20°C/min [36] *source: DSM Resins BV*

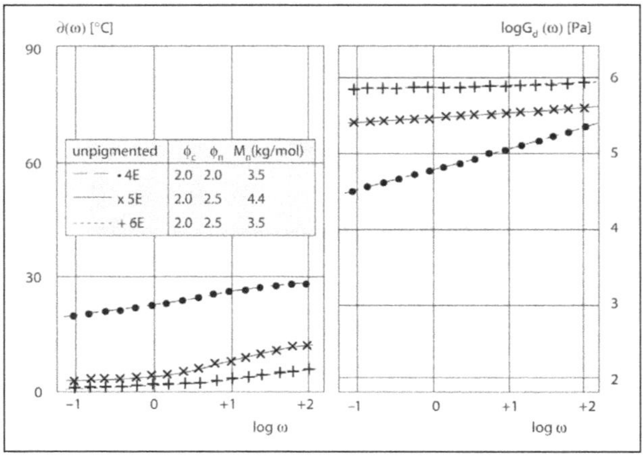

Figure 3.17: DMA isotherms at 200 °C for polyester/epoxy hybrid powder coating after completion of the curing reaction [36] *source: DSM Resins BV*

or during the measuring stage. It is obvious that powder coating behaves as a viscoelastic material since the phase angle dependent on the catalyst concentration and applied frequency lies between 35 and 90°. Otherwise a purely viscous material is characterized by δ= 90° and G_d/ω = constant (equal to the zero shear viscosity η_0) and a purely elastic material by δ= 0° and G_d = constant (equal to the equilibrium modulus G_{d0}).

For **monitoring the curing** of the coating, the oven of the instrument is heated to 200 °C linearly at a heating rate of 20 °C/min. The crosslinking is followed by a period of 10 minutes at a constant frequency of log ω = 0.2 by determining the values of the storage modulus. Using equation $G' = (\sigma_0/\varepsilon_0)\cos\delta$ and (3.70) the dynamic viscosity of the system can be easily calculated. This series of measurements shows the initial decrease in viscosity as a result of the heating and subsequent onset of network formation and network build-up as the curing proceeds, as presented in *Figure 3.16.*

Determination of the dynamic modulus after completion of the curing process makes it possible to calculate the **degree of crosslinking** in the completely cured film. For this purpose the final oscillation run should be performed at a different oscillation rate at constant temperature which is higher than the T_g of the system.

From an extrapolation of G' to zero frequency, the equilibrium modulus $G'_{(0)}$ can be obtained and the crosslinking density of the fully cured film can be calculated using Equation 3.71. *Figure 3.17* represents DMA isotherms at 200 °C for three polyester/epoxy hybrid powder coatings prepared from polyester resins that differ in functionality (F_n = 2.0 and 2.5) and molecular weight (M_n = 3500 and 4400). The relatively high value of δ and the frequency dependence or G_d indicates that the reaction between linear polyester and linear epoxy resin (4E) does not lead to a permanent network. However, the increase of the functionality to 2.5 causes the formation of the permanent networks (5E and 6E). The lower molecular weight of polyester 6E results in a higher value for the dynamic modulus, indicating that the molecular weight between the crosslinking points (M_c) in this case is lower.

The use of DMA to study the influence of the ratio between the functional groups and the catalyst concentration in the powder coating composition, discussed later in Chapter 4, is another indication of the potential use of this technique in powder coating.

3.2.4 Real Time Fourier Transform IR spectroscopy (RT-FTIR)

For radiation curable powder coatings Real Time Fourier Transform Infra Red spectroscopy (RT-FTIR) is a reliable method for **monitoring the rapid photo-polymerization** reactions in the entire mid-IR region (2.5 to 50 μm).

With RT-FTIR the reactivity of UV curable powder coatings can be studied under defined radiation conditions. It is also possible to measure the degree of conversion as a function of time, during and after UV radiation.

Infra red spectroscopy can be applied in transmission as well as in reflection and is based on **absorption of individual infra-red frequencies** (IR bands), which are characteristic for chemical bonds in a compound.

With Fourier Transform IR-ATR (attenuated total reflection) a coating, which is brought in contact with a crystal (with high refractive index), can be examined up to a depth of about 5 μm. Herewith one makes use of the fact, that the totally reflected IR light (at the crystal/coating interface) is reduced ("attenuated") by absorption in the coating. This attenuation takes place by interaction (absorption) of the IR light absorbing coating with the in the coating exponentially decreasing "evanescent wave" of the IR light. With Fourier Transform IR-PAS (photo-acoustic spectroscopy) any IR light absorbing substance can be examined. This technique is based on absorption of IR frequencies by a coating. The absorption takes place at various film depths, dependent on the optical absorption coefficient β (in cm^{-1}) of the individual IR bands. The value 1/β is here the "optical penetration depth" (in cm) and indicates the IR light absorbing depth in the film.

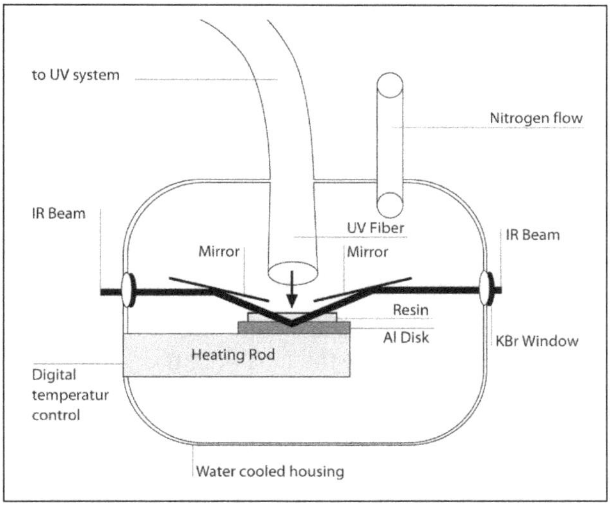

Figure 3.18: Set-up for Real Time Fourier Transform Infra Red measurements [40] *source: Udding-Louwrier et al.* [41]

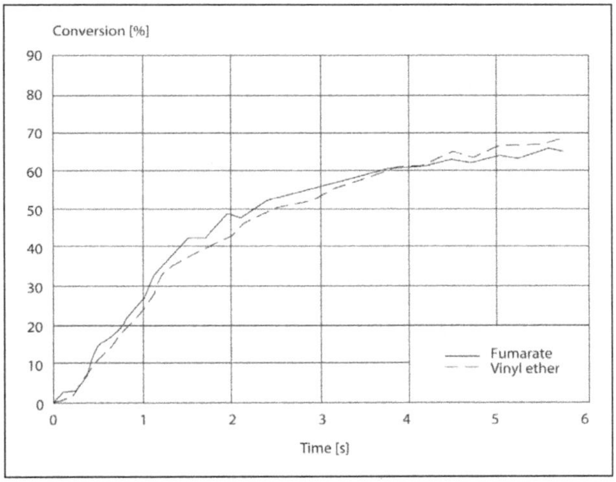

Figure 3.19: RT-FTIR conversion of double bonds of an experimental fumarate-vinyl ether binder system [41] *source: [41]*

The absorbed IR energy is than transformed into heat that flows through the coating, placed in a closed PAS-cell. Heat waves, that reach the surface of this coating, give then their thermal information to the Helium gas in the cell, where "acoustic" waves are formed, which are detected by a microphone.

Resuming, depth dependent polymer specific optical absorption (IR light) is transformed into thermal energy, which is then transported to the film surface and lead, as acoustic waves, to a detector (microphone). Detection depths with PAS are in the order of 5 to 50 μm.

In combination ATR and PAS are two **depth profiling techniques**, which give spectral information on various film depths. With ATR spectral information can be obtained about the "surface" as well as the "sub-surface". Complementary, PAS provides spectral information about the "bulk".

DSM Research [40, 41] has designed a special RT-FTIR-ATR apparatus with a heated plate and a "transflection" (transmission reflection) set-up for the sample arrangement, in which the **UV-curing** of a molten powder coating can be studied. *Figure 3.18* [40] illustrates the set-up of this apparatus.

Though this technique gives information about the curing reaction, it is limited, as it does not allow direct translation to the results of the curing of a UV powder coating on a real IR-UV line (see Chapter 3.1.9 and 3.4.5), which works usually with different UV lamps [41]. Another limitation of the RT-ATR-FTIR technique is its inability to account for differences in film thickness, since the IR beam only penetrates the upper surface of the sample at the coating/crystal interface.

It does, however, allow the monitoring of different types of monomers that are polymerizing. That means it provides the following possibilities:

- the capability to measure radiation curable systems;
- monitoring reactions at various temperatures;
- monitoring of photo-curing reactions under various atmospheres;
- monitoring simultaneously the rates of consumption of different monomers during photo-curing, which is illustrated in *Figure 3.19*.

3.2.5 References

[1] *Widmann, G.*, Proc. of a Symposium in Polymerteknisk Selskab, May 7, 1980
[2] *Borchardt, H.J., Daniels, F.*, J. Am. Chem. Soc., 79, 41, 1956
[3] *Prime, R.B.*, in Thermal Characterization of Polymeric Materials, Ed. *E.A.Turi*, Academic Press, New York, 1981, p. 532
[4] *Ozawa, T., J.* Therm. Anal., 2, 301, 1970
[5] *Fava, R.A.*, Polymer, 9, 137, 1968
[6] *Barton, J.M.*, J. Makcromol. Sci., Chem. A8 (1), 53, 1974
[7] *Manche, E.P., and Carrol, B.*, in Physical Methods in Macromolecular Chemistry, Ed. *B. Carrol*, Vol. 2, *Marcel Dekker*, Inc., New York 1972, p. 309
[8] *Mohler, H.*, FARBE&LACK, 86, 211, 1980
[9] *Henig, A., Jath, M., Mohler, H.*, FARBE&LACK, 86, 313, 1980
[10] *Zicherman, J.B., Holsworth, R.M.*, J .Paint Technol., 46, 55, 1974
[11] *Peyser, P., and Bescom, N.D.*, J. Appl. Polym. Sci., 21, 2359, 1977
[12] *Olcese, T., Spelta, O., Vargiu, S.*, J. Polym. Sci. Symp. N53, 1975, p. 113
[13] *Gabriel, S.*, J. Oil Col. Chem. Assoc., 58, 52, 1975
[14] *Klaren, C.H.J.*, J. Oil Col. Chem. Assoc., 60, 205, 1977
[15] *Linde van der, R., Belder, E.G.*, Proc. VIIth Int. Conf. Org. Coat. Sci. Technol., July 1981, Athens, Greece, p. 359
[16] *Ghijssels, A., Klaren, C.H.J.*, Proc. XIIth Fatipec Congress, 1974, p. 269
[17] *Franiau, R.P.*, Paintindia, September 1987, p. 33
[18] *Riesen, R., Wyden, H.*, Mettler TA3000 Seminar, October 5-7, 1982
[19] *Riesen, R.*, STK Meeting, Basel, Switzerland, 30 Sept. 1981
[20] *Gargallo, L., Russo, M.*, Makromol. Chem., 176, 2735, 1975
[21] *Stamhuis, J.E., and Pennings, A.J.*, Polymer, 18, 667, 1977
[22] *Chevenard, P., Wache, X., De La Tullaye, R.*, Bull. Soc. Chim. Fr., 11, 48, 1944
[23] *Duval, C.*, Anal. Chem. Acta, 2, 92, 1948
[24] *Wendlandt, W.W., and Gallagher, P.K.*, in Thermal Characterization of PolymericMaterials, Ed. *E.A. Turi.*, Academic Press, New York 1981, p. 4
[25] *Bauer, D.R., and Dickie, R.A.*, J. Coat. Technol., 54, 57, 1982
[26] *Freeman, E.S., Carrol, B.*, J. Phys. Chem., 62, 394, 1958
[27] *Scharp, J.H., Wentworth, S.A.*, Anal. Chem., 41, 2060, 1969
[28] *Reich, L., Levi, W.*, in Macromolecular Reviews, Eds. *Peterlin, A., Goodman, M., Okamura, S., Zimm, B.H., Mark, H.F*, Vol.1, Wiley-Interscience, New York, 1963, p. 173
[29] *Friedman, H.L.*, J. Macromol. Sci., Chem., A1, 57, 1967
[30] *Sharp, J.H., Wentworth, S.A.*, Anal. Chem., 41, 2060, 1969
[31] *Coats, A.W., Redfern, J.P.*, Nature, 201, 68, 1964
[32] *Flynn, J.F., Wall, L.A.*, J. Res. Natl. Bur. Std., 70, A487, 1966

[33] *Zsako, J.*, J. Phys. Chem., 72, 2406, 1968
[34] *Cameron, G.G., Fortune, J.D.*, Eur. Polym. J., 4, 333, 1968
[35] *MacCallum, J.R., Tanner, J.*, Eur. Polym.J., 6, 1033, 1970
[36] *Scholtens, B.J.R., Van der Linde, R., Tiemersma-Thoone*, P.J.M., DSM Resins Powder Resins Symposium, Noordwijk 1988
[37] Adapted from *Skrovanek and Schoff*, Prog. Org. Coat., 16, 135, 1988
[38] *Treloar, L.R.G.*, The Physics of Rubber Elasticity, 3rd ed., Clarendon, Oxford, 1975
[39] *Ikeda, S.*, Prog. Org. Coat., 1, 205, 1973
[40] *Dias, A. A., Hartwig, H., Jansen, J.F.G.A.*, Proc. Radcure Coatings & Inks, Harrogate 1998, paper 15; or Surface Coatings Intern. 83, (8), 382, 2000
[41] *Udding-Louwrier, S., Baijards, R., De Jong, E.S.*, Proc. 25th Intern. Waterborne, High Solids and Powder Coatings Symp. New Orleans, Feb. 1999
[42] *Knappe, S.*, Farbe + Lack, 111(8), 34-37, 2005

3.3 Crosslinkers for powder coatings

After binders, crosslinkers used in thermosetting powder coatings are the most important part of the coating composition. They exert an influence on the production, storage, application and exploitation properties of the coatings such as extrusion temperature, storage stability, curing time, curing temperature, flow behavior, impact resistance and hardness, weathering and chemical resistance etc.

It is sometimes very difficult to distinguish between the crosslinker and the binder, because both constituents at the last instance make the complete binding system. Very often the low molecular weight compound present in a minor amount and participating in the curing reaction is considered as a crosslinker. However, in some cases, as in the polyester/epoxy hybrids it is a question of what is the crosslinker and what is the binder, especially in the 50:50 systems. In such a case the choice of the right word is more or less a semantic problem. Therefore bisphenol A epoxy resins and acid functional polyesters, which, depending on someone's preference, can be classified in a group of the crosslinkers, will be omitted in this section, since they will be discussed in Chapter 3.4 as binders for powder coatings.

3.3.1 Crosslinkers of the epoxy type

3.3.1.1 Triglycidyl isocyanurate (TGIC)

Triglycidyl isocyanurate is one of the most important curing agents for carboxyl functional binders, specifically for carboxyl terminated polyesters. In Europe this used to be the most popular crosslinker for polyester powder coatings for **outdoor use**, only recently replaced by HAA systems. The formula of TGIC is given below:

Different methods of preparation of TGIC such as isomerization of triglycidyl cyanurate obtained from cyanuric chloride and glycidol or transesterification of cyanuric tosylate with glycidol are described in the patent literature. However, only the **epichlorohydrin** route has found practical importance.

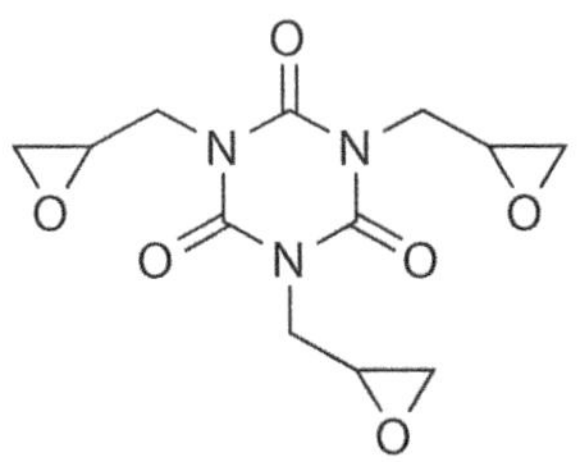

TGIC

The industrial preparation of TGIC from cyanuric acid and epichlorohydrin is extensively described in the patent literature [1-3]. The synthesis is performed by heating cyanuric acid in a large excess of epichlorohydrin in the presence of ammonium or phosphonium halide as catalyst. The subsequent dehydrohalogenation of the formed halohydrin by base gives TGIC as a product, while the excess of epichlorohydrin is distilled off. The obtained crude product is purified by recrystallization.

Equation 3.72

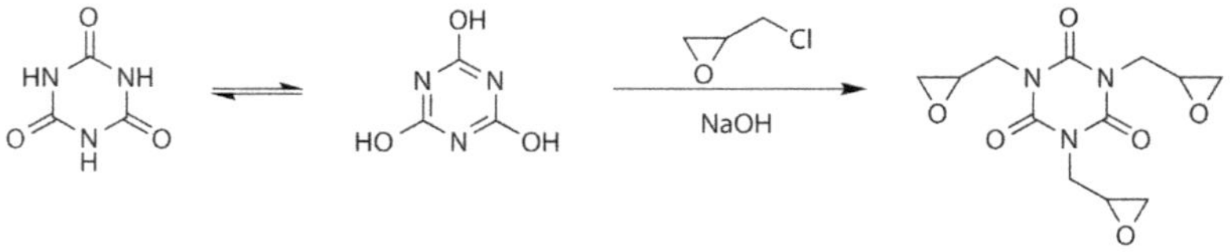

Because of the racemic blend of epichlorohydrin used in the syntheses, by fixing one glycidyl group per molecule, a product that is a 50:50 blend of the two stereoisomers should be obtained. Since in the case of TGIC three glycidyl groups are fixed on one molecule, eight different diastereoisomers can be obtained. Because of the symmetry of the molecule of third order, three pairs of the diastereoisomers cannot be distinguished, thus leading to the following **four isomeric forms**: RRR/SSS and three pairs of RRS/SSR, RSR/SRS and SRR/RSS. Statistically, the composition with respect to the ratio of different pairs will be 1:3 or 25:75. It has been found that TGIC produced in the way described above, is a blend containing 75 % of so-called form A with a melting point of 103 to 104.5 °C, and 25 % of form B with a melting point of 156 to 157.5 °C [4, 5]. *Merck* and *Loutz* have found that the amount of form B with a higher melting point in **commercially available TGIC** from different suppliers varies between 13 and 29 % [6]. They have isolated and purified forms A and B and have shown that they behave very differently when cured with carboxyl functional compounds. Form B is more reactive, and with an equivalent amount of adipic acid it has a gel time at 180 °C of 38 seconds, in comparison with 73 seconds for form A. On the other hand, in a coating formulation with carboxyl functional polyester, form B shows a longer **gel time**, a slower viscosity increase during curing, and requires a longer curing time to obtain good mechanical properties than form A. At first site these are contradictory results, explained by the particularly low solubility of form B in organic media, even at 200 °C, which considerably influences the kinetics of the curing process. This could also be an explanation for the surface and transparency anomalies that are frequently observed in **TGIC/polyester powder coatings**, especially in the case of dark colored paint formulations.

The presence of the B form with a higher melting point in the commercial types of TGIC can considerably influence the curing behavior of the powder coatings. It will be emphasized even more in low TGIC containing powder formulations, which are in principle more critical. For the time being, there is no strong evidence that TGIC will isomerize itself during storage to thermodynamically more stable, but unfortunately not desirable form B.

The theoretical **epoxy equivalent weight** of TGIC is 99, but in practice due to partial self-condensation during its production and certain hydrolysis during storage, the epoxy equivalent value is somewhat higher, being in the range of 105 to 110. This **low epoxy equivalent value** is a rather important characteristic of TGIC. Low molecular weight materials in powder coating compositions dramatically decrease the glass transition temperature (T_g) of the system. This adversely affects the storage stability of the coating. Because of its low equivalent weight, the concentration of TGIC in powder coating formulations is relatively low, usually being not higher than 10 %. Therefore, TGIC affects the T_g of the coating composition to a level that is acceptable in practice. However, it should be kept in mind that every weight percentage of TGIC in the powder formulation decreases the T_g of the coating by approximately 2 °C.

Compared to the bisphenol A types of epoxy crosslinkers TGIC is much more reactive. Powder coating compositions containing TGIC as a crosslinker do not contain curing catalysts in many cases. If a fast curing system has to be formulated, the type and amount of catalyst should be chosen very carefully in order to avoid pre-reaction during the extrusion of the coating, or problems connected with the chemical stability of the formulation during storage.

Coatings in which TGIC is used as a crosslinker exhibit outstanding outdoor durability, and with a good choice of the binders and pigments, systems with a **service time longer than ten years** can be obtained.

Mutagenic potential of TGIC

Some years ago, modern toxicological studies showed some evidence of **mutagenic potential** concerning TGIC. Therefore, the European Union has decided, that the symbol T (Toxic, symbolized by a skull and crossbones), accompanied by the relevant Risk Phrase, must be used for TGIC labeling with effect from May 1998. In 2011 a case of occupational asthma caused by triglycidyl isocyanurate was reported by *Sastre* et al. [7]. A new method for the determination of triglycidyl isocyanurate from powder coatings in occupational hygiene samples by gas chromatography was developed by *White* [8] with an estimated detection limit of 0.002 μg/ml.

TGIC alternatives

In the meantime, intensive research work has been dedicated to the development of technically viable "TGIC alternatives", without any concern for mutagenic potential.

From the reports in the literature on TGIC alternatives, which belong to the group of crosslinkers of the epoxy type, several should be mentioned here:

AMOCO have developed epoxy functional polyesters, to be cured with carboxyl functional polyesters. These products perform well. However for toxicological reasons they have been withdrawn. In 1995 Shell published a patent, which describes the use of epoxy functional polyesters.

They use **tertiary carboxylic functional polyesters**, which are converted into alpha-branched diglycidyl esters. In this way very stable systems can be developed, which have attractive toxicological characteristics and very good outdoor durability, but very poor impact resistance when combined with acidic polyesters. Initial attempts by Shell to improve the impact resistance succeeded, at the cost of the storage stability. But in 1998 they came with a "Shell ODD" outdoor durable system (consisting of an α,α-branched diglycidylester and a "special" acidic polyester), which meets easily the "Qualicoat"-Class 2 (superdurable) specification.

For Class 1 the flexibility (reverse impact) needs however improvement. The accelerated weathering test (QUV-A) shows a greater loss of gloss, after 4000 hours exposure, than a corresponding TGIC system. So far, only semi-commercial products (pilot plant quantities) have appeared on the market for testing.

Three other "TGIC alternatives", which also belong to the group of "Crosslinkers of the Epoxy type", have been introduced by Ciba, DSM Resins and Estron in the 1990's. As these crosslinkers are already available in commercial quantities, they will be discussed separately in Chapters 3.3.1.2, 3.3.1.3 and 3.3.1.4. Recently BASF [9] claimed **urethane** and **urea epoxies** based on the reaction of polyisocyanates and glycidyl alcohols. These special epoxy hardeners show a higher reactivity and thus are suitable for low temperature curing [10].

3.3.1.2 "Solid solutions" of glycidylesters

Already in 1992 several major powder producers were concerned about the hazards of TGIC during powder manufacture and decided to abandon the use of TGIC in the near future.

Therefore, Ciba, (now Huntsman), one of the main producers of TGIC under the trade name "Araldite PT 810", came soon with a TGIC alternative. This crosslinker, described by Ciba as a **polyfunctional epoxy hardener**, is based on several esters, prepared from polyfunctional carboxylic acids, which after glycidylation, can be used as curing agents for carboxylic functional polyesters, In 1994 this curing agent came available (commercially) under the official trade name "Araldite PT 910". According to *Glottis* and *Cutting*[11] "Araldite PT 910" is a solid solution (known from solid state physics) of 1 part (by weight) of the liquid TML (= triglycidyl-trimellitate) in 3 parts (by weight) of the crystalline DGT (= diglycidyl terephthalate). That means: an **optimal functionality between 2 and 3** is obtained here by mixing a solid crystalline compound (terephthalate) with a liquid one (trimellitate) in a ratio leading to a crystalline solid solution.

This concept is so far new in the powder coating technology. The chemical structures of DGT and TML are:

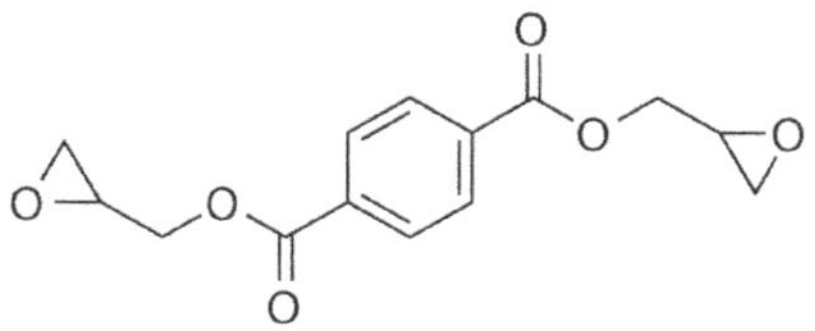

Diglycidylterephthalate (DGT)

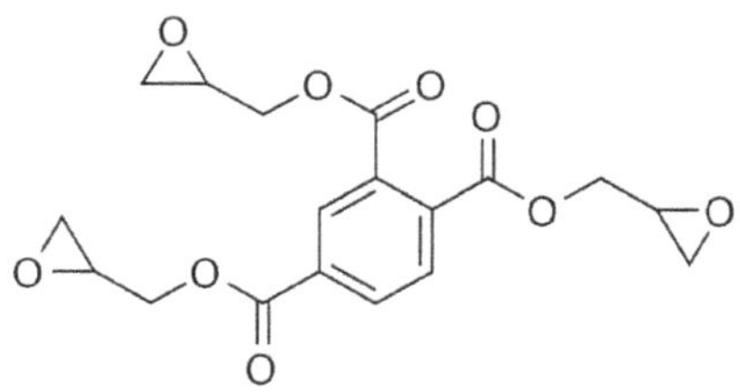

Triglycidyltrimellitate (TML)

Though, from a chemical and technological standpoint "Araldite PT 910" is one of the most obvious solutions for a TGIC alternative (similar crosslinker chemistry) and trials have indicated the lack of a mutagenic potential for both glycidylester compounds, powders based on "PT 910" need to be labeled, due to the sensitization characteristics (positive Ames test) of this curing agent. Therefore, labeling (only in the European Union) should warn with the symbol Xi and the risk phrases R 36/38 and R 43 (irritating to eyes and skin; may cause sensitization by skin contact). Coating powders, containing more than 1 % of "Araldite PT 910", have to be labeled with the risk phrase R 43 only.

However, the **EU legislation** offers the possibility to determine the necessity of R 43 labeling of coating powders as potential skin sensitizers on the basis of experimental test data (in accordance with EU Directive 88/379/EEC, as adapted by Directive 93/18/EEC)[1].

Powder storage stability is affected quite severely by the level of "PT 910" (even more than with TGIC). A 93/7 polyester/"PT 910" ratio has still a good stability (if the polyester has been designed for cure with "PT 910"), but a 91/9 ratio already means an overkill of crosslinker in terms of powder storage stability.

Since the functionality of "PT 910" is lower than that of TGIC, the mechanical properties are more sensitive to the cure schedule and the resin type used. The lower reactivity of "PT 910" can be compensated for either by adding an accelerator (or catalyst), or by using a more reactive polyester.

With the current raw materials it is possible to synthesize such polyester resins, which also have high glass transition temperatures, resulting in good storage stability.

[1] *It is (and remains) within the responsibility of thr formulator to satisfy himself, that his products are properly labeled in accordance with EU Directive 88/379/EEC, as adapted by Directive 93/18/EEC*

Coatings, in which "PT 910" is used as a crosslinker, are comparable to TGIC powders in terms of outdoor durability. Premixing of a "PT 910" system needs more attention than premixing of TGIC systems, due to the high melt viscosity of the polyester resins for "PT 910"-cure and the different melting behavior of "PT 910". These two factors result in a **different behavior of a "PT 910" system** in an extruder. Higher shear forces are required to ensure optimal dispersion in the extruder.

Good premixing and dispersion of "PT 910" systems are very important to obtain good film flow (or leveling), non-hazy appearance and good mechanical properties. These properties depend more on the processing conditions than with TGIC powder systems.

3.3.1.3 Aliphatic oxiranes

Due to the dry blending of the various powder components, before extrusion, crosslinkers have to be solid materials. This is a restriction for those people searching for new non-toxic TGIC alternatives.

A new approach, using liquid aliphatic oxirane crosslinkers was introduced by DSM Resins in the 1990's. This resulted in a complete binder system, launched on the market under the name "Uranox" [12-18].

The aliphatic oxirane (AO) powder coating system uses carboxyl functional polyester resin as binder and an aliphatic oxirane crosslinker, being a derivative of epoxidized fatty acids, which are renewable resources. Aliphatic oxiranes can be classified as almost **non-toxic** with an oral LD50 higher than 20,000 mg/kg [19], they are negative in the mutagenicity tests, do not cause skin irritation, and do not show chronic toxicity effects [20]. They are listed in a number of regulations, issued by the **FDA** for use in food packaging applications [21]. That means powder based on "Uranox" does not have to be labeled, which also means there are no restrictions with respect to waste powders.

The production process of the complete binder system comprises the synthesis of a carboxyl functional polyester resin and an aliphatic oxirane crosslinker in two separate reactors.

The reactivity of the aliphatic oxirane groups is much lower, compared to the glycidyl groups present in the epoxy hardeners normally used for powder coatings. This ensures a safe blending operation on the one side, but on the other side a proper catalyst is required, to compensate for this low reactivity. In order to avoid undesired prereaction, no curing catalyst is added to the "Uranox" system during the blending of resin and crosslinker. Therefore, a special catalyst, delivered in the form of a master batch has been developed, thus allowing a curing schedule of 10 minutes at 200 °C. To further enhance reactivity without compromising storage stability *Senatore* et al. [22] of the Eindhoven University **microencapsulated** the **liquid** epoxide in polyvinylpyrrolidine. Although curing kinetics now depended on the de-encapsulation no significant higher reactivity could be achieved (curing 20 min at 180 °C), showing a storage stability of two weeks at 40 °C. Due to economical reasons, "Uranox" has been withdrawn from the market in 2006.

3.3.1.4 Glycidyl methacrylate (GMA)

Some years ago Estron (in the U.S.) and Worlée (in Europe) introduced a **polymeric glycidyl methacrylate** with the trade name "GMA 300".

According to our own experience, it is a suitable TGIC alternative. In contrast with "Araldite PT 910", no specially designed polyesters are needed and the normal TGIC-designed polyesters can be used, most suitable are polyesters with acid values of 22 to 26 mg KOH/g.

As "GMA 300" has an epoxy equivalent weight of approx. 300, the polyester/"GMA 300" ratio should be 88:12.

Coatings, in which "GMA 300" is used as crosslinker, are comparable to TGIC powders in terms of outdoor durability. The film hardness is however less than with TGIC.

3.3.2 *Polyisocyanates*

From a number of methods available for preparation of isocyanates, only phosgenation of amines and the phosgene-free urea process have found a real commercial importance.

Equation 3.73 $$RNH_2 + COCl_2 \longrightarrow RNCO + 2\,HCl$$

Equation 3.74

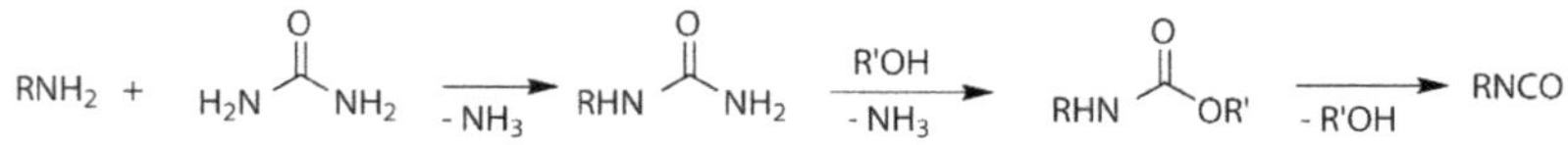

Polyisocyanates used as crosslinkers for powder coatings can be of the aromatic or aliphatic types. Aromatic isocyanates, especially the derivatives of toluene diisocyanate are very attractive from a price point of view. On the other hand, the main drawback associated with the use of **aromatic isocyanates** is their **tendency to yellow** on exposure to sunlight. This was assumed to be due to the oxidation of some terminal aromatic amino groups derived from these isocyanates [23].

The isocyanate groups in the crosslinkers used for the manufacture of **polyurethane powder coatings** must be protected in order to avoid the crosslinking reaction at room temperature and to provide good storage stability of the coating. Another reason for this protection is to reduce the health and safety hazards to a level that allows relatively safe handling of these materials during production and use of the coatings. These crosslinkers are synthesized by the reaction of diisocyanates or their derivatives with protective species, commonly referred to as blocking agents to produce isocyanate-free adducts that are stable in the presence of moisture and in blends with polymers containing reactive hydroxyl groups at normal storage conditions.

Although there are many different combinations between the considerable number of available isocyanates and different blocking agents, only the **derivatives of isophorone diisocyanate (IPDI) blocked with caprolactam** have a major commercial importance. Despite its low price, toluene diisocyanate has found limited applicability only in cheap polyurethane coatings for indoor use because of its low yellowing resistance and light stability.

Commercially available types of IPDI derivatives will be presented in this section, together with isocyanates which are not yet commonly used in practice for powder coating production, but which have potentials to be commercialized in the near future.

3.3.2.1 *Caprolactam blocked IPDI derivatives*

Caprolactam blocked isophorone diisocyanate derivatives are the most widely used crosslinkers for polyurethane powder coatings. They possess a suitable melting range to be used as hardeners for powder coating manufacture by the standard extrusion procedure. This is combined with an **acceptable deblocking temperature** and **curing rate** which is suitable for the common curing schedules when the system is catalyzed properly.

There are three important commercial IPDI caprolactam blocked adducts on the market produced by Bayer SE and Evonik Industries: "Crelan UI" (Bayer) and crosslinkers "Vestagon B1065" and "Vestagon B1530" (Evonik Industries), the last one being the most popular. The NCO content of this crosslinker is 15 % with less than 1 % of free NCO groups. All of them have melting temperature in the range between 75 and 90 °C and a glass transition temperature of about 50 °C. The deblocking reaction takes place at 175 °C for "Crelan UI2 and "Vestagon B1065", and somewhat lower (160 to 170 °C) for "Vestagon B1530". Evonik

Industries [24] have found, that the deblocking temperature can be lowered to 140 °C by using 1,2,4-triazole as blocking agent instead of caprolactam. The chemical structure of the caprolactam blocked IPDI crosslinkers is as follows:

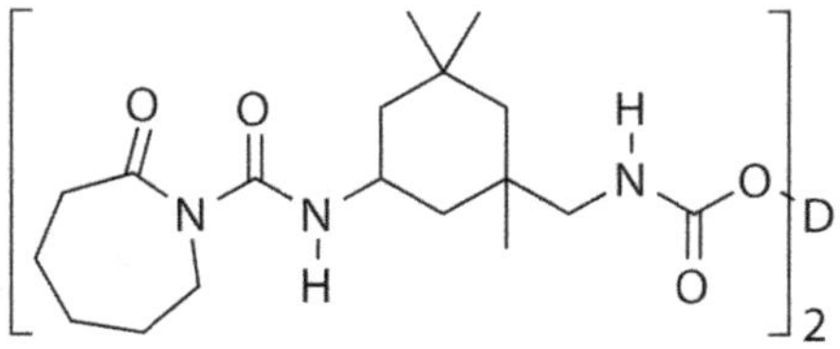

where D is a residue of a diol.

Different patents cover the **production** of caprolactam blocked IPDI oligomers suitable as crosslinkers for polyurethane powder coatings. In a patent of Hüls [25] (now Evonik Industries) 1 mole of ethylene glycol is reacted with 4 moles of IPDI followed by blocking with 6 moles of caprolactam. The product has a melting point of 70 to 80 °C and T_g of 35 to 46 °C. An alternative way to produce caprolactam polyisocyanates starting from a polyamine and carbonyl biscaprolactam has been published by *Loontjens* et al. [26].

Instead of diols, diamines can be used to extend the molecular weight of the IPDI blocked oligomers, in this way introducing urea groups into the oligomer structure [27]. Oligomer obtained by blocking IPDI with caprolactam followed by reaction with isophorone diamine is characterized by a melting point of 80 to 84 °C and 16.3 % of a blocked NCO content.

Isocyanates with a higher functionality than two can be obtained by **trimerization of diisocyanates** to isocyanurates [28].

Equation 3.75

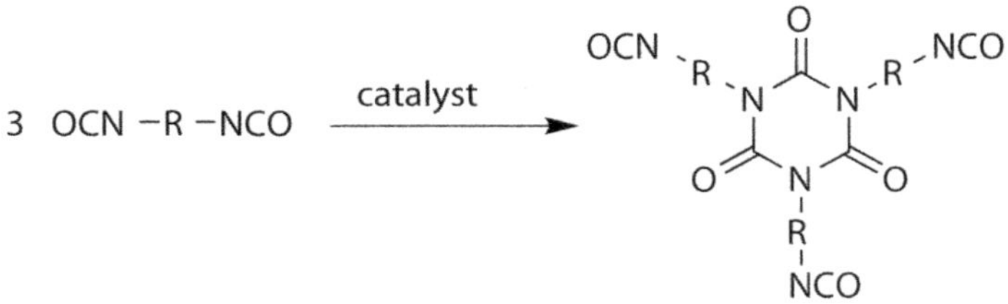

By the correct **choice of isocyanate**, such as 2-methylpentane-1,5-diisocyanate, 2-ethylbutane-1,4-diisocyanate or 3-isocyanatomethyl-, 3,5,5-trimethylcyclohexaneisocyanate (IPDI) separately or in a mixture, a suitable catalyst for trimerization, such as triethylene diamine propylene oxide and caprolactam as a blocking agent [29], a curing agent containing 10 to 16 % blocked NCO groups with a glass transition temperature between 60 and 93 °C can be obtained.

Oligomers of **IPDI with higher functionality** than two can be obtained alternatively by using blends of diols and triols in the preparation of the final product. A patent of UCB (now Cytec) [30] discloses a two-stage process for the production of caprolactam blocked IPDI oligomer. In the first step, 1 mole of IPDI reacts with 1 mole of caprolactam at 70 to 90 °C in the presence of Sn-octoate as catalyst, giving a partly blocked intermediate product. The latter reacts in the second step at 90 to 140 °C with the equivalent quantity of blend comprising 30:70 to 50:50 diol/triol mixture until full consumption of the isocyanate groups.

An interesting **hardener** which is also a derivative of isophorone diisocyanate is described in a patent assigned to Bayer [31]. The hardener is a condensation product of isophorone diisocyanate, caprolactam and 2,2-bis-(hydroxymethyl)-propionic acid in a molar ratio 2:2:1. The condensation reaction performed in toluene gives a crystalline colorless product with an acid value of 70 and melting point between 120 and 128 °C. In combination with TGIC (or other epoxy-group containing compounds) as crosslinker, hydroxyl functional polyester as binder, pigments and a suitable flow additive, a powder coating can be obtained, having 60°gloss below 20 % after curing. It is essential to keep the ratio between the NCO content and the acid number of the hardener between 0.1 and 0.3 in order to obtain matt films. A powder coating formulation, in which this ratio is e.g. 0.8, has a gloss of 77 %.

3.3.2.2 Uretdiones

Uretdiones are heterocyclic four membered rings formed by the reversible reaction of two isocyanates. In this form the isocyanates are temporarily deactivated and upon heating they split back into the reactive isocyanates. Although other diisocyanates [32] can be transformed into uretdiones, too, IPDI based dimers are the only one which found an industrial application in powder coatings. Blocking agent-free polyisocyanate hardener suitable for producing polyurethane powder is internally blocked IPDI, obtained by a self-condensation process in basic media. In the first step a dimerization of IPDI is performed at room temperature in the presence of tri-dimethylamine phosphine or other suitable catalysts to produce an uretdione of IPDI of the following formula:

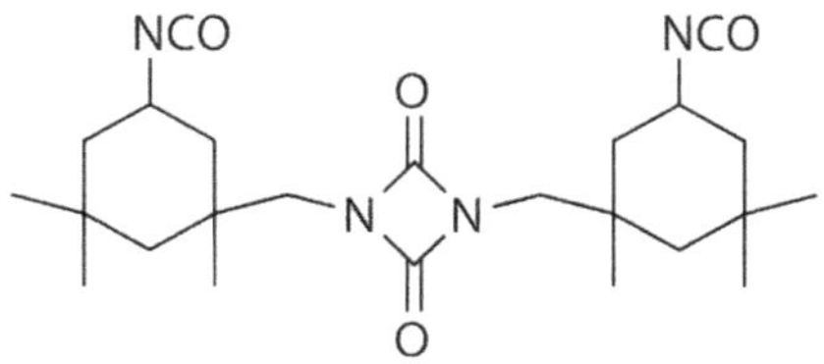

IPDI and its uretdione are in a thermodynamical equilibrium, which is in favor for the dimer at room temperature as can be seen in the *Figure 3.20.* Although at 120 °C the uretdione should have reformed the original IPDI monomers, this process is kinetically hindered. This is the reason why uretdione based powder coating formulations conventionally are crosslinked at 180 °C and higher, unless special catalysts are used, described later in this chapter.

The intermediate uretdione is reacted with diols to produce the final product. The basic structure of the IPDI uretdione hardener is the following:

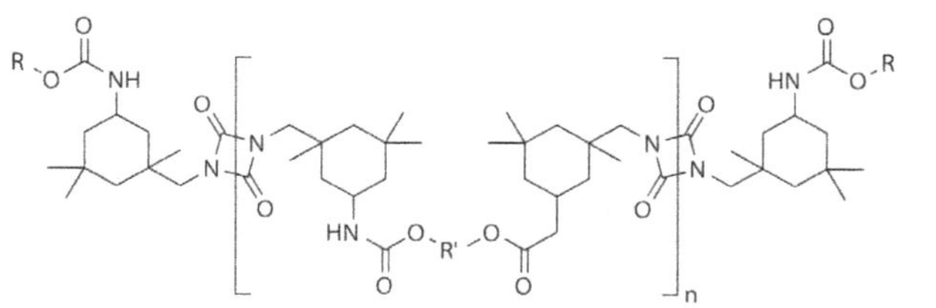

where n is at least 1 and R' is a divalent aliphatic, cycloaliphatic, arylaliphatic or aromatic hydrocarbon group from a diol. R is either the residue of a monoalcohols, or more common, of a diol. Usually R is identical to R'. The process according to [33] proceeds in two steps. Dependent on the IPDI-uretdione/diol ratio, a range of crosslinkers can be obtained with molecular weights between 500 and 4000, melting points between 70 and 130 °C and a low free NCO content. When heated to the stoving temperature of the coating, 98 % of the material converts back to moieties with free isocyanate groups.

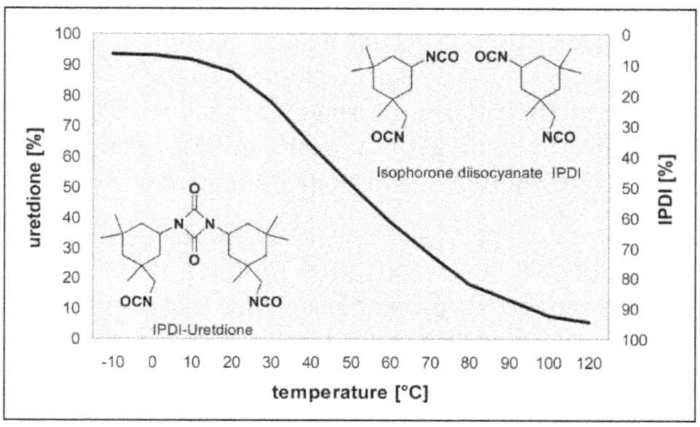

Figure 3.20: Thermodynamic equilibrium of IPDI and IPDI-uretdione

Other patents of Chemische Werke Hüls AG (now Evonik Industries) [34] disclose the possibility of producing practically **isocyanate free uretdiones** of IPDI, using monoalcohols or monoamines as blocking agents for the free-end isocyanate groups. In one of the examples a polymeric IPDI uretdione obtained in the way described above is blocked with 2-ethyl hexanol to reduce the free isocyanate content below 1 %. During the curing, 98 % of the uretdione converts back to free isocyanate which serves as a crosslinker for hydroxyl functional polymers.

A commercial product of this type is introduced and sold on the market by Evonik Industries under the name of "Vestagon BF 1540". The content of free NCO groups is very low (below 1 %) while the total NCO content is 15.4 %. The melting point of the crosslinker is in the range between 105 and 115 °C. Since the usual extrusion temperature does not exceed 120 °C, practically no pre-reaction takes place while the melt is compounding with the binder.

An obvious advantage of IPDI uretdione compared to blocked isocyanates is the **absence of a side product** which leaves the system during curing.

Therefore from a technical point of view, polyurethanes cured by "Vestagon BF 1540" could be a very good alternative to TGIC powder coatings. But, though "Vestagon BF 1540" has been commercially available for more than 20 years, it has never found a well-established place on the market. This can be attributed to the fact, that "Vestagon BF 1540" has shown deficiencies of (too) low reactivity and poor resistance to solvents and chemicals (due to its low functionality of< 2). "Vestagon BF 1540" contains both diols and also certain monofunctional alcohols for irreversible capping of terminal isocyanate groups, which result in the formation (under curing conditions) of monofunctional hardener fractions. These fractions lead to chain termination in the crosslinking reaction with an OH-functional polyester resin and by consequence to a functionality of less than 2. More recent developments by Evonik Industries [34] lead to the introduction of "Vestagon BF 1320", which contains only diols. The use of diols instead of monofunctional alcohols for irreversible capping of terminal isocyanate groups avoids chain-terminating. As a result, each linear polymeric hardener chain receives a terminal OH functional unit. Thereby the **overall functionality** of the individual hardener portions in the "BF 1320" hardener **is raised to 2.0**.

Wenning [35] compared the performance properties of the three **internally blocked crosslinkers** "Vestagon BF 1540", and "Vestagon BF 1320" (former "Vestagon BF 1300" and "1310"), when used in conjunction with polyesters, having OH numbers of 26 and 55 and both pigmented with titanium dioxide at 40 %. "Vestagon BF 1540" was cured for 10 minutes at 200 °C, while "Vestagon BF 1320" was cured for 8 minutes at 200 °C. Compared to the "Vestagon BF 1540" control formulations, films crosslinked with "Vestagon BF 1320" gave very good film properties. Though film flow and gloss levels of all three systems were similar, impact resistances (direct and reverse) for "Vestagon BF 1540" were poor, but the "Vestagon BF 1320" based formulations showed excellent mechanical properties. The improved cure, obtained with "Vestagon BF 1320", results in improved resistance to detergent solution and hot water, with more than 80 % gloss retention after 500 hours of immersion in a 1 % detergent solution at 74 °C or 500 hours immersion in a 90 °C water bath. UV stability and weather resistance is also shown to be comparable to polyester/TGIC powder systems.

The use of a **catalyst** to accelerate the formation of urethane bonds is recommended. Dibutyltindilaurate (DBTL) has proven to be a useful accelerator. When the "BF 1320" systems is catalyzed with 0.15 weight-% DBTL (based on the total formulation), it is possible to completely cure the films at 180 °C, with a reduction in stoving time of between 15 and 20 %.

It should be noted here, that "Vestagon BF 1320" themselves contain already DBTL. Perhaps needless to say, that the internally blocked "Vestagon BF 1320" adduct systems have similar emission levels to "standard" polyester/TGIC powder coating systems, in the range of only 1 % weight loss during cure. The latest developments in this field led to "Vestagon BF 1320" and "BF 1321" (the tin-catalyst free version), which are in particular suitable for the popular **heat transfer printing**. This technology is based on the transfer of a desired pattern on a powder coated substrate. The foil with the printed pattern is pressed on the substrate and heated to initiate a diffusion process of the pigments onto the powder coating. This way the metallic window or door profiles obtain e.g. a wooden appearance.

DuPont recently filed a patent [36] describing the manufacture of **carboxy functional uretdione** based hardeners for higher crosslink densities and flexibility even and lower film thickness.

Some years ago Bayer launched also their first **polyuretdione** crosslinker under the trade name "Crelan VP LS 2147", which is characterized by a low T_g and thus by an extremely low melt viscosity and excellent flow properties. Compared to other polyuretdione hardeners, such as "Vestagon BF 1320", the melt viscosity of "LS 2147" is nearly ten times lower, which could be obtained by modification of the composition of the diols. Such a low melt viscosity and the 30 °C lower melting range allow very moderate extrusion conditions (as a normal temperature of 100 °C is now possible). It has to be taken into account, though, that these kind of low T_g hardeners have a detrimental effect on the physical storage stability of powder coating formulations.

Freudenberg et al. [37] and *Meier-Westhues* et al. [38] have given typical formulations based on "LS 2147", using polyesters with OH values of 50 and 30 respectively. Coatings cured at 180 °C were shown to comply with the GSB and Qualicoat standards for coatings on aluminum. This rather **high stoving temperature** is necessary to achieve cleavage of the four-membered ring of the uretdione into two NCO groups, when using stannous octoate as a catalyst. Therefore Bayer directed their research to improving the reactivity of IPDI-uretdione based powder coatings, which is low, compared to other systems, such as polyester/TGIC powders.

Organotin compounds have been used in polyurethane systems to accelerate the reaction between the NCO groups, produced by cleavage, and the OH groups of the polyester. It is now known, that both the NCO and OH groups can coordinate with the tin atom and that the reaction to form urethane occurs in the ligand sphere of the tin, requiring less activation energy [39]. However, organotin catalysts do not have a significant effect in uretdione systems, as tin complexes have no influence on the first reaction, which leads to the necessary cleavage of the uretdione.

Bayer researchers have found, that the addition of cyclic N,N,N-trisubstituted amidines makes it possible to reduce the **cleavage temperature** by up to 30 °C, depending on the amount of catalyst used [40]. This means, that it is now possible to use a stoving cycle of 15 minutes at 150 to 160 °C for these blocking agent-free polyurethane powder coatings. Such a high level of reactivity meets that of the most reactive polyester/TGIC systems. With polyurethane powder coatings this was only possible, up till now, by using butanone oxime- or 1,2,4-triazole blocked crosslinkers.

DSC analysis of cured films based on IPDI-uretdione based hardeners provides clear evidence of this increased reactivity. For example, the peak enthalpy of reaction of the non-catalyzed system is 206 °C. Addition of 1 % stannous octoate reduces this enthalpy point to 198 °C. When 0.5 % of the amidine catalyst is also added, the peak enthalpy is decreased by a further 35 to 163 °C. Lower levels of amidine give reactivities between the high and low peak enthalpy values.

The mechanism of **amidine catalysis** is not fully understood, since it only brings about the cleavage of the uretdione ring, while tin catalysts, as mentioned earlier, only accelerate the subsequent reaction between the NCO and OH groups. A direct reaction between the OH groups and the uretdione to form allophanate, such as occurs in the case of aromatic uretdiones of low molecular weight, could not be found by the Bayer inventors. Instead, the cyclic amidine molecule attaches itself to the four-membered ring of the uretdione forming a temperature sensitive six membered ring.

A disadvantage of the use of the before mentioned **amidine catalyst** is its **yellowing tendency**. Therefore *Lehmann* and *Gedan-Smolka* [41] of the German Institute of Polymer Research, Dresden, decided to compare different types of catalysts and to examine the influence of 1 % catalyst additions to a combined system of a uretdione containing hardener and 1,8-octanediol (as a model of a simple dihydroxy compound) in a molar ratio of 1:2. Best catalytic effects were detected with metal-acetylacetonates. Further tests have shown, that the catalyzed reaction between uretdione and OH-groups, using **zinc(II)-acetylacetonate, generates an allophanate structure** at temperatures below 150 °C, with a second reaction (measured by DSC) at temperatures above 160 °C, converting the allophanates into urethanes.

These results were used to formulate a powder coating, based on a hydroxyl functional polyester with OH value of 30 mg KOH/g, to be cured by IPDI-uretdione based hardeners, and catalyzed by 1 % zinc acetylacetonate. DSC measurements showed, surprisingly, a reduction of only 10 to 20 K, in contrast to the results with the model tests. This anomalous behavior is caused by the carboxylic end groups of the polyester resin (acid value 5 mg KOH per g), which deactivate the catalytic effect. To overcome this, a small amount of a suitable epoxy compound was added, such as “Araldite PT 910”, to block the carboxyl groups during stoving. Benzyltrimethyl ammonium chloride was added to act as a catalyst for the reaction between epoxide and carboxylic groups.

Best results are obtained at the 1 % level of zinc-acetylacetonate, when stoved at 130 °C for 30 minutes or 140 °C for 15 minutes. In both cases *Erichsen* indentation values of more than 9 mm are measured. The most suitable reverse impact (80 inch · lbs) is found at 15'/150 °C.

This invention (protected by patent applications of the Dresden institute [42, 43]) opens realistic curing temperatures of 130 to 140 °C, which, after further optimizing of the binder component and the uretdione hardener, can even be lowered [44]. Use of a crystalline resin helps to **minimize** the **melt viscosity** of the formulation to ensure good flow even at low temperatures [45]. It has to be taken into account that in the case of allophanate generation the uretdione hardener hasn’t the usual functionality of 2, but more or less 5 to 8, which in some cases might **interfere with** a sufficient **leveling**. Bayer [46] recommended 0.5 to 2.0 % carboxylic acid containing resins for better flow, due to the retarding effect of acids on the crosslinking reaction. The same company filed a patent [47] claiming that flow can be improved by the addition of monoalcohols, too, due to the decreased network density. Thermoforming of aluminum sheets, which change their crystal structure at higher temperatures, was successful with a catalyzed uretdione based powder coating crosslinked at a temperature below 160 °C [48].

In 2006 *Hsieh* et al. [49] presented a non-yellowing catalyst for uretdione based formulations. A rather high concentration of 5 % of this non-disclosed catalyst is claimed to lead to a curing time of 20 min at 150 °C.

Zinc acetylacetonate leads to allophanates at lower temperatures, consuming the double amount of precious uretdione crosslinker. Therefore Degussa [50, 51] (now Evonik Industries) introduced **alternative low temperature curing catalysts for uretdiones**, generating predominantly urethane bonds. These catalysts consist of tetraalkylammonium hydroxides or carboxylates preferably in combination with an epoxide which enhances the reactivity significantly. It is supposed that the reaction of both components at higher temperature leads – by ring opening of the epoxy group – to an alcoholate which acts as the true activating agent (*Figure 3.21*). The required epoxy group can even be covalently attached to

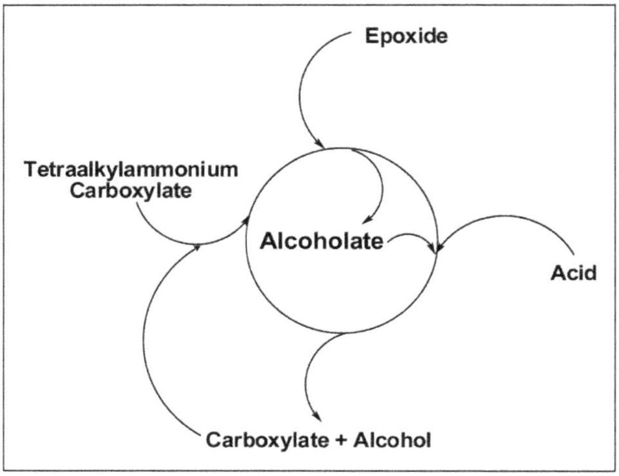

Figure 3.21: Reaction cycle catalyst activation [54]

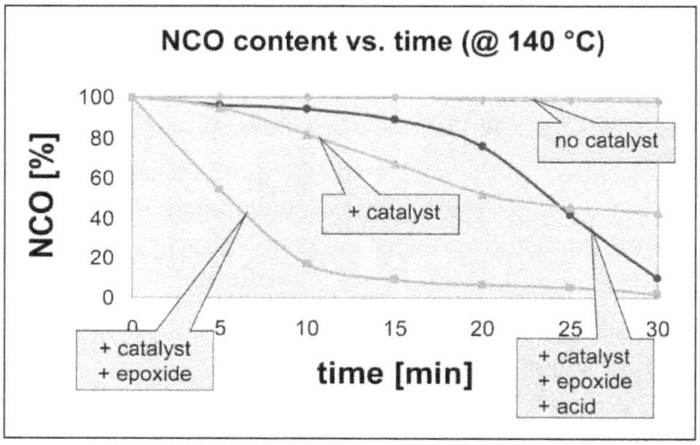

Figure 3.22: NCO content of uretdione based formulations vs. time dependent on the catalyst composition [54]

the uretdione hardener [52]. Thus a combination of an OH-polyester, a crosslinker ("Vestagon BF 1320", Evonik Industries) and 0.5 % of tetraethylammonium benzoate was applied on aluminum sheets and crosslinked at temperatures 120 to 140°C [53]. Molding at room temperature showed no microcracks or decreasing gloss. These weatherstable materials could be applicable for automotive or electronic industry. For **prolonged storage stability** (2 to 3 weeks at 40 °C) of these low temperature curing formulations high T_g (>60 °C) polyesters resins should be used [54], in particular with secondary hydroxyl groups. The NCO content of a formulation depending on the addition of catalyst, epoxy component and acid is shown in *Figure 3.22.*

3.3.2.3 TDI derivatives

Toluene diisocyanate is the **cheapest diisocyanate available** on the market. The commercial product is a colorless liquid with a boiling point of 120 °C at 10 mmHg, and it is 80:20 blend of the 2,4 and 2,6 isomers.

2,4-TDI 2,6-TDI

The NCO group in position 4 is 8 to 10 times as reactive as that in position 2 at room temperature. However, with increasing temperature the reactivity of the ortho group increases at a greater rate than that of the para group. At 100 °C the reactivity of both isocyanate groups is the same.

The difference in the **rates of reactivity** between the isocyanate groups is a specific advantage of TDI since it allows preparation of blocked adduct with a much better defined structure than would be the case if the reactivities were equal. Namely in a reaction with an equimolar amount of caprolactam the main product will be half- blocked TDI with a predominantly free isocyanate group in the 2nd position. This intermediate can then be used for condensation with different polyols or polyamines to produce caprolactam blocked TDI oligomers with a well-defined structure.

Although the price and its suitability for preparation of blocked oligomers with tailored structures and properties are strong points of TDI, its **use in powder coatings is very limited** because of its sensitivity to UV light and emphasized yellowing. Powder coatings for indoor application where the whiteness of the coating is not of primarily importance are typical products where TDI blocked derivatives are used as crosslinkers.

3.3.2.4 TMXDI and other polyisocyanates

Two new isocyanates, m- and p-tetramethylxylene diisocyanates (m- and p-TMXDI), developed by American Cyanamid Company (now Cytec) are potentially interesting materials for making crosslinkers for powder coatings. They have **characteristics typical of aliphatic diisocyanates**, even though they contain aromatic rings. This is because the NCO groups are shielded from the ring by protective methyl groups [55]. They are made by a non-phosgenation process [56, 57]. The structural formulas of m-TMXDI and p-TMXDI are the following:

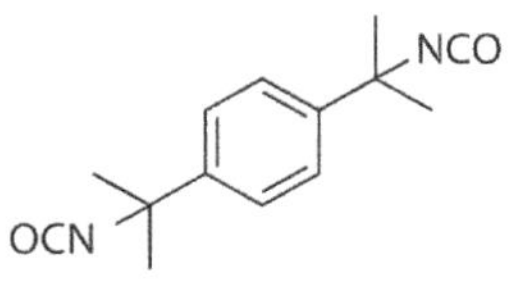

p-TMXDI

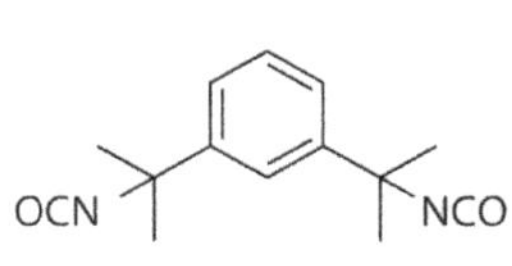

m-TMXDI

m-TMXDI is a liquid at room temperature with a melting point of -10 °C and boiling point of 150 °C at 4 mbar. **p-TMXDI** has the same boiling point, but it is a solid material at room temperature with a melting point of 72 °C.

The reactivities of m- and p-TMXDI with hydroxyl groups are somewhat lower compared to the other aliphatic isocyanates.

A large number of catalysts are examined for improving the reactivity of the TMXDI isomers in the NCO-OH reaction. It is found that the amine catalysts are relatively ineffective as compared with metal salts such as dimethyltin dilaurate, or tetrabutyl diacetyl distannoxane.

TMXDI isomers cannot be used as crosslinkers for powder coatings in the free form. They have to be **blocked** in order to prevent storage instability and viscosity increase during processing of the powder coatings. In order to increase the functionality and to decrease the vapor pressure of m-TMXDI it has been reacted with trimethylol propane to produce a crosslinker with a functionality of three. The use of the trimethylol propane adduct of m-TMXDI in two component acrylic coatings has been reported by *Fiori* and *Dexter* [58]. Since the m-TMXDI/TMP adduct is relatively rigid, it is necessary to select binders that have a lower glass transition temperature and, therefore, are softer and more flexible, in order to compensate for the rigidity of the crosslinker.

An advantage of TMXDI isomers is their relatively lower **toxicity** compared to the conventional isocyanates. For example, it has been found that m-TMXDI does not produce pulmonary sensitization at inhalation doses that cause acute irritation effects [59]. This behavior is assigned to the tertiary character of the isocyanate group in m-TMXDI in contrast to the primary and secondary isocyanate groups in the conventional isocyanates. The use of TMXDI in powder coatings has been on a very low level, up to now.

Although **caprolactam blocked isocyanates** are the most widely used for powder coatings, the possibility of using other blocking agents or other isocyanates, is also described in the patent literature. A patent of Hüls [60] (now Evonik Industries) describes a new flexibilizing hardener, based on a blend of trimerized IPDI and trimerized HDI (hexamethylene diisocyanate), blocked with caprolactam, a so-called "**hybrid hardener**", that improves the flexibility of acrylic urethane and super-durable polyester urethane binders. Products formulated with these binders and cured with the "hybrid hardener", show **improved mandrel bend flexibility** and very good stone-chip resistance. They are, therefore, recommended for automotive clear coats.

DuPont recently filed a patent [61] describing crystalline **HDI-based blocked polyurethane** hardeners. They claim that these crystalline components lead to better flow and higher flexibility. This effect can be enhanced even more, if the corresponding resins are crystalline, too [62].

Sullivan and *Weinberg* [63] claim performance advantages of **H_{12}MDI** as a crosslinker for polyurethane powder coatings. H_{12}MDI is bis-4-isocyanatocyclohexyl-diisocyanate. Due to its cycloaliphatic rings the H_{12}MDI monomer provides **increased flexibility** and **toughness**, compared to HDI and IPDI, and slightly higher reactivity, requiring lower stoving temperature than IPDI, because H_{12}MDI has two secondary isocyanate groups, while IPDI has one secondary and one less reactive primary isocyanate group. Due to the sound performance increased use in powder coatings is predicted.

The patent of *Takeda Yakuhin Kogy* [64] describes the use of 1,3,5-tris(isocyanatomethyl) benzene or 1,3,5-tris (isocyanatomethyl)cyclohexane of the formulas given below, blocked with methyl ethyl ketoxime.

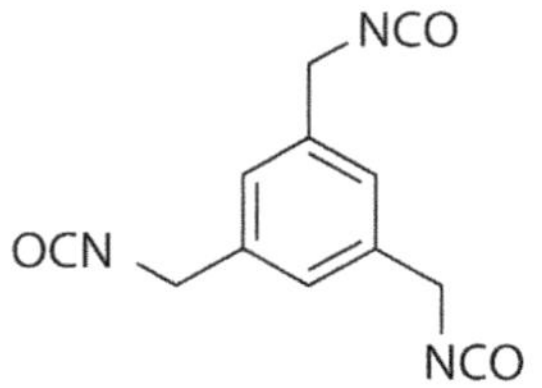

1,3,5-tris(isocyanatomethyl)benzene

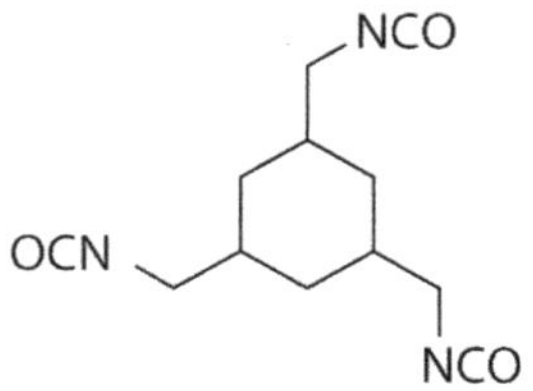

1,3,5-tris(isocyanatomethyl)cyclohexane

It is claimed that in powder composition using hydroxyl functional polyester as a binder, the product gives a coating film free of yellowing and blistering, combined with good outdoor durability, chemical resistance, mechanical properties and storage stability. Moreover, the **low deblocking temperature** of the ketoxime blocked isocyanates allows curing of the coating at 160 °C.

An Eastman Kodak patent [65] discloses the use of adducts of 1,4-cyclohexane bis-methylisocyanate with compounds containing active hydrogen as crosslinkers for powder coating compositions. Specifically, the patent claims the use of oximes such as acetone oxime, for example, which enables a powder coating composition to be made with a **very low curing temperature** of 140 °C.

Cooray and *Spencer* [66] have reported a dissociation temperature of **pyrazole blocked aliphatic isocyanates** of 120 °C. Although there is no evidence that these crosslinkers have been used in powder coatings, they could be potentially very interesting hardeners for producing low temperature curing powders. Recently Bayer [67] patented a crystalline pyrazole blocked nonane triisocyanate. This crosslinker is trifunctional and has a low melting viscosity.

A solid ketoxime blocked isocyanate being a reaction product of acetone oxime and 1,4-cyclohexane bis(methylisocyanate) suitable for manufacturing powder coatings with low curing temperatures is described in the mentioned patent of Eastman Kodak [65].

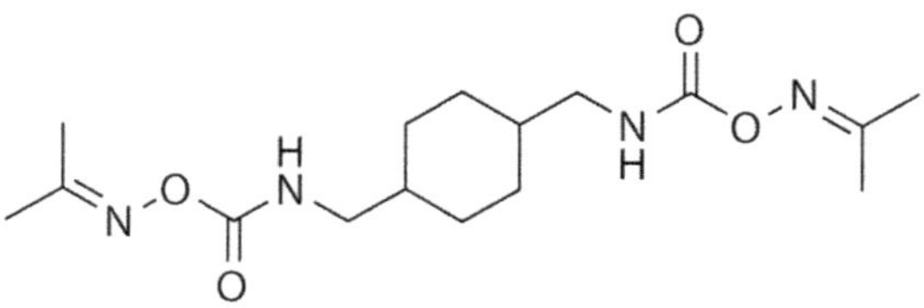

It is reported that in combination with hydroxyl functional polyester gel time of 18.5 minutes at 140 °C is obtained. In the presence of tin catalyst the gel time is reduced to 7.5 minutes, which does not affect the stability of the system.

3.3.3 *Polyamines*

3.3.3.1 *Dicyandiamide and its derivatives*

Dicyandiamide (DICY) or 1-cyanoguanidine is most widely used **crosslinking agent** for epoxy powder coatings. It is a white solid crystalline compound which is produced from calcium cyanamide according to the following reaction:

Equation 3.76

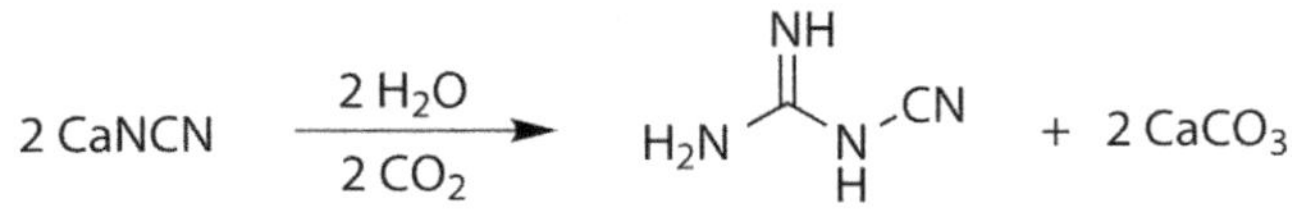

The amino hydrogens present in the structure of DICY can react with the oxirane rings of epoxy resins. The curing mechanism of DICY/epoxy resin is complex involving addition reactions with both epoxide and hydroxyl groups. DICY also catalyses considerably the epoxide-hydroxyl reaction promoting the self-crosslinking of the epoxy resins, having an overall functionality of 5 to 6. Therefore, in the case of DICY and its derivatives there is **no need for strict stoichiometric** proportions between the curing agent and the binder in the coating composition for obtaining infinite network formation, as is the case with most crosslinker/binder systems in powder coatings.

The **melting point** of DICY (211 °C) is much higher than the extrusion temperature during the compounding process. On the other hand its solubility in the binder is relatively low. Therefore, the reaction rate between the amino and epoxy groups during curing depends on the availability of reactive groups on the surface of the DICY particle. For this reason commercial grades of DICY are usually delivered in fine micronized form with the particle size below 75 microns [68, 69]. The lower availability of reactive groups because of the high melting point and low solubility of DICY must be compensated by a slight excess of hardener in the coating composition. Even the excess of DICY does not improve the reaction rate. For obtaining acceptable curing schedules, a proper **catalyst** has to be used. The catalysts are incorporated in the hardener during its manufacture. Most of the DICY producers supply the so-called "accelerated dicyandiamides" providing different levels of reactivity. In this way curing schedules of 15 min at 155 °C can be achieved [70] that do not affect the storage stability of the system very much; stability is only slightly reduced due to the insolubility of DICY in the binder.

Although the **reaction rate** can be controlled and improved by proper choice of a catalyst, several problems still can be addressed to DICY/epoxy resin system. The **insolubility of DICY** creates inhomogeneity problems which are manifested in low gloss, poor flow and in extreme cases non-uniform appearance of the film after curing. DICY itself is a water soluble compound, and often increases the water sensitivity of the cured films. For these reasons, DICY has been chemically modified to produce a family of curing agents which give better flow, high gloss and very low water sensitivity. These so-called **substituted dicyandiamide curing agents** whose general formula is presented below, are characterized by good solubility in the epoxy resins, providing in this way a homogeneous powder coating composition, that does not suffer from the typical disadvantages of pure DICY systems.

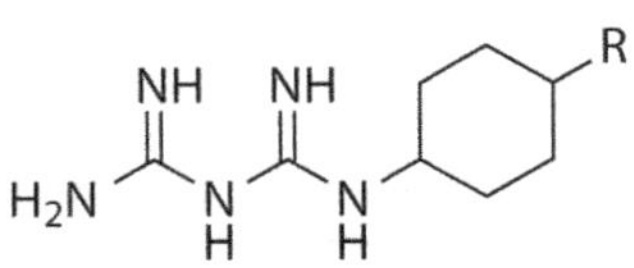

Since the solubility of the substituted DICY in the binder is much better than DICY itself, it will affect the **storage stability** of the system. Therefore, a good compromise has to be found between the desirable curing schedule and acceptable storage stability by choosing a suitable amount of the correct catalyst. In properly formulated coatings, curing schedules of 15 minutes at less than 160 °C can be obtained combined with an acceptable storage stability.

3.3.3.2 *Modified aromatic and aliphatic polyamines*

The majority of primary and secondary aliphatic and aromatic amines, even those that are solids at room temperature, cannot be used as hardeners for epoxy powder coatings because of the extremely high **reactivity** of the system. This creates considerable problems during the processing in the extruder, and dramatically decreases the **storage stability** of the coating. Therefore even a (usually not easy to handle) dryblend process has been proposed in a patent [71]. However, several types of amine derivatives have been developed which provide the necessary stability when combined with solid epoxy resins in a powder coating composition.

Certain aromatic diamines can be used as hardeners for epoxy powder coatings. Methylene dianiline, for example, the formula of which is presented below, has been claimed as a hardener for epoxy powder coatings [72].

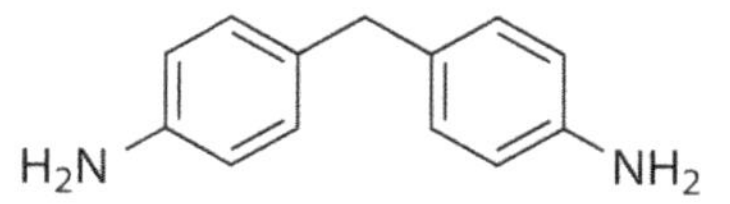

It is a solid compound at room temperature with a melting point of 90 °C and an active hydrogen equivalent of 49.5 %. Epoxy powder coatings formulated with this hardener possess excellent chemical resistance, particularly to mineral acids and bases, and good retention of the properties at elevated temperatures.

It is not common practice to use low molecular aromatic diamines as crosslinkers for powder coatings even when they are solid at room temperature, because of their relatively high vapor pressure at elevated temperatures combined with **considerable toxicity**. Therefore, they are usually modified by a reaction with monoglycidyl phenolic ethers [23, p.677] in order to increase their boiling point or by diglycidyl ethers making resinous adducts [73].

In general, **aromatic polyamines** and their derivatives give systems with excellent reactivity, extremely good chemical resistance, but with a much emphasized tendency towards yellowing. Another weak point of aromatic amines and their derivatives is their very questionable physiological activity because of **toxicity**. This makes them very inconvenient hardeners for a system that is claimed to be environmentally friendly.

A group of aromatic amine curing agents that are mono-salts of **aromatic polycarboxylic acids** and **cyclic amidines** (e.g. 2-phenyl-2-imidazoline) are described in several patents of Hüls (now Evonik Industries) [74, 75]. These hardeners are obtained by reaction of equimolar amounts of pyromellitic or trimellitic anhydride previously hydrolyzed with water to obtain their acid, and 2-phenyl-2-imidazoline. Combined with epoxy resins, pigments and flow agents in a powder coating composition, after curing they produce **matt films** with excellent flow and a very low gloss of only 9 % (at an angle of 60°).

The mechanism of this “chemical” matting can be explained as follows: The amidine is not present in the amount necessary for the full hardening of the epoxy resin; only a pre-crosslinked, but still flexible and glossy coating would form first by reaction with the amine. During a further rise in temperature, the less reactive polycarboxylic acid will also react with the epoxy resin. Due to the polyfunctionality of the acid hardener, this **second crosslinking** would be expected to occur with a strong contraction of the molecules. But this is counteracted by the fact, that the epoxy resin was already pre-crosslinked and the chain molecules possessed only restricted mobility.

A further crosslinking can therefore not be compensated by a uniform shrinkage of the film thickness. By consequence the inhibition of shrinkage will lead to uniformly distributed points of retraction in the film surface and thereby cause the matting effect.

Two matting hardeners, belonging to this group of amine mono-salts, used to be sold by Hüls (now Evonik Industries) under the trade names “Vestagon B 55” and “B 68”.

The **"double curing" mechanism**, as a means of producing matt powder coatings, can also be applied to polyester-based powder coatings. See therefore the separate Chapter 3.6.

Aliphatic polyamines that differ from DICY, produced by condensation of aliphatic polyvalent amine and bisphenol A epoxy compounds, are described in the patent literature. For example, the condensation product of isophorone diamine and diglycidyl ether of bisphenol A with low molecular weight ("Epikote 828") in the ratio of 2:1 with a melting point of 55 to 57 °C is used in combination with high molecular weight epoxy resins as the curing agent [76]. It is claimed that the hardener can be mixed in the melt with an epoxy resin homogeneously, without gelation, providing extremely short curing times.

In the patent of Yuka Shell Epoxy [77] a method of producing dicarboxy-polymethylene diamine of formula $H_2N\text{-}(CH_2)_n\text{-}OCO\text{-}R\text{-}OCO\text{-}(CH_2)_n\text{-}NH_2$ is described involving esterification of aminododecanoic acid and aliphatic diols with two to six carbon atoms. A modification of the process also involves diols derived by condensation of low molecular weight epoxy resins with phenol followed by esterification of the formed secondary hydroxyl groups with amino acids. In this way, different diamines having a melting point in the range between 50 and 140 °C can be obtained.

3.3.4 Polyphenols

Conventional thermosetting phenolic resins of both resol and novolac type can be used in principle as hardeners for epoxy powder coatings. They are prepared by step-growth polymerization of phenols with aldehydes. This involves electrophilic substitution of the aromatic ring to form a phenol alcohol, followed by condensation reaction between the methylol groups leading to the formation of a high molecular compound. Phenol and formaldehyde are by far the most important raw materials for production of **phenol-formaldehyde resins** in general, but other types of phenolic compounds such as cresol and diphenylol propane are also reported to be used for the manufacture of phenolic resins for powder coatings [78, 79].

Although the condensation reaction between phenol and an aldehyde had already been described by Bayer in 1872 [80], the commercial production of these resins started in 1910 after the work of *Baekeland* who developed formulations and an economical method for manufacturing molded parts using phenol formaldehyde resins as binders.

Two different types of product can be obtained by condensation of phenol with formaldehyde depending on the reaction conditions. The reaction between phenol and formaldehyde in alkaline conditions or slightly acidic conditions at a pH above 5, results in a formation of mono, bis and tris alcohols, as well as in many other rather complicated compounds:

Equation 3.77

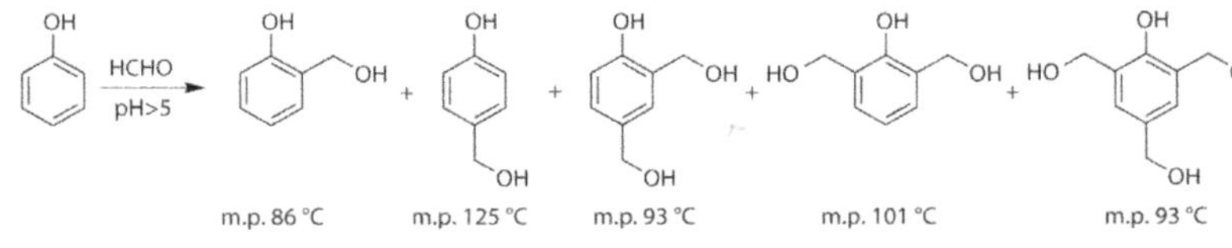

Under alkaline conditions, the formed methylol phenols undergo a condensation reaction forming a polymer. Based on the experiments with model compounds it has been proposed that the following two reactions dominate in this stage of the process [81, 82]:

Equation 3.78

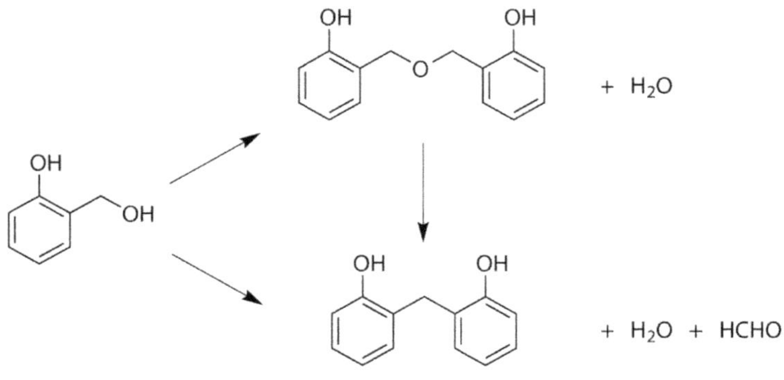

Yeddanapalli and *Francis* [83] have shown that under strong alkaline conditions the reaction yielding diphenylmethanes with splitting formaldehyde as a by-product is the dominating one. However, under weak acidic or neutral conditions, the condensation proceeds through a formation of dihydroxy benzyl ethers.

Under strongly acidic conditions, it has been proposed that the reaction between phenol and formaldehyde is an electrophilic substitution [80, p.47] leading to diphenylmethanes, too.

Depending on the molar ratio between the starting raw materials and the reaction conditions, **two different types of phenol formaldehyde resins** can be produced.

In an excess of formaldehyde and under alkaline conditions the products are branched resins commonly known under the name **resols**. The mole ratio between formaldehyde and phenol is between 1:1 and 3:1. Catalysts used commercially include sodium hydroxide, sodium carbonate, alkaline earth oxides and hydroxides, ammonia and tertiary amines.

Under acidic conditions and a slight excess of phenol (formaldehyde/phenol mole ratios between 0.75 and 0.85) **novolac** types of linear phenol formaldehyde resins are obtained. Oxalic acid is the most widely used catalyst giving low colored products.

The relatively cheap and well-established process of production of phenolic resins in combination with conventional raw materials used for their production, which belong to the typical commodity chemicals, makes them very attractive crosslinkers from an **economical** point of view. On the other hand, bisphenol A epoxy resins when crosslinked with phenolic hardeners produce coatings with exceptionally high chemical, solvent, detergent and boiling water resistance. These properties can not be achieved to such an extent by any other binder/crosslinker system.

Several disadvantages are encountered when phenolic resins are used as crosslinkers for powder coatings. The **yellowing** resistance of the system is rather poor which makes them less suitable for white powder coatings. The presence of the aromatic ether group dramatically decreases their UV resistance. Therefore their use is exclusively limited to indoor application. On the other hand, the curing reaction between oxirane rings of the epoxy resin and phenolic hydroxy groups is rather slow, giving cured films that are in general brittle and have poor impact resistance.

The latest developments in the novolac phenolic resins specially designed to be used as crosslinkers for powder coatings and the new catalysts for increasing the curing rate have overcome some of the problems mentioned above. As a result commercial crosslinkers have appeared that are characterized by the following properties:

- A high degree of compatibility with the bisphenol A epoxy resins.
- The same melting range as that of the epoxy resins, thus providing excellent processing properties of the coating.

- A reactivity that allows **curing at low temperatures**, or the formulation of very fast curing systems (5 min at 180 °C).
- Sufficient latency to ensure safe compounding at high temperature.
- Sufficient film flexibility to meet normal customer requests.

Crosslinkers of the phenolic type can be used for decorative purposes as well as for functional coatings, providing excellent corrosion, solvent and chemical resistance.

Crosslinkers containing phenolic functional groups can be obtained by condensation of epoxy resins with a molar excess of phenolic compounds. However, this is a rather expensive process of making phenolic crosslinkers. An interesting modification of such a process in order to make the system flexible by introducing aliphatic chains is described in a patent assigned to Shell Oil Company [84]. According to this patent phenolic curing agents specifically suitable for use in powder coatings, are prepared by reacting dihydric phenol (diphenylol propane), diepoxide (low molecular weight diglycidyl ether of diphenylol propane) and dicarboxylic acid (adipic acid for example). The phenolic functionality is in excess, exceeding the difference between the epoxy and acid functionality. In one example of the patent a mixture of 2 moles of bisphenol A epoxy resin, 2 moles of bisphenol A and 1 mole of adipic acid are reacted in the presence of tetramethyl ammonium chloride as catalyst for 1 hour at 140 °C. In these conditions the whole amount of carboxyl groups has been consumed in the reaction with the oxirane rings. The second reaction between the epoxy and phenolic groups is then carried out at 160 °C for 2 hours. A clear brittle product with a phenolic hydroxyl content of 230 meq/100 g with a melting range between 75 and 90 °C is obtained.

Most of the epoxy resin producers have developed phenolic crosslinkers specially matching their resins. All of the **commercially available phenolic crosslinkers** have almost the same hydroxyl equivalent weight, ranging between 240 and 260 and relatively low softening point between 65 and 75 °C [85–87]. They are characterized by a very low melt viscosity, which is at 150 °C usually below 1 Pa s, in most cases between 100 and 400 mPa·s. The producer usually incorporates the curing catalyst during the manufacture of the phenolic crosslinkers. They are delivered to the market with different degrees of reactivity, very often offering extremely short curing schedules of 1 min at 230 °C to 15 min at 150 °C. Despite the high reactivity, it is possible to formulate powder coatings with acceptable storage stability.

3.3.5 Acid anhydrides

Acid anhydrides can be used as **curing agents** for powder coatings having epoxy resins or hydroxyl functional polyester as binders. In the case of epoxy binders, acid anhydrides containing only one anhydride ring can lead to an infinite network formation, because after the ring opening esterification of the anhydride group, the resulting carboxyl group can react with the epoxy group of the binder. In the case of hydroxyl functional binders, only di- or polyanhydrides can be suitable curing agents.

In principle all low **molecular weight** acid anhydrides such as phthalic anhydride, maleic anhydride, and trimellitic anhydride can be used for curing epoxy resins, or resins containing epoxy groups.

Trimellitic anhydride (TMA) was widely used as a hardener for epoxy composites. In 2008 the world wide consumption of **TMA** was over 200 kt. In applications such as powder coatings, powder particles, which are always present in the air surrounding the spraying booths, and the large open surface to be coated, which emits a considerable amount of trimellitic anhydride during curing, restricted the use of trimellitic anhydride as a curing agent because of its **toxicity** and **irritancy**. The same holds for the other potentially useful hardener pyromellitic dianhydride (PMDA).

Preparation of dianhydrides from trimellitic anhydride and polyols or amines, as curing agents for epoxy compounds, is described in a patent of the General Electric Co. [88]. Resinous dianhydrides can be obtained, starting from trimellitic anhydride and diols or diamines. The process proceeds at sufficiently high temperatures (230 to 300 °C), to remove water, resulting in an intermolecular trans-esterification. During the reaction a tetracarboxylic acid is formed as an intermediate product, which under vacuum at 260 °C undergoes transformation to a dianhydride. A modification of this patent [89] involves a one-step procedure for obtaining the same products. Common commercially available glycols can be used as diols. The process can be performed at much lower temperatures up to 200 °C, starting from dialkyl ester of different diols (ethylene glycol diacetate, for example) and trimellitic anhydride. The formation of the desired compound goes through a process of acydolysis, taking place on the carboxylic acid radical, which leaves the anhydride radical intact [90].

A patent of Rhone Poulenc Ind. [91] describes a powder coating composition comprising polyester resins with terminal hydroxyl groups with a hydroxyl value of 150 to 250 and a softening point of 70 to 130 °C. The crosslinking agent is a reaction product of trimellitic anhydride and diol, having a softening point between 60 and 130 °C. The dianhydride curing agent is prepared by reacting trimellitic anhydride with C_2-C_{10} aliphatic diols at 180 to 250 °C, with a mole ratio for anhydride/diol of 2:1. The curing process is claimed to be rather short, taking 5 to 30 minutes at 140 to 200 °C.

It is reported that pure ethylene glycol bistrimellitate (EGBM) is a crystalline material with melting point of 166 °C [92] and, therefore, under normal extrusion conditions complete homogenization with polyester resins is not possible. Reduction of the melting point of **EGBM** can be achieved by making oligomers with higher molecular weight. In this way the crystallinity of the product is destroyed, and a material with a melting range below 120 °C can be obtained.

The high solubility of the amorphous oligomer of EGBM, however, creates another problem because of its high reactivity with the hydroxyl groups during processing. By careful control of the temperature below 120 °C and short residence time in the extruder, relatively safe processing can be performed without **gelation** problems. It is reported that this material increases (by roughly five times) the viscosity of the powder coatings during 10 minutes of processing at 120 °C compared to the initial viscosity. It is considered that this is an acceptable value for obtaining powder coating compositions that exhibit good flow [92].

Although the technical problems connected with the use of acid anhydrides as crosslinkers for powder coatings are solved, **toxicity considerations** still remain a permanent obstacle for wider acceptance of these systems by the market. It is reported by *Schmidt* and *Fink* [93-95], that for automotive clear topcoats, based on epoxy functional acrylates, a polyanhydride of an aliphatic dicarboxylic acid (such as dodecane dicarboxylic acid) as crosslinker gives very promising results. This clear topcoat should equal or outperform both the one and two component solventbased clear topcoats in many aspects, being inferior only with respect to the solvent resistance.

This polyanhydride is a partially crystalline, non-toxic compound with a melting point of 80 to 90 °C, acid number of 300 mg KOH/g and mean molecular weight of 2000.

3.3.6 Amino resins

Amino resins were used for a short period in the beginning of the 1970's as crosslinkers for poly-ester powder coatings. The appearance of TGIC/polyester systems swept them from the market, so there is **no evidence** that at this moment there are **commercial systems** based on amino resins as crosslinkers. In the solventbased coatings, however, amino resins play an extremely important role in formulating paints for industrial applications, which include the market for household equipment, general industrial applications, coil coating, can coating, tube coating and automotive coatings. These crosslinkers offer an **outstanding price/performance ratio** combined with relatively low toxicological and health hazard considerations. This is a good enough reason to include them in the group of crosslinkers for powder coatings, in order at least to stimulate powder coating chemists to consider them as a possibility, something which has been neglected for years.

Melamine, benzoguanimine and glycoluril are the main raw materials for condensation with formaldehyde to produce crosslinkers with potential use in powder coatings:

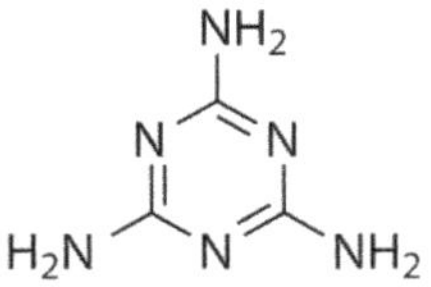

Melamine

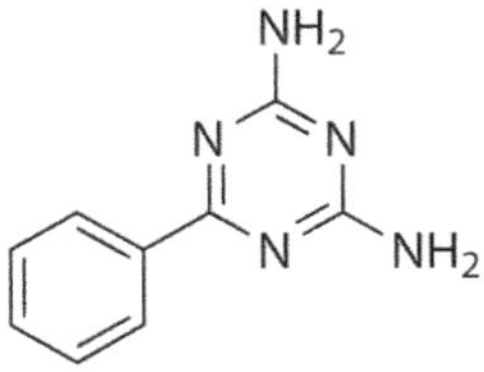

Benzoguanamine

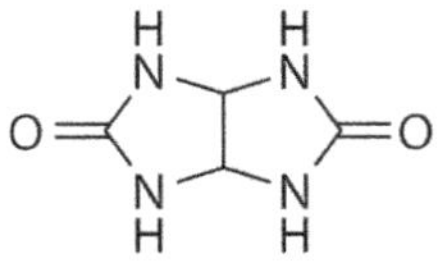

Glycoluril

Two main reactions are involved in the formation and cure of amino resins. The first is a simple addition of formaldehyde to the amino compound forming a methylol intermediate. This reaction is also known as methylolation and may be catalyzed by either acids or bases. The methylol compounds formed are relatively stable under neutral or basic conditions. In the second step in the production of amino resins a condensation reaction between the methylol intermediate and amino compound proceeds, forming a methylene bridge and liberating a molecule of water as a by-product. Unlike the methylolation reaction, the condensation reaction is catalyzed by acids only (see also Chapter 3.1.7).

Because of their high reactivity leading to unstable products and generally poor compatibility with most of the binders used in the coating industry, methylol products of melamine, benzoguanimine, urea or glycoluryl are not used as crosslinkers. Etherification (or the so-called alkylation) with an aliphatic alcohol is the usual modification, which overcomes the stability and compatibility problems.

The reactions of methylolation, condensation and etherification are only an oversimplified presentation of the chemistry involved in the manufacture of amino resins. In fact it is a rather complicated process in which, in general, all of the three reactions proceed simultaneously, involving very complicated kinetics. Therefore, especially for melamine and benzoguanimine resins, it is very difficult to assign an exact mechanism of the resin formation. For details concerning the kinetics of the process which determines for a great extent the type of resins obtained and their properties, a considerable number of published research papers can be consulted [96-115].

The evaluation of products known under common name of HMMM types of melamine resins (hexamethoxymethylmelamine) indicated a remarkable superiority over the butylated melamine. HMMM enamels have excellent gloss, without the characteristic haze of the butylated melamines, and in addition improved flexibility, adhesion, impact resistance with equal or better chemical resistance and exterior durability [116]. At the same time, HMMM resins impart equal film hardness at half the level of butylated melamine resins. Since melamine resins are in general low molecular compounds with a relatively low glass transition temperature, they **decrease the T_g of the powder coating formulation** remarkably. Much lower amounts of HMMM resins in the coating composition allow the formulation of powder coatings with a T_g high enough to provide good storage stability. PPG published a patent [117] recommending carboxylic acid modified aminoplast resins, thus increasing the T_g and providing better storage stability.

Fully methylolated grades of methylol melamine are produced on a commercial scale with a solids content higher than 98 % and degree of polymerization ranging between 1.35 and 1.7 [116]. At room temperature they are highly viscous to waxy solid materials, which makes them **rather unsuitable for use in a powder coating** composition having in mind the present production technology.

Another weak point of HMMM is the **liberation of methanol** during curing. Since the boiling point of methanol is rather low, its rate of leaving the film during curing is very rapid. In thicker films this causes surface defects, usually manifested in the form of blisters. Therefore its use is limited to thin film applications. On the other hand BASF has published a patent [118] claiming oxazolidine blocked aminoplasts, which do not generate volatile byproducts upon curing.

An interesting way of obtaining solid melamine crosslinkers for powder coatings is described in a patent of *Sherwin-Williams* [119]. According to the patent, 1.8 to 2.2 moles of monohydroxyl phenol compounds are condensed with 1 mole of hexamethoxymethyl melamine in the presence of oxalic acid as catalyst. The methanol, which is a by-product of the condensation process, is removed by vacuum, after which a high viscous material is obtained, having upon cooling a softening point of 106 °C and a melting point of 116 °C. It is claimed that such hardeners can be used in combination with hydroxyl containing binders without having surface defect problems in coating layers up to 25 μm. At curing temperatures ranging from 95 to 205 °C films with good hardness, flexibility and gloss can be obtained. Recently DSM [120] showed an alternative way to increase the T_g of melamine. In a patent they described the use of an excess of formaldehyde in a reaction with HMMM followed by an alkylation step with an alcohol leading to a modified amino-formaldehyde resin with a T_g of 42 °C.

Fully **methylated monomeric benzoguanimine formaldehyde condensates** are solid compounds at room temperature, with a melting point around 90 °C depending on the purity. This makes them very suitable for incorporating in the powder coating composition using conventional extrusion procedure preceded with dry blending of the coating constituents. Benzoguanimine resins as a type of aminotriazine resins are **very popular crosslinkers** for can coatings, giving excellent sterilization resistance to the cured films. They are also widely used in the electro deposition coatings. The benzoguanimine structure contributes superior detergent and chemical resistances. As well as splitting off methanol during curing, which could cause the same blistering problem as in the case of HMMM, another weak point of the benzoguanimine resins is their rather poor exterior durability, which restricts their use to indoor applications only.

Toluenesulphonamide modified HMMM resins are described [121] as effective crosslinkers for hydroxyl functional polyester resins in powder coatings. Cured films have excellent solvent and impact resistance. A disadvantage is also here, that methanol is set free during stoving.

Tetramethylmethoxy glycoluril (TMMGU) with the formula presented below is another potential crosslinker for powder coatings [122-126].

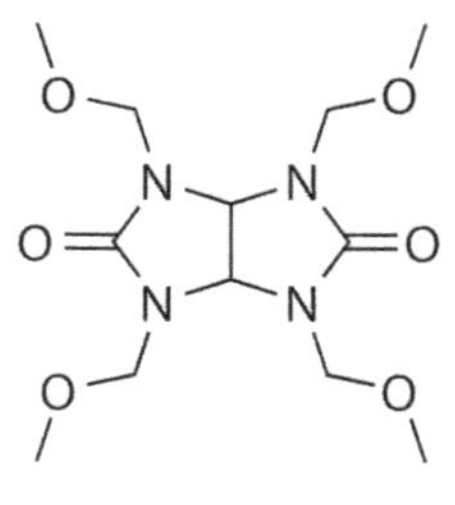

Tetramethylmethoxy glycoluril

TMMGU is a solid at room temperature with melting point of about 90 °C. It has been reported that coatings containing glycoluril resins have excellent mechanical properties, corrosion resistance, humidity resistance and good accelerated durability [112]. During baking, films with glycoluril-formaldehyde resins release much lower amounts of formaldehyde than melamine-formaldehyde containing films do. Compared to melamine resins, glycoluril-formaldehyde resins require stronger acid catalysis to achieve fast crosslinking reaction. At low stoving temperatures some of the titanium pigments depending on the nature of their surface treatment show inhibition in the cure response. This is especially the case of alumina-treated rutile titanium dioxide pigments. In case of alumina silica zinc-treated rutile titanium dioxide pigments the inhibition of curing is not that emphasized. By increasing the catalyst level it is possible to overcome the cure inhibition by the pigments, but as can be expected, the increased catalyst level has an adverse effect on the humidity resistance of the coatings.

TMMGU has been introduced on the market by Cytec (formerly Cyanamid) under the name "Powderlink 1174" and is mainly used to formulate attractive, outdoor durable and reproducible wrinkle finishes (by the use of special catalysts). In principle it could also be used as a TGIC alternative. During baking it releases methanol, which may be a limitation. Another disadvantage is, that "Powderlink 1174" lowers the T_g of the powder drastically, which leads to a limited storage stability.

A **biobased melamine powder coating** crosslinker has been claimed by PPG [127] generated by the reaction of methoxymethylmelamine formaldehyde copolymer and tall oil rosin.

Tris(alkoxycarbonylamino(triazine) (TACT), a new formaldehyde-free melamine crosslinker is recommended by Cytec [128] for curing of OH-polyesters as well as COOH-polyesters. One form of TACT is a monomeric mixture of four tris components, having butyl and methyl carbamate groups. Depending on the exact butyl/methyl ratio, the melting point of the TACT varies between 130 °C and 150 °C. Its equivalent weight is 120 to 130. There is no indication of toxicity concern. Coatings cured by TACT possess good environmental etch resistance.

Polyguanidines have been studied [129] as precursors for polycarbodiimides crosslinkers in polyester powder coatings. Polyguanidines are **"blocked" polycarbodiimides**, which thermally decompose into carboxylic acid-containing polymers to form polycarbodiimides, which then react with the carboxylic acid moieties of the resin to form a crosslinked film.

The crosslinking efficiency of polyguanidine depends on the deblocking efficiency of that polyguanidine to the polycarbodiimide, and the number of guanidine moieties per polymer chain. Increasing the number of guanidine moieties per polymer chain increases the crosslinking efficiency at the expense of image clarity. Increasing the level of a polyguanidine in carboxylic acid-containing coating decreases gloss. This is in contrast with polyester/TGIC coatings, where an increased level of TGIC improves gloss.

It was found, that coatings, containing polyguanidines, yellow more than coatings, containing the highest level of TGIC. An explanation is that the more intense yellow color for coatings crosslinked by polyguandines results from the presence of amino groups or the amine deblocking groups on the crosslinker. It was also found, that polyester/TGIC coatings have better **outdoor durability** than coatings, crosslinked with polyguandines.

3.3.7 Hydroxy alkyl amides (HAA)

At the end of the 1980's a new β-hydroxy alkyl amide (HAA) type of crosslinker (being a reaction product of one mole adipic acid and two moles diethanol amine) was introduced on the market by Rohm & Haas under the name "Primid XL 552". Some years later the sales and production of this crosslinker were taken over by EMS-Chemie, because they saw HAA as a possible **TGIC-alternative**. In 1996 EMS launched a second HAA under the trade name "QM1260". In 1998 Hüls-Creanova (now Evonik Industries) introduced also a HAA crosslinker under the trade name "Vestagon EP-HA 320".

The right chemical names for both HAA are:

N,N,N'N'-tetrakis (2-hydroxyethyl)adipamide (trade name "Primid XL-552" or "Vestagon EP-HA 320") and N,N,N'N'-tetrakis (2-hydroxypropyl)adipamide (trade name "Primid QM-1260" or "Vestagon EP-HA 376"). 320 and 376 refer to the molecular weight of these HAA.

The chemical structure of "Primid XL-552" or "Vestagon EP-HA 320" is as follows:

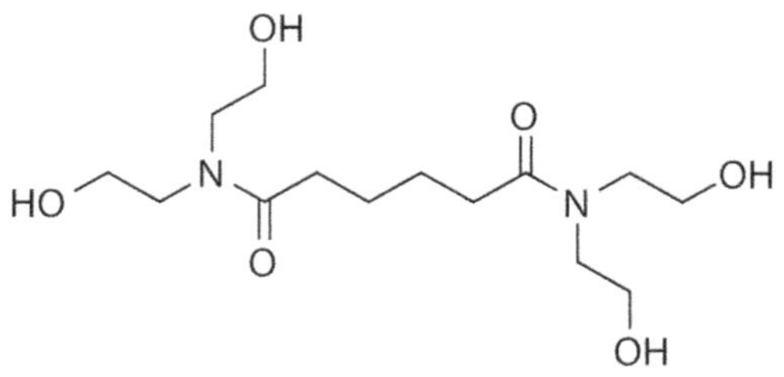

N,N,N'N'-tetrakis (2-hydroxyethyl)adipamide

The crosslinking mechanism is based on an esterification between the ß-hydroxyalkyl groups from the HAA and the carboxyl groups from the polyester, with water as a reaction by-product. In comparison with esterification of normal aliphatic alcohols, where temperatures of above 225 °C or catalysts are necessary, ß-hydroxyalkylamides already react with acid groups at relatively low temperatures, starting at 150 °C.

The **increased reactivity of ß-hydroxyalkylamides** with regard to carboxyl groups is due to the formation of an intermediate oxazolinium structure [130]. This mechanism was first put forward by *Wicks* et al. [131] and later confirmed by *Stanssens* et al. [132], using H-NMR-spectroscopy. *Wicks* also discovered, that in comparison to esterification with aliphatic alcohols (where the aromatic carboxylic acids reacted more quickly than their aliphatic counter parts), the reaction of aromatic carboxylic acids with ß-hydroxyalkylamides was slower than that with the aliphatic acids. Unfortunately the reaction progress can't be detected by DSC due to the evaporation of water. Therefore dielectric analysis (DEA) [133] has been recommended for kinetic studies. Microscopy tools have been used to follow the micro structure during film formation, thus enabling to predict the macroscopic appearance of the cured powder coatings.

The reaction of polyester with HAA appears to be unaffected by a wide range of conventional catalysts. The reason for this is that the **thermodynamic equilibrium** of the intermediate structure at the respective temperature is located totally on the "wrong" side, being pushed to the side of the reaction products predominantly by the evaporation of water. However, Merck [134] claimed that silane treated nanoscale zinc oxide is able to decrease gel time of HAA-formulations significantly. Whereas the pure system showed a gel time of 565 s, 1 % of nanoscale zinc oxide led to 235 s, 9.5 % even to 15 s. According to the same patent no influence of pure zinc oxide, acetate or stearate on the gel time could be detected. The observed result might rather be attributed to the crosslinking effect of the alkoxy silanes than to a catalytic activity.

Since the reaction of HAA can't be catalyzed, the only way to **adjust the reactivity** is to use different resins, e.g. carboxylated polyester with higher branching. Normally this would lead to a poor appearance due to a higher melt viscosity and shorter leveling time. Therefore the resin manufacturer had to find ways to decrease the melt viscosity of these highly branched resins containing HAA cured powder coatings, by designing resins with special raw materials or by using powerful additives. Find more to this topic in Chapter 3.4.2.2.

Due to their chemical and physical properties both HAA types are very well suited for use in powder coatings, according to *Kaplan* [130]. At first glance they have a similar structure. The only chemical difference is that the second product has additional methyl groups in α-position relative to the hydroxyl groups. Both HAA are white, crystalline solids with melting points around 120 °C.

The main differences are the **better color retention** of the additional methyl groups containing HAA compound in direct-fired gas ovens and its improved hydrolytic stability.

As has been mentioned, the esterification reaction between carboxyl-terminated polyesters and hydroxyalkylamides is carried out through an oxazolinium intermediate product and no catalysts or accelerating agents have yet been found [130]. This means that other components such as pigments, fillers and additives have no influence on reactivity during the manufacturing process. By consequence, this gives **good batch to batch reproducibility** and the reactivity remains at the same level for all colors. In practice this means a considerable time reduction during formulating, as no adjustment of the reactivity for different colors is necessary. From a toxicological point of view HAA is completely safe and can be considered as a **non-toxic material**, which needs no labeling.

HAA-containing powders will at first benefit most from the obliged TGIC hazard labeling for the following reasons:

- the hardener is relatively well-known (HAA exists nearly 30 years);
- the total binder system is - from a cost point of view - very attractive;
- the low curing temperature (160 °C is possible);
- the absence of labeling legislation.

However, several inherent disadvantages of HAA-based powders restrict their use. The liberation of water as a by-product of the curing reactions can lead to surface defects **(pinholing) in thick layer applications**. Therefore, the end users need to be trained in handling HAA-based powder coatings. Furthermore, HAA-based powder coatings are relatively sensitive to water and humidity under severe test conditions (e.g. elevated temperatures), though "QM-1260"/"EP-HA 376" gives here an improvement.

In addition HAA cannot be catalyzed, which narrows the formulating latitude for different applications, making it rather difficult to adjust the curing speed. But this "inertness" has also its advantage (see before).

Due to the presence of the amide groups, the yellowing (mainly in direct-fired gas ovens) and overbake resistance of HAA are not as good as with TGIC. Here too, "QM-1260"/"EP-HA 376" shows an improvement.

In general, **powder stability is good**, because HAA does not have a significant effect on the glass transition temperature (T_g), as it is insoluble in the polyester. The powder storage stability is therefore largely determined by the resin glass transition temperature. But this insolubility has also a drawback: it makes it impossible to formulate clear HAA-based coatings.

HAA-based powders are comparable to TGIC powders in terms of outdoor durability, if the right HAA-designed polyesters are chosen.

It should be emphasized, that HAA-based powders are more critical in terms of processing properties than TGIC powders. Therefore, extrusion characteristics should be determined carefully and premixing in particular is an important parameter, determining the final coating properties (such as good film flow).

As has been said, the main disadvantage of polyester and ß-hydroxyalkylamides based powder coatings is the split-off of water during the crosslink reaction, which can form gas bubbles, resulting in blisters and pinholes in the thicker coating area.

Degassing can be improved by slowing down the cure reaction, i.e. keeping a low melt viscosity for a longer time. Until ten years ago it was thought, that the reaction of HAA-based polyester powder coatings was not adjustable [130]. However, at the PCE 2002 it was shown by DSM resins [135] that special construction of the resin structure allows to adjust and to reduce the reaction speed, resulting in longer gel time, improved degassing behavior and thicker blister-free coatings of up to 180 μm, without sacrificing other properties such as gloss, impact resistance and durability.

Also at the PCE 2002 EMS introduced "Primid SF-4510", their newest ß-hydroxyalkylamides hardener. Due to its longer gel time and its unique rheological behavior, it features extremely good flow properties, improved degassing behavior, and is particularly suited for high gloss powder coatings, for coatings on degassing (porous) substrates such as cast aluminum and hot-dip galvanized steel, as well as for thick coat applications [136]. According to the manufacturer "Primid SF-4510" can be seen as a super flow hardener for outdoor resistant powder coatings. The chemical composition of this new HAA is based on benzoic acid and diethanolamine.

Recently AkzoNobel filed a patent [137] describing the reaction product of ε-caprolactone and diethanolamine as a lower cost substitute of conventional hydroxyalkylamides. Similar to "Primid SF-4510" it is difunctional and requires a branched carboxyl functionalized resin for crosslinking.

At the ECS 2011 Evonik Industries introduced a new cycloaliphatic hydroxyalkyl amide "Vestagon EP-HA 368" which can be used as a special one shot matting hardener [138]. This novel compound will be described further in Chapter 3.6 "Matt and semi-matt powder coatings".

3.3.8 References

[1] Ciba Ltd., GB Pat. 814 511, 1960
[2] Shell, NL Pat. 67.16.630, 1969
[3] Ciba, FR Pat. 1 570 826, 1970, GB Pat. 1 250 840, 1972
[4] *Budnowski, M.* Angew. Chem., 80, 851, 1968
[5] *Budnowski, M.* Kunststoffe, 55, 641, 1965
[6] *Merck, Y., Loutz, J.M.*, Proceed. XV Congress A.F.T.P.V., 1983, p. 183
[7] *Sastre ,J.; Carnes, J.; Garcia del Potro, M.; Manso, L.; Aguado, E.; Fernandez-Nieto, M.*, International archives of occupational and environmental health, 84(5), 547-9, 2011
[8] *White, J.*, Health and Safety Laboratory, Annals of Occupational Hygiene, 48(6), 555-563, 2004
[9] a) BASF, US 20040068060, 2004
b) *Edwards, P.E., Striemer, G., Webster, D.C.*, JCT Research, Vol. 2, 7, 517-527, 2005
[10] Ems-Chemie AG, WO 2008061687, 2008
[11] *Gottis, P.G., and Cotting, J.A.*, XXIII FATIPEC Congress book, Vol. B, Brussels, 1996, p. 216
[12] *Stanssens, D.A., Van de Werff, A.J., Molhoek, L., Van der Berg Jeths, R.*, Int. Waterborne, High-Solids and Powder Coatings Symp. New Orleans 1994, p. 653
[13] *Molhoek, L., Van de Werff, A.J., Stanssens,D. and Van der Berg Jeths, R.*, XXII FATIPEC Congress book, Budapest 1994, p. 1. 119
[14] *Misev, T.A.*, R&D Activities of DSM Resins on Powder Coatings, "Uranox" Seminar Amsterdam, 27-28 March 1996
[15] *Van der Linde, R.*, The Reaction of Aliphatic Oxirane with Carboxylic Acids, "Uranox" Seminar, Amsterdam, 27-28 March 1996
[16] *Stanssens, D.*, Coming to the Aliphatic Oxirane, "Uranox" Seminar, Amsterdam, March 1996
[17] *Goemans, C.*, et al. Aliphatic Oxirane Binder, "Uranox" Seminar, Amsterdam March 1996
[18] *Goemans, C*, et al. XXIII FATIPEC Congress book, Vol. B, Brussels, 1996, p. 324
[19] *May, C.A.*, Epoxy Resins, Chemistry and Technology, 2nd Ed., Chapt. 11, 936-945, *M. Dekker* Inc., New York/Basel, 1988
[20] *Gardiner, Th.H.*, et al., Epoxy Compounds, Chapt. 7, in Patty's Industrial Hygiene and Toxicology, 4th Ed., Vol. II, part a: Toxicology. Editors *G.D. & F.E. Clayton*, John Wiley & Sons, Inc. New York, 1993
[21] FDA Regulations CFR 21: §178.3740, §175.300, §177.1650, §177.1210
[22] a) *Senatore, D.; Ten Cate, T. A.; Laven, J.; Van Benthem, R. A. T. M.; De With, G.*, 234th ACS National Meeting, Boston, MA, US, August 19-23, 2007
b) Stichting Dutch Polymer Institute, WO 2008092686, 2008
[23] *Paul, S.*, in Surface Coatings, Science and Technology, John Wiley & Sons, Chichester, 1985, p. 267
[24] *Guida, R. and Denkinger, P.*, Powder Coat. Formulation & Production Conf. Indianapolis, Sept. 1998
[25] Chemische Werke Hüls AG, DE Pat. 3.004.876, 1980
[26] *Loontjens, T.; Plum, B.; Kidd, T.; Scholtens, B.;* European Coatings Journal, 10, 52-58, 2004

[27] Chemische Werke Hüls AG, DE Pat. 3.143.060, 1981
[28] *Nicholas, L., Gmitter, G.T.,* J. Cellular Plastics, 1, 85, 1965
[29] Chemische Werke Hüls AG, DE Pat. 3.322.718 and DE Pat.3.322.719, 1983
[30] UCB, DE Pat. 2.542.191, 1974
[31] Bayer AG, DE Pat. 3.232.463, 1982
[32] Bayer, WO 2004005364, 2004
[33] Chemische Werke Hüls AG, EP 0.045.994, 1980
[34] Chemische Werke Hüls AG, EP 0.045.996 and EP 0.045.998, 1980
[35] *Weiss, J.-V. and Grenda, W.,* XXIII FATIPEC Congress book, Vol. D, Brussels 1996, p. 182
[36] a) DuPont De Nemours, WO 2010068732, 2010
b) DuPont De Nemours, US 20100144927, 2010
[37] *Wenning, A.,* et al., Powder Coating Europe 98, Congress Papers, Amsterdam, Jan 1998, Paper 10, p. 147
[38] *Freudenberg, U.,* et al., Proc. 4th Nuremberg Congress, April 1997
[39] *Meier-Westhues, H.U.,* et al., FARBE&LACK, 103, 4, 140 (1997)
[40] *Van der Wely, F.W.,* J. Polym. Sci., 19, 381 (1981)
[41] a) *Gedan-Smolka, M., Lehmann, D. and Lehmann, F.,* FARBE&LACK, 106, 12, p. 62 (2000)
b) *Gedan-Smolka, M.,* Congr. Papers PCE 2002, Nuremberg, Jan. 2002, p. 145
c) *Edelmann, M.; Gedan-Smolka, M.; Lehmann, D.,* Progress in Organic Coatings, 57(3), 251-258, 2006
[42] IPF -Dresden, DE Pat. 19856877 (1998) and WO 00/34391
[43] IPF-Dresden, DE Pat. 19856878 (1998) and WO 00/34355
[44] *Gedan-Smolka, M., Çetin, S., Lehmann, D. and Komber, H.,* FARBE&LACK, 103, 11, p. 48 (1997)
[45] a) Degussa AG, WO 2005105880, 2005
b) *Gedan-Smolka, M.; Taeger, A.; Lilie, B.; Krug, D.; Riske, T.,* FATIPEC Congress, 29th, 99-108, 2008
[46] Bayer Materialscience, WO 2005095482, 2005
[47] Bayer Materialscience, EP 1903066, 2008
[48] a) *Gedan-Smolka, M.; Lehmann, D.; Lehmann, F.; Edelmann, M.,* FATIPEC Congress, 27th, 1, 291-301, 2004
b) *Gedan-Smolka, M.; Demmler, M.; Edelmann, M.; Topfer, A.,* European Coatings, 82(19), 19-28, 2006
c) *Edelmann, M.; Gedan-Smolka, M.,* Proceedings of the International Waterborne, High-Solids, and Powder Coatings Symposium, 35th 183-192, 2008
[50] a) Degussa AG, EP 1475399, 2004
b) *Spyrou, E.; Loesch, H.; Weiss, J.-V.,* Farbe + Lack, 111(5), 24-27, 2005
c) Degussa AG, US 20050090627, 2005
[51] *Grenda, W.; Spyrou, E.; Weiss, J.-V.; King, C.,* International Waterborne, High-Solids, and Powder Coatings Symposium, 32nd, 270-281, 2005
[52] Degussa AG, EP 1526146, 2005
[53] *Gedan-Smolka, M.; Edelmann, M.; Demmler, M.,* Farbe + Lack, 112(11), 22-27, 2006
[54] a) *Spyrou, E.,* Farbe + Lack, 113(7), 22-27, 2007
b) Evonik Degussa GmbH, WO 2008138855, 2008
c) *Spyrou, E.,* JOT, Journal fuer Oberflaechentechnik, 48(2), 22-23, 2008
[55] *Arendt, V.D., Logan, R.E., Saxon, R.,* J. of Cellular Plastics, Nov./Dec., p. 376, 1982
[56] *Alexanian, V., Forgione, P.S., Chang, L.W.,* American Cyanamid Co., US Pat. 4.379.767, 1983
[57] *Singh, B., Chang, L.W., Forgione, P.S.,* American Cyanamid Co., US Pat. 4.439.616, 1984
[58] *Fiori, D.E., Dexter, R.W.,* Intern. Waterborne, High-Solids & Powder Coatings Symp., February 5-7, 1986, New Orleans
[59] The Pharmacologist, 28, 3, 1986
[60] Hüls AG, DE Pat. 19729242 (1999)
[61] DuPont de Nemours, WO 2008079233, 2008
[62] DuPont de Nemours, WO 2009079536, 2009
[63] *Sullivan, C. and Weinberg, J.,* Powder Coating 2002, Indianapolis, Sep. 2002
[64] Takeda Yakuhin Kogy, GB Pat. 2.080.305, 1980
[65] Kodak, WP 8.300.328, 1981
[66] *Cooray, B., Spencer, R.,* Paint & Resin, October 1988, p. 18
[67] Bayer MaterialScience, US 20050245718, 2005
[68] "D.E.H. 40 Epoxy Curing Agent" (EU 3566-E-680, Dow Epoxy Resins in Powder Coatings, Edition 1981, Dow Chemicals)
[69] "Epikure 143 FF", Data Sheet EK 1.4.10, issued 1992, Shell Chemicals, now distributed by Overlack Global/ Chemical GmbH, as "Epikote Curing agent 143"
[70] "Epikote Powder Coatings - General Principles", Bulletin EK 2.3 (1992) p. 6-7
[72] Hardener HT 972, Ciba-Geigy Corp. Broshure CR 8895M81, 1981
[73] Hardener HU 261, Ciba-Geigy Corp. Broshure CR 8615C90, 1981
[74] Chemische Werke Hüls, EP 0.044.030, 1980

[75] Chemische Werke Hüls, DE Pat. 3.311.404, 1983
[76] Nitto Electric Industry, Jap. Pat. 57.042.760 1980
[77] Yuka Shell Epoxy, Jap. Pat. 58.118.818, 1982
[78] Sumitomo Durez, Jap. Pat. 59.226.066, 1983
[79] Shell Oil Company, US Pat. 4.214.068, 1980
[80] *Knop, A., Pilato, L.A.*, in Phenolic Resins-Chemistry, Applications and Performance, Springer Verlag, Berlin, Heidelberg, 1985, p. 1
[81] *Megson, N.J.L.*, Phenolic Resin Chemistry, Butterworths, London1958
[82] *Martin, R.W.*, The Chemistry of Phenolic Resins, J. Wiley & Sons, New York, 1956
[83] *Yeddanapally, L.M., Francis, D.J.*, Makromol. Chem., 55, 74,1962
[84] Shell Oil Company, US Pat. 4.214.068, 1980
[85] Dow Epoxy Resins in Powder Coatings, Ed. 1981, EU 3568-E-680, EU 3569-E680 EU3570-E-680
[86] Shell Chemicals, "Epikure 171 and Epikure 169", Data Sheets EK 1.4.16 and EK 1.4.15, 1992
[87] Ciba-Geigy Corp., Brochure CR 853A5C81, 1981
[88] General Electric Co., US Pat. 3.435.002, 1967
[89] General Electric Co., US Pat. 3.632.608, 1967
[90] General Electric Co., GB Pat. 1.032.372, 1962
[91] Rhone Poulenc Ind., DE Pat. 2.558.295, 1974
[92] *Walz, G., Kraft, K.*, Proc. 4th. Inter. Conf. in Org. Coat., Sci. and Technol., Athens 1978, p. 56
[93] *Schmidt, H.*, Proc. 13th Intern. Conf. "Powder Coatings - What's Next?", Paper 13, Brussels, Nov. 1993
[94] *Schmidt, H.*, European Coatings Journal (ECJ), 9/1994, p. 574
[95] *Schmidt, H., and Fink, D.*, Surface Coatings International, 2, p. 66 (1996)
[96] *Okano, M, Ogata, Y.*, J. Am. Chem. Cos., 74, 5728, 1952
[97] *Wohnsiedler, H.P.*, Ind. Eng. Chem., 44, 2679, 1952
[98] *Ugelstad, J., and Jonge, de J.*, Rec. Trav. Chem., 76, 919, 1957
[99] *Gordon, M., Halliwell, A., and Wilson, T.*, S.C.I. Monograph. 20, p. 187-198, Society of Chemical Industry, London, 1966
[100] *Gordon, M., Halliwell, A., and Wilson, T.*, J. Appl. Polym. Sci., 10, 1153, 1966
[101] *Sato, K.*, Bull. Chem. Soc. Japan, 40, 2963, 1967
[102] *Sato, K.*, Bull. Chem. Soc. Japan., 40, 724, 1967
[103] *Aldersley, J.W., Gordon, M., Halliwell, A., and Wilson, T.*, Polymer, 9, 345, 1968
[104] *Sato, K.*, Bull. Chem. Soc. Japan., 41, 7, 1968
[105] *Anderson, I.H., Cawley, M., and Steedman, W.*, Br. Polym. J., 1, 24, 1969
[106] *Berge, A., Gudmindsen, S., and Ugelstad, J.*, Eur. Polym. J., 5, 171, 1969
[107] *Berge, A., Kraeven, B., and Ugelstad*, J., Eur. Polym. J., 6, 981, 1970
[108] *Sato, K., and Naito, T.*, Polym. J. Japan, 5, 144, 1973
[109] *Sato, K., and Abe, Y.*, J. Polym. Sci., Chem. Edn., 13, 263, 1975
[110] *Tomita, B.*, J. Polym. Sci., 15, 2347, 1977
[111] *Berge, A.*, 3rd Intern. Conf. in Organic Coat. Sci. and Technol., Athens, 1977
[112] *Tashiro, T., and Oiwa, M.*, J. Polym. Sci., Polym. Chem., 19, 3, 645, 1981
[113] *Ubdegraff, I.H., and Suen, T.J.*, "Condensations with Formaldehyde", Chapt. 14 in Polymerisation Processes, Ed. C.E. Schidhenecht, Wiley-Interscience, NewYork, 1977
[114] *Drumm, M.F., and Le Blank, J.R.*, "The reactions of formaldehyde with phenols, melamine, aniline and urea", Chapt.5, Ed. D.H. Solomon, *Marcel Dekker*, New York, 1972
[115] *Brydson, J.A.*, Plastic Materials, 2nd Ed., Chapts. 19-23, Van Nostrand Reinhold, New York, 1970
[116] *Kirsch, A.J.*, in 50 Years of Amino Coatings Resins, The Twinchell Company, Philadelphia, 1986. p. 10
[117] PPG Industries, US 20040043223, 2004
[118] BASF, US 20040054083, 2004
[119] *Hsieh, B.; Ravichandran, R.; Abi-Karam, F.; Barczak, P.*, Proceedings of the International Waterborne, High-Solids, and Powder Coatings Symposium, 33rd 51-61, 2006
[119] The Sherwin-Williams Co., US. Pat. 4.189.421, 1980
[120] DSM, WO 2006065123, 2006
[121] *Higginbottom, H.P.*, Bowers, G.R. et al., Proc. Lucerne Conf. (Patsis) 1991, p. 129-198
[122] *Parekh, G.G.*, J. Coat. Technol., 51, 658, 101, 1979
[123] *Parekh, G.G.*, US Pat. 4.118.437, 1978
[124] SCM Corp., US Pat. 4.444.941, 1981
[125] SCM Corp., EP 0.021.770, 1979
[126] SCM Corp., US Pat. 4.271.277, 1979
[127] PPG Industries, US 20040225076, 2004
[128] *Essenfeld, A., and Kuang-Jong Wu*, 23rd Intern. Waterborne, High Solids and Powder Coat. Symp., New Orleans, 5-7 Feb. 1997

[129] *Taylor, J.* et al., J. Coat. Technol. 67, (846), 43-52 (July 1995)
[130] *Kaplan, A.*, Congress Papers PCE 98, Amsterdam, Jan. 1998, Paper 9, p. 117
[131] *Wicks, Z.W.*, et al., J. Coat. Technol. 57, (726), 51 (July 1985)
[132] *Stanssens, D.*, et al., 18th Internat. Conf. in Organ. Coat., Sci. and Technol., Athens, July 1992
[133] a) *Williams, F.; Stachyra, Z.; Armengol, J. M.; Grau, E.; Montserrat, S.; Roman, F.*, International Waterborne, High-Solids, and Powder Coatings Symposium, 32nd 253-269, 2005
b) *Roman, F.; Montserrat, S.*, Progress in Organic Coatings, 56(4), 311-318, 2006
[134] Merck, WO 2009015713, 2008
[135] *Stuiver, H., Overkamp, R., Hofkamp, A. and Schutte, M.*, Congress Papers Powder Coatings Europe (PCE) 2002, Nuremberg, Jan. 2002, p. 239
[136] *Franiau, R., and Pledger, T.*, Congress Papers PCE 2002, Nuremberg, Jan. 2002, p. 341
[137] AkzoNobel Nobel Coatings, US 20080114132, 2008
[138] *Grenda, W., Spyrou, E.*, Congress Papers European Coatings Congress (ECS), Nuremberg, 113, 2011

3.4 Industrial thermosetting powder coatings

3.4.1 Epoxy powder coatings

Introduction

A successful introduction of thermoplastic polyethylene powder coatings at the beginning of the 1950's, epoxy powder coatings were introduced in the market in the late 1950's as the first thermosetting powder coatings.

The epoxy group can be introduced into the polymer backbone by different methods, but only a few of them are of commercial importance. The epoxidation of unsaturated compounds by catalytic oxidation with oxygen is one of the most economical methods for production of ethylene oxide, but it has never found a real commercial application in the production of epoxy resins. An easy and commercially widely used method for epoxidation of unsaturated compounds is that of using peroxy acids or esters of peroxy acids:

Equation 3.79

The ease of preparation of peroxy acids by treating the corresponding acids with hydrogen peroxide has led to the development of **in situ** epoxidation methods. In this way separate preparation of peroxy acids and the difficulties connected to their storage and loss of active oxygen due to decomposition are eliminated.

The method of epoxidation with peroxy acids is widely used on an industrial scale for obtaining epoxidized drying oils. Although it is not the typical method for production of epoxy resins used in the powder coating field, it is used for production of liquid epoxy resins of low molecular weight to solid epoxy resins of high molecular weight by epoxidation of polybutadiene, polyisoprene, polycyclopentadiene, poly(isoprene-styrene), poly(isoprene-isobutylene), unsaturated hydrocarbon resins, poly(cyclopentadiene-styrene-methylstyrene), etc.[1, 2].

The most important route for preparation of epoxy resins is conversion of compounds containing activated hydrogen (e.g. in 3.80 alcohols) to epoxides by the hydrohalogenation **reaction with epichlorohydrin**, followed by dehydrohalogenation with a suitable base:

Equation 3.80

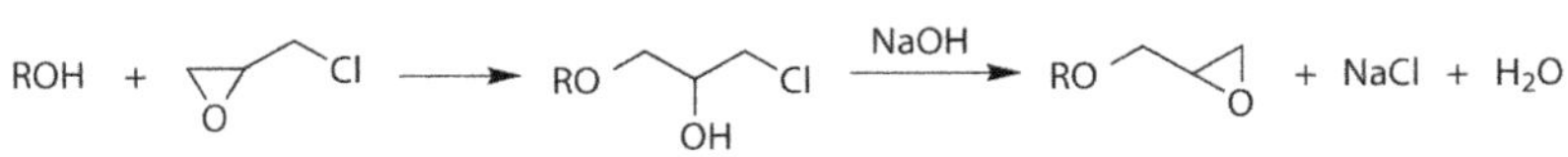

The active hydrogen containing compounds used in this process which gives glycidyl types of epoxies are phenols, alcohols, carboxylic acids, amines, mercaptanes, silanols and certain phosphorus compounds.

By far the most important representatives of binders for powder coatings obtained in this way belong to the group of **bisphenol A** and novolac types of epoxy resins. Several very important low molecular weight epoxy crosslinkers are also obtained through the epichlorohydrin route as it was already described in Chapter 3.3.1. The overall reaction for the manufacture of bisphenol A epoxy resins can be presented in a simplified way as follows:

Equation 3.81

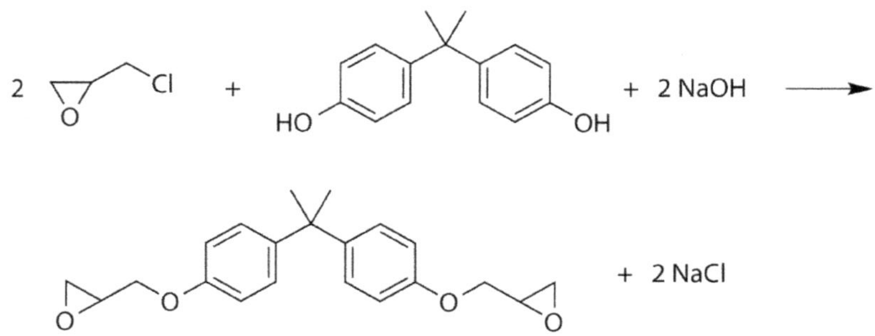

However, because of the simultaneous nature of the process with respect to all of the reactions involved, the newly formed epoxy groups may be attacked by the phenoxy anions instead of epichlorohydrin, leading to the formation of **oligomers with a higher molecular weight** according to the following overall scheme:

Equation 3.82

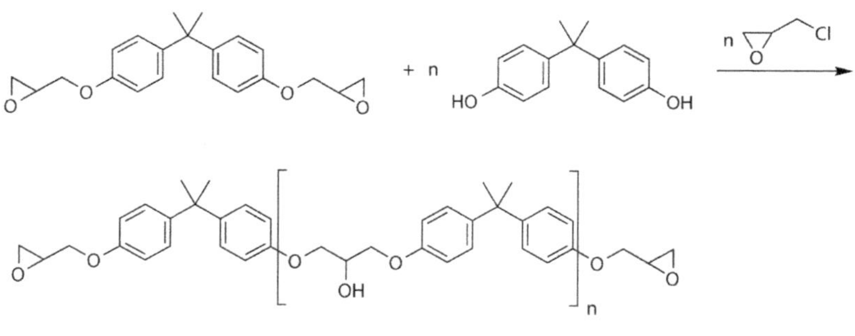

This reaction cannot be avoided, but its extent can be minimized by using a high excess of epichlorohydrin over the stoichiometric ratio.

Another problem that occurs during the manufacture of bisphenol A epoxy resins is the **hydrolysis of epichlorohydrin** leading in the subsequent reaction with the sodium hydroxide to the formation of glycidol:

Equation 3.83

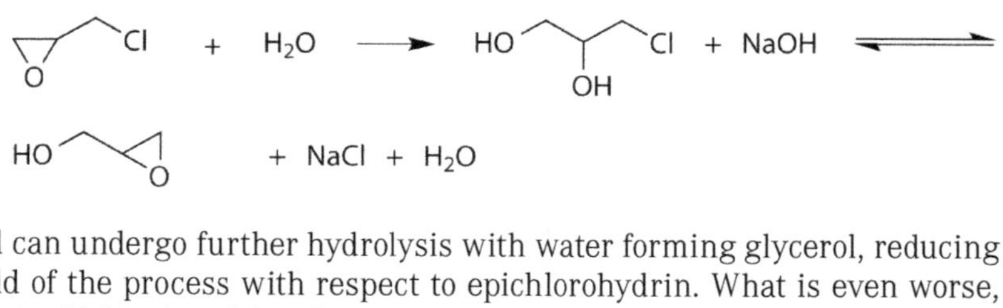

The glycidol formed can undergo further hydrolysis with water forming glycerol, reducing in this way the yield of the process with respect to epichlorohydrin. What is even worse, glycidol tends to react with bisphenol A to form a resin with two hydroxyl end groups. These end groups are not reactive when the resin is applied in a powder coating composition, thus adversely affecting the coating properties.

Diglycidyl ethers of bisphenol A are high viscous liquids at room temperature. With increasing molecular weight the melting point of the resins increases too, and those of interest for production of powder coatings usually have a molecular weight higher than 900, which corresponds to a melting point higher than 70 °C.

The **epoxy equivalent weight** (EEW) defined as the weight of the resin containing one gram equivalent of epoxide, is an important parameter for the powder coating chemist. Very often terms such as weight per epoxide (WPE) or epoxy molar mass (EMM) are used interchangeably instead of EEW.

Although the first bisphenol A type of epoxy resin was synthesized in the laboratories of I.G. Farbenindustrie in 1934, the real commercialization of these new polymers began after the Second World War based on the work of *Castan* in Switzerland and *Geenlee* in the United States. A great number of patents issued in the more than 50 years long industrial history of the epoxy resins has been reviewed several times in the literature [3-6].

From the reaction scheme of the synthesis of epoxy resins it is clear that the molecular mass of the resin can be adjusted by proper ratio of bisphenol A and epichlorohydrin. Very often, however, especially when solid epoxy resins have to be obtained, a condensation of low molecular weight epoxy resin with bisphenol A is performed in the presence of a suitable catalyst:

Equation 3.84

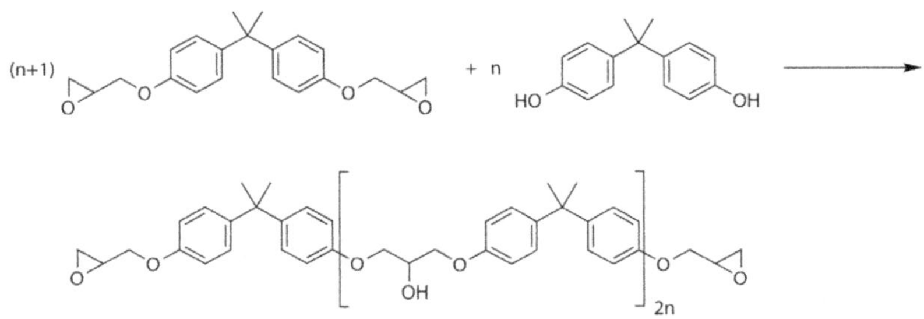

Depending on the ratio between the low molecular weight liquid epoxy resin and bisphenol A, resins with a degree of polymerization up to 30 can be obtained. However, in the coating industry, epoxy resins with degrees of polymerization higher than 12 are not normally employed [7]. The three **major producers of epoxy resins** nowadays (2012) are Dow, Hexion (now Momentive) and Huntsman.

Derivatives of epoxy resins

A modification of bisphenol A epoxy resin with an improved flexibility is described in the patent of Asahi Chemical Ind. [8] where bisphenol A epoxy in molar excess resin is reacted with a long chain (C_{10}–C_{25}) aliphatic dicarboxylic acid of a saturated or unsaturated nature. It is claimed that the powder coating having these modified resins as binders, cured with DICY, imidazole derivatives, dihydrazides of dicarboxylic acids, solid aromatic amines or acid anhydrides exhibits much better flexibility compared with that of the unmodified epoxy resins.

Aliphatic glycidyl ethers prepared from aliphatic alcohols are not used as sole binders for powder coatings, because they are liquid at room temperature. The condensation of aliphatic polyols with epichlorohydrin is performed in two steps. The epichlorohydrin and the polyol are reacted in the presence of Lewis acid to produce the corresponding chlorohydrin. In the second step, the process proceeds with dehydrohalogenation by means of sodium aluminate or other base in a water immiscible solvent. Although alone they are not suitable as binders for powder coatings, they can be used in combination with other resins as additives which provide flexibility.

Improved flexibility is also given by **elastomer-modified epoxy resins.** *Drake* [9] reviewed the various possibilities, and studied in particular the use of "CTBN" (carboxyl-terminated polybutadiene/acrylonitrile polymers) as such an elastomer. Improved toughening and flexibility, which result in powders with better impact resistance, bending, thermal shock and anti-chip properties, are possible by elastomer-modification. According to BASF [10] fibrous fillers help to increase the chip and corrosion resistance of foamed epoxy based powder coatings for automotive underbody components.

In the mid 1960's new multifunctional resins of the **epoxy-novolac type** were introduced in the market. The chemistry of these resins is very similar to that of bisphenol A type. Starting materials are novolac resins obtained by condensation of phenol, bisphenol A or cresol with formaldehyde in acidic media, followed by hydrohalogenation with epichlorohydrin and subsequent dehydrohalogenation with sodium hydroxide. The epichlorohydrin/ phenolic hydroxyl mole ratio should be about 6 to assure complete epoxidation and to avoid crosslinking and possible gelation [11].

Functionality of the epoxy resins

The difference between bisphenol A epoxies and novolac epoxy resins is in their functionality. While bisphenol A epoxy resins have a functionality of 2 or in the case of low molecular representatives even 1.9 [7], the functionality of the **novolac epoxy resins** can be easily adjusted by the resin manufacturer, and it is usually higher than two. This characteristic is very important in applications where a high density of reactive groups in a binder system is needed. Such cases are coatings with high solvent resistance combined with very high reactivity and high performances at elevated temperatures.

Hybrids epoxy resins

Epoxy powder coatings were used at the beginning for functional purposes only, but soon their potential as decorative coatings was recognized and their market share of the thermosetting powder coatings increased to more than 40 %. Nowadays, functional and decorative epoxy powder coatings are **ousted by the so-called "hybrids"** (using carboxyl functional polyesters as the main binder and epoxy resins as crosslinkers), but still enjoy 10 % of the market share. The hybrids are now representing more than 50 % of the market share.

Curing agents: dicyandiamide

The curing agents commonly used in epoxy powder formulations are solid aliphatic amines, solid aromatic amines and their resinous adducts, solid acid anhydrides and their resinous adducts, solid phenolic resins, polybasic organic acids and acid functional polyester resins.

Dicyandiamide (DICY) is a typical representative among the aliphatic solid amines used as curing agent in the production of epoxy powder coatings (see also Chapter 3.3.3.1). DICY has a reasonably high melting point of 211 °C and during the extrusion of the powder coating it does not melt to give a homogeneous mixture with the epoxy resin on a molecular scale. Therefore, the commercial grades of DICY are finely ground powders with a particle size below 75 µm which provide an acceptable homogeneous distribution of the fine particles of the hardener during the extrusion step. Since it is very difficult to obtain an optimum distribution of the hardener, in many cases it is necessary to use more DICY than stoichiometric quantities, in order to ensure a full cure. Another reason for not having strict stoichiometric proportions of the resin and the curing agent of this class is the curing mechanism of DICY which is complex, involving addition reactions with both the epoxy and hydroxyl groups and catalysis of the epoxide/epoxide reaction. The **optimum concentration** is dependent on the types and concentrations of pigments present, the particle size of the curing agent, the degree of dispersion of DICY in the resin and the desired stoving schedule. The best level in a particular formulation, therefore, has to be established by trials.

Catalysts for epoxy powders

The curing speed of powder coating with pure DICY is relatively low, even at comparatively high temperatures, so that stoving schedules of 30 minutes at 185 °C or 20 minutes at 200 °C are required. To obtain an acceptable curing time at temperatures that are normally used in the stoving lines in industry, **use of an accelerator is necessary.** Boron trifluoride/amine complexes have been suggested as latent catalysts for epoxy powders [12]. A wide range of decomposition temperatures between 70 and 300 °C can be obtained dependent on the choice of amine complexing agent. **Imidazoline derivatives** such as 2-methylimidazoline [13, 14] are also reported as catalysts for epoxy/DICY systems. **Tertiary** amino compounds and **quaternary ammonium** salts are also very suitable catalysts for accelerating the curing reaction. A few years ago General Electric [15] claimed guanidines in a patent as catalysts. Very common are substituted ureas as catalysts, which are decomposing at elevated temperatures to liberate an alkylamine accelerating the cure of the powder coating formulation. Commonly the producers supply DICY already containing accelerator in the form of a dry blend. The curing speed of the powder coating can then be adjusted by incorporating in the powder formulation different ratios of accelerated and non-accelerated DICY. A typical formulation of this kind is presented in *Table 3.4* [16].

Barletta et al. [17] investigated the mechanical strength and the wear resistance of epoxy coatings. They found a strong increase of the viscosity after a conversion degree of lower than 0.1 and the gelation around 0.45 to 0.50.

A patent of Hitachi Chemical [18] describes a method for obtaining the powder coating composition in which dicyandiamide is incorporated by dissolving it in novolac phenol formaldehyde resin. In this way the homogeneity of the system is improved, resulting in higher gloss.

Better **gloss** can be obtained with the substituted grades of DICY which have higher compatibility with the epoxy resin. Water resistance of the epoxy powders cured with substituted DICY is also improved, since these curing agents have lower solubility in water than DICY itself.

In general, it is difficult to obtain **matt films** with powder coatings. An interesting method to produce matt finishes has been given by Wagemakers (now DuPont Powder Coatings) [19]. They either dry blend an epoxy powder (not containing a curing agent), with a finely ground accelerated DICY, or with a ready epoxy powder, containing the accelerated DICY. In the first case they obtain a 45° gloss value of only 6 %; in the second case, by dry blending of equal parts, a gloss value of 12 %. Changing the weight ratio of hardener-free powder/hardener containing powder to 30/70 gives a semi-matt film with a 45° gloss value of 20 %.

Table 3.4: Epoxy powder coating formulations with dicyandiamide as curing agent [16]

Material	Parts by weight	
	A	B
"D.E.R. 663" (1)	47.4	47.4
"D.E.R. 673MF" (2)	10.0	10.0
"D.E.H. 41" (3)	1.3	2.6
DICY	1.3	–
TiO_2 rutile	40.0	40.0
	100.0	100.0
Bake schedule		
[min at 180 °C]	12 to 15	10 to 12

(1) Bisphenol A epoxy resin (Dow Chemical Company)
(2) Bisphenol A epoxy resin (Dow Chemical Company) containing 5 % of flow additive, discontinued
(3) Accelerated DICY (Dow Chemical Company), discontinued

Combination of powder coatings

A similar method has been proposed by *Klaren* [20]. He has found that by dry blending of two epoxy/DICY powder coatings with different reactivities and high gloss, a gloss reduction can be obtained by mixing them in various ratios. For example, when 2 parts of an epoxy powder coating with DICY as curing agent having a gel time of 6 to 30 minutes (dependent on the temperature) are dry blended with 1 part of powder coating having the same composition, but being catalyzed with a tertiary amine, thus having a gel time of 1 to 6 minutes, a 60° gloss of only 18 % can be obtained after curing for 15 minutes at 180 °C. Increasing the amount of the high reactive powder coating brings about a gloss increase up to 68 %, although the individually cured powders have a 60° gloss of 100 %. A proper **combination of the fast and slow curing powder coatings** gives a range of films from matt (20 to 25 % gloss) to silk-like (60 to 65 % gloss), that have at the same time good mechanical properties.

In another patent, assigned to Wagemakers (now DuPont) [21], epoxy powders with matt effects are obtained by incorporating (already in the premixing stage) a combination of **two curing agents**, namely at least 2 % of trimellitic anhydride (TMA) and at least 5 % of sulphaminic acid (both calculated on total powder).

By changing the ratio sulphaminic acid/TMA the 45° gloss can be adjusted to any desirable value, ranging from 40 to 5 %. Increasing the sulphaminic acid content leads to a lower gloss. Since the use of TMA and other free acid anhydrides is completely abandoned nowadays (for toxicological reasons; see Chapter 3.3.5), this matting method is no longer practiced.

A combination of DICY with an adduct of abietic acid and maleic anhydride is described in a patent assigned to Dow Chemicals [22]. The curing agent comprises 50 to 80 wt.% of the adduct and 50 to 20 wt.% dicyandiamide. In one of the examples in which the ratio between the different curing agents is 80/20 with a bisphenol A epoxy resin with epoxy equivalent weight of 785 as binder, a gloss of only 15 % (60° angle) has been indicated.

Epoxy powder coatings cured with DICY types of curing agents show excellent chemical and solvent resistance. The resistance towards atmosphere containing sulphur dioxide is extremely good. Water sensitivity is substantially improved when substituted DICY's are used as crosslinkers. When properly formulated, these powder coatings offer **outstanding impact, abrasion** and **scratch resistances**. A very good point is the excellent adhesion without using any primer, which is in general a typical characteristic of most epoxy coatings. The availability of accelerated grades of DICY enables the powder coating formulator to adjust the curing behavior of the coatings to various stoving conditions with considerable flexibility. Curing schedules of 8 to 15 minutes at 150 to 180 °C are commonly used in practice.

Yellowing resistance

Like all powder coatings based on bisphenol A epoxy resins as the main binder, coating formulations containing DICY suffer from sensitivity towards UV light. In the accelerated weathering tests they have shown only 7 % of the initial gloss after 192 hours (Atlas "Weather-O-meter" carbon-arc type), while at the same conditions, acrylic, polyurethane and polyester/TGIC powder coatings retain 98 % of the initial gloss [23]. The yellowing resistance of **DICY cured epoxy powder coatings** is rather poor. On a relative scale range where 10 indicates heavy, and 0 no yellowing, epoxy/DICY systems are among the worst, having a mark of 10 [23].

Because of these two weak points of epoxy/DICY powder coatings, their application is restricted mainly to **functional coatings**, and decorative coatings for indoor use, where the yellowing resistance is not a requested property. The main application of epoxy/DICY systems is certainly the coating of pipes for **subsurface pipelines**, where the epoxy powder coatings provide high corrosion resistance. Even better adhesion and corrosion resistance under hot and humid conditions is claimed by DuPont [24] to be achieved by brominated epoxy resins. According to another patent [25] of this company the same objective, in particular for cathodic disbondment protection, is pursued by the addition of alkanolamines, like trihydroxymethyl aminomethan. A similar approach is followed by Dow [26], by incorporating compounds like 2-amino-2-hydroxymethyl-1,3-propanediol into the epoxy hardener.

Corrosion resistance

Zinc dust is frequently used to enhance **corrosion resistance** in liquid coatings, but *Schuetz* et al. [27] could not observe any significant correlation between zinc content (from 5 to 60 % and thus lower than the usual >80 % in liquid coatings) and corrosion protection in powder coatings. *Harris* [28] claimed that **combination of zinc pigments** with different particle shapes (lamellar flakes and spherical dust), can lead to a significant reduction of zinc metal content without compromising the outstanding corrosion protection.

Blistering and delamination

Nguyen et al. [29] investigated the degradation of fusion bonded epoxy powder coatings on steel substrates. In particular the scratched panels showed anodic blistering near the scratch and cathodic delamination around the blisters when immersed in a $Ca(OH)_2$ saturated sodium chloride containing solution, imitating a marine concrete environment. Unscratched panels showed only delamination, which recovered after drying. In a recent patent [30] of Dow it is claimed that epoxy resins containing both urethane groups and oxazolidinone rings (the reaction product of epoxy and isocyanate) are even more suited for **high temperature stable fusion bonded pipes**, due to the higher onset of the glass transition temperature. Similar benefits are claimed to be accomplished by oxazolidonone containing compounds manufactured by the reaction of a mixture of aliphatic and aromatic epoxy resins and isocyanates [31]. Again the same company described a different approach [32] to obtain high T_g epoxy powder coatings by manufacturing bisphenol diglycidyl ethers using cyclododecanone instead of acetone (the A in bisphenol A) in the bisphenol synthesis.

Curing agents: other amines

Amine derivatives containing primary or secondary amino groups are generally unsuitable as crosslinkers for powder coatings because of their high reactivity with epoxy resins. Even this is the case with those amines that are solids at normal temperature. This creates certain problems during processing of the coating, or considerably reduces the storage time. However, aliphatic polyamines which differ from DICY are described in the patent literature, and some of them are commercially available. For example, the **condensation product of isophorone diamine and diglycidyl ether of bisphenol A** with a low molecular weight ("Epikote 828") in the ratio of 2:1 with a melting point of 55 to 57 °C can be used in combination with high molecular weight epoxy resins as the curing agent [33].

Aliphatic polyamine for preparation of corrosion resistant powder coatings is described as a reaction product of aminododecanoic acid and condensate of bisphenol A epoxy resin with an equivalent amount of phenol [34]. It is recommended as a hardener for epoxy powder coatings for pipes, drums, tanks etc.

Table 3.5: Low curing powder coating with amine hardener [35]

source: Shell Chemicals

Material	Parts by weight
"Epikote 1055" (1)	100.0
Flow agent	0.5
"Vestagon B 31" (2)	6.0

(1) Bisphenol A epoxy resin (Momentive)

(2) Amine curing agent (Evonik Industries, discontinued)

Powder coatings with amine derivatives as hardeners are characterized by extremely short curing cycles at relatively low temperature. The formulation presented in *Table 3.5* [35] based on a commercially available amine curing agent, introduced, but now abandoned, by Degussa-Hüls (now Evonik Industries), "Vestagon B 31", a cyclic amidine, refers to coatings with curing schedules of 15 min at 140 °C, 20 min at 130 °C and 30 min at 120 °C. Since the proportion of curing agent is small, the pre-blending is usually done by high-speed mixers in order to ensure efficient dispersion and homogeneity of the hardener throughout the powder in the next compounding step. It is interesting that in spite of the short curing schedules, the stability of the powder coating is satisfactory.

Curing agents: phenolic resins

Epoxy powder coatings with **phenolic resins as hardeners** are also widely used in practice. In principle, common thermosetting solid phenolic resins can be used as crosslinkers

Table 3.6: Epoxy powder coating formulations with phenolic resinous crosslinkers *source: Momentive*

Formulation	A	B	C
"Epikote 1003" (1)	100		
"Epikote 1004" (1)		100	
"Epikote 1007" (1)			100
"Epikure curing agent 168" (2)	53.3	45.6	22.9
2-methylimidazole	0.52	0.45	0.22
"Modaflow Powder III" (3)	1% w/w delivery form on total binder		
"Kronos 2310" (4)	40% w/w on total binder		
Powder and coating properties			
Gel time at 180 °C, s	80 to 90	95 to 115	125 to 160
Cure time at 180 °C, min	9	12	16
Flow	good	fair	poor
Gloss 60° [%]	95	94	88
Whiteness E313, index	74	72	76
GT adhesion, 0/best 4/poorest	0	0	0
Erichsen slow penetration [mm]	9.8	9.9	9.9
T-bend, T	0	0	0
Impacts, reverse & direct [lb.inch]	>160	>160	>160
Abrasion [mg]	46	53	58
Dynamic solvent resistance [min]	4	4	4
Caustic resistance, 10 % NaOH	gloss and adhesion not affected after 75 days		
Acid resistance, 10 % H_3PO_4	gloss slightly and adhesion not affected after 45 days		
Salt spray resistance			
Blisters, middle and edge	no blisters observed		
Rust creep [mm]	average 3 to 5		
Tape test, middle and edge [mm]	average 3 to 5 at middle; 0 at edge		

(1) Bisphenol A epoxy resin (Momentive)
(2) Phenolic based curing agent (Momentive)
(3) Flow additive (Monsanto)
(4) Titanium dioxide (Kronos Titan)

for formulating epoxy powder coatings. The initial problems with respect to the dark color of the phenolic resins, low reactivity and brittleness have been successfully overcome by the new generation of resinous phenolic hardeners. Coatings with high crosslinking density, high chemical, solvent and boiling water resistance, and good mechanical and decorative properties can be obtained. The **yellowing tendency** is still an obstacle for universal use of these epoxy powder coatings, and like the other epoxy powder coatings they are suitable only for indoor application. In comparison with DICY systems, much shorter curing times can be obtained when suitable catalysts in proper amounts are used. In most cases, the catalyst is usually incorporated in the hardener, which provides better homogeneity of the system. *Table 3.6* contains three guide formulations for powder coatings. The short curing time is quite indicative.

Novolac types of epoxy resins can also be used in combination with phenolic hardeners. In this case, an even **higher crosslinking density** can be obtained, because of the higher functionality of the novolac epoxy resins. A composition suitable for insulation coating for electrical and electronic parts with good moisture resistance, high temperature electrical characteristics and heat-cycle resistance is described in Jap. Pat. 59226065 [36] and comprises 24 % novolac epoxy resin with an epoxy equivalent weight of 190 and a melting point of 75 °C, 16 % phenolic resin of novolac type with a melting point of 78 °C and a hydroxide equivalent of 220, 58.8 % silicon dioxide as a filler, 1 % carbon black as pigment and 0.2 % imidazole as catalyst.

A distinguishing feature of epoxy powder coatings cured with phenolic resins is an exceptionally high gloss combined with good flow. They have very good mar resistance, chip and impact resistance and good adhesion without an adhesion promoter. Resolution [37] claimed enhanced hot wet adhesion to metal substrates, in particular pipes, if amino alcohols are added to the powder coating formulation. According to PPG [38] semi-mat films can be achieved by using a mixture of incompatible epoxy resins in combination with a compatibilizer, e.g. an epoxy amine adduct. With regard to possible contacts with food and skin, they meet the normal toxicity standards. Relatively low baking temperatures can be used for curing, permitting application on more or less heat-sensitive substrates.

Typical **applications of the decorative epoxy/phenolic powder coatings** include appliances, wire goods, bicycle frames, hot-water radiators, tanks, fire extinguishers, toys and tools, electronic components, housing for electronic instruments, heavy-duty electrical cabinets, control boards, fuse boxes, automotive armatures, etc.

Curing agents: acid anhydrides

Curing of the epoxy resin can be done with acid anhydrides or polycarboxylic acids. The **use of acid anhydrides is almost completely abandoned** nowadays, for toxicological reasons. A somewhat lower risk with respect to human health exists when resinous anhydrides, described in Chapter 3.3.5, are used. Polycarboxylic acid anhydrides, described in Chapter 3.1.2, react with epoxy groups in a complex stepwise manner. The efficiency of curing with acid anhydrides is not very sensitive to the stoichiometry, as is the case with polyacids, since in principle only one hydroxyl group is needed to initiate the reaction. However, this advantage is not big enough to compensate for the heavy irritation of the mucous membranes which is quite common when acid anhydrides are used as hardeners in epoxy powder coatings. On the other hand, acid anhydrides are susceptible to hydrolysis by atmospheric moisture which creates certain **problems during storage** of the curing agent itself, or the powder coating. Therefore today, the use of the anhydrides as curing agent either in the form of low molecular weight compounds, or resinous anhydrides of higher molecular weight is very rare.

Table 3.7: Clear epoxy powder coatings with TMA adduct as curing agent[39] *source: Shell Chemicals*

Material	Parts by weight
"Epikote 1003", "1055" (1)	100.0
"Epikure F 50" (2)	10.5 to 11.5
Stannous octoate	0.2 to 0.5
Curing schedule	
at 180 °C [min]	10

(1) Bisphenol A epoxy resin (Momentive),
(2) trimellite anhydride adduct (Momentive), discontinued

Table 3.8: Epoxy powder coating with dihydrazide curing agent[42]

Material	Parts by weight
"D.E.R.-663U" (1)	70.0
Azelaic dihydrazide	5.4
Barium sulfate	84.0
Phthalocyanine green	0.1
Chromium oxide	0.5
Titanium dioxide	3.0
Flow additive (2)	0.7
2,4,6-tris(dimethylamino-methyl)phenol as catalyst	0.7

(1) Bisphenol A epoxy resin (Dow Chemical Comp.)
(2) Flow additive "Modaflow" (Monsanto)

A special feature of polycarboxylic acid anhydride curing agents in epoxy powder coatings is the very rapid curing rate that can be achieved by means of suitable catalysts such as metal salts of organic monoacids. The crosslinked film has excellent hardness, good electrical properties, excellent resistance to solvents and good yellowing resistance on overbake. Curing times of 20 to 30 seconds can be achieved on preheated metal surfaces at 250 to 275 °C in fluidized bed applications, or 10 minutes at 180 °C in electrostatic spraying applications. A typical epoxy powder coating formulation using a trimellitic anhydride adduct is given in *Table 3.7*[39].

Curing agents: others

Polycarboxylic acids are not often used as sole crosslinkers for epoxy powder coatings. From a technological, economical and toxicological point of view, the acid functional polyesters which will be discussed in detail in the following sections have certain advantages. These coatings, known under the name hybrids, hold a dominant position among the epoxy powder coatings. They have almost completely replaced the use of polyacids as curing agents. However, some patent applications describing systems based on epoxy resins and polyfunctional acids can be found. For example, matt or semi-matt epoxy powder coatings can be obtained by curing bisphenol A epoxy resins with salts of trimellitic acid (or other aromatic polycarboxylic acids) and cyclic amidines, e.g. 2-phenylmethyl-2-imidazoline[40]. Two matting hardeners, belonging to this group of amine mono-salts, used to be sold by Hüls (now Evonik Industries) under the trade names "Vestagon B 55" and "B 68". See also Chapter 3.3.3.2.

A patent of Mitsubishi Electric Corp.[41] describes a powder coating obtained by dry blending of acid anhydrides including phthalic, maleic, trimellitic or pyromellitic, polyethylene or polypropylene glycols with molecular weights of 4000 to 40,000 and epoxy resins with epoxy equivalent weight of 100 to 5000.

Epoxy powder coatings with **dihydrazides of dicarboxylic acids** as curing agents which are used for the **protection of pipes** in pipelines are claimed in a patent of the 3M Company[42]. The binder is bisphenol A type epoxy resin, while the dicarboxylic acid dihydrazides are of the following general formula:

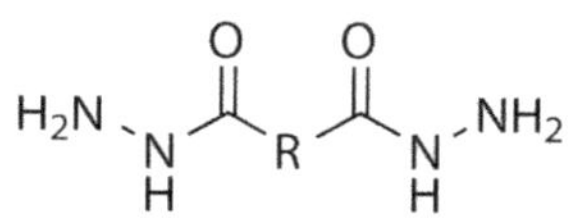

Although in principle any dicarbonic acid can be used, dihydrazides of aliphatic diacids with more than four atoms between the two carboxyl groups are preferred for providing better flexibility to the cured product. Tertiary amines are suggested

Table 3.9: Major uses of epoxy resin powder coating systems [48] *source: Shell Chemicals*

Epoxy resin plus	Domestic equipment	Metal furniture	Building components and fittings	Automotive components	Agricultural equipment	Radiators, boilers	Pipes	Containers	Concrete reinforcing bars	Electrical equipment
Accelerated DICY	•	•		•	•					•
Substituted DICY	•	•		•	•					
Amine derivatives	•	•	•	•		•				
Polyester resins	•	•	•	•	•					
Anhydride adducts				•		•	•	•		•
Phenolic resins			•	•	•		•	•	•	•

as catalysts to speed up the reaction, together with quite high amounts of barium sulfate as filler. An example from this patent is given in *Table 3.8* as an illustration. An exceptionally good boiling water and cathodic disbondment resistance at room and elevated temperature (60 °C) is claimed as an outstanding achievement. In a recent Japanese patent [43] the addition of boric acid esters, like 2,2'-oxybis[5,5-dimethyl-1,3,2-oxaborinane], is recommended for high reactivity and good storage stability of the epoxy formulation.

Marx et al. [44] describe a **phenolic hydroxyl functional arylether resin** that can be combined with epoxy resins. Powder coatings based on this combination show very good flow and can be cured as low as 30 min at 140 °C. The most common arylether is poly-(2,6-dimethyl-1,4-phenylene ether) offered by Sabic. *Guo* [45] stresses the advantages of these resins in thermosetting epoxy powder coatings, including high T_g, toughness, resistance against chemicals and water, low shrinkage and last but not least flame retardancy. For better solubility he recommends low intrinsic viscosity telechelic poly(arylene ether).

Dow filed a patent [46] that claims the combination of epoxy resins and blocked isocyanates. The curing reaction, taking part below 150 °C, includes the formation of oxazolidones (addition of isocyanate and epoxy group), isocyanurate (trimerization of isocyanate) and urethanes (isocyanate reacting with the hydroxyl groups of the epoxy resin).

Weather-resistant, high-gloss epoxy powder coatings can be produced, according to *Krüger* [47], by using cycloaliphatic epoxy resins, prepared from epoxidized cycloalkenes, such as limonene diepoxide. *Table 3.9* [48] gives a survey of the major uses of the epoxy powder coatings cured with different curing agents.

Several typical **advantages** of the epoxy powders over the other coating systems are the excellent chemical and solvent resistance, combined with excellent boiling water resistance. The presence of free OH groups uniformly distributed along the resin chain provides excellent adhesion on metal substrates. This is the reason why epoxy coatings were applied 2006 for corrosion resistance of the world longest bridge (30 km) crossing the Hangzhou gulf [49]. Other outstanding features of the epoxy powder coatings are excellent impact, abrasion and scratch resistance. For use in electrical applications (rotors, stators, electrical housing etc.), they have the necessary dielectric properties.

Weak points of the epoxies are their inferior yellowing resistance and poor outdoor durability due to the presence of the aromatic ether group which is extremely sensitive to UV light. Chalking on outdoor exposure is a phenomenon typical for epoxy powder coatings. Although once it has reached a certain level additional chalking does not occur, the use of epoxy powder coatings has been restricted to functional areas and interior decorative applications.

3.4.2 Polyester powder coatings and related

Introduction

Polyester (PES) powders **dominate the powder coatings market** worldwide and find much more use then related polymer resins like polycarbonates or polyamides. The latter were recently reintroduced into the market by DSM [50] for powder coating formulations, bearing amino, carboxyl or hydroxyl functionality. By adding a second incompatible compound they achieve a certain degree of a matting effect. Polyurethanes based on polycarbonates are claimed to have a better weatherstability compared to polyester containing coatings [51].

The first attempts to make pure polyester powders date from the end of the 1960's. In a patent of Hoechst AG [52] a physical blend of two different types of powdered polyester resins containing carboxyl and hydroxyl functional groups is used as a powder coating with curing time of 15 min at 220 °C. In one of the examples of the patent, a melt mixing of both polyesters with titanium dioxide as a pigment is powdered after cooling, resulting in a system with a curing schedule 15 min at 220 °C. This process of producing powder coatings closely resembles the powder coating technology of today. The curing reaction involves esterification between the carboxyl and hydroxyl groups from both binders, which unfortunately yields water as a by-product causing surface defect problems. Another disadvantage is a rather long curing time at high temperatures. Two groups of saturated polyesters containing carboxyl or hydroxyl functional groups represent almost 100 % of the resins used for production of thermosetting polyester powder coatings.

In principle all reactions for the preparation of esters can be used for manufacturing polyester resins. Of course, not all of them can be performed on an industrial scale because of technical or price reasons. The industrial preparation of polyesters for powder coatings mainly involves direct polyesterification of polybasic acids and polyfunctional alcohols. Depending on which component is in molar excess, the polyester can be **terminated with carboxyl or hydroxyl groups**. The numbers of moles of different constituents in the resin formulation in the case of hydroxyl functional polyesters are interrelated by the following equation [53]:

Equation 3.85 $$\sum N_p = N_{a2} + 2N_{a3} + 1$$

where $\sum N_p$ is the total number of moles of diols, and N_{a2} and N_{a3} are the number of moles of diacids and triacids. For carboxyl functional polyesters a similar equation is valid:

Equation 3.86 $$\sum N_a = N_{p2} + 2N_{p3} + 1$$

where $\sum N_a$ is the total number of moles of the diacids, and N_{p2} and N_{p3} are the number of moles of diols and triols in the resin formulation.

Functionality of the resins can be easily adjusted by incorporation of a certain amount of a branching component which can be triol, tetraol or triacid. The techniques for calculating formulations for polyester resins are beyond the scope of this book. The interested reader can consult a considerable number of published articles and monographs [54-70].

A **big advantage of the polyester resins** is the possibility of adjusting the place of functional groups in the polymer chain to a desired position which has an enormous influence

Table 3.10: Hydroxyl functional polyester resin for polyurethane coatings [7, p.137]

source: Selective Industrial Training Associates/ SITA Ltd

Ingredient	Parts by weight
1. Water	15
2. Neopentyl glycol	436
3. Terephthalic acid	552
4. "Fascat 4201" (1)	1
5. Isophthalic acid	117
6. Sebacic acid	25
7. Trimethylol propane	28
8. Xylene	20
Final characteristics: Acid value = 2.2 mg KOH/g Hydroxyl value = 35 ± 5 mg KOH/g Melting range = 110–115 °C	

(1) Catalyst (Elf Atochem S.A.)

upon the mechanical properties of the cured films. This is one of the reasons why polyesters have superior mechanical properties compared to the acrylics, for example, where the functional groups are distributed at random.

Thermosetting polyester resins for powder coatings are produced on the same equipment that is used for production of polyesters for solventbased coatings, or high molecular weight polyesters. Although the azeotropic method of processing can be used, it is rather an exception in practice, since the subsequent stripping off of the solvent is unnecessary complication. The fusion method of cooking followed by removal of the esterification water at the end of the process by means of a vacuum is the most suitable procedure. As in the case of high molecular weight polyesters, the resin is poured onto a cooling belt provided with a rotating crusher which gives the final flake size of the crushed resin.

A standard method of preparation of hydroxyl functional polyester resin for polyurethane powder coating with the reaction details is given in the book of *Oldring* and *Hayward* [7, p.137].

Example of preparation of a polyester resin

Ingredients 1 to 4 in *Table 3.10* are charged together in the reactor and esterified in a temperature interval from 180 to 245 °C with removal of the reaction water until an acid value below 2 mg KOH/g is reached. Then, components 5 to 7 are charged and esterification continues. In order to facilitate the removal of reaction water, xylene is added to the reaction vessel, and esterification is performed according the so-called azeotropic method until an acid value below 3 mg KOH/g is obtained. Xylene is removed from the system by vacuum distillation, and the resin is discharged on the cooling belt.

Different patents describe methods for making hydroxyl functional polyester resins for polyurethane powder coatings with minor variations with respect to the acid value, hydroxyl value, functionality of the resin, and the choice of raw materials [71-74].

A somewhat different cooking procedure is used with dimethyl terephthalate as a starting raw material in place of terephthalic acid. In such cases the reaction goes in two different steps. In the first step dimethyl terephthalate is transesterified in an excess of glycol with complete removal of methanol as a reaction by-product. In the second step, the polymerization reaction is carried on by means of high vacuum, removing the excess of the glycol from the system, thus shifting the reaction equilibrium in favor of a high molecular compound [75, 76].

An interesting approach using **crystalline polyesters** to obtain good storage stability of the powder coating and to have at the same time low melt viscosity providing good flow properties to the coating has been developed in the laboratories of Eastman Kodak Co. [77, 78]. In the case of branched polyesters, a conventional method of preparation is used, starting from dimethyl terephthalate as a source of dicarboxylic acid moieties, which is transesterified and later on polymerized under vacuum with linear glycols (1,6-hexanediol, 1,4-cyclohexanedimethanol or ethylene glycol) and trimethylolpropane or trimellitic anhydride as branching components. The polyesters have a molecular mass between 700 and 3000, OH number from 30 to 160 and melt viscosity at 160 °C of 50 to 3000 mPa·s, preferably between 50 and 100 mPa·s. The curing agent is caprolactam-blocked isophorone diisocyanate oligomer. The same method is used in the case of linear polyesters, where terephthalic acid and 1,4-cyclohexane dicarboxylic acids in the form of dimethyl esters are transesterified with 1,4-butanediol as a single polyol component. The amount of 1,4-cyclohexane dicarboxylic acid should be at least 50 mol-% in order to obtain a melting temperature below 160 °C. The main advantage that is claimed is the excellent flow as a result of a very low melt viscosity.

A recent patent of Hexion [79] describes amorphous resins grafted with a few percent of crystalline resins. They are claimed to be superior regarding flow and UV resistance in thermosetting powder coatings compared to melt mixtures of the same amorphous and crystalline resins.

PES for hybrids and TGIC

Carboxyl terminated polyesters first developed by DSM Resins BV at the beginning of the 1970's are especially popular binders in Europe [80]. Two main systems are widely accepted by the market: **TGIC/polyester for outdoor applications**, where good gloss retention and yellowing resistance is a strong request, and so-called hybrids composed of carboxyl terminated polyester and epoxy resin of bisphenol A type as a crosslinker. The **hybrids have poor outdoor durability** and yellowing resistance, but excellent boiling water resistance and alkali resistance, which makes them very suitable for household equipment applications. Although the same resins can be used in principle for both systems, there are certain small differences between the polyesters for curing with TGIC and curing with bisphenol A type epoxy resins.

The **functionality** of the epoxy resins of bisphenol A type is two compared to three in the case of TGIC. Therefore, the functionality of the polyester resins for hybrid systems in principle should be higher than the functionality of those intended to be crosslinked with TGIC.

The glass transition temperature of polyester resins for TGIC containing powder coatings should be at least 60 °C in order to achieve good storage stability. The storage stability of polyester/epoxy hybrids is not very critical because of a high enough T_g of the bisphenol A epoxy crosslinkers. Therefore, the glass transition temperature of polyester resins for hybrids can be approximately 10 °C lower compared to TGIC systems. This makes the choice of raw materials for producing polyester resins for hybrids somewhat easier.

With a good approximation, it can be considered that the **T_g of the powder coating** decreases by 2 °C for each weight percentage of TGIC. Since T_g of 40 °C is considered to be the lowest value that one can allow for having good storage stability of the coating, the amount of TGIC in the coating formulation in which the polyester resin is with a T_g of 60 °C usually cannot exceed 10 % by weight on the basis of the binder. Consequently, the acid value of the polyesters for TGIC curing is almost never higher than 50 mg KOH/g (which corresponds roughly to 10 % TGIC). In the case of polyester resins for hybrids the acid value can go up to 80 mg KOH/g.

Table 3.11 presents **typical formulations of polyester** resins for TGIC and hybrid systems for which the details of the manufacturing procedure are given by *Oldring* and *Hayward* [7, p.137].

The **production** of carboxyl functional polyesters can be realized in a one-step cooking procedure, having a proper ratio between the reactants, which ensures a certain molecular weight, combined with the desired acid value. In many cases, however, the process of production of polyester resins with carboxyl functionality is performed in two steps. The reason is much easier process control, especially in the case where the resin is of a branched character. The first step ends up with a hydroxyl functional polyester resin with a certain molecular mass and hydroxyl value. In the second step a calculated amount of dicarboxylic acid is added, and the esterification is carried on until the desired acid value has been reached [81].

Table 3.11: Polyester resins for TGIC and hybrid systems [7, p.137]

source: SITA Technology Ltd

Ingredient	Parts by weight	
	Hybrid	TGIC
Water	28	30
Neopentyl glycol	364	530
Terephthalic acid	423	711
Isophthalic acid	–	88
Adipic acid	41	–
Pelargonic acid	–	58
Trimellitic anhydride	141	43
"Fascat 4201" (1)	3	5
Final characteristics:		
Acid value [mg KOH/g]	80 ± 5	35 ± 5
Melting range [°C]	110 to 120	100 to 110

(1) Catalyst (Elf Atochem S.A.)

A similar process uses dimethyl terephthalate instead of terephthalic acid in the first step, continuing with esterification with dicarboxylic acid in the second step to obtain a carboxyl functional polyester resin suitable for curing with epoxy containing compounds [82].

Addition of acid anhydride to hydroxyl functional polyester at the last step of the process is also a common method for preparation of carboxyl functional polyesters. Trimellitic anhydride was often used for this purpose [83], but nowadays **TMA-free polyesters** are preferred for toxicological reasons.

A somewhat different process is described in the patent of Goodyear Tire & Rubber Co. [84]. In one of the examples of the patent a polyester resin is prepared by transesterification of dimethyl terephthalate with neopentyl glycol in the presence of a small amount of trimethylolpropane. Under vacuum in the next step, the excess of neopentyl glycol is removed and hydroxyl terminated polyester is obtained. The latter is endcapped with carboxyl groups by esterification with isophthalic acid in excess of carboxyl groups in respect of the total resin formulation. **Isophthalic acid enhances the UV resistance** of the powder coating system at the expense of elasticity and costs.

Similar techniques starting with terephthalic acid or blend of terephthalic and isophthalic acid instead of their dimethyl esters are used in many patents [85-90]. A method that differs from the previously described processes involves a depolymerization of high molecular mass polyester resin with polycarboxylic acids [91] resulting in carboxyl functional polyester with lower molecular mass than the starting material. The polycarboxylic compounds are selected among di-, tri- and tetravalent carboxylic acids. A modification of this method uses diols, triols and tetraols as depolymerizing agents [92] having as the end product an hydroxyl functional polyester resin of lower molecular mass. The molecular weight of the final product and its functionality depend on the amount of the depolymerizing agent and the functionality of the same.

T_g values of polyester resins

A polyester resin suitable for production of powder coatings with a very low amount of TGIC in the coating formulation is described in the patent of DSM Resins BV [93]. The resin is obtained by fusion esterification of neopentyl glycol, 1,4-cyclohexanedimethanol, 1,6-hexane diol, trimethylolpropane, terephthalic acid and adipic acid. Products with average molecular masses between 4500 and 12,500 with acid values of 10 to 26 mg KOH/g and T_g of 40 to 85 °C are obtained which are suitable for making powder coatings containing 1.4 to 5.3 % by weight of TGIC.

A similar patent [94] describes a method for obtaining polyester resins with somewhat higher T_g values (45 to 85 °C), which permits the use of diglycidyl terephthalate as crosslinker in amounts of 2 to 9 % by weight, still giving powders with good storage stability.

AMOCO co-researchers [95] have found, that the T_g of polyester resins can be increased (giving powders with improved storage stability at 40 °C) by replacing small amounts (5 to 10 mol-%) of terephthalic acid by 5-tert.-butyl isophthalic acid (tBIA), without affecting the melt viscosity. At higher levels of tBIA (20 mol-%) there was a decrease in mechanical properties.

Johnson and *Sade* [96] describe the influence on T_g and melt viscosity for a number of glycols and diacids, introduced for use in polyesters in the 1990's.

Among them are 1,4-cyclohexanedicarboxylic acid (1,4-CHDA) and 1,3-CHDA as diacids, hydrogenated bisphenol A (HBPA) and hydroxypivalyl hydroxypivalate (HPHP) as glycols. 1,4,-CHDA reduces both the T_g and the melt viscosity of the resin, compared to an analogous resin based on terephthalic acid, while HBPA has the opposite effect. The melt viscosity is also reduced by 1,3-CHDA, but without a significant reduction of the T_g of the resin. HPHP reduces both the T_g and melt viscosity of resins in which it is used as a partial replacement for neopentyl glycol. Due to its extended chain length and higher molecular weight, as compared with neopentyl glycol, it reduces the aromatic content of the resin and improves weatherability and flexibility.

Polyester resins from biobased materials

Due to environmental reasons **renewable sources** for coating raw materials come into focus more and more recently. *Noordover* et al. [97] from the University of Eindhoven reported the synthesis of polyester resins for powder coatings starting from biobased materials like isosorbide and succinic acid. They observed a similar weatherstability for isosorbide containing powder coatings as for as conventional terephthalic acid-based poly(ester urethane) coatings [98]. The Batelle Institute [99] patented bioderived compositions like soy based dimer diol, isosorbide, ethylene glycol and terephthalic acid derivatives for powder coating resins.

Polyester resins and crosslinkers

A general weak point of polyesters is their **low resistance to saponification**. By the proper choice of raw materials used in building up the backbone of the polyester resin, this property can be improved to a considerable extent. Under normal conditions of hydrolysis, resistance to ester cleavage increases with increasing alkyl substituent at the alpha and beta positions on both alcohol and acid. Glycols like neopentyl glycol, trimethylpentanediol, 2,2-diethyl-1,3-propanediol, 2-methyl-2-butyl-1,3-propanediol, "Ester diol 204" (an ester of hydroxypivalic acid and neopentyl glycol) and cyclohexyldimethylol are the raw materials that can be used for this purpose.

It is interesting that a patent describing a process of making polyester powder coatings in which bisphenol A epoxy resin is used as crosslinker had already been filed in 1961 [100]. The coating in one of the examples of the patent intended to be used for electrical insulation purposes is a physical blend of powdered carboxyl functional polyester resin and powdered epoxy resin. It is applied in a fluidized bed on preheated aluminum wire and cured for 2 hours at 205 °C, which is certainly an unacceptable curing schedule for practical reasons.

A period of 10 years was necessary to realize the advantages that are offered by this system over the pure epoxy powder coatings. In 1970, *Huneke* [101] reported that the gloss retention of powder coatings based on blends of epoxy and polyester resins is considerably improved in comparison with pure epoxy powders. However, nothing was said about what kind of polyester resins was used and what curing reaction takes place.

A patent of Unilever NV from 1971 [102] refers to the powder coating composition manufactured by melt extrusion of carboxyl functional polyester resin and epoxy resin of **bisphenol A type**. The coating is applied by the electrostatic spraying method on a metal surface and is cured for 10 min at 200 °C. Later it was realized that the gloss retention and yellowing resistance are not the only advantages that the polyester powder coatings can offer. Excellent mechanical properties, attractive visual appearance, easy tailoring of processing, application and end properties, and an acceptable price/performance ratio are some of the attributes that contributed to these systems becoming the most popular among a wide range of powder coatings offered on the market.

Epoxy containing PES

An interesting modification of polyester resins leading to polymers which contain epoxy groups, thus being suitable to be crosslinked with carboxyl functional polymers or low molecular polybasic acids, is protected by two patents of Hitachi Chemicals [103, 104]. The process involves a reaction of epoxy resin of bisphenol A, hydrogenated bisphenol A type or TGIC with monoalkyl ester of terephthalic acid (typically monomethyl terephthalate) in the first step. In the second step, this product is transesterified with a hydroxyl functional polyester resin having a hydroxyl value between 20 and 200 mg KOH/g and an acid value lower than 20 mg KOH/g, stripping off the methanol. The resulting polyester contains epoxy groups and can be used as a binder or crosslinker in a powder coating formulation together with carboxyl functional materials.

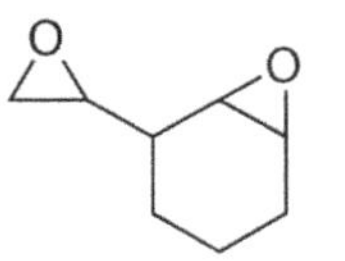

Polyester resins terminated with epoxy groups can be produced by reaction of carboxyl terminated polyester resins in an excess of low molecular weight diepoxide.

Example of a polyester resin terminated with epoxy groups

The patent literature [105] describes such a process using vinyl-cyclohexane-dioxide of formula given left and linear carboxyl terminated polyester composed of neopentyl glycol, phthalic anhydride and hexahydrophthalic anhydride, with a molecular weight of 1800 and an acid number of 62.4 mg KOH/g. Of this polyester 900 g were reacted with 140 g of vinyl-cyclohexane-dioxide to obtain a polyester resin with a melting range between 90 and 95 °C, an acid value of 3.8 and an epoxy equivalent weight of 1035.

Melamine based curing agents

The **sensitivity of the bisphenol A epoxy resins to UV light** restricted their use only in polyester powder coatings for indoor applications. The first attempts to make polyester powder coatings suitable for exterior use were naturally concentrated on the well-known curing chemistry from the solventborne coatings which employ melamine resins as curing agents. More than 20 years of practical experience which had already been built up with the **melamine solventborne coatings** was a guarantee for good outdoor performance of these materials. The use of hexamethoxymethyl melamine (HMMM) as crosslinker for hydroxyl functional polyesters in powder coating formulation is described in a few patents from the early 1970's [106, 107]. Several problems caused this system to **disappear very soon from the market**. The liquid nature of the melamine resins restricts the production of the polyester-melamine powder coatings to the Z-blade kneaders. When this system was developed, extruders were already dominating production equipment for powder coatings with all of the advantages over the Z-blade kneaders which will be discussed in Chapter 5. The second problem facing powder manufacturers and users was surface defects in the form of blisters, due to liberation of methanol during curing. In order to avoid it, the system should not be catalyzed, which resulted in a slow curing rate, requiring a curing schedule of 30 min at 200 °C.

Two different curing agents of a similar type like the melamine based crosslinkers available in a solid form were also developed in the 1970's: **tetramethoxymethyl benzoguanimine** and **tetramethoxymethyl glycoluril** [108]. Although the processing problems were overcome in this way and normal extrusion equipment could be used for compounding, the surface defects and high prices prevented successful commercialization of these powder coatings.

In the meantime, the development in the direction of polyester resins curable with epoxy compounds and blocked isocyanates was so fast and successful, that it swept **polyester/melamine systems** out of the market when they had only just appeared. It was probably the main reason why further research on the amino-hydroxyl systems has ceased, although the biggest part of solventborne coatings for industrial application employs this curing chemistry.

In the last 35 years, three different types of polyester powder coatings established quite strong positions in the market. **Polyester/epoxy powder** coatings well known by the name "hybrids" came as a result of attempts to improve the yellowing resistance of the pure epoxy powder coatings. **Polyester/TGIC** systems dominating the market for exterior durable coatings were developed almost at the same time. **Polyester/isocyanate** or polyurethane powder coatings were the natural consequence of the successful introduction of the solventborne polyurethane coatings. These three types of polyester powder coatings are discussed in further detail in the following sections.

3.4.2.1 Interior polyester powder coatings

"**Hybrids**" is the usual name for powder coatings with polyester resin as a binder and bisphenol A epoxy resin as crosslinker. They are the main representatives of powder coatings used for **interior applications**. It has always been a problem as to how to classify the hybrid systems. The word crosslinker usually refers to the component of the coating which is used in minor amounts to produce a network, reacting with the functional groups of the main binder. In the case of hybrids, the polyester and epoxy resin are used in comparable amounts. In the so called 50/50 systems their quantities in the powder coating formulation are equal. Therefore it is difficult to distinguish between the resin and hardener in this case. Because of better yellowing resistance, the contemporary tendency is towards coatings that employ less epoxy resin than polyester. Let us use this argument as a justification for classifying the hybrids in the group of polyester powder coatings.

Since epoxy resins as curing agents are in most cases two functional, the main binder must have a higher functionality than two. Deviations from the stoichiometric ratio can lead to a marked deterioration of the film properties, such as solvent resistance, boiling water resistance and impact resistance.

Polyester resins for powder coatings are usually characterized by a certain acid number which can be used to calculate the necessary amount of epoxy partner in the coating formulation, knowing its epoxy equivalent value. The following formula can be used for this purpose:

Equation 3.87 $$EC = R \cdot AV \cdot EEW/56100$$

where:

EC = amount of epoxy resin [g]
R = amount of polyester resin [g]
AV = acid value of the polyester resin [mg KOH/g]
EEW = epoxy equivalent weight of the epoxy resin [g]

The reaction speed between the epoxy groups coming from bisphenol A epoxy resin and the carboxyl groups of the polyester is **rather slow**. Normally, the hybrid powder coating should be catalyzed in order to achieve an acceptable curing time at baking temperatures between 160 and 200 °C which are mostly used in practice. It is a rule that the polyester resin manufacturers incorporate the **catalyst in the resin** during its production, providing in this way the most uniform distribution of the catalyst in the powder coating. This is very important, because it is rather difficult to obtain a uniform distribution of very small amounts of catalyst during the extrusion of the powder coating. When additional amounts of catalyst have to be added, a masterbatch technique should normally be used. Catalysts normally used to speed up the reaction rate are tertiary amines or quaternary ammonium salts. In order to get the best results regarding of reactivity and storage stability sometimes **latent catalyst**s are used. A quite new type of these latent catalysts is inclusion catalysts. Such catalysts are based on a complex of included so-named “guest” molecules within the crystal lattice of so-named “host” molecules for example, tetrakishydroxyphenyl ethane, described in a patent of DuPont [109].

But also crystalline amic acids, derived from carboxylic acid anhydrides and diamines, possessing both primary and tertiary amine groups, have been described [110] as latent amine catalysts for hybrid powder coatings. The amic acids exist as **“zwitterions”**, in which the tertiary amine group is protonated, which minimizes both their solubility in the powders and their catalytic activity. On heating, the amic acids undergo thermally induced intramolecular cyclization to imides. The non-zwitterionic cyclic imides contain tertiary amine groups, which catalyze the carboxylic acid/epoxy reaction, leading to cure of the powder coating. An advantage of the crystalline amic acids is that master batching is not needed. Recently BASF [111] introduced latent basic catalysts which form amidines by intramolecular cyclization upon heating. This way they claim to achieve a superior ratio of storage stability to reactivity.

PPG [112] described a polymeric catalyst containing tertiary amine groups, which are bonded to the polymer by a diisocyanate spacer.

Gedan-Smolka et al. [113] developed **two-stage curing mechanisms**, starting from the concept of combined curing. These may enable the processes of film formation and curing to be separated, even in the case of thermosetting powder coatings, such as hybrids. By means of a model system, consisting of epoxide/carboxylic acid/phenolic component, in which the epoxide/acid reaction is faster than the epoxide/phenolic one, they found, by DSC and 1H-NMR analyses, that this two-stage reaction took completely place at 130 °C. That means that by combination of two different curing mechanisms, a significant lowering of the curing temperatures of both individual reactions was realized. This result leads to an optimistic view for application of the two-stage cure of powder coatings, including hybrids, at lower temperatures.

Another way of obtaining low temperature curing epoxy/polyester hybrid powder coatings is reported by *Bergmans* et al. [114]. They claim to have developed two new polyester resins that can be cured at temperatures as low as 120 °C, when combined with low molecular weight type 1 (EEW = 450-550) or type 2 (EEW = ~600) epoxies. Only the polyester resin that gives matt finishes is described in detail. This polyester contains tertiary carboxylic acid groups with a strong intrinsic acidity. Such carboxylate anionic species are excellent nucleophiles for the reaction with epoxide groups and favors esterification addition reactions as the rate determining step, with only negligible other side reactions. The resin should be used with epoxy resins in a stoichiometric molar ratio and the system should be catalyzed with a basic catalyst to promote cure at 120 °C for 30 minutes. In this way 60° gloss values of <10 are obtained. Most **suitable catalysts** include quaternary phosphonium salts, hindered tertiary amines or imidazoles.

The reduction of the curing temperature, thus enlarging the "curing window" is described by *Bayards* et al. [115a]. In their extensive study of 50/50 hybrid powder coatings they mention **six factors that affect the curing window** namely:

- type of (polyester) resin,
- crosslinker,
- catalyst,
- cure window,
- manufacturing method and
- formulation.

The effect of catalyst, epoxy resin, filler and polyester resin architecture on the curing window is shown by means of reversed impact and coating glass transition temperature (coating T_g) building up. Strongest decrease in stoving temperature is observed with increasing catalyst concentration, which depends on catalyst type. The concept ΔT_g (coating T_g minus powder T_g) is used to find a relation with the reversed impact values. Only with the epoxy resin a certain trend is found in this relation. The 50/50 hybrids showed a ΔT_g of 19 to 25 °C. *Bayards* et al. conclude their study with the presentation of a 70/30 hybrid powder that has a curing window of 30 min at 140 °C and/or 10 min at 155 °C.

Barletta et al. describe in their excellent article [116] the surface appearance, the scratch and the wear resistance of hybrid powder coatings. They found a dramatic increase in viscosity at a **conversion degree** of 0.1, gelation at 0.45 and scratch resistance at values higher than 0.55, corresponding to a curing time of at least 20 min at 150 °C.

In order to **improve the leveling properties** of the coating and to reduce the surface defects, suitable additives, whose roles will be discussed later, are usually incorporated. The leveling additives are almost never used as such. Normally, a master batch containing 10 % of the additive in the main binder is used in order to obtain uniform distribution of the additive. However, certain additives like benzoin or polyvinylbutyral resins do not require master batching. The same holds for different types of colloidal silicas which are used as thixotropic agents.

The ratio between polyester and epoxy resin depends on the acid value of the polyester resin and epoxy equivalent weight of the crosslinker. The ratio between the resins is often used to name these systems, such as 50/50, 70/30 etc. the first number indicating the amount of polyester and the second the amount of epoxy resin in the powder formulation.

Table 3.12 [115b] contains different formulations of hybrid polyester powder coating as an illustration. The films after curing are characterized with König pendulum hardness above 200 s, slow Erichsen penetration over 8 mm, reverse impact of 160 inch·lbs and crosshatch adhesion of Gt0. Formulation for 60/40 systems contains choline chloride as catalyst, which is FDA approved, so it can be used to cover surfaces that come in direct contact with food.

Ehsani et al. [117] investigated similar polyester epoxy resin ratios with rheology measurement and DSC and found out that 50/50 systems are much faster than 70/30 since the gelation time increases from 98 to 230 s. A similar slowing down effect was observed by an increase of the filler concentration, which leads to a longer required extrusion time, too.

Higher amounts of epoxy resin in general improve chemical and boiling water resistance of powder coatings, but on the other hand have a negative effect on the yellowing resistance and weatherability. Therefore, systems with low epoxy content have been developed. A patent describing a hybrid that uses only 15 % bisphenol A epoxy resin in hybrid formulation [118] claims excellent yellowing resistance of the cured film combined with excellent flow and impact resistance.

Table 3.12: Polyester hybrid powder coatings [115b]

source: DSM Resins BV

System	50/50	60/40	70/30
"Uralac P 5127" (1)	300	–	–
"Uralac P 5263" (2)	–	360	–
"Uralac P 2450" (3)	–	–	420
"Araldite GT 7004" (4)	300	240	180
"Kronos 2310" (5)	300	300	300
Benzoin	4.5	4.5	4.5
"Modaflow P. III" (6)	9.0	9.0	9.0
Choline chloride	–	3.0	–
Stoving cycles:	15/180	10/180	15/180
[min/°C]	10/200	5/200	10/200

(1) Saturated polyester-acid value of 69–79 (DSM Resins BV)
(2) Saturated polyester-acid value of 48–58 (DSM Resins BV)
(3) Saturated polyester-acid value of 32–38, standard is Uralac P 5071 (DSM Resins BV)
(4) Epoxy resin-epoxy equivalent weight of 715–835 (Ciba, now Huntsman)
(5) Titanium dioxide -rutile type (Kronos Titan GmbH)
(6) Flow control agent (Monsanto)

Two other ways to **improve yellowing resistance of hybrid powders** are described, one by *Thames* et al. [119], the other by *Miller* et al. [120]. *Thames* replaced the carboxylic polyesters at 2 to 6 % equivalent use levels by organosilane diacids. As a result, film flow, corrosion resistance, yellowing and weather resistance were improved, while flexibility and impact resistance decreased with increasing amounts of organosilane diacids.

Miller described two carboxyl functional Ciba (now Huntsman) hardeners, namely "RD 91-411" and "91-412", which have been discontinued without replacement in the meantime.

According to *Frischinger* et al. [121], **high flow ("HF") hybrid powder coatings** can be obtained by the use of a "modified" epoxy resin, which is characterized by an **unconven-**

Table 3.13: Comparison of normal and high flow ("HF") hybrid powder coatings [121]

Epoxy resin	"Araldite GT 7004" (1)	"Araldite GT 6703 ES" (2)	"Araldite GT 6703 ES" (2)	"Araldite XB 6750" (3)
"Araldite" epoxy resin	49.8	49.8	48.7	54.8
"Uralac P 5127" (4)	45.2	45.2	–	45.2
"Uralac P 5998" (5)	–	–	46.3	–
"Araldite GT 3032" (6)	5.0	5.0	5.0	–
Benzoin	0.3	0.3	0.3	0.3
Titanium dioxide - rutile	50	50	50	50
Stoving cycle [min/°C])	←	15/200		→
Flow (internal rating)*	11	4	2	2

**2=best, 18 = worst*
(1) = epoxy resin - EEW. 715–750 (Huntsman)
(2) = novel, modified HF epoxy resin - EEW 690–740
(3) = epoxy resin with flow agent (replaced by Araldite GT 6750)
(4) = saturated polyester resin-acid value 69–79, general purpose resin (DSM Resins BV)
(5) = saturated polyester resin-acid value 69–79, slow cure type, discontinued (DSM Resins BV)
(6) = masterbatch of flow agent in Araldite epoxy resin (replaced by "Araldite GT 2874-1")

tional low melt viscosity in relation to its epoxy content. According to its epoxy content, it can be considered as a type 3½ resin (EEW = ~850). Its softening point is typically that of a type 2 to 2½ (EEW = ~700) epoxy resin. But its melt viscosity is nearly 30 % below that of a type 2 (EEW = ~600) epoxy resin. This anomaly enables the formulation of "HF" hybrid powder coatings with outstanding flow and very good overall performance.

As the viscosity of the epoxy component is always much lower than that of the polyester resin, most profits of the low viscosity of the novel epoxy resin can be obtained by formulating 50/50 hybrid powders. Hybrid powders with a lower epoxy content, e.g. 70/30 systems, show poor mechanical properties. According to *Frischinger*[121], this is partly attributed to the difficulty to obtain a homogeneous mixture upon extruder processing of two resins with more different melt viscosities. Moreover, the reactivity and crosslinking density may be critical. *Table 3.13*[121] gives different formulations of 50/50 hybrids using the novel epoxy resin and other epoxy resins. The films of the powder coatings, formulated in *Table 3.13*, are all characterized by 60° gloss values of more than 95, slow Erichsen indentation of over 8 mm and reverse impact of 160 inch·lbs at a film thickness of 60 μm.

BASF[122] described in a patent a multi layer epoxy based powder coating which can be cured simultaneously.

End use of powder coatings

At present, the most important end uses of polyester/epoxy hybrids cover the following: household equipment such as refrigerators, deep-freezers, washing machines, cookers, and other kitchen equipment, metal furniture, ceiling panels for the building industry, shower cabinets, automotive components, agricultural equipment and machinery, engineering and electrical parts.

3.4.2.2 Exterior polyester powder coatings

Introduction

Powder coatings for exterior application have to fulfill the typical request addressed to all coatings being exposed to water in the form of rain and snow, frost, ultraviolet radiation from sunlight, aggressive industrial atmosphere containing sulfur dioxide etc.

Water and humidity resistance of polyester powder coatings is determined mainly by the nature of the main binder containing ester groups prone to hydrolysis. The **new generation of polyester resins** built up of monomers, which provide improved hydrolytic resistance, in most cases satisfies the normal requirements with respect to the water and humidity resistance. On the other hand, they are characterized by considerable resistance to UV radiation.

The right choice of crosslinker plays an important role in formulating polyester powder coatings with good outdoor performances. The crosslinker should also be resistant to UV light, which excludes the bisphenol A epoxy resins containing ether oxygen. Unfortunately, aliphatic and cycloaliphatic epoxy compounds with good UV light resistance are liquids at room temperature. This makes them unsuitable to be easily compounded into powder coatings using the present melt extrusion technique. Only a few of them are solid at room temperature, but only two have a commercial use; **triglycidyl isocyanurate (TGIC)** and diglycidyl terephthalate (DGT), the former had been by far the more important for many years. But since 1998 it is more and more **replaced in Europe by its less- or non-toxic alternatives**. See later.

Curing agents: TGIC

The first patent protecting the use of TGIC as crosslinker for powder coatings was filed in 1969[123]. According to the invention, 77 to 95 % hydroxyl functional powdered polyester resin with a particle size below 300 μm is dry blended with 5 to 10 % powdered TGIC and 6.5 to 13 % powdered acid anhydride. The powder blend is applied by electrostatic spraying, or by the fluidized bed technique on a metal object which is then cured at 270 °C for 4 minutes, or only 4 seconds at 350 °C. The curing reaction goes in two steps. In the first step, the anhydride reacts with the hydroxyl groups of the polyester resin giving as a result carboxyl functional polyester. In the second step, TGIC reacts with the formed carboxyl groups, crosslinking the systems and building up a network. Although it is claimed that the obtained films have a good appearance, an obvious disadvantage of this system is the inhomogeneity of the coating on a molecular scale, since the constituents are only blended in a dry form.

One year later, a process of melt extrusion of carboxyl functional polyester resins with TGIC as crosslinker, was disclosed in a patent application from 1970 [124]. Very soon in 1972, a polyester/TGIC powder coating for protection of aluminum extrusions and claddings in outdoor architecture was used in Switzerland for the first time [80]. More than twenty-five years of experience in manufacture and application of TGIC powder coatings for outdoor application, have confirmed the reliability of this system for providing long term protection. Up to about 2000, polyester/TGIC systems have undoubtedly dominated the exterior market for powder coatings.

TGIC as a low molecular weight compound decreases dramatically the **glass transition temperature** of the coating. It is a rule that for every weight percentage of TGIC, the glass transition temperature decreases for 2 °C. The amount of TGIC normally used in a standard powder coating formulation is 7 %, although for economical reasons, there is a tendency to reduce the TGIC content to lower values. A patent of DSM Resins [125] describes a system containing only 4 % TGIC. The resins that are used have low acid values between 10 and 26 mg KOH/g and molecular weights between 4500 and 12,000. The high molecular weight of the polyester, and consequently the high shear modulus, decrease the grinding efficiency of the powder coating which could be considered as a small disadvantage.

Hoppe [126] refers to a system containing 5 % TGIC which still has a sufficient crosslinking density, excellent mechanical properties and good flow.

The **amount of TGIC** in the powder coating depends on the acid value of the polyester resin. The best results are obtained when there is a stoichiometric amount of epoxy groups coming from TGIC with respect to the carboxyl groups of the polyester. The following formula can be used to calculate the stoichiometric amount of TGIC, knowing the acid value of the polyester resin:

Equation 3.88 $$TGIC = R \cdot AV \cdot 107/56100$$

where

TGIC	=	amount of TGIC [g]
RR	=	amount of polyester resin [g]
AV	=	acid value of the polyester resin [mg KOH/g]

Table 3.14 illustrates three formulations of powder coatings with different TGIC contents [115b]. The normal curing conditions for polyester/TGIC systems existing on the market at present are 10 min at 200 °C object temperature. Usually it is not necessary to catalyze the system, since the reactivity of TGIC itself is high enough to obtain the desired curing cycle. If faster curing schedules are required, or lower curing temperatures are requested, an incorporation of **suitable catalyst**, as in the formulation 93/7 (*Table 3.14*) is necessary. This is especially the case when aluminum extrusions with plastic inserts have to be coated. Then, the temperature of the object should not exceed 180 °C. *Merfeld* et al. [127] investigated highly reactive TGIC based formulation with up to 0.8 % of choline chloride and benzyltrimethylammonium halogenide, respectively,

as a catalyst. Using special precatalyzed polyesters they were able to cure the formulations at 30 min, 120 °C, but at the same time they observed **undesired prereactions during the processing**. In addition gloss, surface quality and chemical resistance of the coating were compromised, pointing out the **limits of high reactivity**. In this publication other commercially available powder coating systems have been evaluated for low temperature curability, too.

The **yellowing resistance** of TGIC powder coatings at overbake is extremely good; 30 min exposure at 220 °C usually does not give a color change (ΔE) of more than 0.2 to 0.3 NBS.

The first polyester/TGIC powder coatings offered a dramatical improvement in **outdoor durability**, compared to that of the hybrids. While the hybrids show a 50 % reduction of the initial gloss after 50 to 200 hours in the QUV tests dependent on the amount of the epoxy resin in the coating formulation, TGIC systems exhibit 50 % retention of the initial gloss over a period of 500 to 600 hours in the QUV test. In mild climates, such as in Western Europe, these systems perform very well, but exposure in tropical and subtropical climates causes a rather rapid loss of gloss, particularly with dark shades. When exposed to Florida natural weathering, the 60° gloss value of a medium brown (RAL 8014) powder coating decreases to 50 % of its initial gloss after about 2 years. This relatively fast gloss reduction is attributed to photochemical degradation of the polyester resin, caused by light, oxygen, heat, and moisture, generating a photo-induced oxidation of the polyester. The oxidized polyester is further hydrolyzed by action of moisture and heat [128]. According to *Maetens* [129] the CH_2-groups at the β-position of the polyester carbonyl group are first oxidized into anhydrides and subsequently hydrolyzed. *Montserrat* et al. [130] investigated the behavior of hybrids, HAA and TGIC based polyester powder coatings upon thermal aging by DSC and found the latter to show the highest stability.

Table 3.14: Typical polyester/TGIC powder coating formulations [115b] *source: DSM Resins BV*

System	90/10	93/7	95/5
"Uralac P 2220" (1)	900	–	–
"Uralac P 5010" (2)	–	558	–
"Uralac P 4800" (3)	–	–	570
"Araldite PT 810" (4)	100	42	30
"Kronos 2160" (5)	500	300	300
Flow agent (6)	15	9	9
Benzoin	7.5	4.5	4.5
"Araldite DT 3126-2" (7)	–	18	–
Curing schedule	12/180	15/160	15/180
[min/°C]	8/200		10/200
Pendulum hardness (König), [s]	200 to 220	190 to 210	–
Slow penetration (Erichsen)	>8 mm	>8 mm	>8 mm
Reverse impact [inch*lbs]	160	160	160
Cross-hatch adhesion	Gt0	Gt0	Gt0

(1) Saturated polyester-acid value of 48–55 (DSM Resins BV)
(2) Saturated polyester-acid value of 33-38, standard is "Uralac 5201" (DSM Resins BV)
(3) Saturated polyester-acid value of 25-30 (DSM Resins BV)
(4) TGIC (Huntsman)
(5) Titanium dioxide -rutile type (Kronos Titan GmbH)
(6) Flow control agent (Monsanto)
(7) Catalyst (Huntsman)

Super-durable polyester resins

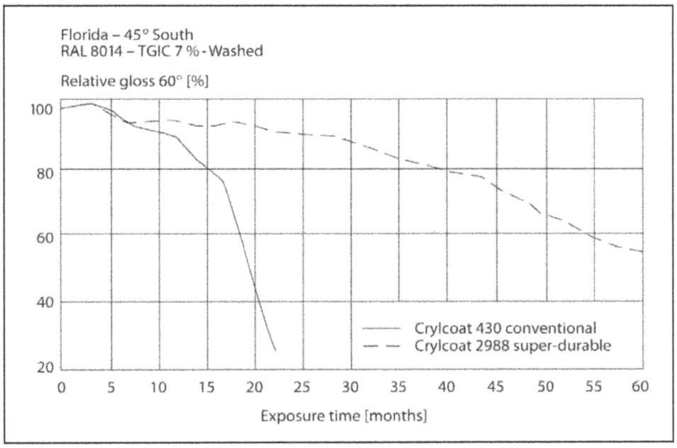

Figure 3.23: Results of exposure tests in South Florida for a conventional and a superdurable polyester powder coating[128]

A major improvement of the UV and hydrolysis resistance can be achieved by replacing terephthalic acid (TPA) by its isomer isophthalic acid (IPA) as the main diacid constituent in the polyester. These so-called **"super-durable" polyester resins** were shown to last at least four times longer [128] in Florida than the conventional TPA-based polyesters (see *Figure 3.23*). Unfortunately, the replacement of the linear TPA by the ring-forming IPA results in a **marked loss of flexibility** of these super-durable polyester powder coatings, which exclude them from use in special applications like coil coating and post-formed metallic substrates, but also from use in architectural coatings, specified by e.g. GSB RAL RG 631 (Germany), Qualicoat Europe or British Standard BS 6496.

15 years ago it has been reported [131], that a special carboxyl terminated polyester resin has been developed by UCB Chemicals (now Cytec) [132], which is launched as a **"flexible super-durable" polyester resin** with outstanding outdoor resistance together with excellent impact and T-bending resistances.

The University of Mumbai [133] in India used successfully **recycled** carboxyl terminated polyester resins from the depolymerization of polyethylene terephthalate (PET) in TGIC formulations.

Properly formulated and cured polyester/TGIC powder coating gives film with excellent hardness, flexibility, impact resistance, abrasion resistance and scratch resistance. The adhesion properties of TGIC systems are in general good, and it is not necessary to use an adhesion promoting primer.

Weak points of polyester/TGIC systems are the flow and leveling of the coatings. The orange peel effect is more emphasized than with other powder coating systems. Considerable work has been done to improve the flow properties of the TGIC powder coatings resulting in new generation resins with lower viscosity and slower viscosity build-up during curing, leading to improved flow and a less emphasized orange peel effect [134].

Zhou, Thames et al. [135] report, that replacing part of the neopentyl glycol (NPG) in polyester resins by 1,3-propane diol (PDO), leads to improved film flow (due to a lower melt viscosity), and a higher T_g, if up to 50 % of the NPG is replaced by PDO.

Evonik-Degussa (now Evonik Industries) [136] recommend a mixture of crystalline and amorphous polyester resins to lower the melt viscosity and thus enhance leveling. In order to avoid undesired haze the crystalline resins should have a melting point from 60 to 120 °C and a melt enthalpy from 70 to 120 J/g. No more than 20 % of these crystalline resins should be used.

Polyester/TGIC powder coatings are mainly used for **protection of outdoor architectural elements**. The surface is in most cases aluminum, but it can also be galvanized steel. In general, the adhesion of the coating on a galvanized steel surface is somewhat inferior in comparison with aluminum. The usual film thickness deposited in one layer is 60 to 80 µm.

Another market segment where TGIC powder coatings are used is for metal furniture for outdoor use (garden furniture) and automotive trim parts such as wheel rims, fenders, bumpers, window frames, windscreen wipers, bicycle frames, motor-cycles etc.

Curing agents: TGIC alternatives

Polyester powder coatings using crosslinkers other than TGIC are also described in the patent literature. A patent of DSM Resins [137] gives a powder coating composition using carboxyl functional polyester resin with an acid value between 10 and 30 mg KOH/g as a main binder and 2 to 9 % **diglycidyl terephthalate** (DGT) as a crosslinker. The use of diglycidyl hexahydrophthalate is also described in the same patent. The outdoor durability of DGT/polyester powder coating is comparable with that of the TGIC systems. However, a disadvantage of DGT is its higher epoxy equivalent weight than TGIC. Therefore, higher amounts of DGT by weight have to be used in order to maintain a stoichiometric ratio between the epoxy and carboxyl groups. Consequently, the decrease of T_g of the powder coating is stronger and therefore polyester resins with higher T_g's have to be used in combination with DGT as crosslinker. This narrows the choice of raw materials which can be used for making the polyester resin.

Table 3.15: Typical polyester/"Araldite PT 910" powder coating formulations [115b] *source: DSM Resins BV*

Component	I	II
	Parts by weight	
"Uralac P 3495" (1)	558	558
"Araldite PT 910" (2)	42	42
"Accelerator DT 3126" (3)	–	6
"Kronos 2160" (4)	300	300
Flow control agent (5)	9	9
Benzoin	4.5	4.5
Stoving cycle [min/°C]	10/200	6/200 or 15/170
Substrate	Q-panels, type AL-46, 0.8 mm	chromated aluminum
Film thickness [µm]	50	50
Gloss 60°	94	94
Erichsen slow penetration [mm]	>8	>8
Reverse impact [inch · lbs]	60	60
Cross-hatch adhesion	Gt 0	Gt 0
Pendulum hardness, König [s]	190 to 210	190 to 210

(1) Polyester resin (DSM Resins BV), acid value 25 to 30 mg KOH/g, T_g = 66 °C.
(2) TGIC-alternative (Huntsman)
(3) Catalyst (Huntsman)
(4) Titanium dioxide (Kronos Titan GmbH)
(5) "Modaflow Powder III" (Monsanto); or "Resiflow PV 5", (Worlée); or "Byk 365 P", (Byk-Chemie)

A crosslinker, comparable with DGT, is described by *Gottis* and *Cotting* [138]. It is a **"solid solution"** of 1 part (by weight) of the liquid triglycidyl trimellitate (TML) in 3 parts of the crystalline DGT. That means: an optimal functionality between 2 and 3 is obtained in this way, by mixing a solid crystalline compound (DGT) with a liquid one (TML) in a ratio leading to a crystalline solid solution. In 1994 this crosslinker came to the market under the trade name "Araldite PT 910" and can be considered as an alternative to TGIC, which shows some evidence of mutagenic potential.

A somewhat more reactive type of "PT 910" was later introduced as "Araldite PT 912".

Both "PT 910" and "PT 912" need, however, **labeling**, due to their irritating properties. "Araldite PT 910" is described in detail in Chapter 3.3.1.2. *Table 3.15* shows two formulations of high-gloss white powder coatings, to be cured with "Araldite PT 910". According to *Buysens* [139] a duplex system consisting of a zinc and a TGIC free polyester powder coating ensures an **excellent corro-**

sion protection even in under aggressive conditions. Dry rural areas (C2) show a corrosion rate of 0.1 to 0.7 µm zinc/year, whereas industrial or maritime areas with high humidity and high salt level, respectively (C4), lead to a corrosion rate of 4 to 8 µm zinc/year.

Curing agents: β-hydroxyalkylamides (HAA)

Some other crosslinkers have also been introduced in the market as TGIC-alternatives. The most important types are β-hydroxyalkylamides (HAA) with the trade name "Primid" or "Vestagon EP-HA" (see Chapter 3.3.7). As there has been mentioned, the curing reaction between a carboxylated polyester and a **β-hydroxyalkylamide (HAA)** is not a polyaddition (as with TGIC), but a poly-condensation reaction, which means, that water is set free during stoving. And, due to the presence of amide groups, the yellowing (mainly in direct-fired gas ovens) and overbake resistance of HAA-based powder coatings are not as good as with TGIC. Furthermore, HAA-based powder coatings are relatively **sensitive to humidity**, water and detergent solutions under severe test conditions (e.g. elevated temperatures).

Therefore the following formulation guidelines for HAA-based powder coatings are recommended [140] by EMS:

- **Prevent micro-pinholing with thick coatings**
  Due to the polycondensation reaction and the polar nature of HAA, resulting in differences in surface tension between HAA and the non-polar polyester, powder coatings based on HAA tend to be more sensitive to micro-pinholing than traditional powders, based on TGIC. This limitation can be greatly reduced by selection of an optimized HAA-designed polyester resin, combined with good formulating practices. Numerous degassing additives are available, which allow the formulation of HAA-based coatings micro-pinhole-free up to a film thickness of more than 100 µm. The strongest degassing agent is benzoin. A minimum of 0.2 % (calculated on total powder) is recommended as a starting point for all formulations. In white and pale shades this level should not be exceeded, if there is a concern regarding overbaking. In such cases the degassing properties of benzoin can be enhanced by the use of "supplementary" additives (or extenders), such as polyolefin waxes, hydrogenated castor oil derivatives, or crystallized aluminum hydroxide. The use of supplementary degassing additives may have adverse effects on other properties, such as glass transition temperature (T_g) and/or storage stability. Therefore pretesting is always recommended.
- **Select a HAA-designed polyester resin**
  Resins, specifically developed for curing by HAA, are available from nearly all well-known resin manufacturers in Europe. UCB (now Cytec) [141] recommended an isophthalic acid rich carboxyl functional polyester, with linear chain aliphatic (poly)ether diols or longer chain aliphatic diols [142] to provide good flexibility in addition to superior weather-stability. A Japanese patent [143] claimed the use of melamine modified resins to decrease the water sensitivity of HAA systems, thus enabling them to show full adhesion after 500 h at 40 °C in water. Rohm and Haas [144] described the benefits of phosphoric acid modified polyesters on adhesion on metal substrates. They recommended in particular NPG based polyesters since they show no dehydration, due to the lack of a β-H atom. DSM filed a patent [145] describing branched polyester resins, which are partly defunctionalized with a monoacid like tert. butyl benzoic acid. These resins are claimed to need less costly hardener, show higher reactivity and still have sufficient flow. According to the same company [146] branched polyester resins terminated with aromatic diacids require a curing time of 12-15 min at 155 and show no blooming (a haze on the paint surface caused by condensation of gaseous products produced during the curing – in particular visible at dark colors). Hexion [147] introduced **branched polyesters for low temperature curing**

(15 min at 150 °C or 30 min at 140 °C) containing dimer fatty acids to ensure sufficient flow. Cytec [148] offers lower temperature curing polyesters for HAA and glycidyl methacrylat based powder coating formulations, too, especially for outdoor applications.

- **Avoid overbaking discoloration**
  In sensitive colors, e.g. whites and pale shades, use a maximum of 0.2 % benzoin. When this is not sufficient, use non-yellowing supplementary degassing additives (see above).
- **Improve direct-fired gas oven stability (if required), using "external" stabilizers**
  Powder coatings are sensitive to discoloration during the stoving process. This problem is aggravated by the use of direct-fired gas ovens, due to reactions between the coating and the combustion products (mainly nitrogen oxides NO_x). The degree of discoloration depends on the type and color of powder coating and the degree of stabilization. A number of HAA-designed resins are already available, which are specifically "internally" stabilized for direct-fired gas oven applications. However, discoloration can also be reduced (but not completely prevented) by incorporation of "external" stabilizers (e.g. "Irgafos PEP-Q") in the powder coating formulation. Recommended level is 0.2 to 0.6 % on total powder. In a recent patent [149] DSM recommends transition metal halogenides like copper bromide. Note, that these stabilizers have a maximum effectiveness level, whereby exceeding this level has no further benefit. Excess stabilizer may even result in discoloration. Here too, pretests are always necessary. It should be remarked here, that commonly used heat stabilizers and anti-oxidants (such as organophosphites) are not suitable for direct-fired gas oven applications. Sometimes an appropriate "co-stabilizer", such as sodium benzene phosphinate, may be used to enhance overbake stability, without adversely affecting gas oven stability [140].
- **Improve detergent and corrosion resistance, if required**
  In accelerated corrosion tests, HAA-based coatings may be susceptible to color change and gloss reduction. But for high-gloss pigmented HAA-based powder coatings holds: Detergent and corrosion resistance can be brought to the level of polyester/TGIC systems by using resins, specifically designed for curing by HAA.
- **Optimize the properties of HAA-cured powder coatings by formulating with crystallized aluminum hydroxide, if high gloss is not required**
  EMS Chemie [140] has found, that, apart from its improving effect on detergent and corrosion resistance, a milled grade of crystallized aluminum hydroxide (mean particle size of 10 μm) reduces gloss of HAA-based powder coatings very effectively, but in contrast to other fillers (or extenders), it has a distinct positive effect on film flow. It also reduces the sensitivity of HAA-cured polyester powder coatings to micro-pinholing, as its acts as a degassing agent. A fourth benefit of the aluminum hydroxide is, that it increases the T_g of the powder and by consequence its storage stability). Kansai Paint [150] claimed that addition of strontium carbonate-modified aluminum phosphate anticorrosive agents improve water sensitivity and corrosion resistance of HAA coatings.
- **Check the tendency of "static build-up" of the powder**
  HAA-based coating powders show sometimes a static build-up during handling or spraying, which is demonstrated by reduced free flowing properties. If this occurs, the powder must be given a "secondary treatment" [151] with a highly dispersed silica or aluminum oxide or it must be reformulated by incorporation of a small amount of an antistatic agent.

Table 3.16 illustrates a formulation for a high-gloss white, gas oven stabilized, HAA-based powder coating, which is also pinhole-free in thicker films.

The use of hydantoin epoxies containing two to four epoxy groups as crosslinkers for carboxyl functional polymers in powder coatings is described in a patent, assigned to DuPont [152]. It is claimed that the outdoor durability especially in the case where carboxyl acrylics are used as main binders, should meet the requirements of automotive clear topcoats.

In a patent of Bayer AG [153] the use of triglycidyl-triazolidine-3,5-dione as a crosslinker for powder coatings for outdoor use is disclosed. Comparable or slightly better gloss retention on outdoor exposure is claimed compared to the powder coating compositions where TGIC is used as a crosslinker.

Table 3.16: Typical formulation for a high-gloss white polyester/HAA powder coating [140]

Component	Parts by weight
"Crylcoat 7630" (1)	560
ß-Hydroxyalkyl amide (2)	30
"Resiflow PV 88" (3)	8
Benzoin	2
"Kronos 2160" (4)	400
Curing schedule [min/°C]	10/180
Substrate	Q-panel, R-36, steel 0.8 mm
Film thickness [µm]	60
Gloss 60° [%]	93
Erichsen slow penetration [mm]	>8
Reverse impact resistance [inch·lbs]	160
Cross-hatch adhesion	Gt 0
Pencil hardness, Wolff-Wilborn Type 291	H

(1) Polyester resin (Cytec), acid value 31-36 mg KOH/g, stabilized for gas fired ovens
(2) "VESTAGON HA 320" (Evonik Industries) or "Primid XL-552" (Ems)
(3) Flow agent (Worlée)
(4) Titanium dioxide, rutile type (Kronos Titan GmbH)

Curing agents: acid anhydrides

Hydroxy functional polyester powder coatings cured with acid anhydrides exhibit good outdoor durability. *Walz* and *Kraft* [154] compared the gloss retention of powder coating based on saturated polyester cured with modified ethylene glycol bis trimellitate (MEGBM) with that of epoxy/DICY and polyester/epoxy hybrid powder coatings. Results presented in *Figure 3.24* indicate clearly the potentials of the polyester/anhydride system for exterior use.

An advantage of polyester/anhydride powder coatings is the fast curing rate which allows stoving schedules of 10 min at 180 °C to 30 min at 140 °C. Powder coatings in which the polyester resin is based on isophthalic acid need higher stoving temperatures or longer stoving times to bring about complete curing. Despite the short curing cycles and high reactivity, storage and processing stability of polyester/anhydride powder coatings are good enough from a practical stand point. In *Table 3.17* a typical formulation of a polyester/anhydride powder coating is presented [154].

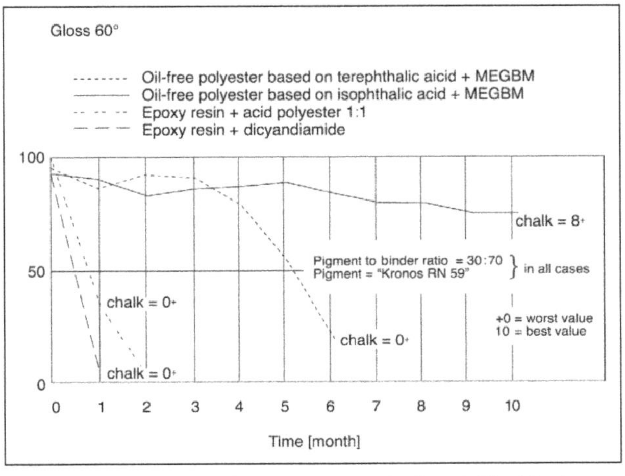

Figure 3.24: Results of exposure tests in South Florida for different powder coating systems [154] source: *International Conference in Organic Coatings, Sci. and Technol., Athens/Greek*

Polyester/anhydride systems have not gained market shares compared to their performances. The main reason is the **toxicity of the hardeners** which are heavy respiratory and mucous membrane irritants. In some countries their use is even pro-

Table 3.17: Powder coating composition based on saturated polyester resin and ethylene glycol bistrimellitate hardener (EGBM) [154] *(Source: International Conference in Organic Coatings, Athens)*

Component	Parts by weight
Oil-free polyester	51.7
Modified EGBM	17.2
Titanium dioxide	29.6
Flow agent	1.5
Stoving schedule [min/°C]	10/200
Gloss 60°	94
Impact resistance [inch·lbs]	160
Processing stability expressed as viscosity increase during 10 min processing at 120 °C (η_{10}) compared to initial (η_0); (η_{10}/η_0)	4

hibited. Although the so-called resinous anhydrides or bistrimellitates are less problematic from a toxicological point of view, they did not contribute very much to a greater acceptance of these powder coating systems by the market.

It is reported by *Schmidt* et al. (see Chapter 3.3.5), that for **automotive clear topcoats**, based on epoxy functional acrylates, a polyanhydride of an aliphatic dicarboxylic acid (such as dodecane dicarboxylic acid) as crosslinker, gives very promising results. The clear coat seems to equal or outperform both the one and two component solventbased clear topcoats in many aspects, being only inferior with respect to solvent resistance. This polyanhydride is a partially crystalline, non-toxic compound with a melting point of 80 to 90 °C, acid number of 300 and a mean molecular weight of 2000.

Instead of hydroxyl functional polyesters other OH-polymers can be used as powder coating resins. The T_g of polyethers is too low, but polycarbonates do qualify in every respect. Although they have remarkable properties they haven't found a broad application, yet, presumably do to cost reasons. BASF [155] claimed that hyperbranched polycarbonates show better flow and higher reactivity compared to conventional compounds.

Curing agents: polyurethanes

The full success of polyurethane solventborne coatings was naturally followed by trials to use the same curing chemistry in the powder field. To prevent the premature reaction of the isocyanate groups with the functional groups of the polyester resin it is necessary to use blocked isocyanates as crosslinkers. Blocked IPDI and TDI derivatives with ε-caprolactam as blocking agents are the most commonly used crosslinkers for polyurethane powder coatings. The toxicity and very poor weatherability and yellowing resistance of TDI restrict its use to indoor cheap polyurethane powders. **IPDI derivatives** which are described in detail in Chapter 3.3.2 are for the time being the **most important commercial crosslinkers**. Caprolactam blocked IPDI oligomers release the blocking agent during curing. Therefore, for thick layer application these crosslinkers are not the best choice because of the possible surface defects in the form of blisters. On the other hand, the blocking agent contributes very much to the flow improvement of the polyurethane powder coatings; namely, the curing reaction does not start until deblocking takes place. This results in a permanent viscosity decrease during the increase of the temperature at the stoving period of the coating. This is not compensated by a viscosity increase as a result of a network formation and an increase of the molecular weight as is the case with epoxy or polyester/TGIC systems. Moreover, once the blocking agent is released, it only remains for a short time in the film before it evaporates, acting as a solvent, contributing to a further decrease of the melt viscosity of the coating. Therefore, polyurethane powder coatings have **the best flow properties** among the thermosetting powders. In many cases, the flow and leveling properties of the polyurethane powder coatings can be even compared with those of the solventborne coatings. The presence and release of caprolactam, on the other hand, is a nuisance from an ecological point of view, slightly spoiling the image of the powder coatings as environmentally friendly materials. However, the released amount of caprolactam per square meter of coated surface is much less than the amount of solvents emitted by the liquid coatings.

Recently Bayer [156] described an unusual blocking agent free polyurethane powder coating system based on a mixture of micronized trimeric IPDI (<10 μm) and crystalline polyesters. Curing is completed after 15 min at 100 °C generating a mat finish. No data are disclosed concerning the storage stability.

A new **polyester-urethane system** is described by *Williams* and co-workers at Cray Valley Iberica [157]. The system is based on blocked hexamethylene diisocyanate trimer (HDT), hydroxylated polyesters and catalysts. The blocking agent consists of high boiling point esters of p-hydroxy benzoic acid instead of caprolactam. The solid blocked HDT is commercially available as “Rhodacoat P 8500”. The loss of weight of “P 8500” is lower than for the ε-caprolactam blocked curing agent, because evaporation of the blocking agent is only initiated at temperatures beyond 210 °C, although the initial deblocking temperature is approximately 110 °C. As the T_g of “P 8500” is low (30 °C), polyester resins with high T_g are needed for good or sufficient storage stability. It is claimed, that weather resistant coatings can be formulated from gloss to matt finishes. Low gloss can be obtained by incorporating a special catalyst. With such a system it is possible to formulate a reproducible low gloss in a “one-shot” process.

Polyurethane powder coatings using uretdione oligomers of IPDI can be considered as coatings containing **internally blocked isocyanates**. As a result there is no release of blocking agent during curing. This is an advantage from an ecological point of view, but it must be admitted that the flow properties and gloss of these coatings are somewhat inferior compared to caprolactam blocked systems.

Some years ago Bayer introduced a polyuretdione crosslinker with the trade name “Crelan VP LS 2147”, which is characterized by an extremely low melt viscosity and excellent flow properties. Due to the low T_g of the crosslinker a detrimental effect on the storage stability of the powder coating formulation has to be taken into account, though.

Somewhat earlier Hüls (now Evonik Industries) succeeded in increasing the reactivity of their “Vestagon BF 1540” and launched their improved uretdione hardener under the trade name “Vestagon BF 1320”. “BF 1320” as well as “LS 2147” is described in detail in Chapter 3.3.2.2.

IPDI derivatives like all aliphatic or cycloaliphatic isocyanates have a slow reaction rate. A proper **catalyst is a prerequisite for adjusting the curing schedule** of the system to the most common requests in practice. Tertiary amines can be used for this purpose. However, metal catalysts are by far the most effective for this system. It has been found that compounds containing tin, bismuth, zinc, titanium, iron, antimony, zirconium, cobalt, nickel, aluminum and many other metals can in principle be used as catalysts for NCO/OH curing. However, catalysts of commercial importance are the organotin compounds such as dibutyltin dilaurate (DBTL), dibutyltin oxide (DBTO), stannous octoate (SnOct) and bismuth neodecanoat. It is common practice for the resin producer to incorporate the catalysts in the resin during the production process, in order to ensure the most homogeneous distribution of the catalyst in the system.

For replacing the toxic organotin compounds *Blank* [158] recommends Bi-, Al- and in particular **Zr-chelates**. The latter is claimed to be most effective and is even faster than DBTL.

Polyesters used for polyurethane powder coatings usually contain hydroxyl and to a minor amount carboxyl functional groups. The ratio between them depends on the stoichiometry between the reactive monomers in the formulation of the polyester resin, and on the degree of conversion of the carboxyl groups during esterification. Both hydroxyl and carboxyl groups can react with the polyisocyanate crosslinker. In an excellent article *McLaferty, Figlioti* and *Camilleri* discuss the influence of the hydroxyl/carboxyl ratio in the polyester resin upon the curing characteristics of IPDI and TDI based polyurethane coatings [159]. The same is done and discussed for uretdione based polyurethanes by *Bronk*[160]. Evonik-Degussa (now Evonik Industries) [161] recommends a mixture of crystalline and amorphous hydroxyl functional polyester resins to lower the melt viscosity and thus enhance leveling and flexibility. In order to avoid undesired haze the crystalline resins should have a melting point from 60 to 120 °C and a melt enthalpy from 70 to 100 J/g. No more than 20 % of these crystalline resins should be used.

The effect of end-capping carboxyl groups in hydroxylated polyesters is described by *Wu, Padaki* and *Maetens*[162]. As hydroxylated polyesters are formulated with an excess of hydroxyl groups, a very high part of the carboxyl groups is consumed, and the resin chains are completely "hydroxyl end-capped" and can be seen as "fully condensed". If the acids have not completely reacted, then the resin will have both acid and hydroxyl end-capping and such resins are mentioned "under-condensed". The objective of their study was to verify if a small amount of carboxylic acid is capable of catalyzing the hydroxyl-isocyanate (OH/NCO) reaction during polyurethane formation. It was found, that residual carboxylic end groups in hydroxyl polyester resins do not accelerate the curing. Some other conclusions from this study are: under-condensed resins give more overbake yellowing than condensed resins and less salt spray resistance. The residual carboxyl groups in under-condensed polyester resins, which may cause worse film properties, can be partially consumed by an epoxy curing agent, but not by isocyanates.

Ratliff et al. [163] compared different powder coatings systems regarding corrosion resistance on a variety of steel substrates. Although the pretreatment of the substrate was crucial **polyurethane formulations** turned out to show **the best overall performance**.

Typical formulations of polyurethane powder coatings using caprolactam blocked IPDI oligomer and internally blocked uretdione derivative of IPDI are presented in *Table 3.18.*

Table 3.18: Polyurethane powder coatings for outdoor application *Source: Evonik Industries*

Material	Parts by weight-%			
	A	B	C	D
"Uralac P 1580" (1)	41.4		46.8	
"Crylcoat 690" (2)		55.4		49.7
"Vestagon B 1530" (3)	17.1	8.1		
"Vestagon BF 1320" (4)			21.7	8.8
"Kronos 2160" (5)	40.0	35.0	30.0	40.0
"Resiflow PV88" (6)	1.0	1.0	1.0	1.0
Benzoin	0.5	0.5	0.5	0.5
Stoving schedule [min/°C]	12/200	20/200	12/200	15/200
Impact dir [inch·lbs]	80	20	50	20
Stoving loss [%]	2.5	1	0	0

(1) Hydroxyl functional superdurable polyester resin (DSM Resins BV)
(2) Hydroxyl functional superdurable polyester resin (Cytec)
(3) Caprolactam blocked IPDI derivative (Evonik Industries)
(4) Uretdione derivative of IPDI (Evonik Industries)
(5) TiO_2-rutile (Kronos Titan GmbH)
(6) Flow promoting agent (Estron)

Polyurethane curing agents: matting

A double curing mechanism as a means of producing **matt polyurethane coatings** employs carboxyl/epoxy curing together with an hydroxyl/isocyanate curing reaction [164]. A special hardener, "Crelan KL5-2668", containing carboxyl and blocked isocyanates as functional groups is made, reacting 2 moles of isophorone diisocyanate with 2 moles of caprolactam and 2 moles of 2,2-bis-(hydroxymethyl)-propionic acid. In combination with TGIC as co-crosslinker, hydroxyl functional polyester, pigments and suitable flow additive, a powder coating can be obtained having gloss below 30 % after stoving. It is essential to keep the ratio between the NCO content and the acid number in the hardener between 0.1 and 0.3 in order to obtain matt films. The control powder coating formulation, in which the ratio mentioned above is 0.8, has a gloss of 77 %.

A typical disadvantage of this matting system is its sensitivity to traces of pretreatment chemicals, causing glossy spots or stripes.

In 1997 Bayer [165] launched a method for a **non-blocked matt polyurethane powder coating system**, which can also be manufactured in a "one-shot" process. This means the troublesome dry blending (or co-grinding) of two powders can be avoided. The principle behind this system is the combination of an internally blocked poly-uretdione hardener with a long-chain dicarboxylic acid, such as dodecanedioic acid. The uretdione based hardener reacts with the hydroxylic polyester, and the dicarboxylic acid with a hydroxyalkyl amide. The parallel crosslinking reactions occur at very different rates and therefore interfere with each other, causing a kind of film disturbance and by consequence very matt films with 60° gloss values of about 10 to 15 %. The degree of matting can actually be controlled via the ratio of uretdione/dicarboxylic acid. With increasing uretdione content gloss increases. The powder formulator can therefore easily produce the required degree of matting by adjusting his formulation. The cured matt films yield good mechanical and chemical resistance properties, and a 60° gloss value of about 14 %.

Based on the same idea of the double curing mechanism, Creanova (now Evonik Industries) came in 1998 with an amine-modified IPDI-uretdione adduct [166]. Surfaces with reproducible matting degree can be obtained, giving 60° gloss values of about 11 % and excellent film flow.

In 1999 another attempt was made by *Thometzek, Freudenberg and Meier-Westhues* of Bayer [167] to obtain really deadmatt films (gloss only 1 %). They used a hardener, "Crelan VP LS 2181/1", which has two different functionalities in the molecule. These two polar groups are again isocyanate and carboxyl, and they allow two parallel curing reactions. The blocked isocyanate reacts with a hydroxylated polyester, and hydroxyalkyl amide or epoxy resins are used to co-react with the carboxyl group.

With this system the degree of gloss is unaffected by the curing time and/or by variations of the curing temperature (which should be about 180 °C).

It is also possible to combine the blocked "Crelan VP LS 2181/1" hardener with a uretdione crosslinker to vary the gloss between 1 and 85 %. In these formulations the film leveling, mechanical properties and chemical resistance of the coatings seem to be excellent. If this method gives reproducible results at low gloss levels, then it will find a market place for low gloss façade finishes.

Table 3.19: Comparison of outdoor resistant powder coatings [80] *source: Product Finishing Magazine, Cincinnati, Ohio*

Property	Acrylic (GMA)	Polyurethane	Polyester/TGIC	Polyester/HAA
Weathering resistance	+++	++	++	++
Mechanical properties	–	++	++	++
Corrosion resistance	•	+	+	•
Storage stability	-	++	++	++
Leveling	++	++	+	+
Edge covering	•	+	+	+
Resistance to overcuring	+	+	++	•

+++: outstanding ++: very good +: good •: average –: bad

Curing agents: summary

Table 3.19 gives a comparison between three different **powder coating systems for outdoor application** [80]. It is interesting that the acrylic coatings which will be discussed in the following chapter do not range higher than the polyester/TGIC systems with respect to weathering resistance, although in the liquid systems, acrylics are by far the best ones for outdoor use. This corresponds with the results reported by *Gulpen* and *Van de Werff*, who performed weathering tests on different powder coatings and compared them with the conventional wet systems [23]. Although the comparison in *Table 3.19* is qualitative only, and some objections can always be addressed, it gives a good indication of expectations from various outdoor powder coating systems. The table is also of interest, when compared with the present-day situation.

3.4.3 Acrylic powder coatings

Introduction

It is interesting that, despite the fact that acrylic solventborne coatings are the best thermosetting materials for outdoor application, in the powder field they do not have the importance which might normally be expected. Market analysis shows that even in Japan, which has always been the biggest market for acrylic powder coatings, their **market share is declining**.

Several reasons can be considered as relevant for this position of the acrylic powders. The mechanical properties of acrylics, which are generally inferior, compared to the other coating systems, are even worse in the case of acrylic powder coatings. Surprisingly, the weathering resistance, which is the strongest point of the acrylics, was not as good as with the solventborne acrylic coatings. The latest developments indicate that the **weathering properties** of pure acrylic powder coatings have improved very much [168] and it might be expected that in the near future much more attention will be paid to this system. This could be especially the case if improved flow properties, combined with environmental compliances contribute to more extensive use of powder coatings in the automotive field.

Nowadays, these expectations have been realized by the introduction of the acrylic based automotive **primer-surfacer** and acrylic **clear topcoat**. Details will be given at the end of this chapter.

Polymethyl acrylate and polymethyl methacrylate produced by free radical polymerization were the first acrylic polymers manufactured on an industrial scale. The products under names of “Plexigum” and “Perspex” were introduced in 1927 by Röhm&Haas AG in Germany and Imperial Chemical Industry in England [169]. Although acrylic ester polymers can be obtained by anionic polymerization, thermal polymerization in the absence of catalyst, photo-polymerization and radiation-induced polymerization, **free radical polymerization** is the predominant synthetic route in industrial manufacture.

Since powder coating resins are 100 % solid materials, bulk polymerization of acrylic esters should be the most attractive production route. However, technical problems with respect to the removal of the exothermic heat during polymerization, restricts the use of this method almost exclusively to manufacturing articles by casting. Although the bulk polymerization process using vinyl versatate as a co-monomer is a known process, there is no evidence in the literature for practical use of this production method in manufacturing powder coating resins.

Polymerization in solution is carried out by adding the monomers to an organic solvent in which both the monomer and the polymer are soluble. Typical catalysts are organic peroxides, hydroperoxydes and azo catalysts. The molecular weight of the polymer is controlled by selection and concentration of the initiator, the type of solvent, monomer concentration and temperature. When the polymerization is accomplished, the solvent is removed either by vacuum distillation or by a spray drying technique.

Emulsion polymerization is widely used for the manufacture of acrylic dispersions for waterborne coatings. The monomer blend is dispersed in the water phase by means of agitation and suitable surfactant. Polymerization takes place in the water phase initiated by water soluble catalyst. Emulsions with a particle size between 0.01 and 5 μm are easily obtained. The precipitation and drying of such small particles is rather a difficult process. Therefore, the suspension polymerization process, by which particles with a diameter of a few millimeters are obtained, is preferred.

Suspension polymerization is carried out in a similar manner to emulsion polymerization. The monomer added to water is dispersed by agitation in droplets of a few millimeters. Unlike emulsion polymerization, reaction is initiated by a catalyst soluble in the monomer blend. As a result, the polymerization takes place entirely in the monomer phase. To overcome sticking of the droplets at certain critical moments when polymerization is partially complete, a small amount of protective colloids such as polyvinyl alcohol is added to the batch. Removal of the particles by filtration, their purification by washing with water and drying are the following steps after polymerization has been completed. The product is in the form of small beads which are easy to handle and process.

Typical monomers used for production of acrylic resins are acrylic and methacrylic esters of C_1–C_8 alcohols. Different functional groups can easily be introduced by **co-polymerization** with monomers such as acrylic and methacrylic acid, hydroxyl ethyl acrylate and methacrylate, hydroxyl propyl acrylate and methacrylate, glycidyl methacrylate, acryl amide, methylol acryl amide, butoxymethyl acryl amide etc. In addition, acrylic and methacrylic esters co-polymerize readily with practically all types of vinyl monomers. This versatility provides a possibility for development **resins with a broad variety of properties**.

The **importance of the glass transition temperature** as an essential parameter determining the stability of powder coatings is discussed in Chapter 4. Glass transition temperatures of acrylic homopolymers vary in a broad range, as presented in *Table 3.20*[170]. By co-polymerizing acrylic monomers, polymers with intermediate glass transition temperatures can be prepared. An empirical equation proposed by *Fox*[171] can be used to predict the glass transition temperature of the copolymer from the known values of the homopolymers:

Equation 3.89:

$$\frac{1}{T_g} = \frac{W_1}{T_{g1}} + \frac{W_2}{T_{g2}} + \frac{W_3}{T_{g3}} + \ldots \frac{W_n}{T_{gn}}$$

where $W_1, W_2 \ldots W_n$ are the weight fractions of each acrylic monomer in the copolymer, and T_{g1}, $T_{g2} \ldots T_{gn}$ are the glass transition temperatures of the respective homopolymers.

Table 3.20: Glass transition temperatures of various acrylic homopolymers [170] *source: John Wiley & Sons, Inc.*

Polymer	T_g [°C]	Polymer	T_g [°C]
Polyacrylates:		Polymethacrylates:	
methyl	8	methyl	105
ethyl	–22	ethyl	65
n-propyl	–51	n-propyl	33
n-butyl	–54	n-butyl	20
2-ethylhexyl	–55	n-octyl	–20

Reactivity of acrylic monomers differs considerably. Consequently, wide differences in the reactivity ratios of many co-monomer mixtures can be expected.

This leads to heterogeneity in the composition of the chains an in the distribution of the monomer units along the chain. Although methods for controlling heterogeneity of the copolymer based on programmed addition of the co-monomers have been developed, it is believed that this is one of the main reasons for rather **poor mechanical properties** of acrylic powder coatings.

In polycondensation types of polymers, such as polyesters, the functional groups can be located at the chain ends and the distance between the branching points in the chain can easily be regulated by choosing right ratio between the reacting monomers. In the case of acrylic resins, it is not easy to choose the position of the functional groups. They are distributed at random not only along the chain, but also among the macromolecules, thus having chains with very low functionality or no functionality at all, and chains with very high functionality. Since the part of the chain with the so-called dead end having no functional groups does not contribute much to the impact resistance (chains with low or no functionality), and, on the other hand, the impact resistance is inversely proportional to the crosslinking density for the same molecular weight of the crosslinked polymer (chains with high functionality), the overall result is crosslinked film with low flexibility and poor impact resistance. Modern methods to functionalize acrylic resins specifically at the end positions, like ATRP (atomic transfer radical polymerization), are still to expensive for a use in powder coatings.

Curing agents: carboxylic acids

Acrylic powder coatings which are used in practice are almost exclusively based on **epoxy functional acrylic resins cured with long chain dicarboxylic acids as hardeners**. The long aliphatic chain of the crosslinker, e.g. dodecanedioic diacids (DDDA), provides flexibility and impact resistance to the cured film, but still far below the values which are usually obtainable with the other powder coating systems. In most cases the impact resistance of the acrylic powder coatings does not exceed 30 inch·lbs. This is still several times lower compared to the values of polyester and polyurethane powder coatings.

A typical **acrylic powder coating with glycidyl functionality** and dicarboxylic aliphatic acid as hardener is described in the patent of Glidden Co. [172]. Acrylic resin prepared by suspension polymerization of ethylene unsaturated monomers, including glycidyl monomers, is formulated into a powder coating together with a stoichiometrical amount of 1,10-decanedicarboxylic acid (with respect to the epoxy/carboxyl ratio), flow agent and titanium dioxide in the case of pigmented coatings, or without pigments to produce a clear coating. The coatings are prepared by extrusion of the components at 80 °C, and pulverization in the presence of liquid nitrogen to keep the temperature below 20 °C during the milling operation.

A patent of DuPont [173] discloses a system in which the glycidyl functional acrylic resin containing 12 to 20 % glycidyl methacrylate as co-monomer, the main crosslinking agent being aliphatic dicarboxylic acid with 10 to 20 carbon atoms, blocked aliphatic diisocyanate, hexamethoxymethyl melamine or tetramethoxy methyl glycoluryl as co-crosslinkers in minor amounts and UV stabilizer of benzotriazole type. The unpigmented composition can be used as an automotive clear topcoat characterized with high gloss, smooth surface and excellent weathering resistance, but the **mechanical properties are not good enough** for this purpose.

Better flexibility of the cured films is claimed in a disclosure of *Labana* and *Chang* [174] where glycidyl functional acrylic resins are cured with **carboxyl terminated polyethers** obtained by termination of hydroxyl functional polyalkylene glycols with diacid anhydride.

Other systems having acrylic resin with carboxyl functional groups as the main binder and epoxy functional low molecular weight crosslinkers are also described in the patent literature. In a patent of Bayer [175], **carboxyl groups bearing acrylic resin** is used in combination with **triglycidyl isocyanurate** to produce clear or pigmented powder coatings which are claimed to have good mechanical properties, gloss and weather resistance.

Curing agents: polyurethanes

Self-crosslinking acrylic powder coating can be produced using acrylic resin containing hydroxylalkyl acrylic co-monomers which have been reacted before or after co-polymerization with other unsaturated monomers with partially caprolactam blocked isocyanates [176].

Blocked isocyanates are also used as crosslinkers for acrylic powder coatings. An example is a powder coating composition containing hydroxyl functional acrylic resin as the binder and caprolactam blocked adduct of hexanediol and isophorone diisocyanate [177]. After curing at 180 °C for 20 min, films with good flexibility, gloss and chemical and solvent resistances are obtained. Nowadays (2012) "Joncryl 587" (S.C. Johnson) is the **market leader** for OH-terminated acrylics, which can be crosslinked by polyurethane hardeners.

A powder coating composition containing hydroxyl functional acrylic resin and blocked isocyanate prepolymer is described by *Yousuf* [178] and **compared with** two **polyester based polyurethane powder coatings**. Obvious advantages of the acrylic system are better solvent resistance and gloss retention during the accelerated weathering tests. However, the direct and reverse impact resistances of the acrylic coating are much worse compared to the polyester powders.

Several attempts have been made to benefit from the good mechanical properties of the polyester resins and good weathering resistance of the acrylics, combining them into one system. A patent of Dainipon Ink and Chemicals, Inc. [179] refers to a system containing 70 to 90 % carboxyl functional polyester resin and 10 to 30 % glycidyl functional acrylics. The average number of carboxyl groups per molecule of polyester resin is very low, largely below 2, in order to obtain good flow properties. Acrylic resin itself has a rather low molecular weight between 500 and 3000, having a relatively high content of glycidyl methacrylate of at least 20 wt-%. It is claimed that after curing at 160 °C for 20 minutes, smooth high glossy films with good impact resistance and flexibility are obtained.

A **combination of polyester and acrylic resin** is described in a patent of Nippon Paint [180] and comprises 65 to 95 wt-% of polyester resin with an hydroxyl value of 200 to 100 mg KOH/g and acid value of 1 to 20 mg KOH/g, 4 to 30 wt-% blocked isocyanate and 2 to 30 wt-% acrylic resin containing hydroxy and glycidyl groups with an hydroxy value of 5 to 150 mg KOH/g, a glycidyl group content of 0.0035 equivalents/g and molecular mass of 3000 to 50,000 with T_g = 40 to 85 °C. It is claimed that the coating has an improved resistance to staining and solvents.

A system developed by S.C. Johnson & Son, Inc. is based on a **blend of hydroxyl functional acrylic resins and hydroxyl functional polyesters** with blocked isocyanates as crosslinkers. An improved outdoor durability is claimed as a result of the positive influence of the acrylic co-binder in the coating composition [181].

In both systems, the amount of acrylic resin is rather low. In the case of Dainipon Ink, the low amount of acrylic resin comes as a result of the low epoxy equivalent weight. Higher epoxy equivalent weights will lead to glycidyl acrylates with higher molecular weight, resulting in limited compatibility with the polyester. Although in both systems there is an improvement in the weathering resistance, the effect is not as high as would be expected if higher amounts of acrylic resin were used in the coating formulation. In the absence of a real solution to the problems related to the inferior mechanical properties of the acrylic systems, even when acrylic polyester blends are used, the so-called „**dual coat dry-on-dry**" powder coating system has been proposed offering excellent physical film performance properties [182]. The first coat is a thin application of a polyester-based powder coating. The topcoat, applied directly over the polyester basecoat, dry-on-dry without previous curing, is based on acrylic resin. In this way the advantages of both acrylic and polyester coatings are combined into one system.

The **dry-on-dry system** can be certainly considered as an improvement of the performances of the acrylic powder coatings. However, the performances of the coating and the reproducibility of the results depend to a large extent on the skill of the operator. **One coat system** will not suffer from such a disadvantage. A solution of the compatibility problems, leading to acrylic rich polyester/acrylic hybrids, could be a way to substantially improve the weathering resistance of powder coatings for outdoor use.

This way was more or less followed by *Dumain* [183] and *Iwamura* [184] of Reichhold Chemicals Inc, who describe a novel curing mechanism, involving the development of a bi-functional polyester with both hydroxyl and carboxyl functionality. The carboxyl groups are cured with a glycidyl methacrylate (GMA) acrylic resin, specifically designed for low gloss powder coatings. This acrylic resin has an epoxy equivalent weight of 480.

The OH groups of the polyester are cured with a blocked isocyanate crosslinker. The **matting effect** relies here too upon the **difference in curing speed** between the acid/epoxy reaction and the hydroxyl/isocyanate reaction. This polyester/acrylic hybrid is claimed to give smooth, ultra low-gloss films with excellent mechanical properties and durability. In addition to providing matte finishes, this polyester-acrylic hybrid technology can be modified to produce high gloss coatings, by simply replacing the acrylic resin with another grade of GMA acrylic resin (with an epoxy equivalent weight = 600). These high gloss finishes are claimed to have the same excellent mechanical properties and durability. By mixing the two GMA acrylic resins in various ratios, it is possible to produce a full range of gloss, from matte to high gloss.

Iwamura [171] also describes a novel **acrylic-polyester hybrid**, consisting of a bi-functional (carboxyl and hydroxyl) isophthalic (IPA) type polyester, which is cured with a bi-functional (glycidyl and hydroxyl) GMA-acrylic hardener and a blocked isocyanate. This bi-functional acrylic (with an epoxy equivalent weight of 380 and a hydroxyl value of 50 mg KOH/g) combined with an IPA-rich polyester is claimed to show a still **better balance of durability and mechanical properties** than a „conventional“ mono-functional acrylic/bi-functional polyester hybrid system.

Curing agents: bisphenol A epoxy resins

Another system, developed by S.C. Johnson & Son, Inc., is based on a **combination of a carboxylated acrylic resin** and a type $2^1/_2$ (EEW = ~700) **bisphenol A epoxy resin** [185]. Such acrylic-epoxy hybrids offer a nice set of properties. They combine the typical properties of acrylic resins (such as UV resistance, hardness, chemical resistance) with typical epoxy resin properties (such as flexibility and chemical resistance). A comparison of acrylic-epoxy hybrids with the „normal“ polyester-epoxy hybrids, shows, that the impact resistances of the acrylic hybrids are not as good as with polyester hybrids, but this will be sufficient for most applications. The films pass the conical mandrel and Erichsen tests. Acrylic-epoxy hybrids clearly outperform polyester-epoxy hybrids in all the other properties, which have been tested by *Kerssen* [185] of S.C. Johnson.

Markets, that currently utilize polyester-epoxy hybrids, but need more hardness, scratch and wear resistance, can benefit from acrylic-epoxy hybrids.

An advantage of the acid (or carboxyl) functional acrylics is, that they are more "forgiving" than the GMA acrylics, concerning **contamination**. According to *Kerssen* contamination by acid acrylics can easily be controlled by „**good house-keeping**“.

Automotive applications

Powder coatings have a rather long history in automotive applications. However, for long time all applications have been limited to underbody and interior trim components, and later on to exterior trim parts and steel and aluminum car wheels. Because of aesthetic reasons (mainly related to the poor flow) and special application requirements, the car producers were reluctant to use coating powders as full-body coatings.

Automotive applications: primer-surfacer

Though Honda in Japan started in the 1980's with the application of a powder coating **primer-surfacer** on their car bodies, the auto car producers in the U.S. were more careful, regarding the speed of introduction of full-body powder primer-surfacers. The transition to powder coatings on body components began in 1992 when General Motors started (after ten years of building up experience and know-how) using an acrylic based primer-surfacer and an acrylic trim coating at the company's Shreveport plant. Since then, GM has installed several additional plants, using the acrylic based primer-surfacer.

The system applied to car bodies generally consists of a cathodic electrodeposition primer to provide corrosion protection; a primer-surfacer to protect the electrocoat primer from UV radiation and provide a smooth surface for the basecoat; a basecoat to impart color and metallic appearance; and a clear topcoat, giving additional UV resistance and scratch and environmental etch resistance.

Miller and *Kerr* [186] summarize the **qualities offered by the acrylic powder primer-surfacer** as follows:

- overall surface smoothness, improving the appearance of the succeeding layers
- chip and corrosion resistance for the entire vehicle body
- essentially no VOC (= volatile organic compounds) emission
- almost 100 % utilization and no waste disposal problems

Specifically the combination of high utilization rate and elimination of paint sludge, collected in the spraying booths, makes powder primer-surfacer an extremely interesting paint from an economical point of view, and can be considered as a **major driving force** for changes in the automotive industry. In the period 1991 to 1995 General Motors and Chrysler introduced full-body acrylic powder primer-surfacer in more than 10 production sites for passenger cars. In 1996 Eurostar-Works in Graz (50:50 joint-venture of Chrysler and Steyr-Daimler-Puch) have introduced powder primer-surfacer as a full-body anti-chip coating for their passenger cars [187]. It can be concluded, that the developments in the 1990's secured the position of the powder primer-surfacer at the car manufacturers in the USA and this trend is followed now in Europe.

According to *Green* [188] of the Ferro Corp, an **acceptable acrylic resin for use in automotive primer-surfacer powder coatings** is comprised of 15 to 35 % glycidyl-methacrylate (GMA), 5 to 15 % butyl methacrylate, and the balance is a combination of methyl methacrylate and styrene. Besides good gloss and film leveling, the balance of properties includes package stability, good fluidization and transport properties, and a wide „cure window".

In 2007 about 3 million cars and light trucks received a powder primer-surfacer treatment. The film thickness started with 75 μm and has been reduced to 50 μm. Due to cost reasons further reduction is to be expected [189].

To prevent filiform corrosion at aluminum wheels Rohm and Haas claimed GMA systems based on a mixture of hydrophobic methacrylates like bicycloalkyl methacrylate and non hydrophobic starting materials like methyl methacrylate in a patent application [190].

Automotive applications: topcoat

The automotive clear topcoat was considered for years as a too difficult field for powder coatings to enter in. But the permanent pressure for reduction or elimination of the VOC in the painting process, the economical advantages offered by the powder coating technology and, last but not least, the proven success with the acrylic powder primer-surfacer, initiated extensive research work at both powder paint producers and car manufacturers. Ford, General Motors and Chrysler have formed the so-called Low-Emission Paint Consortium (LEPC) to jointly develop new technologies for application of powder clear topcoats on automotive bodies [191-194].

Volvo, BMW, Volkswagen, Citroen, Audi, Jaguar and Ferrari are among the European car manufacturers experimenting with powder clear topcoats used over metallic base coats [195]. The system, that has been **chosen by all car manufacturers**, is a blend of epoxy functional acrylic copolymer, containing glycidyl methacrylate (GMA) as co-monomer and dodecane dicarboxylic acid or its (poly)anhydride as crosslinker. Due to cost reasons instead of dodecane dicarboxylic acid a mixture of adipic acid and sebacic acid is recommended in a patent [196]. *Schmidt*[197] reports rather positive experience with his automotive acrylic powder clear topcoat. It equals or outperforms both the one and the two component solventbased clear topcoats in many aspects, being only inferior with respect to the solvent resistance. See also Chapter 3.3.5.

While **acrylic resins** are primarily chosen for their UV and **outdoor resistance**, (typical **automotive requirement**: "CAM 180" (Xenon Weather-o-meter): > 4000 h and > 4 years Florida 5° South) they must also possess a number of **additional properties**. *Kinza* [198] summarizes these properties as follows:

- stoving schedule 25 minutes at 140 °C, (max. 145 °C, depends on base coat), but „usable" paint films already from 20 minutes at 120 °C
- color must be clear waterwhite and no yellowing should take place during cure
- extremely clean resin and powder, as small dirt particles are visible
- good intercoat adhesion
- good mar and chip resistance
- good environmental etch resistance
- high gloss (at 20° >85 %) and distinctness of image (DOI)
- good degassing properties for the pre-dried, but non-cured waterborne base coat
- excellent film flow on pre-dried waterborne basecoat
- film thickness max. 70 μm (for acceptable film flow) at the moment, but should be reduced in future
- good recoatability with repair paint
- powder must give good application qualities and perform well in respect to fluidization, impact fusion, transfer efficiency, response to tight cure schedules and freedom from excessive volatile byproducts during cure;
- good chemical resistance to e.g. petrol, motor oil, water, acids and alkalis, brake fluids, bird droppings, tar, resin droplets from trees.

From the three acrylic resin systems (hydroxy acrylic + isocyanate, acid acrylic + epoxy functional hardener, GMA acrylic + acid hardener), the latter is preferred, due to the required low stoving temperature, film appearance and lack of by-products. To optimize the resin formulations, chemists have a number of variables at hand, such as monomer composition (low styrene content), molecular weight, degree of functionality and glass transition temperature T_g. But it often occurs, that an improvement in one property causes the decrease of another property and, while increasing molecular weight, T_g and functionality will improve mechanical properties, application properties and chemical resistance, this may result in poorer film leveling and appearance.

According to *Green* [188] a correct balance of molecular weight, monomer composition and T_g, gives GMA acrylic clear powder coatings with excellent appearance properties, good impact fusion resistance and good storage stability. *Green* lifts here the veil of secrecy and says, that this balance is achieved, when the molecular weight is below 2500, the T_g above 80 °C, and the monomer composition is adjusted, so that the melt viscosity is less than 400 mPa·s at 150 °C. Increasing the molecular weight above 2500 has a negative effect on film flow, while a decrease of T_g below 80 °C leads to poor storage stability and worse impact fusion resistance.

It has to be taken into account that GMA/DDDA powder coating formulations, which are currently used for automobile topcoats, have to be **stored in the refrigerator** (<10 °C) and are comparatively expensive. Therefore a market pressure exists to substitute these systems by lower cost alternatives, like HAA or TGIC based formulations. In addition the next generation OEM powder topcoats have to compete with the **higher scratch resistance** of state-of-the-art liquid alternatives, like "Ceramicclear" from PPG and "iGloss" from BASF.

In 2005 Dainippon Ink and Chemicals [199] filed a patent in which orgasiloxane modified GMA based formulations are claimed. The high scratch- and water-resistance of those systems make them in particular suitable for automobile topcoats. In 2008 DuPont [200] filed a patent claiming urethane modified carboxylic acids as hardeners for GMA systems. This way the undesired whitish film formed especially on underbaked GMA clearcoats can be avoided.

Gribble [201] emphasizes, that well formulated acrylic powder clear coats not only meet the distinctness-of-image (DOI) achieved by the current liquid clear coats technology, they additionally have much less variation in the DOI of horizontal versus vertical surfaces. This results in a vehicle with a more uniform, consistent appearance over the entire car body.

Developments and achievements in acrylic powder coatings for automotive clear topcoat are also described by *Ortiz* and *Winters* [202] of Ferro Powder Coatings and *Dornitz* [203] of PPG, while progress with the LEPC project is outlined by *Pearson* [204] of Ford and *Attinoto* [205] of ABB (major plant designer in the LEPC project).

Automotive applications: film thickness

As the mentioned film thickness of powder clear coats (max. 70 μm) is still too high to the opinion of the car industry, powder producers are trying to **reduce** this **film thickness**. For thin films, with high gloss and excellent flow, down to 25 to 30 μm, it is necessary to remove not only all coarse particles (>50 μm), but also the finest fraction (<5 μm) from the powder, for good sprayability. This is of course a time consuming and expensive business.

However, it is observed, that **electrostatic spray guns are now commercially available**, that are capable of applying powders with all particles below 25 μm, and **depositing thin films**, lower than 25 μm.

This will require manufacturers to develop methods for the economic production of these low sized powders. Since this technique has been mastered in the production of toners for the photocopying industry, it should not be beyond the ingenuity of thermosetting powder producers. And indeed, some powder manufacturers have succeeded now in mastering this technique.

They are capable of applying thin and pore-free clear coats with excellent film flow and good powder fluidization properties. Some of these new techniques, such as *Seibert's* "Particle Engineering" [206] and Courtaulds' "Particle Management" [207] are mentioned or described in literature. Other methods are still the secrets of the inventors, but are more or less based on the same principles.

According to *Higgins* [207] "Particle Management" is the term used by Courtaulds to describe their patented discovery of new unique ways of manipulating the behavior of charged powder particles. The behavior of charged particles is fundamental to the electrostatic spray application and is the main determinant of the uniformity with which particles are laid down onto the surface of the object being coated. Particle behavior decides the ease with which powders can be fluidized with air and conveyed evenly through long tube lengths to the spray guns, and that in turn strongly influences the uniformity and consistency of the spray cloud over time. The Particle Management technology involves the use of additives and surface treatment of the particles, to produce very fine powder, which has "fluid" properties, allowing a thinner coating to be applied.

Quite another way to apply thin films is to use **aqueous powder slurries** (APS systems), originally introduced by Ciba-Geigy in the early 1970's. They haven't achieved a broad commercial acceptance, yet, but nowadays this slurry technology is being used by BASF [208] for application of acrylic clear topcoats on the Mercedes A-class in Rastatt, Germany, 800 t in 2007 [189]. The powder particles (3 μm) in the slurry are much smaller than conventional powder coating particles (25 to 50 μm) and allow thus a lower film thickness of 20 to 30 μm or even less by conventional wet paint guns. Special emulgators are claimed to increase the shear stability of these powder slurries making them applicable by high speed rotary bell atomizers (ESTA) [209]. *Lendl* [210] emphasized the easy process to change the color and the low film thickness (15 μm) of powder slurries.

Automotive applications: car manufacturer

BMW is the first car manufacturer in the world to use powder clear coat in standard production. *Wimmer* [211] describes the development steps and initial line experiences at BMW's Dingolfing plant. In 2007 BMW used more than 800 tons of **powder clear coat**, 1.6 kg for each vehicle [189]. Due to their success, BMW will also use acrylic powder clear coat in their new paint shops. The powder is supplied by PPG and DuPont. Although the GMA-powder system is more expensive on a material base than liquid alternatives it is still competitive due to the lower capital costs [189].

Harley Davidson also uses now **acrylic powder clear coats** for their motor bicycles. Other automobile producers, which are or were interested in the use of (acrylic) powder clear topcoats, are: Volvo/Autonova, Renault, Peugeot, Ferrari and Audi. Toyota have selected suppliers of powder clear coats and cars will regularly be coated and tested at pilot plants.

In their paper, related to the new developments in the application equipment for powder coatings for automotive use, *Yanagida* et al. [212] predicted, that powder coatings will replace the conventional and waterborne primer-surfacer and clear topcoating in the future as follows:

	Current type	**Next generation**
Primer	Electrocoat	Electrocoat
Primer-surfacer	Solvent- or waterborne	Powder
(Metallic) basecoat	Solvent- or waterborne	Waterborne
Clear topcoat	Solventborne	Powder

This would reduce the VOC emission from 75 g/m^2 painted surface now to 15 g/m^2.

It should be mentioned, that the "Smart" city car (a development of Mercedes-Benz and Swatch and introduced in 1996) is the first car that is completely powder coated. The coating powder is a novelty, metallic effect powder, which can be applied in one single layer. Neither clear coat protection, nor primer-surfacer is needed! PPG and Wörwag developed this powder, based on pigmented acrylic resins. In 2005 Swatch left this venture.

Rekowsky[213] reported the state-of-the-art of acrylic powder coatings in 2002 under the pregnant title of **"Close to Liquid Coatings"** and mentioned not only the successes with automotive clear coats at BMW, but also the potentials of the new pigmented acrylate powder coatings for industrial use. The expected potentials are based on the following properties of the latest generation of pigmented acrylic powder coatings: super-durable weathering resistance, high gloss, high scratch resistance, ultra smooth film flow, low curing system, low film thickness possible. Rekowsky concludes his article with the expectation, that the success in the automotive industry is only the start for **a bright future for acrylic powder coating systems**.

3.4.4 Unsaturated polyester powder coatings

Until recently, unsaturated polyester resins have found very **limited application**, mainly as binders for in-mould powder coatings for sheet molding compounds (SMC). The principle of curing is a free radical polymerization of the double bonds of the resin initiated with peroxides or other standard initiator/accelerator systems widely used in the field of unsaturated polyester resins for construction purposes. The problem of **oxygen inhibition** which is typical for this type of curing reaction restricts their use for coating surfaces that are in direct contact with air during curing. In the case of in-mould powder coatings this problem is avoided simply because the crosslinking of the powder coating takes place on the surface that is in direct contact with the mould wall.

Polyester resins for **in-mould powder coatings** are not different in principle from the unsaturated polyester resins for glass reinforcement, concerning the resin backbone. Since the coating has to be in a powder form, the use of styrene and other liquid monomers which serve as reactive diluents is excluded, and only solid sources of unsaturation should be considered.

The production of unsaturated polyester resins proceeds by polyesterification of unsaturated diacids and glycols. Although unsaturated glycols and saturated diacids can also be used, because of the availability and price of the former, this is not the case in practice. Most widely used unsaturated diacids are maleic anhydride and its *trans* isomer fumaric acid. Since **unsaturated polyesters** can not be homopolymerized usually styrene is added. Due to the potentially **toxic properties of styrene** it is desirable to minimize the use of styrene in these formulations. Therefore DSM recently filed a patent[214], replacing maleic and fumaric acid by itaconic acid, which is claimed to be polymerizable with a lower amount of styrene. Among the diols, ethylene glycol and 1,2-propylene glycol are the most important. The degree of unsaturation can be easily adjusted by replacing part of the unsaturated diacid with saturated diacid. For this purpose phthalic anhydride has been widely accepted. Isophthalic acid, the *meta* isomer of phthalic acid, has found wide use as the diacid imparting excellent hydrolytic stability of the cured resins. Adipic acid can be used for flexibility improvement. For the same reasons diethylene glycol can also be used sacrificing to a certain extent the hydrolytic stability of the cured material[215]. In that respect 1,6-hexanediol is a better choice, while the other commercially available higher glycol, 1,4-butane diol, has to be esterified carefully, because it forms tetrahydrofuran during esterification[216]. Neopentyl glycol, due to the presence of the two methyl groups in the α-position to the hydroxyl groups, provides excellent hydrolytic stability to the polyesters.

Although **the production process** of saturated and unsaturated polyester resins is the same, there are certain specific differences. During the polyesterification of the maleic acid, there is always a certain equilibrium with the *trans* isomer, fumaric acid[217-219].

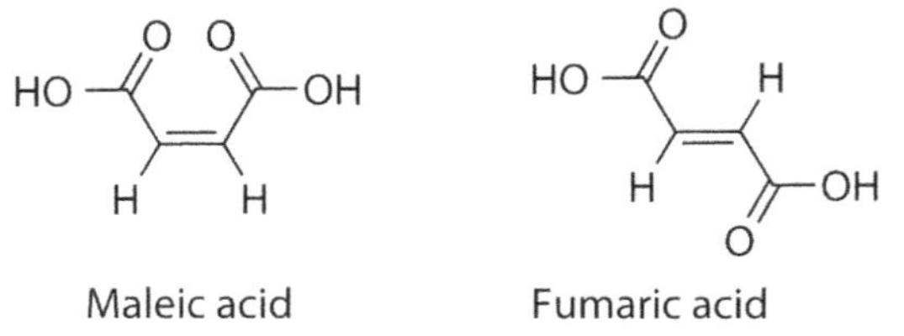

The degree of isomerization depends on the reaction conditions, and the catalysts used during esterification. A high esterification temperature promotes the isomerization. Even the carboxylic groups from the maleic acid are good isomerization catalysts. Therefore, a longer esterification time, due to a slow reaction rate, leads to a higher degree of isomerization.

Fumaric acid is more reactive during free radical polymerization at the crosslinking stage, shortening considerably the curing time of the unsaturated polyester.

Example of preparation of unsaturated polyester

An example of preparation of unsaturated polyester for an in-mould powder coating will be presented as an illustration of the process described by DSM Resins [220].

According the described process, 2672 parts by weight of 1,6-hexane diol, 692 parts of terephthalic acid and 0.45 parts of diphenylol propane are esterified at 230 °C in an inert atmosphere until an acid number of 7.5 mg KOH/g is reached. The reaction mass is cooled down to 170 °C and 1864 parts by weight of fumaric acid are added. The esterification continues at 210 °C until an acid number of 9.3 mg KOH/g is obtained. Before cooling down and discharging, the polyester resin is inhibited with 0.45 parts of hydroquinone. The resulting resin has an unsaturation number of 3.6 mol/kg, a melting point of 99 °C, an acid value of 8 and a hydroxyl value of 52 mg KOH/g.

Crystalline unsaturated polyesters have an advantage over the amorphous types, giving powder coatings that have improved antiblocking properties during storage and better flow after melting before curing takes place [221]. On the other hand, the impact resistance of powder coatings based on amorphous unsaturated polyester is better, and the cured films exhibit higher flexibility and less brittleness. Isocyanate modified unsaturated crystalline polyesters, however, have an improved impact resistance, without sacrificing the stability of the system during storage [220].

Next to the unsaturated binder, in-mould powder coatings contain high boiling point copolymerizable monomers with a functionality higher than two, pigments and additives such as initiators, release agents, flow promoters, stabilizers etc.

The boiling point of the monomers used in combination with the unsaturated polyesters must be in principle 25 °C higher than the temperature at which the powder coating is applied and cured, in order to avoid possible losses during curing, or the appearance of surface defects. The main role of the polyfunctional monomers in the powder coating composition is to **increase the crosslinking density** of the cured material. Suitable monomers used in practice are triallylcyanurate, triallylisocyanurate, trimethylolpropane triacrylate and triallyltrimellitate.

Catalysts which are mostly used for the curing of unsaturated compounds belong to the group of peroxides, such as, hydroperoxides, ketoneperoxides or peresters. They are present in amounts between 0.5 and 1 % by weight with respect to the unsaturated binder. **Conventional accelerators** such as cobalt salts or tertiary amines which are very often used in curing conventional unsaturated resins for construction purposes can be incorporated in the powder coating composition in order to speed up the curing rate. A patent of Morton Thiocol Inc. describes an unsaturated polyester powder coating composition of which the curing rate can be easily altered by a combination of a slow and a fast initiator [222].

Table 3.21: Unsaturated polyester inmold powder coating[220]

Material	Parts by weight
Unsaturated polyester resin	100
Acrylate flow agent	1
Titanium dioxide	10
Triallylisocyanurate	3
p-Benzoquinone	50 ppm
t-Butylperbenzoate	2
Curing schedule [min/°C]	20/120

Although the powder coating can be prepared by physical mixing of finely grounded constituents, the best results are of course obtained by melt mixing of all components, cooling and grinding of the obtained homogeneous blend [223, 224]. In order to avoid prepolymerization during melt compounding, a certain amount of a suitable **polymerization inhibitor** (quinone or catechol compounds) can be added. As a rule, the addition of the catalysts takes place just before cooling. An example of an in-mould powder coating described by DSM Resins [220] is given in *Table 3.21.*

In-mould powder coatings are usually applied by an **electrostatic spraying** technique on preheated walls of the mould at a temperature between 100 and 160 °C. Immediately after application the powder particles melt down and form a continuous layer. In order to ensure that the coating layer can withstand some mechanical load, the coating is allowed to cure partially, before the construction material from which the article is made enters the mould. A complete curing in this stage can cause interadhesion problems between the coating layer and the article that is coated. Unsaturated polyester in-mould powder coatings are mainly used for articles made from reinforced unsaturated polyester resins at elevated temperature and pressure via the SMC or BMC (bulk molding compounds) techniques, or by injection molding.

In a recent patent [225] DSM described the combination of unsaturated polyester and urethane vinylether for **low temperature** curable (<130 °C) in-mould coatings. With a similar formulation containing special initiator, accelerator and inhibitor combinations based on compounds like "Perkadox 26", "Trigonox C", copper and/or cobalt salts and tert-butylhydroquinone DSM [226] was able to achieve ultra low curing temperatures of powder coatings down to 70 °C, while the extrusion step took place at 60 °C!

Dohnt [227] gives a description of Ferro's in-mould powder coating process and summarizes the **advantages** of powder coatings compared to liquid paints. Ferro's powder coating is applied to the interior of the heated mould at approximately 150 °C and then pressed together with the SMC. After 3 to 5 minutes curing, a perfect smooth one-layer coating is achieved on 95 % of the production. According to *Dohnt*, future developments in in-mould powder coating will be focused on lower curing temperatures of around 80 °C, which opens a wider range of applications, and materials other than SMC can then be coated by this process.

The application of thermal curing unsaturated polyester resins is no longer restricted to in-mould powder coatings, as they are now also used for **low bake powders** with curing temperatures of around 125 °C. *Subramanian* and *Sullivan* [228] describe this new development, , using fumaric and allyl modified unsaturated polyesters and which is based on the "**dual cure**" of the carboxyl groups with an epoxy compound (e.g. TGIC) and the C=C double bonds with an organic peroxide.

Free radical curing of the unsaturated groups can also be achieved by metal driers and UV irradiation. Peroxides are fast curing, but can cause problems with processing and storage stability. UV irradiation is also fast and overcomes the stability problems, but has limitations when applied to thick and pigmented coatings (see Chapter 3.4.5). Metal driers (e.g. cobalt driers) are slower to react.

The general expectations of the end-user of low temperature cure (LTC) **powder coatings on MDF board** are: good flow and a smooth coating; good storage stability; excellent scratch resistance; resistance to common household items; reasonable impact resistance; and low gloss.

With these guidelines, a series of unsaturated polyester resins with fumaric and allyl functional groups were prepared. Two of them, each with 4.1 % C=C, acid value of 35 to 45 and hydroxyl number <10, were blended in the ratio 1:1 and pigmented with titanium dioxide. Flow agent, TGIC and benzoyl peroxide were added to complete this dual cure formulation. This formulation is easily extruded on a single screw extruder at 80 to 100 °C. Curing temperature is 125 °C and the coatings have good solvent resistance and pass the 100 double-rub MEK test.

3.4.5 Radiation curable powder coatings

3.4.5.1 UV (and EB) curable powder coatings

Introduction

Unsaturated polyester resins (and other unsaturated compounds) are also used in radiation curable ("radcure") powder coatings. The principle of curing is a **free radical polymerization** of the double bonds of the resin, **initiated with UV light**, and not with peroxides (as described in Chapter 3.4.4). Here a photoinitiator is needed (instead of a peroxide) to form free radicals upon UV irradiation.

Epoxy resins can be polymerized by cationic UV cure, using salts of very strong acids (mostly "onium" salts) as photoinitiators. The curing mechanisms are described in detail in Chapter 3.1.9. *Boneza-Tomaszewski* et al. [229] observed that conventional radiation curable epoxy powder coatings suffer from a low epoxy concentration leading to an insufficient crosslink density. Thus they recommend epoxy-polyester resins with higher epoxy concentrations to overcome this problem. This kind of formulation was applied on MDF showing similar properties than radical polymerizing powder coating systems [230].

UV curable coating powders are suitable for application on **heat sensitive substrates** such as wood, MDF, plastics and paper. This application demands for cure temperatures well below 150 °C and in many cases even below 100 °C.

The acid-epoxy chemistry that is widely used in powder coating can be catalyzed effectively to curing temperatures well below 150 °C. However, with this chemistry the main problem is the **balance between storage stability, melting, flow and cure**. Since a thermally activated (bimolecular) crosslinking reaction is governed by the *Arrhenius* Law (see Chapter 4.6 of this book), it will be virtually impossible to develop a thermally cured system, that is stable (no pre-reaction) during conventional extrusion at e.g. 100 °C, and should give a smooth film with full conversion at e.g. 120 °C.

One of the approaches to solving this "impossible" problem is the use of radiation-activated crosslinking instead of a thermally activated reaction. With the proper choice of the chemistry this can result in a system that allows for the desired separation of melting and film flow from the final cure [231].

In practice of UV curable powder coating the powder particles are first **melted by shortwave IR radiation**, which allows for melting without overheating the substrate. After coalescence of the powder particles and the formation of a continuous film, the film is **cured by UV or EB radiation**. This separation of melting and curing allows for processing of the powder coating in standard equipment, storage of the powder at room temperature and at the same time ensures an efficient curing at a relatively low temperature (*Figure 3.25*).

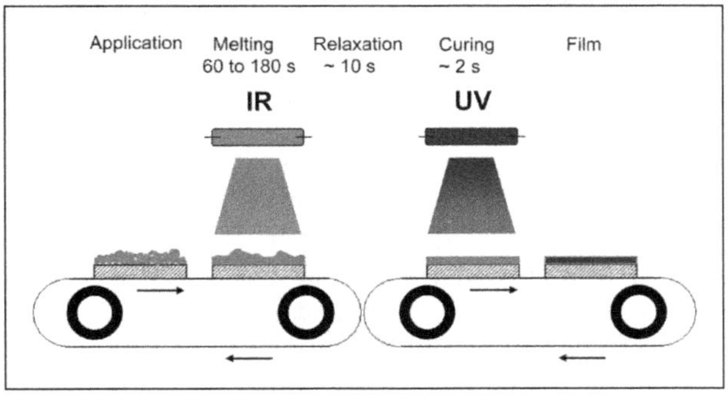

Figure 3.25: UV powder coating process

Additional advantages of such "radcure" powder coatings might be high line speeds, reduced energy costs and less toxicity problems compared to liquid radcure acrylates (due to high molecular weight solid resins) [232]. A recent article [233] compares the raw material costs of a 50 μm UV powder coating (0.9 €/m²) with a 25 μm solventborne coating (2.4 €/m²). In addition a 222 % higher productivity and a substantial lower space requirement are emphasized. Compared to conventional powder coatings, though, UV powders are about 25 % more expensive [189].

The patent literature reveals that chemical companies have been interested in UV powders since 1971. Prior to 1990, there have only been appeared a few patents. But after 1992, a growing number of patents have been published. This shows the growing development of the radiation (UV, but also EB, albeit in a much less degree) curable powder technology. Since 1995 the share of UV powder on all powder publications and patents has doubled.

The UV powder technology is rather complex in terms of the **classification of respective topics**. The chemistry of the binders, the influence of crystalline resins on melt viscosity and storage stability, the use of hyperbranched polymers, different lamp types and curing conditions, the choice and concentration of photoinitiators, new applications and practical results are the main subjects and will be discussed in this order.

Binders: composition

Several **types of solid binder systems for radiation** (in particular UV) **curable powder coatings** have been reported. They include: unsaturated acrylates and combinations with unsaturated polyesters, patented by DSM [234], BASF and Bayer [235]; methacrylated unsaturated polyester, described by UCB (now Cytec) [236]; mixtures of unsaturated polyesters and unsaturated urethane acrylates, described by *Fink* and *Brindöpke* [237] and patented by Hoechst [238]; combinations of amorphous and (semi)crystalline urethane acrylates, described by *Wenning* (Degussa, now Evonik Industries) [239]; *Williams* (Cray Valley) [240] emphasized the versatile use of urethane acrylates for wood applications; acrylated hyperbranched polyesters, described by *Hult et al.* [241]; methacrylated polycarbonates published by *Loewenhielm* et al. [242]; mixtures of unsaturated polyesters with (meth)acrylamide derivatives, described by *Wittig* and *Gohmann* (AkzoNobel) [243]. The already mentioned cationic curing of solid epoxy resins was patented by Ciba-Geigy [244] and some years later described by *Finter* et al. [245].

DSM Coating Resins [246] has developed a non-acrylic system, based on a blend of an unsaturated polyester, which contains maleic or fumaric unsaturation, and a vinyl ether polyurethane crosslinker. The UV curing mechanism of this DSM binder is based on the free radical initiated 1:1 copolymerization of the electron-rich vinyl ether groups with the electron-poor maleate of fumarate groups. See also Chapter 3.1.9.

Binders: crystallinity

In an article, presented at the PCE '98, *Blatter* [247] describes the difficulties to be encountered in the necessary improvements and future developments of the UV powder technology.

One of them is the need to provide sufficient **physical storage stability**, which means, that the glass transition temperature T_g of an amorphous resin should still be above 50 °C and this restriction has a major influence on the minimum temperature for UV curing of the powder coating. A suitable low viscosity, to achieve a smooth film, will only be attained at temperatures of at least 50 °C above the T_g. The recommended minimum curing temperature for UV curable powders is, therefore, mainly determined by the physical storage stability requirement, but also by the molecular mobility of the UV powder. It has been shown by Photo Differential Scanning Calorimetry experiments, that crosslinking proceeds only after UV irradiation at elevated temperature. Irradiation of the powder with UV light at room temperature possibly only leads to partial consumption of the photoinitiator. At room temperature, the molecular mobility is too low to allow for network formation. Above 80 °C, according to *Udding-Louwrier* et al. [232], the molecular mobility is high enough to ensure a good curing. To adjust T_g and melt viscosity Bayer [248] claimed the addition of reactive diluents like trimethylolpropane triacrylate or dipentaerythritol pentaacrylate to the UV powder formulation in a patent application.

A possible solution to this fundamental problem of powder storage stability could come from the use of **crystalline resin** systems with a melting point above 40 °C. These crystalline resins will have a much lower T_g and thus a much lower melt viscosity above the melting point, compared to today's amorphous resins at the same temperature.

Complete crystalline, unsaturated polyesters have been developed and patented by UCB Chemicals (now Cytec) [249] and give very low melt viscosities at 100 °C (e.g. from 0.2 up to 6.0 Pa·s, instead of 2500 Pa·s for amorphous unsaturated polyesters at 100 °C). Improved flow and/or lower fusion temperatures (<90 °C) are already achieved by these formulations.

As known to those skilled in the art, crystallinity is promoted by:

- Monomers, used for **poly-condensation**, which contain an even number of carbon atoms.
- Use of **cyclic monomers**, which are symmetrically substituted with high symmetry, such as e.g. para-substituted phenyl rings; examples are terephthalic acid, cyclohexane dicarboxylic acid, dimethylolcyclohexane. For example in a Chinese Patent [250] a composition of hexanediol, terephthalic acid, trimellitic anhydride and glycidyl methacrylate is recommended.
- Use of these **linear symmetric monomers** alone (preferred by UCB [249]), which also means only one diacid and only one diol, and thus not mixtures of various diacids or diols.
- The **crystalline nature** of polyester resins makes them non-sticky solids at room temperature. Lower surface hardness, associated with these crystalline resins, appears to be a function of the degree of unsaturation in the resin. At higher levels of unsaturation, the hardness is excellent (therefore UCB (now Cytec) used crystalline polyesters with methacrylic end groups), but the film is lacking in flexibility. Trials in combination with amorphous resins may improve this flexibility.

Wenning (Degussa, now Evonik Industries) [239] combines amorphous and (semi)-crystalline urethane acrylates to get **tailor-made binders**, which exhibit excellent adhesion, good flexibility and high hardness both in clearcoats and in pigmented formulations [251] on different substrates.

Production of semi-crystalline UV powder coatings needs some precautions. In the case of DSM's MA-VE system, they are summarized by *Bayard* [252]. Due to the semi-crystalline nature the extrusion temperature may not exceed 70 °C, and the extrudate temperature may not be higher than 75 °C. This temperature is above the T_g, but below the melting range of the crystalline parts. Above the melting temperature the melting of the crystalline vinyl ether leads to a strong reduction in viscosity, which makes extrusion virtually impossible. This also means that at present it is not (yet) possible to manufacture coating powders from completely crystalline materials in conventional powder production equipment (extrusion).

During grinding the temperature may not exceed 40 °C, since otherwise lump formation may take place. Both *Kersten* [253] and *Paulus* et al. [254] warn, that, after extrusion, insufficient recrystallization of the crystalline component in a semi-crystalline system, lowers the T_g of the powder and leads to a **worse storage stability**. Here the non-recrystallized parts act as a "plasticizer". *Spyrou* [255] describes the benefits of appropriate crystalline resins, which show a higher degree of recrystallization. As a rule of thump 20 % of crystalline component lead to a decrease in melt viscosity of 80 %. The **recrystallization degree can be enhanced** further by heating (tempering) the mixture a few hours at a temperature between the T_g of the amorphous and the melting point of the crystalline component. The storage stability can be increased even more by addition of special waxes, e.g. polyamide waxes.

After curing the crystallinity is often undesired, since it leads to reduced flexibility and opaque films. Therefore DSM [256] patented special urethane acrylates based on hexamethylene diisocyanate, branched diols and hydroxyacrylates. Those resins tend to be crystalline before curing, but amorphous afterwards.

Williams et al. [257] investigated the **network formation** of di- and trifunctional urethane acrylates. They used a variety of test methods to follow up the behavior of these mixtures during melting and curing including photo-rheology and photo-DSC. Best results were obtained for mixtures of both types of resins. Aromatic diisocyanates based urethane acrylates lead to a decrease of reactivity but in the same time to an increase in film hardness.

Binders: dendrimers

The use of dendrimers and hyperbranched polymers, as sources for (semi)-crystalline resins or for resins (and/or crosslinkers) with very low melt viscosity, has been studied and described by *Malmström* et al. of the Swedish Royal Institute of Technology [241, 258] and by *Van Benthem* of DSM Research [259].

Dendritic macromolecules, such as dendrimers and hyperbranched polymers, have become the subjects of increasing interest amongst polymer chemists over the last decade, in their desire to synthesize new and improved polymers.

The concept, that these materials will give improved properties, is based on the assumption that dendrimers (with their spherically shaped molecules) are inherently different to linear polymers and will give **significant changes** in **physical** and **physico-chemical properties**. Some of these qualities have been realized, for dendritic polymers offer high functionality, broad variety of functional groups and unique rheological behavior. These properties make them ideally suitable for film forming applications. The melt viscosity in relation to their molecular weight is kept low, due to their compact morphology hampering chain entanglements, while on the other hand the large number of functional endgroups enables efficient crosslinking. For coating practice this means the combination of good flow, excellent chemical properties, durability and good mechanical properties.

The difference between dendrimers and hyperbranched polymers can be described [259, 260] as follows: Dendrimers have a regular structure and form monodisperse materials, while hyperbranched polymers have a less regular structure and form polydisperse materials. Synthesis of dendrimers is very time-consuming and therefore expensive, while hyperbranched polymers are often synthesized in one-pot reactions, using commodity raw materials, and make them more readily suitable for commercialization.

For **UV curable powders**, functionalization of the hyperbranched polyesteramide with acrylic or methacrylic acid is necessary. This can be done effectively in a one-pot two-step procedure. Such UV curable hyperbranched resins excel in curing speed, as compared to conventional acrylic or methacrylic oligomers, and also in film hardness.

Cheng et al. [261] introduced C-18 alcohols into hyperbranched UV curable resins, thus inducing crystallinity with a melting point of 120 °C. This way they achieved a remarkable storage stability of the powder coating formulations of one week at 45 °C. Only a few coatings data are published, therefore it is difficult to evaluate the influence of the long alkyl chains on appearance and mechanical properties of the film.

Photoinitiators

In UV curable powder systems the (choice of) photoinitiator plays an important role. The photoinitiator, adapted to the powder coating technology, should have the following properties [236, 262]:

- solid, with melting point below the working temperature of the melting stage;
- non-volatile and non-yellowing, when the binders are melted;
- high activity;
- it should not contribute to a lowering of the glass transition temperature, which otherwise may cause poor storage stability and caking;
- chemically stable at the fusion temperature, which means no thermal initiation of the curing reaction;
- no liberation of volatile radicals, and no discoloration by these radicals;
- insensitive to oxygen, since it could react to form peroxy radicals at the surface of the film, and these are very poor initiators of photo-polymerization.

It has been found [263] that a photoinitiator of the α-hydroxyketone (AHK) class, especially α-hydroxy-acetophenone (α-HAP) meets these requirements, when tested in a clear coat. The AHK initiator especially resists oxygen inhibition and shows lowest yellowing. For **pigmented UV powder coatings** good results are reported [264], also in thick coatings, with a novel class of photoinitiators, the so-called bis-acylphosphine oxides (BAPO).

A combination of HAP and BAPO gives good surface cure (HAP) and through cure (BAPO), which can be explained as follows: HAP ensures good surface cure, due to the low sensitivity of its radical scission products towards oxygen, while BAPO gives good through cure (or cure in depth) for two reasons [265]:

- Its absorption peak in the visible region of the spectrum enables this initiator to absorb the **long wavelength light** that penetrates **deeper** into the surface of the coating.
- The capacity of BAPO to photobleach on exposure promotes **cure at depth** and efficiency.

A 75/25 blend of AHK and BAPO is commercially available as "Irgacure 1800" (ex Ciba). Pigmented formulations (containing 2 wt.% "Irgacure 1800") have shown the possibility of MDF application of the MA-VE binder system in coatings, containing 15 to 20 wt-% TiO_2 pigment, with film thickness of 100 μm! Below 15 % TiO_2 the **hiding power** is not acceptable [231].

Other AHK/BAPO ratios are also used. For white powder coatings, containing 20 % rutile TiO_2, ca. 2.5 % of a 1:1 blend of BAPO/AHK (or HAP) is recommended for optimal whiteness and good cure [266]. Higher BAPO/AHK ratios of 2:1 up to 5:1 give a slightly better cure at the cost of a greater residual yellowness [265]. In their Nuremberg 2001-lecture *Bayards* et al. [252] recommend a BAPO/HAP ratio of 2:1 for a white powder, containing 15 % TiO_2.

Rawlins [267] describes the influence of the **amount of photoinitiator** on the curing process. If too little photoinitiator is used, insufficient UV light is absorbed, or a wrong combination between photoinitiator absorbance and UV lamp output takes place. By consequence, unreacted functional groups or excess photoinitiator remains in the film, which have a negative effect on the outdoor durability. When an excess of photoinitiator is used, this will eliminate the residual functional groups, but initiates too many photoinitiation sites, resulting in low molecular weight and crosslink density and, therefore, a coating with poor mechanical and chemical proper ties is formed.

The use of **infra red** (medium IR) radiation **to heat up and melt the powder layer**, has shown to be very advantageous, because it allows the coating to heat up in a very short time (e.g. in 1 to 2 min up to 120 °C), keeping the exposure of the substrate to heat at a minimum of e.g. 80 °C. *Schmidt* [268] from Tigerwerk recommended an addition of IR-absorbing pigments like antimony-tin oxide or indium-tin oxide for this step.

It was also remarked by *Blatter*, that it may be difficult to apply UV-curing techniques to pigmented powder systems, because the pigments absorb UV light. This absorption makes the curing of the area below the surface of the coating difficult. However, the right combination of fillers/pigments, photoinitiators (such as BAPO, see before), UV light sources and belt speed can solve or moderate this problem.

Curing conditions and UV lamps

A paper by *Udding-Louwrier* et al. [269] describes the **influence of curing conditions**, like temperature, type of UV light used (bulbs with various wavelengths), the UV dose and the UV peak irradiance, on the coating performance of a UV curable powder coating.

The "**UV dose**" is the total energy (in J/cm^2) that arrives at the substrate, as it passes under the lamp. This dose is inversely proportional to the belt speed under any given light source, and proportional to the number of exposures (or the number of rows of lamps). It is also dependent on the output power of the lamp. The UV dose, which is also defined, according to *Skinner*[270], as the energy, needed to cure the coating to such an extent, that it meets the relevant specifications and performance criteria, varies with the individual formulations. Values of about 1000 mJ/cm^2 (measured with an "IL 390 Light Bug") for 100 μm clear coats, are mentioned by *Udding-Louwrier* et al. [269].

The "**UV irradiance**" is the power (in W/cm^2) of the UV light that is delivered to the surface, as it passes under the lamp. Irradiance varies with lamp output power, the design of the reflector and the distance to the surface. It is a characteristic of the lamp geometry and, therefore, it is independent of belt speed. The intense peak of focused power, directly under a lamp, is referred to as "peak irradiance". Working with lamps, which provide a higher peak intensity of UV light, give a more efficient cure. This is very important, when curing thicker coatings. *Skinner*[270] explains this as follows: Reduction of light energy as it passes through any coating obeys the well-known Lambert-Beer law. That means any energy which is not absorbed in the upper layers of the coating, and is not reflected, is available for absorption by the photoinitiator molecules in the lower part of the coating. Higher irradiance at the surface will give higher light energy within the "body" of the coating, which ensures, that a more uniform degree of polymerization is achieved throughout the depth of the coating. Increasing the peak intensity (or peak irradiance) results in a lower UV dose needed for cure, and thus in a higher belt speed. The UV light source can be either a mercury vapor arc lamp or mercury vapor microwave powered lamp.

The **spectral distribution pattern** (or emitted wavelength) of a UV lamp is a characteristic of the material used in the bulb. So-called "H"-bulbs (mercury type) which emit light in the short wavelength area around 300 nm, are suitable for curing of clear coats, which contain photoinitiators with strong absorption peaks in the short wavelength region (e.g. AHK or

HAP types). The "D"-bulb (iron-doped mercury type) emits medium wavelength UV light. For pigmented coatings, containing photoinitiators of the BAPO type, absorbing in the longer wavelength region (350 to 400 nm),"V"-bulbs (gallium-doped mercury type), emitting light of this wavelength, are used.

To obtain an indication of differences in the **degree of cure as a function of** the above mentioned various **curing conditions**, *Udding-Louwrier, Bayards* et al. [252, 269] determine the glass transition temperature (T_g) of the cured coatings on free films of 60 to 80 μm thickness, using either a modulated differential scanning calorimeter (MDSC) or a differential photo-calorimeter (DPC).

A higher degree of cure results in a higher crosslink density and usually in a higher T_g of the cured coating.

The conclusions of their investigations can be summarized as follows:

- When clear coatings are cured with an H-bulb, the coating Tg's seem to be more related to the UV dose than to the peak irradiance.
- The effect of peak irradiance on the curing of white pigmented coatings is found to be small, in comparison to the effect of the UV dose.
- When coatings are cured at temperatures of 100 °C, or higher, the temperature itself does not seem to have a great effect on the coating T_g. But *Castell* et al. [271] found a decreased conversion and photopolymerization rate at higher temperatures measured by photo-DSC and real-time FTIR-ATR. They believe the reduced conversion can be attributed to chain transfer processes and depolymerization.
- For a properly cured coating, the coating T_g should be at least 55 °C.
- Curing below 90 °C results in a lower conversion of the photo-polymerizable groups and by consequence the final coating T_g is lower.
- Optimal curing of pigmented coating is obtained with a combination of α-HAP and BAPO photoinitiators, while using a UV lamp with sufficient emission at higher wave-lengths.

Difficulties are likely to arise with the curing of three-dimensional objects, because the coating is only cured, where sufficient radiation intensity is reached. Therefore, plant design will be needed to overcome the problem of component shape, according to *Blatter* [247]. However, in recent years successful efforts have been devoted to understanding the requirements for achieving good cure on **complex 3-D parts**, according to *Skinner* [270] of Fusion Systems. In Europe, there are now many production lines using UV curable liquid coatings on 3-D parts, especially in the automotive industry (e.g. on headlamp lenses and reflector housings). Using modular lamps, each mounted in its own housing and fixed to a supporting frame, allows maximum flexibility in the positioning of the lamps. Each lamp has its own power supply. If required, these can be variable power supplies that allow fine tuning of the UV dose being delivered to specific areas of the workpiece. Such lamp arrangements can of course also be used for the UV-curing of powder coatings.

Electron beam curing (EBC) is another possibility for the application of the "radcure" powder coating on heat-sensitive substrates. The same powder coating formulations can be used as for UV curable powders, but with the exception of the photoinitiators, which are superfluous, due to the much higher energies of electron beams. The curing speed of EBC is even higher than with UV radiation and the penetration depth is not hindered by hiding pigments.

The greatest disadvantage of EBC is the very high investment costs for this technology, although less expensive EB equipment has been introduced in the last couple of years. Maybe, this is the reason, that there is no evidence, at the moment, of commercial existence of EB curing of powder coatings. Therefore, EBC will not be given further attention here, but for more information see Chapter 3.1.

New markets for UV powder coatings

Due to the advantages of the UV-curing principle, the following new markets became accessible for powder coatings:

- Special metal applications, which include the coating of **pre-assembled metal parts**, that contain plastic components or other heat-sensitive parts.
- Coil coating, where the **high curing speed** is an advantage. The very high line speeds of more than 100 m/min, used for coil coating with liquid parts, can also be realized now with UV curable powder coatings, using the new Electro Magnetic Brush (EMB) technology [272]. For more details of this technology see Chapter 6.2.
- Wood composites (e.g. MDF) and natural solid wood, which could form a **huge potential for UV curable powder coatings**. Hard board and medium density fiber board (MDF) have become increasingly popular in the furniture industry and UV powder coatings are able to be applied in a single thick coat with a good surface finish and excellent stain resistance [231, 232, 252, 269, 273].
  Natural solid wood surfaces are nearly invariably coated with semi-gloss clear coats to retain the attractive appearance of wood. The irregular surface structure and the high moisture content (degassing) make natural wood the most difficult substrate for powder coatings. Moreover, the electric field of corona charging can cause fiber lifting (also with MDF). Therefore, tribo charging is always recommended.
  The application of a sealer, prior to the powder coating, might offer a solution to the degassing problems. But much more attractive will be (or is) the use of (semi)-crystalline systems with melting points far below 80 °C.
- Coating of plastics. By far the **largest market for plastic coating is the automotive industry**, where there is a trend to coat plastic components (like bumpers) in the same color as the car body. However, plastics are difficult to coat, even with liquid paints, due to poor adhesion. In addition, the low conductivity of most plastics requires pretreatment to make the components sufficiently conductive for powder coating.

Commercial applications

An outline of the results achieved in commercial applications is now presented:

Coating of metal parts is the most advanced area of UV powder application today. Powders can be cured in less than 3 min (total fusion + curing) at 100 to 120 °C, giving a good balance of adhesion, flexibility and chemical resistance. Adhesion to metal is enhanced significantly by addition of acidic phosphorous and acrylate containing **adhesion promoters**. Since these tend to add to the acrylic double bonds, they loose the adhesive properties during the storage of the formulation. Thus Cytec published a patent [274] claiming the use of an additional tertiary amine, avoiding this undesired side effect. In addition to clear coats, a range of colors is currently available, including brown, blue, a black variant and some metallics. Therefore new photoinitiators have been developed. It is one of the key development targets for UV powder producers to offer a full range of colors and gloss values. High gloss [275], semi gloss and textured finishes [276] are already possible. Very low gloss and matt coatings still need more development.

Coating of wood composites with UV powders is more difficult than for metal substrates. However, the right combination of powder, application technology and curing schedule, using infrared heating and high intensity UV lamps, allows pigmented powder coatings on MDF with 100 μm film thickness, good hiding power and a smooth surface (see before). Wood composites, such as MDF, tolerate temperatures up to 120 °C, which ensures very good flow characteristics.

Several colors, including white, are already available and meet most of the **requirements of the furniture industry**, like scratch and stain resistance. Therefore, a big market is envisaged for the coating of MDF. *Drummond* et al. [277] estimate the potential market for MDF powder coating, alone in North America, to be greater than the total powder applications to metal substrates.

In 2006 *Knoblauch* [278] stated that a typical UV powder system is capable of finishing 2 Mio m^2 materials per year, with all surfaces, edges, cutouts and backsides, if required. Since four different UV powder formulators (DuPont, H.B. Fuller, Protech and Tiger/Drylac) were active this time no standard product was available. Typically UV powder shows a pencil hardness of 4H and an outstanding Taber abrasion test (ASTM D4060) of 0.09 mg loss at a 500 mg load after 500 cycles with a CS 10 wheel. The UV powder applicators were and continue to be market builders of the UV powder coating industries.

In a cooperative study by Ciba Specialty Chemicals and DSM Resins [275], *Misev* et al. and *Udding-Louwrier* et al. describe formulations for white rutile TiO_2, carbon black, blue (β)-Cu-phthalocyanine, red diketo-pyrrolo-pyrrole and red iron oxide **colored UV curable powder coatings**. They come to the conclusion that the through-cure increases with the particle size of the pigments. That means, when very fine carbon black is used, film thickness of black UV coatings should not exceed 40 to 50 µm and maximum concentration should be 0.5 to 0.6 % of carbon black [253]. In addition the Ciba-DSM team finds that coatings, containing yellow pigments, like iron oxide yellow and bismuth vanadate, show low reactivity and poor hiding. The curing of these systems is strongly impeded by the **high absorption of yellow pigments in the spectral activity range of the BAPO photoinitiator**. This is confirmed by *Te Walvaart* et al. [279] in a study of selecting pigments for colored UV curable powder coatings.

Natural wood is the most challenging substrate for powder coating. To retain the attractive structure and appearance of wood, clear coats are here preferred, which means that inclusions of gas bubbles or other imperfections (escaping from the wood surface, even at temperatures below 100 °C) are highly visible. Therefore, much development work on the application and curing technology has been (and is) done, to meet the quality, that is expected by the furniture industry. According to *Godan* [280] plywood and solid beech wood can be coated with UV powder, although the surface has to be carefully sanded first.

Souris and *Buysens* [281] give a formulation for a **clear matte finish** for natural hardwood/veneer composite board. This formulation uses a combination of an amorphous and a semi-crystalline resin (ratio 85:15), which lowers the melting temperature considerably. A heating cycle of 30 to 40 seconds at 90 to 100 °C is suitable for coating veneers of beech, ash and oak successfully. Maximum film thickness of 40 µm should not be exceeded, otherwise degassing problems can occur. An UV powder, based on this formulation, has been tested according to the **requirements of the kitchen furniture industry** and showed high qualities, comparable with the high quality laminates/foils currently used in kitchen and bathroom cabinetry. These high qualities include: excellent abrasion, scratch, heat and chemical resistance. Another way to get to mat UV cured powder coatings is described by Rohm and Haas [282]. They mixed free radical and cationic polymerizing resins in addition to the two appropriate photoinitiators to control the gloss.

In 2005 DuPont [283] discovered a **unique way to get to mat UV powder coatings**. After applying, but before melting the powder coating, it is irradiated with a distinctive dose UV light leading to a crosslinking only at the surface and thus adjusting the gloss. Then the coating is fused, melted and cured regularly with UV light. The same principle can even be applied by irradiating the powder coating particles before deposition on the substrate [284]. Adding a hydrophobic UV curable resin like a siloxane urethane acrylate is claimed to lead to easy to clean mat surfaces similar to a Lotus effect [285].

As stated earlier, the experience with UV powders on plastic substrates is rather limited. Adhesion problems and the large variety of plastics with very different properties make this area very difficult for the development of a new coating technology. Nevertheless, here too, progress has been made. So *Souris* and *Buysens* [281] formulated a UV curable **topcoat for PVC sheets**, used for resilient flooring applications. This topcoat is a matte clear coat, again based on a mixture of an amorphous and a semi-crystalline resin. It should fulfill the following properties: excellent scratch, abrasion and chemical resistance, combined with very good coating flexibility. *Wouters* et al. [286] grafted first a benzophenone type photoinitiator and then an acrylate to the surface of polypropylene. This way they almost tripled the surface energy from 26 to 70 mJ/m^2, thus increasing the adhesion of this modified polypropylene surface to a UV powder coating significantly.

Commercial applications for UV curable powder coatings are a reality since 1999, with two installations running successfully in the U.S. and one in the UK [270, 287], two years later followed by a third one in the U.S., one in Canada [288] and a second one in the UK [289]. The first two installations in the U.S. both take advantage of the lower melt and flow temperature of UV powders, to apply coatings to pre-assembled parts, containing heat-sensitive components (electric motors and automotive radiators). The third plant in the USA is the first MDF UV powder line. This company had already quoted for work valued at over $ 1 million and the line was fully operational in the late 2001. The first UK line employed UV powder coatings on MDF furniture pieces (TV tables and other music furniture components). A detailed description of this process is given by *Blatter* et al. [287] and *Romer* [289]. The use of MDF in combination with a UV powder coating has given the customer completely new design possibilities for the furniture and a surface quality that was not available from the "old" technology (chip-board with melamine or PVC-laminate). According to *Blatter* this first commercial application in the UK (and in Europe) has proven, that a **close cooperation between oven supplier and powder producer is necessary**. The powder coating formulation as well as the oven design has to be carefully adapted to the coating job. Optimization of the whole coating system is described in "cooperative" patents by DuPont Powder Coatings AB [290] and Triab/Triline AB [291] and allows a completely satisfying solution for the customer. According to *Hammerton* [292] the water content of the MDF has to be at least 5 %, due to conductivity reasons, but not higher than 8 % since blistering and edge cracking has to be avoided. The density of the MDF should be at least 750 kg/m^3.

The fact, that new industrial lines are in their start-up phase [281], proves that the UV powder concept works and is **more and more accepted as an attractive coating technology**. This view is supported by the existing know-how that has already been built-up during the past years by resin suppliers, powder formulators and equipment manufacturers. Moreover, the fact that the technology is environmentally friendly and cost-attractive can convince more companies to invest in it. *Santandrea* [293] gives a nice overview emphasizing the advantages of UV powder but showing the limits, too. He mentions that only special (higher priced) MDF grades are appropriate for the curing conditions.

Triab's "UV Speed Oven", described by *Karlsson* [294], is used for the above mentioned UK plant and consists of three sections, an IR section, a combined IR and convection section and a UV-curing section. **Combination of the two different heating methods** is the prime feature of this development, because it enables close control of the rate of heat-up and the surface temperature during the melting stage, prior to UV-curing. Heat-up should be fast and uniform to guarantee optimal flow and final surface quality (see Chapter 4.10 of this book) and the "Triab Speed Oven" is designed for this purpose.

The two major players in Europe, concerning UV curable resins, are DSM Resins with their semi-crystalline MA-VE system, based on a blend of an amorphous unsaturated polyester and a crystalline vinyl ether urethane crosslinker; and UCB Chemicals (now Cytec) with amorphous as well as semi-crystalline methacrylated unsaturated polyesters. They are commercially available under the trade names "Uracross" (DSM) and "Uvecoat" (UCB, now Cytec).

Both DSM and UCB (now Cytec) distinguish between resin systems for MDF and systems for application on metals. Therefore, DSM [273] delivers one unsaturated MA-polyester, which is combined either with a short-chain crystalline VE crosslinker (for MDF) or with a longer-chain, more flexible VE crosslinker (for MDF).

UCB (now Cytec) [295] do not deliver blends of resins and crosslinkers, but various separate amorphous and semi-crystalline "Uvecoat" resins, which are either recommended for MDF, textured finishes, metals, or flexible clear coats.

Summary

The state-of-the-art (2012) with UV curable coating powders can be summarized as follows:

Although UV curable powder coatings have proved to be suitable for industrial-scale use for years, their field use is still limited to only a few applications. It is possible, when these powders are considered, that the difficulties involved are more emphasized than the advantages. Problems often cited are: limited pigmenting capability; poor storage stability; and high capital investment costs and a 25 % higher powder coating formulation price. Despite the **slow acceptance of UV powder coatings**, it should not be disregarded, that raw material suppliers and some powder manufacturers have made considerable efforts to make them more acceptable and to further broaden the industrial application possibilities of UV curable powders.

3.4.5.2 NIR curable powder coatings

In 1998 near infra red (NIR) radiation with a high intensity was introduced at the PCE 98 by *Bär* and *Sedlmeyr* [296]. The radiation intensity of the new lamp is several orders of magnitude higher than for conventional short/middle wave infra red lamps. In addition, the NIR radiation can penetrate deep into the powder layer, which leads to a **homogeneous heat up of whole coating in an extremely short period of time** (seconds!). In contrast with UV-curing, separate melting and curing is disposed of. For the working principle of NIR-curing see Chapter 3.1.9.

Two years later, at the PCE 2000, *Bär* and *Sedlmeyer* [297] showed that the energy density of the NIR lamps has been further increased. At the same time *Blatter* et al. [287] investigated the interaction between the NIR radiation and the powder coating, which is described in a patent by DuPont [298]. Their studies lead to the conclusion, that **normal powder coatings will either show incomplete cure** or will be damaged if exposed to high intensity NIR radiation for prolonged periods of time. Only few conventional powder coatings will cure in less than 2 min by NIR radiation. Therefore DuPont [299] claimed the pretreatment of the substrate surface with material which absorb high-energy radiation within a wavelength in the range of 250 to 2,500 nm like carbon, graphite, magnetite, iron oxide, iron oxide black, tin oxide, and antimony oxide. The same company recommended measuring the thermal radiation which is given off by the substrate during curing to detect incomplete hardening of the powder coating formulation [300].

Surprisingly, NIR curing gives coatings with an excellent aspect, which is attributed to the very high heat up rate and the efficient energy transfer. According to DuPont [301] it is recommended to add NIR reflecting pigments like carbon black to the formulation. It is known [302] from conventional (convection) cure, that the heat-up rate is very important for a good flow and smoothness of a powder coating. See also Chapter 4.10. *Blatter* indicates that NIR adapted powder coatings are already available in a full range of colors and gloss levels, including metallics and textures. The properties of their coatings equal at least those of conventional powder coatings and adhesion to metals is even better. To enhance adhesion even more DuPont [303] recommends using a pulsed heating process leading to melting of the formulation close to the substrate first, then removing the heat, followed by irradiating again to crosslink the powder coating formulation at the surface. NIR powder coatings are usually 20 % more expensive than conventional powders [189].

Figure 3.26: NIR-curing line
source: Adphos Innovative Technologies GmbH

The first commercial application of the NIR powder technology was realized in 1999, when aluminum profiles have been coated with NIR adapted powder coatings, with curing times of <10 s. This change-over to NIR technology made it possible for the customer to increase his powder coating line speed considerably, without the need to invest in building extensions. Several other projects of metal coating, including pre-assembled objects (containing heat sensitive parts), have been realized. Adphos reported to have delivered five NIR-curing devices for powder coatings to job coaters in Europe, two in U.S. and three in Asia.

The **benefits of the NIR technology** can be summarized as follows:

- high transmission throughout the powder layer, as powders have low absorption in the NIR region (transparency);
- high reflection at the surface of the substrate (when metal coating);
- curing from the bottom up (increased adhesion);
- extremely high heating rates (emitter temperatures of more than 3000 K), which gives improved film flow;
- extremely shortening of the curing process (within seconds);
- higher efficiency (reduced energy costs per m^2 per part);
- homogeneous heating of the coatings by optimized (NIR adapted) powder formulations (gives improved flow);
- minimal thermal stress of the substrate, without through heating (application on heat-sensitive materials is possible).

Figure 3.26 shows a picture of a typical NIR-curing line.

Are NIR- and UV-curing competing technologies?

The answer is given by *Blatter* et al. [287] by saying: Both NIR and UV radiation curing have similar advantages in terms of curing speed. Some limitations, like the curing of complex 3-D objects, are similar for both technologies. But, in most cases, one of the curing techniques will have distinctive advantages. For instance, NIR radiation is spontaneously

Table 3.22: Comparison of powder curing techniques

	Conventional low T-cure	UV curing	NIR curing
Time for melting and curing	20 to 30 min	2 to 3 min	1 to 20 sec
Maximum surface temperature	140 to 160 °C	100 to 120 °C	<100 to >200 °C, (depending on substrate)
Preferred curing mechanism	polyaddition/polycondensation	chain polymerization	polyaddition
Film thickness limitation	no	<100μm	no
Product range available today	full range of colors, gloss levels, textures	selected colors, various gloss levels, some textures (also matt)	no color limitations, various gloss levels, textures and metallics
Through heating of thick substrates required	yes	no	no (but depending on substrate)
Price value	low	high	medium
Availability of raw materials on heat sensitive substrates	low	high	medium
Potential for non-metal applications	very limited	yes	yes
Limitations on shape of substrate	no	restricted for flat and simple 3-D shapes only	restrict. for flat and simple 3-D shapes only

absorbed by water. Therefore, it will be difficult to use NIR powder coatings on substrates that contain a significant amount of moisture, because this moisture will readily be evaporated and can cause film disturbance. **Coating of wood substrates**, in particular with clear coats, will therefore probably remain a **domain for UV-curing powder coatings**. In addition, UV cured powders provide a superior chemical resistance and hardness of the coating, which may be advantageous for many applications. UV lamps consume much less electrical energy than NIR lamps.

On the other hand, **NIR powder coatings** allow extremely short curing times on metal substrates with **no film thickness limitations**. Red and especially yellow colors, which are difficult to cure in higher thicknesses with UV powders, are no problem for the NIR technology.

Blatter [287] and *Zimmermann* [304] made the following instructive comparison of powder curing techniques *(Table 3.22)*:

3.4.6 Silicone-based powder coatings

For more than fifty years solvent-based silicone resins have been used for the coating of objects, which are **exposed to high temperatures** (>200 °C), such as boilers, engines, exhaust stacks, heat exchangers, mufflers, barbecue equipment, etc.

Due to environmental concerns and regulations, recent efforts by silicone resin manufacturers and powder coating formulators have expanded the utility of silicone-based technologies into the field of powder coatings [305, 306]. Silicone resins are structurally based on a framework of Si-O and Si-C bonds. They provide excellent heat and weathering resistance, but also inertness to chemicals and good electrical characteristics. These properties are due to the **stability of the Si-O-Si bond**, stability of organic radicals attached to the Si atom, (because of the screening effect of the siloxane bond), the formation of stable Si-O-Si bonds within the polymer upon partial disintegration (in contrast to organic radicals) thus preventing film disruption, and the insulating effect of the siloxane layer (formed during thermal oxidation).

Excellent heat resistant properties will be provided by pure silicone polymers, but for most practical applications, the organic-modified silicones meet the requirements for the majority of specifications.

Most silicone resins are produced by the co-hydrolysis of di- and trichlorosilanes. The resulting di- and trisilanol moieties are highly reactive and the majority will immediately condense to form a polymeric resin matrix. Therefore hydrolysis and polycondensation are generally performed in one step. Polycondensation between silanediols yield straight-chain polymers;

Equation 3.90

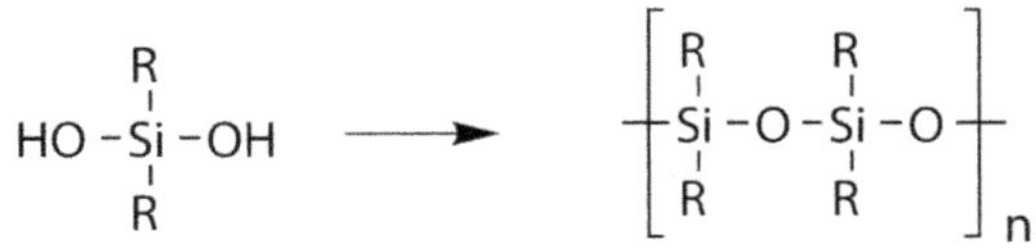

Whereas similar reactions between silanetriols result in the formation of highly crosslinked polymer molecules:

Equation 3.91

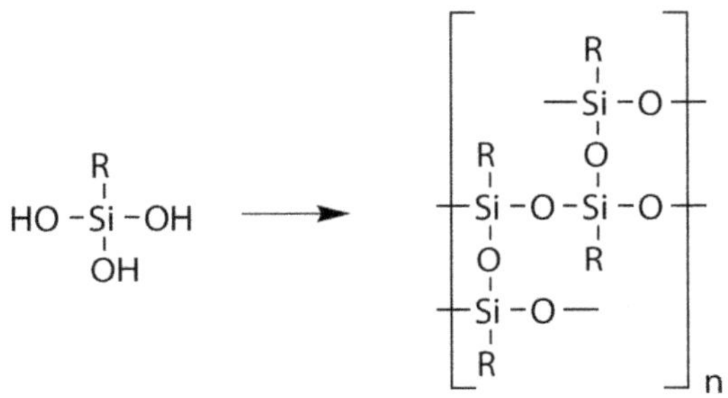

These processes differ from the "familiar" carbon chemistry in two major aspects:

- The strength of the Si-O bond significantly exceeds that of the Si-Si or the Si-C bonds. That means there is a strong tendency for the two elements (Si and O) to combine and to build up polymers, in which the Si-O bond constitutes the structural unit.
- Reactions of carbon compounds tend to produce only five- or six-membered rings, whereas silicon rings of relatively large size may be formed quite readily. It seems, therefore, according to *Groggins* [307], that cyclization competes more seriously with chain-polymer formation in the reactions of the silicon atom, than it does in carbon chemistry.

Synthesis of the before mentioned chlorosilanes can be carried out by means of the well-known *Grignard* reaction, by reacting silicon tetrachloride with alkyl magnesium bromide:

Another, more direct way of preparing chlorosilanes, involving the reaction of silicon with alkyl- or arylchlorides, with the use of copper or silver catalysts is also possible [307]. Both methods produce a mixture of the various chlorosilanes, which leads, after hydrolysis, also to a mixture of the various silanols.

Silanol groups, which remain unreacted during the following polycondensation, can be condensed after application of the coating or co-reacted with organic polymers. Due to the electro-negativity of the Si atom, a hydroxy group attached to a silicon atom is more polar and therefore more acidic than its carbon analog (carbinol). As a result, **reactions between**

silanol functional resins and hydroxyl functional organic polymers will take place at relatively low temperatures (110 to 115 °C).

The cure scheme for a silicone-based powder coating is dependent on the silicone content and the selection of the organic co-binder. A typical cure for a coating, based on 100 % silicone would be 30 min at 230 to 250 °C. Cerium acetylacetonate can accelerate the stoving process [306]. Silicone resins are high molecular weight polysiloxanes, which mostly contain methyl- and/or phenyl groups as organic substituents.

Methyl silicone resins are harder and less thermoplastic than phenyl silicones, but less heat resistant, because of the greater sensitivity to oxidation of the CH_3 groups.

Pure **methyl silicones** have, due to the relative small mass of the methyl group, a lower C-content and a higher SiO_2-content than the phenyl types. A disadvantage is that they are **not compatible with other organic resins**.

Phenyl silicone resins have good heat stability up to temperatures of 250 °C, but like all organic groups, they will decompose at about 400 °C. While they are claimed to be the better choice for heat resistance in the intermediate range (100 to 375 °C), it should be remarked, that when the organic component is oxidized, the losses in film weight of phenyl groups will be higher than those of methyl groups. Additionally, the reactive silanols, sharing a silicon atom with a phenyl group are sterically hindered. This can result in poor cure and reduced physical properties. The **main advantage of phenyl silicones is their compatibility** with other organic resins. Compatibility is essential to ensure miscibility with the resin components, reducing the tendency to cross-contamination problems and film disruption, and at the same time improving physical properties.

An important factor affecting the suitability of a silicone resin is the **crosslink density or degree** of substitution. In the natural state the four valency sites of a silicon atom are bound with oxygen. Silicone-based resins are created by substituting one, two, three or four of these oxygen atoms with a variety of moieties. The number of substituted oxygens per silicon atom is called the "degree of substitution". In the case of conventional silicone-based coating resins, phenyl and methyl groups are most often utilized, and commercially available silicone resins possess degrees of substitution within a range from 1.0 to 1.8. In this range the physical form of the resin can vary from a hard brittle solid to a viscous liquid. Physical properties are also affected by the molecular weight and residual silanol (SiOH) content of the resin.

If highest heat resistance is required, a powder coating will be formulated with a 100 % silicone. This will provide the best thermal properties in applications, where heat resistance has a higher priority than organic compatibility. A resin with a balanced phenyl/methyl ratio and highest allowable degree of substitution should be selected.

For less stringent applications a silicone resin with high phenyl content should be used allow for mixing with organic resins, to reduce raw material costs and to improve the physical properties of the coating.

Selection of the right pigments to achieve maximum temperature resistance is very important. Coatings with short-term exposures (<1000 hours) at up to 350 °C, or long-term exposures up to 225 °C can be produced with TiO_2 and mica. Applications with short-term up to 525 °C or long-term exposures up to 250 °C often include heat resistant black iron oxides. Aluminum, zinc and stainless steel powders are used for high temperature applications, in the short term up to 650 °C or long term at 350 °C.

The inclusion of metal pigments with their high thermal conductivity prolongs the life of the coating by conducting heat away from the coated substrate. Additionally, the inclusion of **metal**

pigments compensates for the inevitable oxidation of the silicone resin at high temperatures by stabilizing the coating as the metal fuses with the resin to form a ceramic coating with stable metal-siloxane bonds. Similarly, ceramic frits are used to create durable coatings, which are stable up to 750 °C [305]. DuPont [308] claimed that the addition of up to 50 % of alkali tetraborate will lead to a heat stability of at least 550 °C, suitable in particular for boilers, ovens, heat exchangers, and cooking utensils.

Enhancements to the thermal stability of the coating can also be achieved by increasing the amount of lamellar fillers (such as mica), or incorporation of anti-oxidants.

Table 3.23 [306, 309] gives typical formulations for a matt black and a glossy white silicone-based powder coating.

In the mid-nineties, also **silicone elastomeric powders** have been introduced [305, 310]. These soft and flexible spherical particles (average particle size 3 to 5 μm) can be incorporated into powder coating formulations to create coatings with unique appearance, texture and feel. With doses of 5 to 15 % they give tough textured finishes. At higher levels suede or soft-feel coatings can be produced. DuPont [311] claims the use of low-melting glass particles to fill voids in the film and prevent adhesion failure.

Dow Corning have commissioned a study by the University of Southern Mississippi. The results of the study are described by researchers from Dow Corning, Dow Chemical and the University of South Mississippi [312]. The study concentrated on three themes: the effect of pigment volume concentration (pvc); the effect of silicone

Table 3.23: Two silicone-based powder coatings

source: Wacker-Chemie

A. Silicone powder coating with good color stability and heat resistance up to 300 °C	
	Parts by weight
"Silres 604" (1)	56.3
"Crylcoat 2617-3" (2)	11.0
ß-hydroxyalkyl amide (3)	0.7
"Blanc Fixe F" (4)	10.0
"Synergy Red 6054" (5)	20.0
"Resiflow PV88" (6)	1.5
Benzoin	0.5
Curing conditions	30 min/230 to 250 °C
Properties: Film thickness	40 to 60 μm
Pencil hardness	4 H
Gloss 60°	74 %
Gloss 60° after 20 h 265 °C	29 %
Crosshatch	GT 1
MEK-resistance	>200 DR
B. Silicone powder coating with corrosion and high heat resistance	
	Parts by weight
"Silres 604" (1)	47.5
"FERRO PK 3095" (7)	15.0
"Glimmer TM" (8)	11.0
"Heucophos SAPP" (9)	13.5
"Talkum EL–10" (10)	12.3
"Resiflow PV 88" (6)	0.3
Benzoin	0.4
Curing conditions	30 min/230 to 250 °C
Properties: Film thickness	40 to 60 μm
Pencil hardness	4 H
Gloss 60°	15.4 %
Gloss 60° after 1 h 500 °C	12.8 %
Crosshatch	GT 1
Impact resistance (dir.)	> 80 inch·lbs
For better free flow 0.1 - 0.3 % "HDK H30" (Wacker Chemie AG) or "Aeroxide Alu C" (Evonik Industries) can be added.	

(1) Methyl polysiloxane (Wacker Chemie AG)
(2) Polyester (Cytec)
(3) "VESTAGON HA 320" (Evonik Industries) or "Primid XL-552" (Ems)
(4) various suppliers
(5) Pigment (Engelhard)
(6) flow agent, (Worlée)
(7) iron oxide (Ferro)
(8) mica filler (ASpanger Mineralwerke)
(9) anticorrosion pigment (Heubach GmbH)
(10) Talcum (Luzenac-Europe)

resin composition, using four Dow Corning silicone resins; and the compatibility and high temperature performance of silicone-polyester at ratios of 70:30 and 30:70.

For all their experiments they used a high temperature resistant spinel black pigment. The conclusions of their studies can be summarized as follows:

- the most **suitable pvc level** is 24 %;
- **film thickness should be kept below 40 μm**, to avoid solvent popping and wrinkling;
- all tested solid silicone resins can be used to formulate **high temperature resistant powder coatings** (a heat exposure test of 1000 hours at 425 °C can be realized using the spinel black pigment);
- it is important to match silicone/organic powder coating blends using polymers of **similar viscosities and reactivities**.

The problem of solvent popping is thought to be caused by water, evolved due to silanol-silanol condensation.

Acrylic-silicone powder coating resins are described by a team of researchers at the Glidden Co. [313]. They blended a hydroxyl acrylic resin with a solid silicone resin and cured the blend with a blocked isophorone diisocyanate adduct. The preferred silicone resin is a hydroxy functional low molecular weight cyclic silicone intermediate ("DC-Z-6018"), having a number average molecular weight of about 600. The hydroxyl functional acrylic polymer and the silicone resin are dry blended in a weight ratio of between 50 and 75 % acrylic resin and between 25 and 50 % cyclic silicone. It is shown by accelerated weathering, that **gloss retention is much higher for the silicone containing materials** (particularly at the 50 % level) than for the acrylic resin without any silicone resin.

The same researchers [313] attempted to make a graft copolymer of silicone-acrylic, but this failed, due to the high reactivity of the silanol hydroxyl (SiOH) groups. But, upon reacting the silanol hydroxyl groups with fluorinated alcohols before grafting the acrylic monomers, a graft copolymer of silicone-acrylic can be successfully made. It is believed, that the weathering resistance of this fluorinated silicone-acrylic copolymer should be excellent.

In a patent [314] of Rohm and Haas the use of trifluoromethane sulfonic acid diethylamine salt as a wrinkling agent in silicone based powder coatings is claimed.

A paper by *Rawlins* and *Thames* [315] reports the synthesis, characterization and evaluation of novel **silicone modified polymers as clear UV curable powder coatings**. These polymers are described as acrylated, silylated diglycidyl ethers of bisphenol A (DGEBA). They are synthesized by reacting acrylic acid with DGEBA polymers in the presence of a base catalyst and free radical inhibitors. Silylation is realized by the addition of 10 mol-% excess of hexamethyl disilazane to the washed acrylated DGEBA. UV curable powders are made by adding two photoinitiators to each of the silicone modified polymers, and melt mixing and grinding them in the usual way. Though the silylated modifications of DGEBA polymers are likely to be of academic interest only [316], there are some interesting facts arising from this study. The most important ones concern the low melt viscosities, achieved with these polymers (without sacrificing storage stability) and the complete freedom from surface defects, such as cratering and orange peel, obtained without the addition of flow control agents.

Wacker [317] reported the use of solid silicone-polyesters in conventional thermoset powder coatings, leading to a higher heat- and UV resistance compared to the pure polyesters. Silicone modification reduces the polymer-T_g by nearly 32 % from the starting polymer.

The use of silicone resins for high temperature resistant powder coatings is still in its infancy and a lot of new developments and applications are to be expected from this interesting field of special powder coatings in near future.

3.4.7 References

[1] *Greenspan, F.P.,* Chemical Reactions of Polymers, Ed. E.M. Fettes Interscience, New York, 1964, p. 152
[2] *Paul, S.,* Surface Coatings: Science and Technology, John Willey and Sons, 1985, p. 219
[3] *Wise, J.K.,* J. Paint Technol., 42, (540), 29, 1970
[4] *Ranney, M.W.,* Epoxy Resins and Production: Recent Advances, Noyes Data Corp., 1977
[5] *Brushwell, W.,* Am. Paint and Coat. J., Nov. 2, 65, 1981
[6] *Brushwell, W.,* FARBE&LACK, 87, (3), 201, 1981
[7] *Oldring, P., and Hayward, G.,* Resins for Surface Coatings, Vol. II, p. 4, SITA Technology, London, 1987
[8] Asahi Chemical Ind., Jap. Pat. 58.187.464, 1982
[9] *Drake, R.,* Proc. 13th Internat. Conf. (PRA), Brussels, Nov. 1993, Paper 4
[10] Basf Coatings Japan, WO 2007138399, 2007
[11] Sumitomo Durez, Jap. Pat. 59.226.066 and 59.226.067, 1983
[12] *Greensitt, E.A.,* Paint Technology, 28, (9), 16, 196
[13] *Kimmo Peltonen,* Thermal Degradation of Epoxy Powder Paints, Institute of Occupational Health, Helsinki, 1986
[14] *Peltonen, K.,* Analyst, 111, 819, 1986
[15] General Electric Company, US 20050261398, 2005
[16] Dow Epoxy Resins in Powder Coatings, Dow Chemicals Brochure Ed. 1987 and 1991
[17] *Barletta, M.; Bolelli, G.; Gisario, A.; Lusvarghi, L.,* Progress in Organic Coatings, 61(2-4), 262-282, 2008
[18] Hitachi Chemical KK, Jap. Pat. 58.087.123, 1981
[19] Wagemakers, NL Pat. 138.009 (1973)
[20] Shell Oil Company, US Pat. 3.824.035, 1974
[21] Wagemakers, NL Pat. 134.229 (1971)
[22] Dow Chemical Europe, EP 0.072.371, 1981
[23] *Gulpen, N.J.H., Werff van de A.J.,* J. Paint Technol., 47, (608), 81, 1975
[24] DuPont De Nemours, WO 2007146388, 2007
[25] DuPont De Nemours, WO 2005023941, 2005
[26] Dow, WO 2009094235, 2009
[27] *Schuetz, A.; Kaiser, W.-D.,* Farbe + Lack, 110(4), 22-24, 26-27, 2004
[28] *Harris, S.,* PPCJ, Polymers Paint Colour Journal, 196(4497), 14, 2006
[29] *Nguyen, T; Martin, J.W.,* Journal of Coating Technology Research, 1(2), 4, 82-92, 2004
[30] Dow, WO 2009058715, 2009
[31] Dow, WO 2011087486, 2011
[32] Dow, WO 2011068645, 2011
[33] Nitto Electric Industry, Jap. Pat. 57.042.760, 1980
[34] Yuka Shell Epoxy, Jap. Pat. 58.118.818, 1982
[35] Shell Chemicals, Epikote Resin Powder Coatings. Manual EK 2.10, 1989, p. 12
[36] Sumitomo Durez KK, Jap. Pat. 59.226.065, 1983
[37] Resolution Performance Products, US 20040147690 2004
[38] PPG, WO 2007002115, 2006
[39] Shell Chemicals, Epikote Powder Coatings - General Principles, Bulletin EK 2.3, p. 9, 1992
[40] Chemische Werke Hüls AG, EP 0.044.030, 1980
[41] Mitsubishi Electric Corp., Jap. Pat. 57.018.765, 1980
[42] Minnesota Mining and Manufacturing Company, US Pat. 3.876.606, 1975
[43] Sumitomo Bakelite Co., JP 2008248100, 2008
[44] *Marx, E.,* et al., Intern. Waterb., High Solids and Powder Coat. Symp. New Orleans 2000
[45] *Guo, H.,* IP.com Journal, 9(9A), 3-4, 2009
[46] Dow, WO 2008147641, 2008
[47] *Krüger, A.,* FARBE&LACK 105, (9), 106 (1999)
[48] Shell Chemicals, Epikote Powder Coatings - General Principles, Bulletin EK 2.3., p. 10, 1992
[49] *Wu, X.,* Reguxing Shuzhi, 21(4), 29-33, 2006
[50] DSM, WO 2008031589, 2008
[51] Sabic Innovative Plastics, WO 2009012391, 2009
[52] Hoechst AG, DE Pat. 1.913.923, 1969
[53] *Misev, T.,* XIX FATIPEC Congress, Proceedings, Vol. III., Aachen, 19-23 Sept., 1988, p. 99
[54] *Patton T.C.,* Off. Digest, 32, (430), 1544, 1960
[55] *Lynas-Gray, J.I.,* Paint Technology, 11, (124), 129, 1946
[56] *Lynas-Gray, J.I.,* Paint Technology, 12, (133), 7, 1947
[57] *Burrel, H.,* Paint Oil and Chem. Rev., 110, 19, 1947
[58] *Wangsness, I.L., Jerabek, R.D, Murphin A.T., and Naponen, G.E.,* Off. Digest, 26, 1062, 1954
[59] *Seaborne, L.R.,* Paint Technology, 19, (208), 6, 1955

[60] *Wiederhorn, N.M.*, Am. Paint J., 41, (2), 106, 1956
[61] *Kraft, W.M.*, Off. Digest, 29, 780, 1957
[62] *Vaughan, C.L.P. and Schmitt, F.E., Jr.*, Off. Digest, 30, (405), 1131, 1958
[63] *Glaser, D.W.*, Off. Digest, 33, 642, 1961
[64] *Patton T.C.*, Alkyd Resins Technology, Interscience Publishers., John Wiley, New York 1962
[65] *Misev, T., Ban, N. and Bravar., M.*, Hemijska Industrija, 33, (5), 177, 1979
[66] *Misev, T.*, Hemijska Industrija, 34, (6), 164, 1980
[67] *Misev, T., Ban, N. and Bravar, M.*, Hemijska Industrija, 34, (7), 179, 1980
[68] *Mleziva, J.*, in Polyestery, SNTL, Praha 1964
[69] *Tysall, L.A.*, in Calculation Technique in the Formulation Alkyds and Related Resins, Paint Research Association, May 1982 ed.
[70] *Misev, T.*, J. Coat. Technol. 61, (772), 49, 1989
[71] Cargil Inc., US Pat. 4.275.189, 1980
[72] Kodak Ltd., WP 8.300.328, 1981
[73] Hüls AG, EP 317 741, 1987
[74] Mitsui Toatsu Chem. Inc., EP 314 447, 1987
[75] Chemische Werke Hüls AG, DE Pat. 3.004.876, 1980
[76] Chemische Werke Hüls AG, DE Pat. 3.322.719, 1983
[77] Eastman Kodak Co., EP 0.070.118, 1981
[78] Eastman Kodak Co., US Pat. 4.352.924, 1981
[79] Hexion Specialty Chemicals, WO 2009109313, 2009
[80] *Bodnar, E.*, Product Finishing, August 1988, p.22
[81] UCB, BE Pat. 0.841.213, 1975
[82] UCB, BE Pat. 0.841.681, 1975
[83] Unilever N.V, US Pat. 4.147.737, 1979
[84] Goodyear Tire & Rubber Co., US Pat. 4.379.895, 1982
[85] DSM Resins BV, NL 8.204.206, 1982
[86] DSM Resins BV, BE. Pat.0.898.100, 1982
[87] DuPont, US Pat. 4.242.253, 1979
[88] Dainipon Ink Chem., Jap. Pat. 58.034.869, 1981
[89] Hitachi Chemical, Jap. Pat. 57.055.920, 1980
[90] Nippon Ester, Jap. Pat.59.011.375, 1982
[91] Nippon Ester, Jap. Pat. 57.135.829, 1981
[92] Nippon Ester, Jap. Pat. 57.126.822, 1981
[93] DSM Resins BV, BE Pat. 898.099, 1982
[94] DSM Resins BV, NL Pat. 8.204.206, 1982
[95] AMOCO-co-researchers, Proc. 25th Intern. Waterborne, High Solids and Powder Coatings Symp., New Orleans, 18–20 Feb. 1998
[96] *Johnson, L.K., and Sade, W.T., J.* Coat. Technol., 65, (826), 19 (Nov. 1983)
[97] a) *Noordover, B. A. J.; van Staalduinen, V. G.; Duchateau, R.; Koning, C. E.; van Benthem, R. A. T. M.; Mak, M.; Heise, A.; Frissen, A. E.; van Haveren, J.*, Biomacromolecules, 7(12), 3406-3416, 2006
b) *van Haveren, J.; Oostveen, E. A.; Micciche, F.; Noordover, B. A. J.; Koning, C. E.; van Benthem, R. A. T. M.; Frissen, A. E.; Weijnen, J. G. J.*, Journal of Coatings Technology and Research, 4(2), 177-186, 2007
[98] *Noordover, B. A. J.; Heise, A.; Malanowksi, P.; Senatore, D.; Mak, M.; Molhoek, L.; Duchateau, R.; Koning, C. E.; van Benthem, R. A. T. M.*, Progress in Organic Coatings, 65(2), 187-196, 2009
[99] Battelle Memorial Institute, WO 2006102279, 2006
[100] Minnesota Mining and Manufacturing Company, US Pat. 3.340.212, 1961
[101] *Huneke von H.*, Metalloberfläche, 24, 9, 315, 1970
[102] Unilever N.V., DE Pat. 2.163.962, 1971
[103] Hitachi Chemical, Jap. Pat. 57055.958, 1980
[104] Hitachi Chemical, Jap. Pat. 57.055.923, 1980
[105] Takeda Chemical Ind., Jap. Pat. 59.230.069, 1983
[106] Scado BV, US Pat. 3.624.232, 1970
[107] UCB SA, DE Pat. 2.352467, 1972
[108] SCM Corporation, EP 0.021.770, 1979 and US Pat. 4.271.277, 1979
[109] DuPont De Nemours, US 20070231580, 2007
[110] *Pappas, S.P., Kunz, V.D. and Pappas, B.C.*, J. Coat. Technol. 63, (796), 39 (1991)
[111] BASF, WO 2010057922, 2010
[112] PPG, WO 2005033170, 2005
[113] *Gedan-Smolka, M., Çetin, S., Lehmann, D., and Komber, H.*, FARBE&LACK, 103, (11), 48 (1997)
[114] *Bergmans, A.*, Polym. Paint Colour J. (PPCJ), 191, (4445), 16 (Oct. 2001)

[115] a) *Bayards, R.A., Groen, H., Koldijk, F., and Beukers, D.*, Proc. 7th Nuremberg Congres, April 2003
b) DSM Resins BV, Powder Coating Resins Brochures 1993, 1999, 2000
[116] *Barletta, M.; Lusvarghi, L.; Mantini, F. P.; Rubino, G.*, Surface and Coatings Technology, 201(16-17), 7479-7504, 2007
[117] *Ehsani, M.; Yousefi, A. A.; Yeganeh, S. S.*, e-Polymers, http://www.e-polymers.org/journal/papers/mehsani_201209.pdf, 2009
[118] DSM Resins BV, BE Pat. 0.898.100, 1982
[119] *Thames, S.F., Panjnani, K.G., Pace, S.D., and Blanton, M.D.*, European Coatings Journal (1994) 10, 705; or same authors + *Cumberland, B.R.*, J. Coat. Technol., 67, (841), 39(1995)
[120] *Miller, R.S., and Joshi, V.*, Proc. 21st Intern. Waterborne, High Solids and Powder Coat. Symp. New Orleans, 9-11 Feb. 1994
[121] *Frischinger, I., Cotting, J., Finter, J., Gottis, P. and Poget, C.*, Proceedings.4th Nuremberg Congress 1997
[122] BASF Coatings, WO 2005120724, 2005
[123] Metallgesellschaft AG, DE Pat. 1.905.825, 1969
[124] Unilever N.V., GB Pat. Appl. 61107-70, 1970
[125] DSM Resins BV, EP 0.107.888, 1982
[126] *Hoppe, M.*, J. Coat. Technol., 60, (763), 58, 1988
[127] *Merfeld, G.; Mordhorst, S.; Koeniger, R.; Acar, A. E.; Molaison, C.; Suriano, J.; Irwin, P.; Warner, R. S.; Gray, K.; Smith, M.; Kovaleski, K.; Garrett, G.; Finley, S.; M., D.; Spicer, M.; Naguy, T.*, JCT Research,2(8), 661-668, 2005
[128] *Maetens, D., Loutz, J., and Merck, Y.*, Proc. 13th Intern. PRA Conf. on Powd.Coat. Brussels, Nov. 1993
[129] *Maetens, D.*, Progress in Organic Coatings, 58(2-3), 172-179, 2007
[130] *Montserrat, S.; Calventus, Y.; Hutchinson, J. M.*, Progress in Organic Coatings, 55(1), 35-42, 2006
[131] *Merck, Y., Maetens, D., Moens, L., and Buysens, K.*, European Coatings J. (ECJ) 12 (1999) p. 18
[132] UCB Chemicals WO 97/20895
[133] *Malshe, V.; Hemantkumar, T.i R.*, Asia Pacific Coatings Journal, 17(4, Suppl.), 15-17, 2004
[134] *Hoppe, M.*, J. Coat. Technol., 60, (763), 53-57, 1988
[135] *Zhou, L., Thames, S.F., Smith, O.W., Smith, C.J., Boon, W.H., and Gwyn, D.*, Paint & Coatings Industry, Vol. XVIII, (2), 52 (2002)
[136] Evonik Degussa GmbH, WO 2008068068, 2008
[137] DSM Resins BV, NL Pat. 8.204.206, 1982
[138] *Gottis, P.G. and Cotting, J.A.*, XXIII. FATIPEC Congress book Vol. B, p. 216, Brussels, 1966
[139] *Buysens, K.*, European Coatings Journal, 6, 18-22, 2005
[140] EMS-Chemie, “Primid” Technical Supplement 7.4 (1996)
[141] UCB, WO 2004083326, 2004
[142] UCB, WO 2004083325, 2004
[143] Nippon Paint Co., JP 2004196952, 2004
[144] Rohm and Haas Company, EP 1852453, 2007
[145] DSM, EP 1873183, 2008
[146] a) *Cuijpers, J.; Buijsen, P.; Posthuma, C.; Verhoef, H. J.; Beukers, D.; Paauwe, J.*, European Coatings Journal, (6), 38-41, 2009
b) DSM, WO 2010094811, 2010
[147] a) Hexion Specialty Chemicals, WO 2010069531, 2010
b) *Capra, A.; Panero, I.; Lorenzo, D.D., Ghisolfi, S.*, European Coatings Show, Proceedings, 108, 2011
[148] *Moens, L.; Bauters, E.; van Muylder, M.*, European Coatings Journal, (6), 25-28, 2010
[149] DSM, US 20090186973, 2009
[150] Kansai Paint Co., WO 2006038491, 2006
[151] *De Lange, P.G.*, XX. FATIPEC Congress, Nice, p. 395, 1990
[152] DuPont De Nemours Co., US Pat. 4.402.983, 1980
[153] Bayer AG, EP 0.024.680, 1979
[154] *Walz, G., and Kraft, K.*, Proc. 4th Intern. Conf. in Org. Coat. Sci. & Technol., Athens 1978, p. 56
[155] Basf, WO 2006089940, 2006
[156] Bayer Materialscience, EP 2159238, 2010
[157] *Williams, F., Armengol, J., Grau, E., Monleon, J., and Schultz, E.*, Proc. 27th Int. Waterborne, High Solids and Powder Coatings Symp. New Orleans, Feb. 2000
[158] *Blank, W.*, XXVI FATIPEC Congress “Quo Vadis - Coatings”, Dresden 2002 in Macromol. Symposia 187, p. 261, Ed. Wiley-VCH, Sept. 2002
[159] *McLafferty, J.J., Figlioti, P.A., and Camilleri, L.T.*, J. Coat. Technol., 58, (733), 23, 1986
[160] *Bronk, J.M.*, Proc. PRA Symp. “Powder Coat. - What’s Next”, Birmingham, Mar. 1999
[161] Evonik Degussa GmbH, WO 2008068073, 2008
[162] *Wu, B., Padaki, S., and Maetens, D.*, Proc. 26th Int. Waterborne, High Solids & Powder Coatings Symp., New Orleans, 10-12 Feb. 1999

[163] *Ratliff, K.; Eastman, J.; Faecke, T.,* Proceedings of the International Waterborne, High-Solids, and Powder Coatings Symposium, 33rd 133-144, 2006
[164] Bayer AG, DE Pat. 3.232.463, 1982
[165] *Freudenberg, U., Meier-Westhues, U., and Laas, H.,* European Coatings Journal 9/1997, 804
[166] *Wenning, A., Weiss, J., and Grenda, W.,* Congress Papers Powder Coatings Europe 98 (PCE 98), Amsterdam, Jan. 1998, Paper 10, p. 147
[167] *Freudenberg, U., Thometzek, P., and Meier-Westhues, U.,* Proc. 26th Intern. Waterborne, High Solids & Powder Coatings Symp., New Orleans, 10-12 Feb. 1999; or same authors: 5th Nuremberg Congr. 1999; or FARBE&LACK, 105, (9), 46, 1999
[168] *R.M. Gallas,* Acrylic Modified Polyester System for Powder Coatings, Johnson Wax Brochure, 1988
[169] *Horn, M.B.,* Acrylic Resins, Reinhold Publishing Corp., New York, 1960, p.15
[170] Adapted from *Luskin, L.S., Myers, R.J.,* in Encyclopedia of Polymer Science and Technology, Ed. *N.M. Bikales,* VolumeI, Interscience Publishers, New York, 1964, p. 299
[171] *Fox, T.G.,* Bull. Am. Phys. Soc., 1, 3, 123, 1956
[172] The Glidden Company, EP 0.256.369, 1986
[173] DuPont De Nemours Co., EP 0.045.040, 1980
[174] Ford Motor Company, US Pat. 3.880.946, 1975
[175] Bayer AG, US Pat. 3.836.604, 1974
[176] BASF Farben & Faser, DE Pat. 3.310.545, 1981
[177] Deutsche Gold und Silber Scheideanstalt, US Pat. 3.867.347, 1975
[178] *Yousuf, M.K.,* Modern Paints and Coatings, June 1989, p.48
[179] Dainippon Ink & Chemicals, EP 0.038.635, 1980
[180] Nippon Paint, Jap. Pat.115516, 1986
[181] *R.M. Gallas,* Acrylic Modified Polyester System for Powder Coatings, Internal Publication, S.C. Johnson & Son Inc., 1988
[182] *Yousuf, M.K., and Pfeffer, W.G.,* Dry-on-Dry Powder Coating Systems, Internal Publication, S.C. Jonson & Son Inc., July, 1988
[183] *Dumain, E.,* Coatings World, Vol. 1, no. 2, Oct. 1998, Part 1, p. 27
[184] *Iwamura, G.,* Proc. 5th Nuremberg Congress 1999, Paper 20
[185] *Kerssen, G.W., and Verlaak, J.,* XXIII FATIPEC Congress book, Vol. B. p. 419, Brussels, June 1996
[186] *Miller, R. and Kerr, J.,* Powder Coating '92, Conf. Proc. p. 79, (1992)
[187] N.N., FARBE&LACK, BASF L+F erfolgreich mit Pulverprimern 102, (8), (1996) p. 16
[188] *Green, D.,* Paint & Coating Industry, Sept. 1995, p. 49
[189] *Linak, E., Kishi, A., Yang, V.,* Thermosetting Powder Coatings, SRI CONSULTING, 2008
[190] Rohm and Haas Company, EP 2098575, 2009
[191] *Cole, G.E.,* Industrial Paint & Powder, 69, (12), 24 (1993)
[192] *Bailey, J.M.,* Industrial Paint & Powder, 70, (7), 12 (1994)
[193] *Koop, J.M.,* Powder Coating, 5, (4), 52 (1994)
[194] *Clark, P.D.,* Proc. Intern. Body Engin. Conf. IBEC '94, 12, 44 (1994)
[195] *Schmidt, H.,* 13th Intern. PRA Conf., paper 13, Brussels, Nov. 1993
[196] Bayer Materialscience AG, EP 1491567, 2004
[197] *Schmidt, H. and Fink, D.,* Surface Coatings International, 2, 66 (1966)
[198] *Kinza, W.,* JOT Conf. EPS '97, Munich, 30-31 Jan. 1997, p. 131
[199] Dainippon Ink and Chemicals, WO 2005087881, 2005
[200] DuPont de Nemours, US 20080299323, 2008
[201] *Gribble, P.R.,* Finishing '93, Conf. Proc., Cincinnati, Ohio, 25-28 Oct. 1993
[202] *Ortiz, C. and Winters, L.,* Proc. 1st Intern. Conf. on Car-Body Powder Coatings, Berlin, 22-23 June 1998
[203] *Dornitz, R.A.,* Proc. 1st Intern. Conf. Car-Body Powder Coatings, Berlin, 1998
[204] *Pearson, R.,* Proc. 1st Intern. Conf. Car-Body Powder Coatings, Berlin 1998
[205] *Attinoto, R.A.,* Proc. 1st Intern. Conf. Car-Body Powder Coatings, Berlin 1998
[206] *Gribble, P.R.,* Proc. 1st Intern. Conf. Car-Body Powder Coatings, Berlin 1998
[207] *Higgins, R.J.,* Powder Coatings - European Supplement of PPCJ and Product Finishing, Jan. 1998, p. 9
[208] *Kreis, W.,* Proc. 1st Intern. Conf. Car-Body Powder Coatings, Berlin, June 1998; and/or PPCJ, 188, (4411), 12 Dec. 1998)
[209] BASF, WO 2007082656, 2007
[210] *Lendl, T.,* JOT, Journal fuer Oberflaechentechnik, 44(5), 40-42, 2004
[211] *Wimmer, W.,* Congress Papers PCE '98, Amsterdam, Jan. 1998, S.27, p. 433
[212] *Yanagida, K., Kumata, M. and Yamamoto, M.,* J. Coat. Technol. 68, (859), 47, (1996)
[213] *Rekowsky, V.,* Congress Papers PCE 2002, Nuremberg, Jan. 2002, p. 137
[214] DSM, WO 2010108963, 2010
[215] *Williams, H.,* Plastic Inst. Trans., 24, 37, 1956

[216] Bayer AG, DE Pat. 1.105.610, 1961
[217] *Feuer, S.S., Bockstahler, T.E., Brown, C.A., Rosenthal, I.*, Ind. Eng. Chem., 46, 1643, 1954
[218] *Hayes, B.T., Read, W.J., Vaughan, V.H.*, Chem. & Ind., 1957, 1165
[219] *Mleziva, J., Vladyka, J.*, FARBE&LACK, 68, 144, 1962
[220] DSM Resins BV, EP 0.188.846, 1984
[221] DSM Resins BV, EP 0.106.399, 1983
[222] Morton Thiokol, EP 309 088, 1987
[223] Stamicarbon BV, US Pat. 4.228.113, 1980
[224] Stamicarbon BV, US Pat. 4.287.310, 1980
[225] a) DSM, WO 2010136315, 2010
b) *Schutte, M.*, European Coatings Congress, Proceedings, 112, 2011
[226] DSM, WO 2010052290, WO 2010052291, WO 2010052293, WO 2010052294, WO 2010052295, WO 2010052296, 2010; WO 2011138431, WO 2011138432, 2011
[227] *Dohnt, W.*, Congress Papers PCE 2002, Nuremberg, Jan. 2002, p. 67
[228] *Subramanian, R. and Sullivan, C.*, Congress Papers PCE 2002, Nuremberg, Jan. 2002, p. 231
[229] *Boneza-Tomaszewski, Z.; Penczek, P.; Bankowska, A.*, FATIPEC Congress, 27th (Vol. 1), 87-95, 2004
[230] *Boneza-Tomaszewski, Z.; Bankowska, A.; Penczek, P.*, Advances in Coatings Technology, International Conference, 6th, Warsaw, Poland, Nov. 23-26, 2004
[231] *Witte, F., De Jong, E.S., and Misev, T.A.*, RadTech Report 10, 13, Sep./Oct. 1996
[232] *Udding-Louwrier, S., De Jong, E.S., and Bayards, R.A.*, Proc. RadTech '98 Chicago (USA), 19-22 April 1998, p. 106
[233] *Schwarb, R.; Knoblauch, M.; Cox,* J., CoatingsTech, 7(7), 44-49, 2010
[234] DSM Resins, WO 93/25596 (1993)
[235] a) BASF, EPA 650978, 650979 (1994) and 678562 (1995);
b) Bayer, EPA 350730 (1990)
[236] *Maetens, D.*, Proc. RadTech '98 - Chicago (USA), April 19-22, 1998, p. 170
[237] *Fink, D. and Brindoepke, G.*, a) Paper 29, 3rd Nuremberg Congress, Mar. 1995;
b) European Coatings Journal 1995, p. 606
[238] Hoechst, EPA 585742 (1994)
[239] *Wenning, A.*, XXVI FATIPEC Congres "Quo Vadis-Coatings", Dresden 2002, in Macromolec. Symposia 187, p. 597, Ed. Wiley-VCH, Sept. 2002
[240] *Williams, F.; Armengol, J. M..; Grau, E.; Chambat, C.; Laurens, C.*, RadTech Europe 05, Oct. 18-20, 1, 533-540, 2005
[241] *Hult, A., Johannson, M., Jansson, A. and Malmström, E.*, Proc. RadTech Eur. '99 Berlin, 8 Nov. 1999, p. 634
[242] *Loewenhielm, P.; Nystroem, D.; Johansson, M.; Hult, A.*, Progress in Organic Coatings, 54(4), 269-275, 2005
[243] *Wittig, M., and Gohmann, T.*, Polym. Paint Col. J. (PPCJ), 184, (4343), 34, 1994
[244] Ciba, EPA 667381 (1994)
[245] *Finter, J., Frischinger, I., Haug, Th. and Marton,R.*, Proc. 4th Nuremberg Congress April 1997
[246] DSM Resins, EPA 636669 (1995)
[247] *Blatter, K.*, Congress Papers PCE '98, Amsterdam, Jan. 1998, p. 91
[248] Bayer Materialscience, EP 1538186, 2005
[249] UCB Chemicals, EPA 739922 (1995)
[250] *Luo, Y.*, CN 1528846, 2004
[251] *Spyrou, E.; Wenning, A.; King, C. L*, UV & EB Technology Expo & Conference, Charlotte, NC, United States, May 2-5, 394-400, 2004
[252] *Bayards, R.A., Antonisse, M.M.G., Udding-Louwrier, S., and Meij, E.*, Proc. 6th Nuremberg Congress, April 2001]
[253] *Kersten, J.*, a) Congress Papers PCE 2002, Nuremberg, Jan. 2002, p. 257
b) FARBE&LACK, 108, (2), 86 (Feb. 2002)
[254] *Paulus, W., Binder, H., Hennig, I., Rieger, J., Schrof, W., and Schwalm, R.*, FARBE&LACK, 107, (10), 46, 2001
[255] a) *Spyrou, E.*, RadTech Europe 05, Oct. 18-20, 2, 269-276, 2005
b) Degussa AG, WO, 2007045609, 2007
[256] DSM, WO 2006082080, 2006
[257] *Williams, F.; Armengol, J. M.; Grau, E.; Decomble, V.; Chambat, C.*, RadTech Europe 07, Nov. 13-15, 2007
[258] *Malmström, E., and Hult, A.*, Proc. 26th Intern. Waterborne High Solids and Powder CoatingsSymp. New Orleans, Febr. 1999
[259] *Van Benthem, R.A.T.M.*, a) Proc. 25th Int. Conf. in Org. Coat., Athens, 1999, p.345;
b) Verfkroniek 73, p. 15 (April 2000)
[260] *Kim, H.J.*, J. Pol. Sci. Polym. Chem., 36, (11), 1685, 1998
[261] *Cheng, X.-E.; Huang, Z.; Liu, J.; Shi, W.*, Progress in Organic Coatings, 59(4), 284-290, 2007
[262] *Biller, K.M. and McFadden, B.*, Industr. Paint & Powder, 1996, nr. 27, p. 22-25
[263] *Witte, F.M.*, European Coatings Journal, 3/96, p. 115
[264] a) *Misev, L., Kunz, M., Strobel, R.*, Proc. RadTech Europe '95, Maastricht, p. 93

b) *Witte, F., Kieft, G.*, et al. Proc. RadTech Europe '95, Maastricht, p. 437
[265] *Laver, H., and Chamberlin, D.*, Proc. PRA Powder Coatings - What's Next ? Birmingham. U.K., Mar. 10, 1999, Paper 8
[266] *Margraf, R., Laver, H., Bender, J., Lehmann, K. and Schmid, O.*, Proc. RadTech Europe '99, Berlin, 8 Nov. 1999, p. 615
[267] *Rawlins, J. W.*, Powder Coating, Aug. 2000, 37-41
[268] Tigerwerk Lack- und Farbenfabrik, WO 2005051552, 2005
[269] *Udding-Louwrier, S., Bayards, R., and Feima, N.*, Proc. RadTech Europe '99, Berlin, 8 Nov. 1999, p. 607
[270] *Skinner, D.*, Proc. RadTech Europe '99, Berlin, 8 Nov. 1999, p. 599
[271] *Castell, P.; Wouters, M.; Fischer, H.; de With, G.*, Journal of Coatings Technology and Research, 4(4), 411-423, 2007
[272] *Binda, P.*, a) Powder Coatings '99 Symp., Birmingham, Mar. 10, 1999; b) Congress Papers PCE 2000, Amsterdam, Jan. 2000, p. 377
[273] *Udding-Louwrier, S., Bayards, R. and De Jong, E.S.*, Proc. 26th Int. Waterborne High Solids and Powder Coatings Symp., New Orleans, Feb. 1999, p. 420
[274] Cytec Surface Specialties, WO 2006037493, 2006
[275] *Misev, L., Schmid, O., Udding-Louwrier, S., De Jong, E.S. and Bayards, R.*, J. Coat. Technol., 71, (891), 37 (April 1999)
[276] *Buysens, K.*, a) Rad Tech Report, 13, (4), 18 (July/August 1999);
b) Proc. RadTech Europe '99, Berlin, Nov. 8th, 1999, p. 622
[277] *Drummond, C.K., Mihalic, S., and Guskov, S.*, Congress Papers PCE 2000, Amsterdam, Jan. 2000, p. 125
[278] *Knoblauch, M.*, Asia Pacific Coatings Journal, 19(2), 12-15, 2006
[279] *Te Walvaart, C., Jahn, R., Lahaye, J., Laver, H. and Schmid, O.*, Congress Papers PCE 2000, Amsterdam, Jan. 2000, p. 309
[280] *Godan, A.*, Surface Coatings International, 91(1), 31-33, 2008
[281] *Souris, Y., Buysens, K.*, Congr. Papers PCE 2002, Nuremberg, Jan 2002, p. 267
[282] Rohm and Haas, US 20040067304, 2004
[283] DuPont De Nemours, US 20050276917, 2005
[284] DuPont De Nemours, WO 2006069149, 2006
[285] DuPont De Nemours, WO 2007059133, 2007
[286] *Wouters, M.; Muixi, P. C.*, Contact Angle, Wettability and Adhesion, 5, 207-227, 2008
[287] *Blatter, K., Strid, M., Thiele, O., and Zimmermann, F.*, Congr. Papers PCE 2000, Amsterdam, Jan. 2000, p. 115
[288] *Mitchell, S.*, Congr. Papers PCE 2002, Nuremberg, Jan. 2002, p. 73
[289] *Römer, J.*, Congress Papers PCE 2002, Nuremberg, Jan. 2002, p. 113
[290] DuPont Powder Coatings AB, WO 97/05963
[291] Triab/Triline AB, WO 97/05964
[292] *Hammerton, D.*, International Wood Composite Materials Symposium Proceedings, 39th 233-248, 2005
[293] *Santandrea, S.*, RadTech Europe 05, Oct. 18-20, 1, 45-50, 2005
[294] *Karlsson, L.*, Congress Papers PCE 2000, Amsterdam, Jan. 2000, p. 47
[295] a) *Buysens, K.*, Congress Papers PCE 2000, Amsterdam, Jan. 2000, p. 253
b) *Zune, C. and Buysens, K.*, European Coatings Journal 05/00 (2000), p. 18
[296] *Bär, K., Sedlmeyr, M.*, Congr. Papers PCE 98, Amsterdam, Jan. 1998, p. 71and Polym. Paint Col.J. (PPCJ), April 1998, p. 18
[297] *Bär, K., and Sedlmeyr, M.*, Congr. Papers PCE 2000, Amsterdam 1998, p. 321
[298] DuPont, WO 99/41323
[299] DuPont De Nemours, WO 2004058417 2004
[300] DuPont De Nemours, WO 2005042648, 2005
[301] DuPont De Nemours, WO 2005078030, 2005
[302] *De Lange, P.G.*, J. Coat. Technol. 56, (717), 23 (Oct. 1984) and in Proc. 6th Int. Conf. in Org. Coat. Sci. & Technol., Athens, 1980 p. 366
[303] DuPont de Nemours, US 20050255238, 2005
[304] DuPont Powd. Coat., Techn. Brochure "Ray-Tec" PowderCoating, by F.Zimmermann
[305] *Witucki, G.L.*, Powder Coating, 8, (8), 19 (Nov. 1997)
[306] *Witucki, G.L.*, Congress Papers PCE 2000, Amsterdam, Jan. 2000, p. 247
[307] *Groggins, P.H.*, Unit Processes in Organic Synthesis. 3rd Ed., McGraw-Hill, New York, 1947, p. 843
[308] DuPont De Nemours, WO 2007126640, 2007
[309] Wacker-Chemie, "Silres", Techn. Brochure, Munich, 1998
[310] Dow Corning, "Silicone Additives", Techn. Brochure, 1991
[311] DuPont De Nemours, WO 2004076572, 2004
[312] *Popa, P., Be A., Storey, R. and Baugh, D.*, Proc. 26th Int. Waterborne, High Solids and Powder Coatings Symp., New Orleans, 10-12 Feb. 1999

[313] Glidden Research Team, Proc. 25th Int. Waterborne, High Solids and Powder Coatings Symp. New Orleans, 18-20 Feb. 1998
[314] Rohm and Haas, EP 1403350, 2004
[315] *Rawlins, J., Thames, S.,* Proc. 26th Int. Waterborne, High Solids and Powder Coatings Symp. New Orleans, 10-12 Feb. 1999
[316] *Harris, S.,* Focus on Powder Coatings, Sept. 1999
[317] Wacker Chemical Corporation, WO 2005063857, 2005

3.5 Additives in powder coatings

Additives play a very important role in the formulation of powder coatings, as they have become an essential part of them for several reasons, which will be discussed in this section. According to *Richart* [1], **the "ideal" powder additive** should meet the following requirements:

- solid, preferably a fine powder with a crystalline melting point or T_g >50 °C;
- chemically non-reactive with binder resin and hardener;
- specific in its function;
- 100 % active;
- effective at low levels of addition, when possible.

Unfortunately, many useful additives are liquid and are therefore difficult to incorporate, unless they are added as a masterbatch in resin or absorbed on a silica carrier. The various classes of powder additives and their functions are described now in the following subsections.

3.5.1 Flow control additives

Their function is to prevent cratering in the powder coating film and to reduce the degree of "orange peel".

Film formation and flow of thermosetting powder coatings are governed by two main parameters, namely **surface tension and melt viscosity** [2]. Surface tension is the major driving force (at a given temperature) to cause flow. The only resistance to flow is the viscosity of the molten coating at baking conditions (see also Chapter 4.7). Therefore, flow control additives can be defined more accurately as "surface tension modifiers".

In principle, there are two basic prerequisites for optimum flow in powder coatings:

1) **High surface tension**, which must however be less than the surface tension of the substrate, to ensure adequate wetting of the substrate.
2) The surface tension must exhibit **local homogeneity** at its interface with the air.

Due to the current method of powder production, this latter requirement is almost impossible to meet without the use of flow agents, since powder coatings always exhibit local inhomogeneity.

Without flow additives powder coatings not only show poor flow behavior, such as the **"orange peel" effect**, but the surface is also prone to pinholing. These effects are due to problems, caused by local differences in the surface tension of the molten powder coating film, which make the molten film move in the direction of the higher surface tension, resulting in a (heavily) textured film (*Figure 3.27a and b,* without and with a flow additive, respectively).

The formation of a powder coating film can be divided into a coalescent stage, followed by a fluid stage. Poor flow originates predominantly in the second stage. In pinhole formation, the local variations in surface tension are so great, that the film is virtually pulled apart. These problems

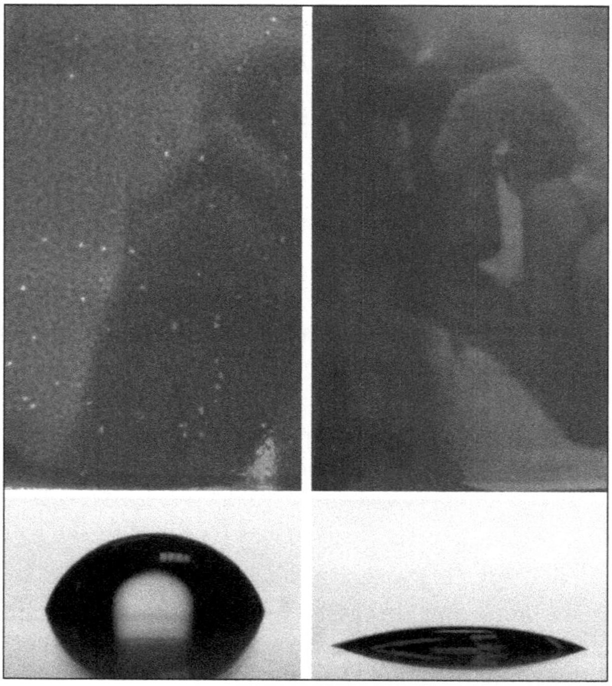

Figure 3.27a and b: Effect of flow additives on appearance and contact angles source: BYK-Chemie

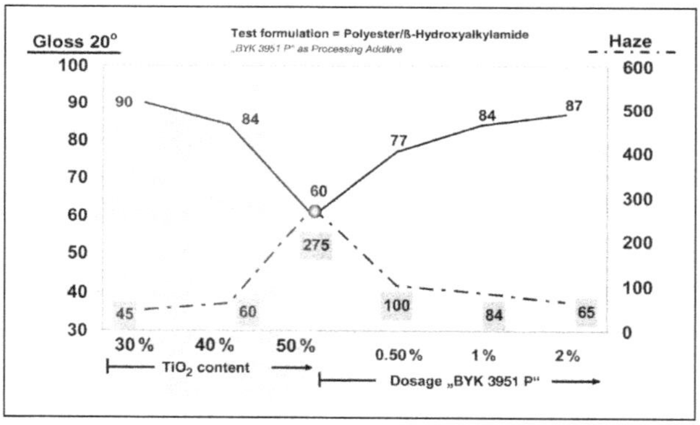

Figure 3.28: Effect of a flow additive on the gloss and the haze of a powder coating formulation source: BYK-Chemie

already occur in the first stage of film formation (coalescence) and then increase in the fluid stage.

Surface defects are independent of substrate and cannot be classified as "substrate wetting faults", but rather as "extreme flow faults".

Flow control additives, currently in use, work by influencing the properties of the interface by smoothing out (or equalizing) local differences in surface tension. This can be explained as follows (see also Chapter 4.7):

By adding a "surplus" of (mostly) a polymeric acrylic additive (which is always done in practice), such an "additional" quantity of **the flow additive migrates up to the surface** of the molten resin (reinforced by its incompatibility in the coating system) to equalize its chemical potential at the surface and in the bulk of the binder. When such an acrylic polymer arrives at the surface, the polar backbone remains in the molten resin (or binder), while the alkyl chains try to escape into the surrounding environment, providing better conditions for a stable equilibrium. Consequently, the surface tension tends to equalize across the surface, because of the greater concentration of an oriented molecular structure all over the surface. Thus this concentration effect "smooths out" local differences in surface tension. *Figure 3.28* shows the effect of a flow additive on the gloss and the haze of a highly filled (50 % TiO_2) powder coating formulation.

At present, flow control additives are mainly homopolymeric or copolymeric acrylates, while certain silicone-acrylate combinations and pure silicones are regarded as "dangerous". There is namely a basic objection to the use of silicones, since earlier experiences with highly active and/or incompatible silicones (such as poly(dimethyl)siloxanes) have caused contamination of entire plants.

It must, however, be admitted, that there are suitable modified polysiloxanes (such as poly ether- and polyester-modified polysiloxanes), which can improve the flow properties of powder coatings, without causing contamination. These silicones are often used in clear coats [1].

Studies by *Wulf, Uhlmann* et al. [3, 4] with polyester-modified methylalkylsiloxanes have shown that the surface tension of powder coatings is too much reduced with these polysiloxanes (even at levels of 0.1 % on binder), to give good flow properties. Strongest reduction takes place, when the polysiloxane contains short alkyl side chains.

Medium or high molecular weight poly-n-butylacrylate is generally used as a homopolymeric acrylate while copolymers of 2-ethylhexylacrylate and various co-monomers are also important [5-7].

An article by *Hajas* et al. [8] presents the results of **a comparative study of flow additives** of different chemical structure and molecular weight. In their study the flow behavior of powder coating films is characterized, using laser-optic measurements of "rippling" (by means of BYK's "wave-scan") as a function of additive concentration. "Wave-Scan" has been designed as a portable instrument for the characterization of "orange peel". It incorporates a laser that scans the surface and detects the pattern of light and dark areas in the same way as the human eye. The merit of the cited article is that the authors have evaluated a range of different flow additives of which they could assess the effect on flow in a quantitative way.

Ming Zhao [9] describes a new acrylic-based flow modifier that not only enhances the flow and leveling during film formation, but also improves "cross compatibility" in powder coating systems. According to a recent patent [10] of BYK monobutyl ether of polypropylene glycols can be used as leveling agents.

Since most flow control agents are (highly) viscous liquids, it is usual to **masterbatch these additives** at 5 to 15 % concentrations in the base resin, or to deposit them on fumed silica at 60 to 70 % additive concentration. This enables the additives to be measured easily and dispersed uniformly in the premixing stage of powder production. As the silica can produce a haze in clear films, a masterbatch in resin is preferred, when the highest level of clarity in clear coatings is desired.

Fluorocarbons [11] are also used as flow control additives at even lower addition levels than silicones. They are not widely used, due to their extremely high cost and solvent content and they are not available as a powdered masterbatch. Advantages of fluorocarbons are that they exhibit excellent recoatability and they are often found to be the only effective flow additive on difficult substrates, such as oily or otherwise contaminated metal [1].

Some high molecular weight thermoplastic polymers are effective in preventing craters in powder coatings. An example is polyvinylbutyral, which was used in the early sixties as "flow" agent. Due to its rather high melting point, it is difficult to disperse in the molten resin during extrusion and therefore it was soon replaced by the liquid polyacrylates.

At the moment polyvinylbutyral of relatively high melt viscosity is only used in functional coatings, as it improves edge coverage without appreciably detracting from the flow properties. Some types of cellulose acetobutyrate (CAB), having low melt viscosity, are used to prevent cratering in powder coatings.

DuPont filed a patent [12] claiming that the use of 2 % of an opacifying agent (e.g. "Thermoplast Black X 70" dye) in an epoxy formulation leads to an increase of the DOI smoothness from 76 to 96.

A unique way to improve the smoothness is described by DuPont [13] in a world patent application: They apply vibration during the application and curing of the powder coating particles resulting in an enhanced leveling.

3.5.2 Degassing additives

The most widely used chemical of this group is benzoin, which functions as a “solid solvent” to keep the film open, long enough for air to escape from the film.

In epoxy resin films it reduces melt viscosity and surface tension. **Benzoin causes yellowing** of white and pale colors, due to its conversion to the highly colored benzil, especially during overbake. A minimum of 0.2 % benzoin (calculated on total formulation) is recommended as a starting point for all colors. In pale and white shades the level of 0.2 % should not be exceeded, if there is a concern regarding overbaking. But up to 0.8 % benzoin can be utilized in black and dark shades.

The degassing properties of benzoin can be enhanced by the use of “supplementary” additives [14] (e.g. some non-fuming, non-yellowing polymeric additives).

There are also a number of **non-yellowing alternatives to benzoin**, which are (nearly) as good degassing agents as benzoin [14]. The use of these alternatives may have adverse effects on other properties. Therefore pretesting is always recommended.

Suitable alternatives to benzoin are: stearic acid, a very effective degassing and flow agent, but it lowers the glass transition temperature T_g drastically, so that its level is restricted by decrease in storage stability; hydrogenated castor oil derivatives, which also improve film flow and lower T_g; various polyethylene and polypropylene wax types (if they have no matting effects) [15-19]; some mono- and bis-amide waxes [17-19].

But also CHDM (= 1,4-cyclohexanedimethanol) dibenzoate [20] and solid surfactants, based on acetylene diol chemistry [21] are known as benzoin alternatives. Recently Troy filed a patent [22] describing the acid catalyzed reaction product of benzoin and glycerol (a ketal) as a non-yellowing alternative to benzoin.

3.5.3 UV absorbers and light stabilizers

They improve the exterior durability of powder coatings and include UV light absorbers for clear coatings, and **sterically hindered amine light stabilizers (HALS)** for pigmented systems [23]. UV absorbers protect coatings from their photodestruction by direct sunlight. They absorb and transform the ultra violet component of the daylight into harmless heat, in this way eliminating the destructive action of the sunlight. Light stabilizers are capable to inhibit reactions of photochemical destruction of polymer molecules, blocking their active centers [24]. But often UVA and HALS are used together, for their synergistic combination gives superior protection to coatings against loss of gloss, cracking, delamination and color change [23].

Recommended concentrations are: 1 to 1.5 % UVA and 0.5 to 2.0 % HALS on binder.

Derivatives of **hydroxyphenyl benzotriazole** are most frequently used as UV light absorbers [23]. Sterically hindered amines (HALS) are almost exclusively derivatives of 2,2,6,6-tetramethyl piperidine.

Zeren et al. [25] mention the development of a new class of UV absorber (trade name “Tinuvin 405”), an advanced hydroxyphenyl triazine, which is advantageous in powder coatings, because of its low melting point (77 °C). This leads to an easier extrusion than with the benzotriazoles, which have a higher melting point (141 °C). Another advantage of the new UVA is, that “Tinuvin 405” is slightly blue-shifted, which minimizes the risk of initial color, when a clear coat is applied over light colored base coat.

In another article *Zeren* [26] evaluates four UVA types, namely a hydroxyphenyl triazine (“Tinuvin 405”), a benzotriazole, a benzophenone and an oxalanilide. With these UVA types the hydroxyphenyl triazine shows the best performance in terms of gloss and color retention, when tested in superdurable polyester and acrylate clear coats.

At the Chinese Szechuan University it was found that nano-rutile TiO_2 improved the UV stability distinctly [27]. *Holzmann* et al. [28] reported the beneficial effect of fatty acid and phosphonate based grafting agent coating of nano TiO_2 on the UV-stability of powder coatings.

3.5.4 Anti-oxidants

For protection against **yellowing in overbake conditions** anti-oxidants (or heat stabilizers) are used. Usually they are combinations of a sterically hindered anti-oxidant and a hydrolytically stable organophosphite [23].

The problem of discoloration (or yellowing) during the stoving process of powder coatings is made worse by the use of direct-fired gas ovens, in which interaction between the coating and the combustion products (mainly nitrogen oxides) can occur. The degree of discoloration depends on the type of powder coating (**HAA cured powders are more sensitive**) and the degree of stabilization. Commonly used heat stabilizers, as mentioned above, are not suitable for direct-fired gas oven applications. Here very special "external" stabilizers are recommended [14] and an appropriate "co-stabilizer" (e.g. sodium benzene phosphinate SBP) may be used to enhance overbake stability, without adversely affecting direct-fired gas oven stability. See also Chapter 3.4.2.2.

Heat stabilizers have a maximum effectiveness level, whereby exceeding this amount (0.2 to 0.5 % on total formulation is recommended) has no further benefit. Excess stabilizer may even result in discoloration. Here too, pretests are always necessary.

3.5.5 Pigment dispersing aids

In (highly) pigmented powders, the addition of materials, which act as **dispersants** and often **reduce melt viscosity**, can **improve gloss**, distinctness of image (DOI) and leveling of the coating. For this purpose modified polyethylene and amide waxes are recommended [16, 17, 19], but a more effective dispersing agent is based on a zinc salt of a hydrophobic sulphonic acid, absorbed on a precipitated silica [29]. The recommended level is 1 to 7 % of this additive (as supplied), based on total pigment weight. This level depends of course on the type and dispersibility of the pigment(s). This latter property can also be improved by surface treatment of the pigment. DSM [30] described the addition of the dispersing aid during or after the synthesis of a polyester resin leading to significant better haze values.

A rather recent development is the use of **pre-dispersed (organic) pigments** (also called pigment preparations), which are pigment masterbatches, containing pigments that are predispersed at high concentration in a low molecular weight (polyester) resin. According to a patent of Degussa (now Evonik Industries) even liquid pigment preparations [31] can be used.

A new approach to resolving the dispersion problem is the use of block copolymer dispersants in powder coatings, developed at the Eindhoven University of Technology by *Van der Linde* and co-workers [32]. They use poly (2-vinylpyridine)-β-poly (ε-caprolactone) copolymers (P2VP-β-PCL) as pigment dispersants (or stabilizers) in powder coatings. Their results with this P2VP-β-PCL show that coatings have excellent gloss and extremely low haze levels. The absence of pigment flocculation is an indication, that the block copolymer has moved to the pigments during extrusion, by diffusion and convection through the resin, and has completely covered the pigment surface. *Maxwell* et al. [33] claimed that **polymeric dispersion agents** can improve carbon black jetness, opacity at lower film build, particularly for TiO_2, and the formulating window with chrome yellow replacements. In addition higher gloss levels are found. According to *Rentschler*[34] the hiding power of pigments is enhanced further by using finer particles, e.g. of TiO_2, which can be substituted up to 30 % by appropriate micronized inexpensive extenders.

In 2008 DSM[35] introduced special polyester resins ("Uralac HiTone") containing an **"anti-bridging agent"** referring to a substance that decreases the bridging of pigment particles by the binder resin molecules and thus enabling the powder coating formulation to hold a higher load (>40 %) on pigments without showing a deterioration of the crosslinked film. This anti-bridging agent should include at least one polar group like phosphoric, sulfuric or maleic acid to link to the pigment and at least one additional functional group to react with the crosslinker.

Tang et al.[36] used methacryloxypropyl trimethoxysilane to coat titanium dioxide for compatibility reasons before manufacturing an acrylate powder coating formulation through a one step minisuspension polymerization.

3.5.6 Antistatic and charge control additives

Electrical surface resistivity of a coating powder and its applied film can be decreased by the addition of antistatic and charge control additives. Though the results produced are similar, the purpose of these two types of additive is different[1]. An antistatic agent improves the ability of the deposited coating to **conduct electrostatic charge to ground**, while a charge control additive will **improve the transfer efficiency** and the ability of the powder to penetrate Faraday cage areas.

With electrostatic (corona) charging systems, particle charging occurs by ionic attachment. Charge efficiency is rather poor, since only about 0.5 % of the air ions, produced by the corona, are attached to the powder particles. The remaining free air ions travel to the workpiece, where they accumulate in the deposited powder layer.

Powder particles charge more efficiently, if their electrical resistivity is lower, so that an antistatic additive will improve charge efficiency[1]. By consequence, this can **reduce spraying time** or allow the charging voltage to be reduced.

The greater part of antistatica belongs to the chemical groups of cationic quaternary ammonium salts or anionic alkyl sulfonates, but also glycerol monostearate is used. Mostly, they are incorporated in the premix (prior to extrusion), but it is also possible to coat them onto the powder particles by postblending. A typical addition level is 1 to 2 %, which is able to lower the resistivity e.g. from 10^{13}–10^{14} Ωm to 10^{9}–10^{12} Ωm. When applying by corona gun, the powder resistivity should be around 10^{12} Ωm[1, 37]. If the powder is too conductive, it will quickly lose its charge and will not adhere to the substrate.

Some antistatics are catalysts for epoxy containing powders or have a tendency to yellowing during curing, so careful selection is always recommended. Glycerol monostearate is non-yellowing.

The successful use of antistatic or charge control agents in powder coatings can best be illustrated by the following comments, which the author received from powder users: "The improved electrical attraction of the powder particles to the workpiece (reduced back-ionization effect) and the easy coverage of all sorts of corners and small openings are a real disclosure!"

3.5.7 Tribo-charging additives

Nitrogen containing compounds can improve the degree of tribo-charging and for this reason some organic pigments, but also β-hydroxyalkylamides are effective in tribo systems.

To prevent the nitrogen containing additives from accelerating cure of (epoxy)powder coatings, tribo-chargeability is normally enhanced by the use of sterically hindered amines or amino-alcohols. Therefore, **light stabilizers of the HALS type** (e.g. “Tinuvin 144”) are often used in small amounts as tribo promoters.

Hoechst [38, 39] recommends **polymeric ammonium compounds** (mol. weight >5000 Dalton) and tetra substituted phosphor-, arsenic- or antimony-compounds as tribo additives. A Polish patent [40] claims that the reaction product of bisphenol A diglycidylether, diisopropyl amine and isophorone diisocyanate achieves a tribo chargeability of 1.9 μA. But also a simple nitrogen compound, such as melamine, is a suitable tribo-additive. The best way of incorporation of these compounds is adding to the premix, followed by co-extrusion.

At an addition level of 0.5 to 1 wt-% (on total powder) of sterically hindered tertiary amines (HALS type) the tribo chargeability of a polyester powder lies between 3.0 and 3.6 μA.

Quaternary ammonium salts, guanamines, benzotriazole show values of 2.1–2.8 μA.

Much stronger tribo chargeability is given by an *oligomeric* sterically hindered tertiary amine (trade name “Eltribo”), developed by *Boneza* and *Penczek* [41] of the Warsaw Industrial Chemistry Research Institute. It strongly improves the tribo chargeability of carboxy polyester coating powders, without influencing the kinetics of the curing process and other properties of the coating. At an addition level of 1 to 3 wt-%, calculated on polyester resin, “Eltribo” enhances tribo chargeability to – not less than – 2.5 mA.

It should be remarked, that, due to the **bad tribo-chargeability of polyester resins**, special “tribomatic” polyesters (with chemically built-in additives) are available. When the tribo-chargeability of such powder coating systems is still insufficient, small quantities (about 0.1 %) of a highly dispersed aluminum oxide powder can be added to enhance the tribo-chargeability [42]. This Al_2O_3 must be co-ground with the extrudate chips.

3.5.8 Anti-caking (free-flow) additives

It is well-known, that silicas and in particular **highly dispersed** (or “fumed” or “colloidal”) **silicas**, which are mixed by postblending with coating powders, belong to the most effective free-flow additives. But also **highly dispersed or colloidal alumina** (aluminum oxide) is very suitable for this “secondary” treatment [43] of coating powders.

These additives are extremely fine, in the range of 10 to 40 nm [44-46], and may be added either to the chips of extrudate and then co-ground, or after the powder has been ground. When co-grinding is chosen, some of this very fine material is carried over into the bag collector, but this technique is preferred rather than blending with the ground powder, since this means an additional production operation. Moreover, co-grinding provides a more stable and permanent effect than postblending, because with co-grinding a kind of mechanical “anchorage” of the silica/alumina is obtained. *Table 3.24* shows an overwiew of different silicas and aluminas for thermoset and thermoplastic powder coatings.

Table 3.24: Silica and alumina additives for different powder coating systems

Additive	Powder coating
“AEROSIL 200” (1)	S
“AEROSIL 380” (1)	S
“AEROSIL R 972” (1)	S, P
“AEROSIL R 812” (1)	S, P
“AEROSIL R 8200” (1)	S, P
“AEROXIDE Alu C” (2)	S, P
“AEROXIDE Alu C 805” (2)	S, P

(1) Silica (Evonik Industries) *S = thermosets*
(2) Alumina (Evonik Industries) *P = thermoplastics*

Figure 3.29: Effect of free flow additives
(Source: Evonik Industries)

Disturbances of the free-flowing properties of powders can, nearly always, be attributed to physical contact of the powder particles.

The way of action of free-flow agents is based on the ability of highly dispersed silica or alumina to coat (or envelop) the primary powder particles effectively as **a thin, "monomolecular" and moveable layer**. Acting as tiny spacers the spherical silica particles produce a kind of "ball-bearing" effect, sliding and tumbling to promote free flow of the powder. Simultaneously they reduce the force of attraction and the friction between the treated powder particles, preventing them to come into immediate contact with each other. This leads to a significant improvement in free-flow and application characteristics.

The amount of free-flow aid required is very low in most cases. Addition of 0.1 to 0.3 wt% (on total coating powder) of a fumed silica/alumina is usually sufficient [43] and improves fluidization, resistance to blocking and impact fusion, and transport through hoses *(Figure 3.29)*.

This "secondary" treatment also gives improved charging characteristics, better spray patterns, reduced moisture sensitivity, lower "angle of repose" and last but not least: improved "free-flow factor" R (as defined and determined according to ISO 8130-5 "Free-flow properties of a powder/air mixture").

It should be mentioned, that **higher levels of fumed silica** (0.3 to 0.7 %) will **act as thixotropic agents** (in the same way as with liquid paints), resulting in improved edge coverage, which is needed in many electrical insulation and functional powder applications, but also often resulting in reduced gloss and film flow (leveling).

Special commercialized "spacer" particles with lower size and density are described by *Zhang* et al. [47] to ensure the free-flow properties of ultra fine powder coatings (10 to 20 μm). Applying this low coating thickness is claimed save up to 50 % powder material without compromising the protective effect.

3.5.9 Mar resistance and slip improving additives

Their purpose is to improve lubricity (slip), scratch (mar) and abrasion resistance, and anti-blocking properties and, therefore, to reduce physical damage during handling or use of the coated object.

As there exists some confusion in literature about these concepts, *Verkholantsev* [48] has given the following **definitions: Slip** is the low frictional resistance of a surface; **scratch** or mar resistance is the resistance to damage by sharp or hard objects; **abrasion resistance** is the resistance to damage by rough and hard objects; **anti-blocking** is that property, which prevents coated surfaces from sticking together. *Verkholantsev*[48] completes his definitions with the following explanatory notes: Scratch resistance is independent of hardness, as some elastomers are highly scratch resistant. Mar resistance implies scratch resistance as well, but it means also the resistance to soiling. Non-blocking means a non-stick between two surfaces, the resistance to adhesion even under pressure pulling the surfaces together.

Abrasion resistance is enhanced by **incorporation of hard particles**, e.g. SiO_2, SiC [49], Al_2O_3, preferably in nano size to prevent optical deterioration. Chinese scientists [50] recommended <5 % of nano-fillers of inorganic oxides, in particular a mixture of SiO_2 and $CaCO_3$, to increase hardness, shock strength, flexibility, adhesion, abrasion and thermal stability of a UV powder coating. PPG [51] described in a patent the use of 0.1 to 0.5 % of aluminum oxide platelets, silicon carbide and diamond powder in powder coatings increasing the retaining gloss after mar from 55 % (reference) up to 96 %. Tigerwerke claimed the use of 3 % polyethylene coated corundum in epoxy powder coatings, increasing the Taber abrasion test resistance from 150 to 200 cycles in a patent [52]. According to a recent patent [53] of the Doerken AG it is advisable to add these hard particles after the extrusion step to avoid damaging of the screws. The same applies for lubricants which would contaminate the screws.

Surface slip, which means low values of the coefficient of friction, and to some extent also the scratch resistance, are provided by the corresponding additives, mainly by the mechanism in which they act as lubricants. Lubricants are additives that are capable to form **a fluid film on the surface**, filling the unevenness by bridging the microscopic "peaks". This allows the aligned bodies to glide through the surface without causing mechanical damage. Liquid lubricating is characteristic of low melting waxes and silicones. More efficient abrasion, mar and block-resisting properties show those additives, which form solid protection films on the surface.

All these additives are agents, which are only partly soluble, insoluble or incompatible in the binder and migrate to the surface of the film during curing, forming a continuous or some-times discontinuous layer [1].

Most common types are polyethylene **waxes**, modified polypropylene waxes, combinations of PTFE and polyethylene powders (also called "polyfluors"), or special silicones [1, 48].

The lower density PE types are tougher and provide better mar resistance than the higher density PE waxes. But the higher density PE powders show better slip properties. All PE types permit recoatability of the powder coating, when used at levels of <0.3 % (on total powder). Moreover, they allow gloss retention and can improve stain and chemical resistance of the coating, because the film is more difficult to wet. In addition, they can be useful components of anti-graffiti coatings, though here the type of binder and the crosslink density play a more important role. Modified polypropylene waxes improve mar resistance, but are not advised as slip aids.

Polyfluors (blends of PTFE and PE waxes) can act as slip aids as well as improvers of mar resistance, depending on the type of polyethylene. The softer the PE wax, the lesser the initial slip and the better the mar resistance. The harder the PE wax in that combination, the greater the initial slip, but the lower the coefficient of friction, which means a lower mar resistance [48, 54].

Fischer-Tropsch waxes, straight chain saturated hydrocarbons with the chemical composition $C_{52}H_{106}$ to $C_{56}H_{114}$ and produced during the hydrogenation of carbon monoxide, provide also improved slip, if added at levels of <1 %. Blends of Fischer-Tropsch waxes with PTFE are used for the same purpose, when solvent resistance of the powder coating is required [18, 54].

Fischer-Tropsch waxes can be considered as high melting paraffins. Due to their lower molecular weight, they give better lubricity than (modified) polyethylenes and are preferred, when fingernail scratch resistance is required.

Low molecular weight hydroxyl functional silicones are sometimes used to improve mar resistance, if added at levels of max. 0.5 % [55].

Special polymer blends, based on a "new" chemistry, have been developed by Lubrizol [17] as slip improvers for use in applications, where loss of gloss and haze formation cannot be tolerated.

3.5.10 Texturizing additives

These products give a structured effect to the film surface, characterized by having a homogeneous surface texture with uniform gloss and color. This effect is achieved by controlling the particle size of the powder, the melt flow and the surface tension of the molten film, by the use of insoluble or incompatible additives [1].

The usual method of influencing the surface tension, is to leave out the flow control agent and to replace it by a small amount of CAB (cellulose acetobutyrate, with a medium melt viscosity), added to the premix, prior to extrusion [56]. CAB lowers the surface tension to a certain degree, resulting in a **controlled cratering effect** where the coating does not flow out completely, but the craters do not go right down to the substrate. Levels of CAB are between 0.005 and 0.5 % on total powder, depending on the desired aspect and the type of powder used. At higher levels of CAB the texture becomes less pronounced or the film becomes even smooth.

The concept of controlled cratering can also be applied, when the cratering additive is post-blended with the finished powder [1]. That means without the use of CAB.

In this method the powder does not contain a flow additive, but the cratering additive is one of the recommended flow control agents (!), used in masterbatch (MB) form (or absorbed on fumed silica). The addition level, using a 5 % finely ground MB, may vary between 0.3 and 1.5 % (on total powder), again depending on the desired aspect and type of powder used.

Addition of 0.5 % MB leads to a wave-like regular texture, without holes or craters. With 0.3 % a more flattened wave-like texture is obtained. The aspect of the texture depends not only on the addition level of the MB, but is also greatly affected by the flow of the base powder. High flow base powders generally provide a broad (wave-like) texture, while low flow base powders give a finer and tighter texture. The texturizing MB can also be pigmented or ground to various particle sizes, to provide a wide range of textures.

The background of this way of texturizing can be explained as follows:

CAB is a more or less **"polluting" agent and necessitates great precautions**: perfect cleaning of all types of premixers, extruders, mills and spraying equipment. To prevent these difficulties, CAB should be replaced by a non-polluting (non-contaminating) raw material. The solution has been found by adding minor (less than normal) quantities of a flow agent as a MB.

An extra advantage of this method is that there are no problems of reproduction with laboratory formulations. A disadvantage of this texturizing process could be an extra extrusion step for the preparation of the MB, unless this is available as a MB of the flow agent in a suitable resin. In this case the MB needs only to be finely ground.

Another method of making textured powder coatings utilizes powders with "controlled" flow and particle size [1]. The powder is prepared with a relatively low "pill flow" in the range of 15 to 20 mm (to be determined according to ISO 8130-11 "Inclined-plane flow test"). If this is ground to a relatively coarse powder and sieved through a 140 mesh screen (which means 100 % >100 µm), the coating will be faintly textured. Adding (by dry blending) a fraction of larger particles of the same powder composition (by grinding the powder to a fineness of about 200 µm and sieving through a 70 mesh screen), will produce a "leatherette" or **"pebble" texture**.

In a Japanese patent [57] it is mentioned that for fine patterning the ratio of surface tension and viscosity should be between 0.25 and 0.40 to provide an optimum thixotropy. The best and easiest way of obtaining low pill flow is to use high oil absorption fillers, like talcs or (sometimes) clays.

A more or less identical way [58] of producing textured coatings is to use powders with a deliberate shortage of flow agent (0.005 to 0.3 %, based on binder). The powder is prepared in the normal way and ground to a particle size of 100 % >100 µm. Then a 10 % master-batch of flow agent (in the same binder) is prepared, ground to a particle size of 200 to 600 µm and added to the coating powder. The addition level is preferably 0.1 to 3 % of this MB. The coarser the MB particles, the stronger the texturing effect.

"Spatter" finishes can also be obtained by dry blending two powders with different particle size. It has been found [59] that a good spatter effect can only be obtained, if one of the powders contains very large particles (up to 500 µm or even more). The other powder should have the normal standard particle size distribution and good film flow properties.

To maintain good application properties (which means: regular distribution of particles on the surface of an object; no influence of spraying distance on the texture effect), the very coarse powder should have a much lower density than the powder with the normal particle size.

This can be explained as follows: Charge acceptance (and with it powder particle distribution in the powder cloud and the electrical field) is improved by increasing particle size. But with particles larger than 100 µm the influence of gravity strongly increases. To compensate both effects as well as possible, the following measure is evident: The small particles must be made specifically heavier and the coarse particles specifically lighter [59].

The two powders can be mixed in a 1:1 ratio or in any other ratio that guarantees that the powder mixture contains particles, of which 20 to 40 wt% lie between 400 and 500 µm.

Polymeric additives, which reduce film flow or are infusible in the binder, can be used to formulate textured powder coatings. Examples are a family of SMA (styrene maleic anhydride) copolymers [60] with melting ranges from 55 to 150 °C, which can give fine and coarse leatherette finishes in epoxy and hybrid powders, without affecting the mechanical properties. The recommended addition level varies between 2 and 7 % on total binder. It should be remarked, that only the "F" series of the range of SMA copolymers produce this texture effect.

PTFE powders in various particle sizes (and also belonging to the group of polymeric additives) can produce different degrees of texture [1, 18]. They are generally supplied as fine powders with particle size between 10 and 15 µm, and are effective at levels of 2 to 3 %. Also combinations of PTFE and "synthetic" (Fischer-Tropsch) waxes, with 52 to 56 carbon atoms in their molecules, and sometimes designated as "modified" PTFE, are used [18], as well as blends of PTFE and polyethylene [16, 17]. Other polymers, which do not melt and/or flow completely or are incompatible, are used for texturizing. They include some grades of coarse polypropylene powders [18], rubbers [54], metallic (preferably zinc) salts of ethylene-acrylic acid copolymers ("ion containing polymers" or "ionomers") [61], nylons and other high melting polymers [1, 54]. These inert polymeric additives do not significantly affect the color or physical properties of the coating. Compared with the "classical" method of texturizing, using only high filler levels, these additives provide better control of the obtained texture and by consequence better reproducibility of the texture effect.

Incompatible thermosetting acrylic and silicone resins may also be used sometimes for texturizing powder coatings [1]. Many of the mentioned polymeric additives also cause a visible reduction of gloss, which not always can be measured, due to the textured surface of the coatings.

Waelde [62] of Troy Chemical describes their "Powdermate 508 TEX", a texture additive that provides reproducible effects in any powder system by incorporating a certain amount of thixotropy at the extrusion stage, and then adding the "508 TEX" to the finished powder. The best way to do this is to co-grind it with the extrudate. The addition level varies between 0.3 and 2 % (on total powder).

A **certain degree of thixotropy** is essential for good reproducibility, and is given by using high oil absorption fillers, such as talcs and colloidal clays. aluminum organic compounds (AOC's) are claimed [63] to be suitable for use as texturizing agents in all powder coating systems. AOC's, derived from alkoxides, have been used for many years in liquid paints to improve through drying of alkyd resins and/or to achieve thixotropy or rheology modification.

Two special types (a solid and a liquid one) have been developed, that exhibit latent reactivity, at elevated temperatures, with powder coating resins, containing carboxyl, hydroxyl and oxirane functionality, to provide crosslinked and/or textured finishes. The solid type is used at 10 to 15 % as sole curing agent for epoxy powders and gives fully cured dull matt (!) films with excellent mechanical properties. When used above 10 %, the coatings have an attractive uniform fine wrinkle finish.

The liquid type is used as a texturizing additive, specifically developed for polyester, hybrid and epoxy powder coatings. It is a chelated derivative of an aluminum alkoxide, absorbed on amorphous silica. Typical usage is in the range of 3 to 8 % (based on resin component), providing fine and uniform semi-matt textures (non-wrinkling).

Nice wrinkle finish powder coatings, based on hydroxyl polyesters, cured with **tetramethoxymethyl glycoluril** TMMGU, can be obtained, when they are catalyzed with several different amine-blocked sulphonic acid compounds, such as dimethylaminomethyl propanol (DMAMP)-blocked p-toluene sulphonic acid (trade name WL catalyst "XL-320") or the diethylammonium salt of trifluoromethane sulphonic acid (trade name "Fluorad FC-520") [64-66].

These **wrinkling additives** should be used at levels of 0.3 to 0.5 % (on total binder). As they are liquids, they should be added as masterbatches in the hydroxylated polyester resin. UCB (now Cytec) [67] sells a 5 % MB of a wrinkling additive, under the trade name "Crylcoat 120".

Basu et al. [68] used DSC and mechanical profilometry to study this wrinkle formation. Evaporation of the unblocked amine from the surface of the coating leads to a compressive elastic stress, which buckles the skin to produce wrinkles. The same research group recently published an article [69] showing that undesired starburst defects in wrinkled powder coatings often are caused by elevations in the film created by non-dispersed particles or impurities.

The influence of **wood grain effects** on filiform corrosion on aluminum substrates was tested by *Fedrizzi* et al. [70]. They found a better corrosion resistance of the wood grain modified samples but on the other hand lower barrier properties against water uptake measured by electrochemical impedance. The authors attribute this apparent contradiction to a destruction of the coating integrity by the corrosion products in the case of the modified powder coating, leading to a slow down of the filiform corrosion process.

A much more elaborative way to texture a powder coating has been reported by *Tripp* et al. [71]. They heated the cured powder coating above the T_g and imparted a texture on it by using a release medium having a replicative surface.

3.5.11 Gloss control additives

Matting additives (or additive matting agents) are defined by *Bate* [54] as products, which have been designed to reduce the gloss of a powder coating, **without interfering with the systems resin stoichiometry**. That means, neither matting hardeners, nor chemical matting systems (which involve mixing of binders or powders with different reactivity) belong to this group of additives.

These matting additives are able to reduce gloss down to 60° values of about 50 to 60 %. For 60° gloss values lower than 50 %, materials (and/or methods) are needed, which cannot be regarded as "additives". See therefore Chapter 3.6.

Two main types of matting additives are available:

- waxes (either incompatible and (non)-melting waxes, or wax/accelerator masterbatches);
- silicas.

The use of both types is restricted to maximum levels of 5 % on total powder, as higher percentages lead to detrimental powder coating characteristics.

Gloss reduction by waxes takes place because during stoving the molten wax migrates and concentrates near the film surface, forming a thin surface layer, enriched with wax. This floating or "bloom" effect can be aided by the incompatibility of the molten wax with the molten coating. Various wax types are used, often in micronized form. So polypropylene powder [17, 18] can produce satin finishes at addition levels of 2 to 5 % (on total formulation). Polyethylene wax [19] and combinations of polyethylene and polypropylene waxes [17] are also often used to reduce gloss.

A non-melting additive is **Teflon** (or PTFE) powder, which is often sold in combination with other waxes [18, 72]. It should be remarked, that waxes are seldom used alone, due to their tendency to produce hazy films with a waxy feel, which is prone to finger printing.

Wax/accelerator combinations are described in a patent by Ciba-Geigy [73], in which they combine waxes with metal (preferably zinc) salts of organic compounds (preferably mercapto benzthiazol) in the weight ratio wax/Zn-compound >5. Typical addition levels are 3 to 5 % of this combination (based on total binder), providing 60° gloss values of 55 to 65 %.

A product, more or less based on this patent, is available as "Ciba Matting Agent DT 125-2". **Silica matting agents** [74, 75] are prepared by a precipitation process, in which sodium silicate and sulphuric acid react together to give a slurry of precipitated silica. After filtration, washing, drying and milling, an untreated silica is formed, consisting of silicon dioxide (SiO_2) in the form of a hydrated amorphous "synthetic" silica.

Suitable **precipitated silicas** have an average particle size of 3 to 10 µm. They should be added at the premix stage and can produce semi-gloss and satin finishes at addition levels of 3 to 4 % (on total formulation). In contrast with the mentioned wax types, precipitated silicas show a greater tendency to affect film flow in a negative way.

Apart from waxes and silicas some other matting additives should be mentioned, *Loar* and *Verlé* [76] describe small particle sized zinc-ionomers (prepared by partial neutralization of an ethylene-acrylic acid copolymer with zinc), which by replacing up to 5 % of the polyester, give 60° gloss values of 55 to 75 % (in HAA-cured polyester powder coatings) depending on the amount of Zn-ionomer. But see also Chapter 3.6.

As has been remarked in Chapter 3.5.10, many texturizing polymeric additives also cause a visible reduction of gloss. A survey of “Waxes for powder coatings”, with all their various functions (not only matting), is given by *Heiling*[77]. It has to be considered, though, that wax might lead to reduction of the adhesion [78].

Wacker [79] described a way not to decrease but to increase the gloss by adding 5 % of an extender (vinyl acetate copolymer) to the powder coating formulation, thus enhancing gloss by 10 %.

3.5.12 Catalysts (accelerators)

The purpose of catalysts is either to accelerate the stoving reaction and to reduce the stoving time, or to lower the curing temperature. As only very small amounts are needed, they are often supplied as masterbatches by the resin manufacturers. The list of available catalysts (including masterbatches) is very long. Many examples are given in Chapter 3.1, 3.3, 3.4 and 4.6, to which the reader is referred.

3.5.13 Miscellaneous additives

In this chapter all other additives are summarized which are used in a minor amount and show a distinctive influence on the properties of a powder coating.

Rohm and Haas [80] claimed acrylic core-shell polymers based on a polymethyl methacrylate shell and a polybutyl acrylate core as **flexibilizers**, in particular for low temperature curing epoxy systems. PPG [81] patented epoxy based polycaprolactones, generated by the ring opening polymerization of ε-caprolactone started by OH-containing epoxies, which are claimed to increase the **impact resistance** of a powder coating. According to Dow [82] block copolymers containing miscible and immiscible parts (e.g. block copolyethers based on ethylene oxide and butylene oxide) are useful as **tougheners** for epoxy powder coatings. *Yu* et al. published articles [83] claiming that calcium carbonate epoxy nano-composits improved **cupping, impact resistance, tensile properties** and **salt spray resistance** of an epoxy powder coating, if the nano-particles are well dispersed. These effects having an optimum at 3 % of calcium carbonate are attributed to interactions between the nano-particles and the epoxy powder coating virtual increasing the compactness of the coating. Recently DuPont filed a patent application [84], describing the use of 2.5 % of hydroxyfunctional methacrylate resins in GMA/ polyanhydride powder coatings formulation, to achieve a reversed impact of 30 inch·pounds.

Nanoscalic silver for **antibacterial** effects is described by *Greiner*[85]. **Antimicrobial** powder coatings are already applied nowadays in certain hospitals (e.g. in Duarte, California) or for the interior of refrigerators (e.g. offered by Bosch and Siemens, Germany) [86].

Tigerwerke [87] patented a process to add nano-particles, in particular indium-tin oxide (a powerful IR absorber), to a polyester in the step of the manufacturing this resin. In this way a perfect distribution in the powder coating is ensured.

According to AkzoNobel [88] high performance intumescent epoxy based powder coatings are formulated by adding two different types of **blowing** agents like phenyltetrazol and ammonium polyphosphate and/or melamine phosphate.

Clariant published a patent [89] in which polar modified (e.g. by maleic anhydride) polyethylene waxes are added to powder coatings to enhance the **adhesion** to sealants.

The **anti-soiling** properties of a fluor-containing polyurethane powder coating are claimed to be enhanced by addition of 4 % of a hydrophylizing mixture of alkylsilicates, like ethyl polysilicate [90].

Windridge [91] added up to 20 % of glass or ceramic hollow microspheres to a powder coating achieving thermal insulation as well as **anti-condensation** properties.

AkzoNobel [92] filed a patent describing the addition of an electroheat-susceptible material in a primer for polypropylene. This way they were able to apply an epoxy phenolic powder coating and to cure it via microwave irradiation with a surface temperature not exceeding 130 °C.

A **drying** indicator like phenolphthalein or thymolphthalein for cementitious powder coatings was added by Beissier [93].

According to Rohm and Haas [94] silicone coated hydrophobic nano-particles like "Aerosil 972" increase the **filiform corrosion** resistance of GMA systems.

A Chinese university group filed a patent application [95] describing the **IR-reflecting** properties of powder coatings containing oxides of iron, yttrium, zinc, silica, aluminum, manganese, nickel and copper. In addition these materials are claimed to be able to transform UV light and visible light into heat.

To achieve higher **film thickness** in fluorpolymer powder coatings, 3M used so called "high build agents" like N,N-diethyldodecane amide, polyethylene glycol and butyrin [96].

An unusual method was applied by *Zhu* et al. [97] to manufacture hydrophobic powder coatings which is claimed to retain its hydrophobicity after mechanical **abrasion**. They mixed 35 % "Aerosil R815S" (Evonik Industries) to 65 % of a TGIC based polyester powder coating. After curing this mixture at 200 °C they grinded and sieved this material used it as a hydrophobic additive for the original powder coating formulation.

Piazza et al. [98] investigated the effect of 2 and 4 % of montmorillonite clay on epoxy powder coatings. They found an increase in thermo stability, barrier properties and adhesion to metal substrates.

A recent patent [99] of BASF describes the use of 1 % iron particles, e.g. made by decomposition of iron pentacarbonyl, in polyamide thermoplastics to improve the **heat aging** resistance.

A **lotus effect** based on a unique double-scale hierarchical (micro- and nano) superhydrophobic structure was generated by dryblending 2.5 % nano-PTFE, 1 % hydrophobic nanosilica, 1 % diatomite (to prevent easy rub off) and an ultra fine (15 to 20 μm) commercially available resin [100].

The addition of 1 % of stearic acid coated calcium carbonate nano-particles led surprisingly to a significant increase of **reactivity** of a hybrid epoxy formulation. *Kalaee* et al. [101] found a drop of the activation energy from 20 to 16 KJ/Mol corresponding to a decrease of the curing time from 12.5 min to 11 min at 170 °C. 5 % nano-particles reduced this time to 6 min.

3.5.14 Pigments and fillers

Surprisingly little is published in the recent scientific literature regarding pigments and fillers especially designed for powder coatings. Conventionally titanium dioxide, iron oxides and carbon black are used as pigments; whereas calcium carbonate, calcium metasilicate, barium sulfate and talcum find a regular application as fillers. An exception is made by aluminum and pearlescent pigments which produce a highly desired metallic appearance [102]: These tend to separate after dry blending into the powder coating formulation leading to an insufficient reproducibility. Therefore metallic pigments are preferred, which are bonded to the resin. In addition aluminum pigments are now offered in pellet form diminishing the risk of metallic dust explosions. **Metallic powder coatings** show an annual growth rate of about

10 %. Still, bright metallic effects are often more easily achieved nowadays by liquid coatings, like waterborne systems. The color of pigmented powder coatings is sometimes influenced by curing time and temperature. There are even organic pigments like dibromanthathron (pigment red 168), which react with the hardener dicyandiamide in epoxy systems [103]. Color matching in powder coatings is a substantial challenge. One of the reasons is a lack of standardization, since nowadays there are four different color systems active in the market: Munsell, Pantone, RAL, and NCS [102].

This chapter will focus predominantly on the recent developments in the field of pigments and fillers.

According to *Albrecht* [104] the encapsulation of metallic pigments with a resin significantly increases abrasion resistance, weathering and chemical stability of the powder coating.

An interdisciplinary team of university and company scientists described the beneficial effect of 5 to 10 nm silica coatings on aluminum metallic pigments on the corrosion resistance [105].

Prokes et al. published an article [106] comparing 33 pigmented epoxy formulations and 15 silicone formulations in terms of physico-mechanical properties and corrosion resistance. Aluminum oxide pigmented epoxy formulation showed a high hardness but a low flexibility. Zinc dust enhanced both the physico-mechanical properties and the corrosion resistance in the area of 0 to 30 % pvc. Silicone based powder coatings showed general a higher hardness than epoxy powders.

Encapsulated titanium dioxide, preferably by methacrylated polycaprolactones, is claimed to lead to less orange peel effect, lower haze, better flow and thus to a higher DOI (distinctness of image) [107].

Barletta [108] reported the beneficial effect of IR-precuring of a bilayer metal flake powder coating, which limits the undesired movement of the decorative pigments. The final step includes a conventional oven-curing.

Tecolor filed a patent [109], describing the use of a bonding agent in a dry mixing process applied on powder formulations and pigments. Upon heating the bonding agent is brought into an adhesive state to improve the agglomerization of powder and pigment particles.

The blanching effect caused by water penetration into tinted powder coatings can be significantly reduced by surface functionalization of the inorganic pigments with alkoxysilanes [110]. *Stahlberg* et al. used wollastonit ($CaSiO_3$) and Barite ($BaSO_4$) with a pvc up to 25 % as fillers in HAA cured polyesters to reinforce the coatings for an application with a high compressive load. In particular above the glass transition temperature the creep and flow of the powder coating is restricted due to the interaction with the inorganic filler.

3.5.15 References

[1] a) *Richart, D.*, Industr. Paint & Powder (Powder R&D Section) Fall 1999, 1, (1)
b) *Harris, S.*, Focus on Powder Coatings, Oct. 1999
[2] *Lange, P.G. de*, J. Coat. Technol., 56, (717), 23 (1984)
[3] *Uhlmann, P., Wulf, M., Juckel, H., Michel, S. and Grundke, K.*, FARBE&LACK, 105, (1999) 12, p. 50
[4] *Wulf, M., Uhlmann, P., Michel, S. and Grundke, K.*, Congress Papers PCE 2000, Amsterdam, Jan. 2000, p. 299
[5] *Grolitzer, M. and Erickson, D.*, J. Coat. Technol. 67, (1995) 845
[6] *Schulte, H.*, Paint & Ink Internat., Nov/Dec (1995) p. 8
[7] *Jacquin, J.D.*, Congress Papers PCE 98, Amsterdam, Jan. 1998, p. 171
[8] *Hajas, J., Juckel, H., and Bubat, A.*, FARBE&LACK, 102 (1996) 9, p. 30
[9] *Ming Zhao*, Proc. ECC, Zuerich, Sept. 2001

[10] BYK-Chemie, WO 2009033529, 2009
[11] Minnesota Min. & Manufact. (3M) Co., Product Info "Fluorad Coating Additives"
[12] DuPont de Nemours, WO 2009117334, 2009
[13] DuPont De Nemours, WO 2011081871, 2011
[14] EMS-Chemie, "Primid", Technical Supplement 7.4.1
[15] DSM Research, EP 471409 (1992)
[16] Byk-Cera, Prod. Info XI 5 "Ceraflour" - micronized waxes (1998)
[17] Lubrizol, Techn. Leaflet "Lubrizol's Coat. Additives f. Powder Coatings" (1998)
[18] Micro Powders Inc., Techn. Leaflet "Micronized Waxes"
[19] Clariant GmbH, Techn. Brochure "Ceridust" waxes
[20] *Arendt, W.*, Proc. 20th Int. Waterborne, High Solids & Powder Coatings Symp., New Orleans, Feb.1993, p. 508
[21] Air Products' "Surfynol 104S" in Chem. Market. Reporter, 15 Apr.1996, 249 (16) 17
[22] Troy Technology Corporation, WO 2008130538, 2008
[23] Ciba Spec. Chem. Brochure "Tinuvin" Light Stabilizers & "Irganox" Antioxydants f. PC's
[24] *Verkholantsev, V.*, European Coatings Journal (ECJ), 3 (2000) p. 58
[25] *Zeren, S., and Renoux, D.*, Congress Papers PCE 2002, Nuremberg, Jan. 2002, p. 165
[26] *Zeren, S.*, XXVI FATIPEC Congress, Dresden, 2002, in Macromol. Symposia 187, p. 343
[27] *Guo, G.; Cao, J.; Duan, X.; Huang, W.; Tu, M., Xiandai H.*, 24(5), 38-40, 2004
[28] *Holzmann, D.; Schoefberger, W.; Holzinger, D.; Schmidt, T.; Knoer, G.*, Monatshefte fuer Chemie, 142(8), 855-860, 2011
[29] King Industries (repres. by Worlée), Product Data Sheet "K-Sperse"
[30] DSM, WO 2006066764, 2006
[31] Degussa AG, WO 2006047238, 2006
[32] *Duivenvoorde, F., Jansen, K., Laven, J. and Van der Linde, R.*, J. Coat. Techn. 74, (931) 49 (Aug. 2002)
[33] *Maxwell, I. D.; Siddiqui, K.; Cartridge, D. J.*, Noveon Specialty Additives, USA, Proceedings of the Annual Meeting Program of the FSCT, 82nd 28/1-28/9, 2004
[34] *Rentschler, T.; Classen, A.; van Beek, H.; Gribble, P.*, Paintindia, 55(8), 73-78, 2005
[35] a) DSM, US 20070248825, 2007
b) *Bolks, J.; Brendeke, S.; Blokhuizen, E.; Vercoulen, P.*, Proceedings of the International Waterborne, High-Solids, and Powder Coatings Symposium, 35th 171-181, 2008
c) *Zimmermann, F.; Blokhuizen, E.; Bolks, J.*, JOT, Journal fuer Oberflaechentechnik, 49(6), 42-44, 2009
[36] *Tang, E.; Cheng, G.; Shang, Q.; Ma, X.*, Progress in Organic Coatings, 57(3), 282-287, 2006
[37] *Hughes, J.*, Electrostatic Powder Coating, RSP Ltd, J. Wiley & Sons, Chichester (1984)
[38] Hoechst AG, EP 476647 (1992)
[39] Hoechst AG, EP 315082 (1991)
[40] Instytut Chemii Przemyslowej, PL 187012, 2004
[41] *Boneza-Tomaszewski, Z. and Penczek, P.*, XXVI FATIPEC Congress, Dresden 2002, Macromol. Symposia 187, p. 417
[42] *Bormann, M.*, Technik heute, (5), 50 (1987)
[43] *Lange, P.G. de*, Proc. XX FATIPEC Congress, Nice, Sept. (1990) p. 395
[44] Degussa, Techn. Brochure Pigmente Nrs. 18, 31, 32 "Aerosil" (1989)
[45] Wacker-Chemie, Techn. Brochure Wacker HDK "Hydrophobic Fumed Silica" (1986)
[46] Cabot, Techn. Brochure "Cab-O-Sil" in Coatings
[47] a) *Zhang, H.; Zhu, J.*, AIChE Spring National Meeting, Orlando, Apr. 23-27, 2006
b) *Zhang, H.; Zhu, J.*, University of Western Ontario, US 2004009340, 2004
[48] *Verkholantsev, V.*, European Coatings Journal (ECJ), 12 (1999) p. 70
[49] Faming Zhuanli Shenqing Gongkai Shuomingshu, CN 1511906, 2004
[50] *Wei, J.; Li, H.; Sun, Z.; Wang, L.*, PMSE Preprints , 91, 631-632, 2004
[51] PPG, US 20050249939, 2005
[52] Tigerwerk Lack- und Farbenfabrik, WO 2008014527, 2007
[53] *G. Reusmann*, Ewald Doerken AG, WO 2010079181, 2010
[54] *Bate, D.A.*, The Science of Powder Coatings, Vol. I, p. 115, SITA London (1990)
[55] Worlée, Techn. Brochure "Worlée Add360" in Additives f. Powder Coatings (1998)
[56] Shinto Paint Co. & Janome Sewing Machine Co., DE Pat. 2.437.731 (1974)
[57] Sumitomo Osaka Cement Co, JP 2004303729, 2004
[58] Libert Frères, BE. Pat. 805 344 (1974)
[59] Herberts GmbH., DE Pat. 4.104.681 (1991)
[60] Atochem, Techn. Bulletin "SMA Copolymers"
[61] Allied Signal, Techn. Leaflet "Aclyn"
[62] *Waelde, L.*, Proc. ICE 2001 Conf., Nov. 2001
[63] Alcan Chemicals, Techn. Data "AOC's"; and *Harris, S.*, PCB, 14, (6), 4 (1991)

[64] *Jacobs, W., Foster, D., Sansur, S., and Lees, R.*, Proc. 21st Intern. Conf. in Org. Coat. Science and Technol., Athens, (1995) p. 259
[65] *Foster, D., Mealmaker, W., and Rivera, C.*, Congress Papers PCE 2000, Amsterdam, Jan. 2000, p. 234
[66] Cytec Industries, Inc. Techn. Bulletin "Powderlink 1174"
[67] UCB Chemicals, Product Info. "Crylcoat 820/120" (1999)
[68] *Basu, S. K.; Scriven, L. E.; Francis, L. F.; McCormick, A. V.; Reichert, V. R.*, Journal of Applied Polymer Science, 98(1), 116-129, 2005
[69] *Reichert, V.; Basu, S. K.; Francis, L. F.; McCormick, A. V.; Scriven, L. E.*, Progress in Organic Coatings, 63(1), 33-37, 2008
[70] *Fedrizzi, L.; Stenico, M.; Deflorian, F.; Maschio, S.; Bonara, P. L.*, Progress in Organic Coatings, 59(3), 230-238, 2007
[71] *Tripp, M. J.; Juers, D. H.*, WO 2011017074, 2011
[72] Degussa-Hüls AG (now Evonik Industries), Info "VESTAGON AD" Matting Additives
[73] Ciba-Geigy, EP 0165 207 (1985)
[74] Crosfield Chemicals, Techn. Leaflet "Crosfield Silicas in Powder Coatings"
[75] Grace, Ltd., Tech. Bulletin "Syloid Matting Agents"
[76] *Loar, D. and Verlé, P.*, Proc. 13th Int. PRA Conf., Brussels, Nov. 1993, Paper 10
[77] *Heiling, G.*, XXVI. FATIPEC Congress, Dresden 2002, Macromol. Symposia 187, 437
[78] *Buyl, F.*, Dow Corning S.A., ASTM Special Technical Publication, STP 1453 (Durability of Building and Construction Sealants and Adhesives), 65-87, 2004
[79] Wacker Polymer Systems, WO 2004072198, 2004
[80] Rohm and Haas, EP 1411093, 2004
[81] PPG, WO 2005095529, 2005
[82] Dow, WO 2006052730, 2006
[83] a) *Yu, H.; Wang, L.; Shi, Q.; Jiang, S.; Jiang, G.*, Journal of Applied Polymer Science, 101(4), 2656-2660, 2006
b) *Yu, H. J.; Wang, L.; Shi, Q.; Jiang, G. H.; Zhao, Z. R.; Dong, X. C.*, Progress in Organic Coatings, 55(3), 296-300, 2006
[84] DuPont De Nemours, WO 2011011780, 2011
[85] *Greiner, W.*, JOT, Journal fuer Oberflaechentechnik, 45(5), 38-39, 2005
[87] Tigerwerk Lack- und Farbenfabrik, WO 2005075548, 2005
[88] AkzoNobel Coatings, US 20060160927, 2006
[89] Clariant, US 20060111492, 2006
[90] Dainippon Toryo Co., JP 2008266361, 2008
[91] *Windridge, D.*, WO 2008068464, 2008
[92] AkzoNobel Coatings, WO 2008055923, 2008
[93] Beissier, WO 2008003672, 2008
[94] Rohm and Haas Company, US 20090192247, 2009
[95] Shanghai University of Engineering Science, CN 101864241, 2010
[96] 3M Innovative Properties, WO 2010080202, 2010
[97] University of Western Ontario, WO 2010000056, 2010
[98] *Piazza, D.; Lorandi, N. P.; Pasqual, C. I.; Scienza, L. C.; Zattera, A. J.*, Materials Science & Engineering, A: Structural Materials: Properties, Microstructure and Processing, 528(22-23), 6769-6775, 2011
[99] BASF, WO 2011051123, 2011
[100] *Mozumder, M. S.; Zhang, H.; Zhu, J.*, Macromolecular Materials and Engineering, 296(10), 929-936, 2011
[101] *Kalaee, M.; Akhlaghi, S.; Nouri, A.; Mazinani, S.; Mortezaei, M.; Afshari, M.; Mostafanezhad, D.; Allahbakhsh, A.; Dehaghi, H. A.; Amirsadri, A.*; et al., Progress in Organic Coatings, 71(2), 173-180, 2011
[102] *Linak, E., Kishi, A., Yang, V.*, Thermosetting Powder Coatings, SRI CONSULTING, 2008
[103] *Herbst, W., Hafner, C.*, Farbe und Lack, 82, 293-411, 1976
[104] *Albrecht, A.; Hirth, U.-A.; Schreiber, B., PPCJ*, Polymers Paint Colour Journal, 198(4531), 26-28, 2008
[105] *Maardalen, J.; Lein, J. E.; Bolm, H.; Hallenstvet, M.; Rekowski, V.*, Progress in Organic Coatings, 63(1), 49-54, 2008
[106] *Prokes, P.; Kalendova, A.*, Journal of Physics and Chemistry of Solids, 68(5-6), 1083-1086, 2007
[107] a) DuPont De Nemours, WO 2010080429, 2010
b) DuPont De Nemours, WO 2010080426, 2010
[108] *Barletta, M.*, Progress in Organic Coatings, 72(3), 498-510, 2011
[109] Tecolor, WO 2011013114, 2011
[110] *Van Beek, H., Fritzen, P., Rohe, B.*, European Coatings Journal, 12, 72-74, 2011

3.6 *Matt and semi-matt powder coatings*

The use of matt and semi-matt powder coatings account for more than **40 % of all powders** for interior and exterior applications. Such low gloss coatings are demanded for both decorative and architectural purposes. Undesired light reflectance can be reduced, when matt or semi-matt coatings are used. In addition, surface defects such as scratches and imperfections in the substrate are less visible on matt surfaces.

Reduction in gloss of a powder coating down to 60° gloss values of less than 50 % is brought about by the development of an irregular and/or micro-textured film surface during the curing reaction.

Various methods have been developed to produce irregular surfaces with their consequent light scattering effect, leading to low gloss readings.

At present there are four main matting methods in use:

- Addition of pigments and/or fillers above the **critical pigment volume concentration** (CPVC) can form irregularities in the film and as a result gloss reduction. This method has limited use, since the high levels of fillers and/or pigments result in poor film flow and in reduction of mechanical properties of the cured film and the overall matting effect is not always appreciated.
- Addition of **incompatible waxes** or plasticizers is a method of creating surface inhomogeneity, because the waxes migrate to the surface during film curing, forming "micro-droplets", which provide a fine and rough surface with a reduced gloss effect. However, there is a tendency of the wax to leach out (or to "exude") from the film at those levels of addition (>5 %) where they are likely to be effective; an effect that limits the application to light colors (where the whitening effect of the micro-droplets is less noticeable) and to 60° gloss values higher than 50 %. See Chapter 3.5.11.
- **Dry blending** of two different powders, each having a different reactivity, or consisting of two incompatible types (based on the "contamination phenomenon" [1]). For improved mixing dry blending is often replaced in practice by "co-grinding" of the extrudates of the two powders. The oldest known examples of this method are described in patents by Wagemakers' Lakfabrieken [2] and Shell [3]. They obtain low gloss values, either by dry blending two epoxy/DICY **powders with different reactivities**, or by dry blending an epoxy powder (without hardener) with an epoxy/DICY powder or with DICY lone. A full description is given in Chapter 3.4.1.
  A well-known example of dry blending two polyester powder coatings with different reactivities is the combination of high gloss powders, based on polyester/TGIC 90:10 and polyester/TGIC 96:4 or on polyester/HAA 90:10 and 96:4 (or 96.5:3.5) [4-6]. The greater the difference in reactivities, the lower the gloss of the coating. In this way 60° gloss values of 25 to 35 % are achievable.
  An alternative is the dry blending of two powders, based on different chemistries with different melt viscosities, surface tension or miscibility properties, leading to film disturbances (often also caused by contamination) and to gloss reduction. Examples are: dry blend mixtures of high gloss hybrid and polyester/TGIC powders, hybrid and pure epoxy powders, polyurethane and polyester/TGIC powders, etc. Some combinations can lead to 60° gloss values of 35 to 40 % [1].
  Gloss reduction caused by **contamination** (or incompatibility) of two powders often gives powders with a damask-like semi-matt finish.
  The suitability of two powders to form a semi-matt effect after dry blending can be estimated by the ISO 8130-12 method *"Determination of compatibility"*.
  The dry blending manufacturing technique is **time-consuming** (and thus costly), as always two powders should be produced. Moreover, it is often difficult to obtain a uniform

(semi-)matt aspect and reproducible gloss values. Dry-blending can be facilitated if the particle size is reduced and the particle size distribution narrowed [7].

- Producing (semi-)matt powder coatings by a **"one-shot" process** and in this way get around the uneconomic dry blending technique. This **"chemically induced incompatibility"** process involves the use of binders, by which at least two chemical reactions with different speeds take place during stoving. It is based on the principle of the "double (or multiple) curing" mechanism, which can consist of premixing either one resin with a dual (or bi-) functional curing agent, or one resin with (a physical mixture of) two curing agents, or two different resins with a single curing gent, or two different resins with two or three curing agents. Here too counts: the greater the differences in reaction speeds, the lower the gloss of the film. In a Japanese patent [8] the required difference in reactivity is claimed to be at least 400 s in geltime. They used a mixture of the blocked polyisocyanates "Vestagon B1530" (ε-caprolactam blocked IPDI derivative, Evonik Industries, geltime 1200 s) and "Vestanat B1358" (MEK-oxime blocked IPDI trimer, Evonik Industries, geltime 800 s) to form a coating with a gloss at 60° of 39 %. *Table 3.25* gives an overview about different techniques.

Since the reactivity of curing reactions is very specific to the different crosslinking mechanism, every resin/hardener composition developed individual one shot matting systems.

Epoxy and hybrid systems

The oldest examples of bi-functional curing agents are homogeneous and well-defined mono-salts of certain aromatic polycarboxylic acids (e.g. pyromellitic acid) and special cyclic amidines (e.g. 2-phenyl-2-imidazoline or 2-methyl-2-imidazoline), described in patents of Hüls (now Evonik Industries) [9, 10]. These salts are the well-known "Vestagon B 55" and "B 68", used as "chemical" matting hardeners for epoxy and hybrid powder coatings.

During stoving the salts decompose into their original products, namely the highly reactive amidine and the much less reactive polycarboxylic acid. Again the matting effect is based on differences in reactivity during stoving.

Schülde [11] explains the matting mechanism as follows (see also Chapter 3.3.3.2):

The amidine is not present in the amount, necessary for full curing of the epoxy resin. Only a pre-crosslinked, but still flexible coating would form first. This coating would be expected to have a glossy surface, typical of curing due to amidines. During a further rise in temperature, the less reactive polycarboxylic acid will also react with the epoxy resin. Due to the marked polyfunctionality of the acid hardener, this crosslinking would be expected to occur with a strong contraction of the molecules. However, this is counteracted by the fact, that the resin was already pre-crosslinked and the chain molecules possessed only restricted mobility. A further crosslinking can therefore not be compensated by a uniform shrinkage of the film thickness. By consequence, the inhibition of shrinkage will lead to **uniformly distributed points of retraction in the film surface** and thereby cause the matting effect.

It can be shown (to confirm *Schülde's* theory), that by adding higher amounts of "Vestagon B55" or "B68", the proportionately higher amounts of amidine, contained therein, lead to full crosslinking of the epoxy resin (already during the heating-up process and at the lower curing temperature, required for the amidine crosslinking), with the formation of a glossy (!) coating, typical for the amidine hardener. When still higher temperatures are reached, a further reaction between epoxy resin and the polycarboxylic acid component cannot take place and therefore the coating remains glossy.

"Vestagon B 55" and "B 68" are only suitable for matting pure epoxy and epoxy containing hybrid powders. "B 68" has a stronger matting effect than "B 55". With "B 68" 60° gloss levels down to 10 to 15 % are possible.

Matt powder coatings, containing "B 55" or "B 68", are distinguished by good film flow and good resistance to hot water and detergents. The absence of any visible surface texture, despite a markedly uniform matt effect, is surprising. The mechanical properties (particularly the impact resistance) are poor, when film thicknesses exceed 50 μm. Both "Vestagons" are sensitive to yellowing on overbake, especially with white and light colors. Stoving temperature must be at least 180 °C, which can be seen as a disadvantage for some applications. Recently Degussa (now Evonik Industries) was able to decrease the curing temperature of B68 to 160 °C without compromising the matting effect by adding a tetraalkylammonium carboxylate as a catalyst [12].

Trials have not failed to improve the mentioned negative properties and simultaneously maintaining the benefits of the "Vestagons".

The following examples of a nearly similar double curing mechanism can be given:

De Cock and *Godts* [13] describe a Shell (now Momentive) matting hardener for epoxy powders, namely "Epikure DX 187 FF" (discontinued by Momentive), which is a mixture of a highly reactive imidazole derivative and a much less reactive polybasic acid. The two-stage cure mechanism is identical to that of the "Vestagons B 55/68". Low gloss coatings with good flow, high film flexibility (good impact resistance) and high whiteness are claimed, when "Epikure DX 187 FF" is used as hardener for epoxy resins. *De Cock* and *Godts* assume that the first stage cure (with the imidazole derivative) is responsible for the flexibility of the coating, while the second stage of the curing cycle (with the polybasic acid) gives the reduction of gloss by a uniform contraction of the flexible film.

Süddeutsche Kalkstickstoff Werke (better known as SKW Trostberg) [14] brought a matting hardener to the market under the name "Dyhard PC 45", which is a physical blend of 45 % 2-phenyl-2-imidazoline (called component A) and 55 % of a triazine derivative (component B). It can be used universally for all pure epoxy and hybrid powder systems, currently in use. SKW give the following explanation of the matting mechanism:

Upon curing, component A starts crosslinking at about 110 °C. Small areas with **local differences in surface tension** are formed. The complete crosslinking is done by reaction of comp. B, starting at about 150 °C. The balanced curing of both components produces small, dot-like "hills" on the surface, giving the visual impression of a matt surface.

"Dyhard PC 45" was widely accepted by the powder coating industry, with the advantages such as very low yellowing even at overbake, low stoving temperature, and good flexibility. However, from time to time some instability of "PC 45" has been experienced, due to the very fact, that it is a physical mixture. Under certain circumstances both components A and B can react with each other, giving a hardener that does not provide a gloss below 20 %. Therefore, SKW decided in 1999 to offer the two components of "PC 45" separately, instead of "PC 45" itself. In this way the powder manufacturer himself can adjust the desired gloss of his products.

The combination of the "Dyhard PC 45" components provides the following extra benefits:

- low 60° gloss levels of 10 to 15 % (adjustable by changing the A/B ratio);
- low curing at 165 °C is possible (at a stoving time of 15 min).

"Dyhard PC 45" is ideal for epoxy and hybrid clear coats. It gives **semi-matt to dull matt surfaces**, together with a haze-free film matrix. Therefore SKW recommend such clear coats as glass coatings. As the "Dyhard PC 45" components-blend produces a high crosslink density, a high coating T_g and high chemical resistance are guaranteed, which opens new applications in clear and pigmented automotive and/or medicinal powder coatings, according to SKW.

Thomas Swan [15] delivers also (epoxy and hybrid powder) matting hardeners, based on physical blends of two curing agents, under the trade name "Casamid". Their range of "Casamids" consists of four types, of which number 1432 gives lowest gloss levels of <5 % in epoxy powders and 10 to 15 % in hybrids.

Ciba [16] uses a polyolefin wax in combination with the zinc salt of an organic anion (preferably a mercapto benzotriazole), which has the ability to accelerate the curing reaction of epoxy/polyester (hybrid) powder coatings. The matting effect is probably due to its poor "solubility" in the binder system, causing **localized acceleration in the curing reaction** and as a result an irregular film formation, giving a micro-textured surface [17]. Therefore, this matting system can also be classified as a double (or dual)-curing system.

This product is commercially available as Ciba "Matting Agent DT 3329-1". It compliments the matting effect of the already cited "DT 125", as it allows 60° gloss levels of 35 to 40 % to be realized, by adding 3 to 6 % on binder. The matting efficiency of "DT 3329-1" is dependent on the choice of the polyester resin, the degree of homogenization and the cure temperature.

"Matting agent DT 3329-1" can be used to formulate semi-gloss hybrid powders with good flow, color stability and good mechanical properties. As it has no adverse effect on weather resistance, it can **also be used as a matting agent for polyester/TGIC (or HAA) systems**.

As DT 3329 contains a waxy component, it is best suited to semi-matting light and not to dark colors, in order to prevent visible leaching-out of the wax.

An example of a matting method based on a multiple curing mechanism, is described by *Chetcuti, Dreher* and *Gottis* [17]. They use a multi-functional, cycloaliphatic carboxylic acid for matting of epoxy and hybrid powder coatings. This product makes it possible to obtain gloss levels from 0 to 50 %. According to *Chetcuti* et al. the **high functionality causes an inhomogeneous crosslink density** throughout the cured film, while the aliphatic nature results in a certain immiscibility with the aromatic groups of the polyester or epoxy resins of the binder. The aliphatic carboxyl groups of this matting hardener react at a slower rate than the aromatic carboxyl groups of saturated films, so that the curing reaction, caused by this hardener, starts, when the film has been partially crosslinked, with the result of a restricted concomitant contraction of the film. It was further found, that the gloss level for epoxide and hybrid powders is independent of the curing conditions, which suggests, that the immiscibility and inhomogeneous crosslinking density, due to the high functionality of this matting hardener, are more important than the differences in reactivity during the curing process.

Schmidthauser [18] discusses the use of low molecular weight SMA (styrene maleic anhydride) copolymers for gloss reduction of epoxy powder coatings. It is again based on a double-curing process that leads to micro-textured films, causing reduced gloss. Double curing takes place by the fast reaction between the epoxy and SMA resins (with their anhydride groups), in combination with the slow reaction between the epoxy resin and dicyandiamide.

The use of SMA resins as gloss reducing agents can be extended to the more complex hybrid systems. In this case the reactivities between the three components, epoxy, polyester and SMA (and/or SMA partial ester) must be carefully balanced to give the different curing rates, necessary for low gloss. This can be done by using catalyzed polyester.

The powders should be formulated with 5 to 10 wt% of an anhydride functional SMA resin, which is more effective in hybrid powders than the partial ester SMA grades. Also an excess of epoxy resin with respect to the polyester should be used. By curing at 200 °C, gloss levels of about 30 to 40 % (at 60°) can be achieved with the recommended "SMA 3000" resin.

Skora and *Green* [19] discuss also a technology, based on the chemically induced incompatibility (or dual cure concept). They incorporate a specifically designed carboxy functional acrylic resin in an epoxy powder that can contain several amine curing agents, to produce reduced gloss.

The **major benefits of this epoxy/acrylic matting resin/amine system** are:

- gloss not affected by stoving temperature (important when there is a temperature gradient in the oven);
- excellent mechanical properties;
- non-yellowing light colors possible;
- wide gloss range available with different amine curing agents.

Three different amine curing agents can be chosen: imidazole-epoxy adduct ("Epikure curing agent P101", Momentive), substituted "DICY", such as ortho-tolyl biguanide (OTB), or an accelerated "DICY", such as "Epikure curing agent P108" (Momentive).

Epoxy coatings, formulated with an imidazole-epoxy adduct and the acrylic matting resin can be cured at temperatures as low as 130 °C. If color and non-yellowing properties are required, then the imidazole adduct should be replaced by a substituted DICY. Each of these amines has limitations, which may eliminate them from some applications. For example, the imidazole adduct has limited shelf life and yellows.

Polyester and polyurethane systems

Matting of polyester and polyurethane powders can also be done by means of a "one-shot" (or "co-extrusion") process, based on the double or multiple curing mechanism, which can consist of combining either one resin with a physical mixture of two curing agents, or one resin with a bi-functional hardener, or two different resins with a single curing agent, or one to two resins with two to three different curing agents, etc. The following examples can be given:

In 1997 Bayer [20] launched a method for the formulation of a "non-blocked" (or more exactly "internally" blocked) matt polyurethane powder coating. The principle behind this system is the **mixture of an uretdione hardener with a long-chain dicarboxylic acid** (e.g. dodecanedioic acid). The uretdione reacts with the hydroxylic polyester and the dicarboxylic acid with TGIC or hydroxyalkylamide (extra added to the formulation). For a matt black coating Bayer researchers combine 45.4 % polyester resin (hydroxyl number 30), 7.5 % of the uretdione adduct "Crelan VP LS 2147", 5.6 % dodecanedioic acid and 4.0 % hydroxyalkyl amide ("Primid XL-552" or "Vestagon EP-HA 320"). They complete the formulation with 1.5 % flow agent, 1.0 % tin-carboxylate (as a catalyst), 1.1 % carbon black and 33.9 % "blanc fixe".

The **parallel crosslinking reactions** occur at very differing rates and interfere with each other, causing a kind of film disturbance and by consequence a semi-matt film with a 60° gloss value of 15 to 20 %. With decreasing uretdione content gloss will be lowered and 60° values of 10 to 15 % are possible.

Based on the same idea of dual curing, Creanova (now Evonik Industries) came in 1998 with a bi-functional hardener, described by *Wenning* et al. [21]. They create a fine and uniform microtexture in the surface of a polyurethane powder coating by incompatibilities in the system between the resin and hardener, by introducing an urea structure. This can be obtained by reacting the uretdione with diamines, instead of diols. The formed bi-functional amine-modified IPDI uretdione adducts can react with a hydroxylated polyester resin, giving reproducible matting degree with 60° gloss values of 11 % and excellent film flow.

In 1999 an attempt was made by Bayer to obtain deadmatt (gloss only 1 % at 60°) polyurethane coatings, described by *Thometzek* et al. [22], now using the bi-functional hardener "Crelan VP LS 2181/1" with an **isocyanate and a carboxyl functionality** in the molecule. These groups allow two parallel curing reactions. The blocked isocyanate reacts with a hydroxyl polyester, and hydroxyalkyl amide or epoxy hardeners (e.g. TGIC) are used to react with the carboxyl group. For example, *Thometzek* et al. combine 50.4 pbw (parts by weight) hydroxyl polyester (with OH number of 30); 10.8 pbw "Crelan VP LS 2181/1"; and 1.8 pbw "Araldite PT 910" (a TGIC alternative). The result is a very matt coating with perfect film flow.

Thometzek et al. [22] also give an example of the combination of two different resins with one single hardener. Using a "one-shot" process, two polyesters with very different hydroxyl numbers and "Crelan VP LS 2147" uretdione hardener are combined to produce polyurethane powder coatings with low gloss. For example, two polyesters with hydroxyl numbers of 30 and 300, respectively, are combined in the ratio 5:1 or 3:1. The difference in reactivity causes defects in film formation during stoving and in a "rough" surface structure. The result is again a (semi-) matt coating with very good mechanical properties, excellent film flow and 60° gloss values of about 15 to 20 %. Recently DuPont filed a patent [23] claiming similar formulations for anti-graffiti applications.

Gloss reduction of exterior durable polyester powder by a "one-shot" process, combining two hydroxyl polyester resins of strongly different OH number, with one isocyanate hardener has already been described. An improvement on this system can be made, according to *Skora* and *Green* [24] by **substituting an acrylic resin** with high hydroxyl value **to replace the comparable polyester** with the same hydroxyl value. Since the OH value of an acrylic resin is easier to control than that of the polyester, a more consistent product is obtained from batch to batch.

Another development, mentioned by *Skora* et al., is a curing agent based on a glycidyl-functional acrylic resin. It is possible to formulate these glycidyl curing agents so, that during the cure of acid-functional polyesters, incompatible domains are generated, giving matt finishes of about 7 % at 200 °C.

These systems provide good storage stability and long-term durability. The gloss is, however, **sensitive to stoving temperature** and care must always be taken with the glycidyl-functional acrylic agent to avoid contamination problems!

Polyester acrylic hybrids

Dumain [25] of Reichhold Chemicals describes a curing mechanism, involving a bi-functional polyester (with both hydroxyl and carboxyl functionality). The carboxyl groups are cured with a glycidyl methacrylate (GMA) resin, specifically designed for low gloss powder coatings. Its epoxy equivalent weight is 480. The hydroxyl groups of the polyester are cured with a blocked isocyanate crosslinker. The matting effect relies also upon the difference in curing speed, here between the acid/glycidyl reaction and the hydroxyl/isocyanate reaction.

This **polyester/acrylic "hybrid"** is claimed to give smooth, ultra low gloss coatings with excellent mechanical properties and durability. See also Chapter 3.4.3.

In his article *Dumain* provides a formulation, which includes catalysts, such as dodecanedioic acid, to optimize film smoothness and crosslink density, and also 2-heptadecyl imidazole to enhance the low gloss stability of the coating.

Iwamura [26], also of Reichhold, describes a novel acrylic/polyester hybrid, consisting of a bi-functional (carboxyl and hydroxyl) isophthalic acid (IPA) type polyester, which is cured with a bi-functional (glycidyl and hydroxyl) GMA-acrylic crosslinker and a blocked isocyanate.

This **bi-functional acrylic** (with an epoxy equivalent weight of 380 and a hydroxyl number of 50) combined with the IPA-rich bi-functional polyester is claimed to show a still better balance of durability and mechanical properties than the "conventional" mono-functional acrylic/bi-functional polyester hybrid system, described by *Dumain*.

Maetens [27] uses the combination of carboxylated telechelic polyesters and an acrylic copolymer hardener, containing glycidyl methacrylate (GMA), which is "statistically" distributed in the polymer chains [28].

This special acrylic copolymer consists of 80 % methyl methacrylate and 20 % GMA and has been brought to the market by UCB Chemicals (now Cytec) under the trade name "Crylcoat 700". With this polyester/acrylic copolymer combination very low gloss (<10 % at 60°) pigmented and clear TGIC-free powder coatings can be obtained through a single extrusion ("one-shot") process. Such coatings exhibit, after curing, good film flow, as well as ageing properties, corrosion- and overbaking resistances (no color difference after stoving at 180 °C. or at 220 °C). However, mechanical properties are poor (particularly impact resistance), which means, that these matt coatings cannot resist sudden mechanical deformation. A better distribution of the functional (epoxy) groups in the "Crylcoat 700" copolymer may improve this, according to *Maetens*.

The high degree of matting is again caused by a strong microstructure of the film surface. As DSC analysis of the curing process shows a very wide reaction exotherm, it is supposed, that there are not two, but multiple reaction speeds, causing a nearly deadmatt film. Therefore, curing with "Crylcoat 700" can be considered as an example of a matting method, based on a **"multiple curing" reaction**.

Partial replacement of "Crylcoat 700" by TGIC or HAA (in a stoichiometric way) allows to manufacture a wide range of gloss levels, which are, however, not always reproducible.

It is remarked by *Maetens*, that no separate extrusion rooms are needed for working with "Crylcoat 700", as the percentage of acrylic copolymer in the binder never exceeds 30 %! The **notorious acrylate contamination phenomenon** has never been observed by him during working with "Crylcoat 700".

Polyester cured with TGIC or HAA

Another example of matting due to **high functionality of a polyester resin** is recently introduced [29] by Cytec. This new "one shot" HAA powder coating system is claimed to allow low and reproducible gloss levels. A similar idea is pursued by EMS [30], using the reaction product of trimellitic acid anhydride and trimethylolpropane as a hyperbranched mat hardener.

Addition levels of 3 to 8 % of hardener (based on binder) are recommended, depending on desired gloss level, type of binder and amount and type of filler used [31]. Calcium carbonate should not be used, as it can react with the acid hardener. Polyesters of a low acid number (e.g. 20) are required, due to the carboxyl functionality of the hardener.

The main advantage of this acid hardener is that zero gloss levels are attainable with pure epoxy powder coatings (showing a "micro-rough" surface) and gloss levels of less than 10 % with hybrid systems, while maintaining the mechanical properties.

Thomas Swan & Co. [32] describe a "one-shot" process, using a reactive matting agent for use in outdoor durable powder coatings, **in particular polyester/TGIC systems** and other systems, curing with a similar reaction mechanism, e.g. glycidyl esters. The trade name is "Casamid 2228". Their matting system is also based on the double curing mechanism. The matting method deviates from other described systems, because matting of polyester/TGIC coatings requires here the ratio of TGIC to polyester to be unbalanced to provide **excess TGIC for reaction with Casamid 2228**.

Gloss levels of 10 to 40 % (at 60° angle) can be obtained by curing with a stoving schedule of 20 min. at 180 °C or 10 min. at 200 °C. Lower temperatures will adversely affect matting.

"Casamid" is neither suitable for polyester/HAA ("Primid") systems, nor for polyurethane powders.

Degussa (now Evonik Industries) [33] described the use of **special semi-crystalline carboxyl terminated polyesters in HAA systems** to lower the gloss down to 33 % at 60°. The same principle can be applied with special OH-functionalized semi-crystalline polyesters in polyurethane formulations.

Loar and *Verlé* [34, 35] use copolymers of ethylene and acrylic acid, neutralized to 25 to 50 % with various divalent cations (preferably Zn), as reactive flatting agents for polyester/HAA powder coatings. The remaining carboxylate functionality of such an "**ion containing polymer**" or "ionomer" makes it ideal for use in powder coatings, based on resins with similar chemistry.

In their experiments they replace 10 % of the polyester resin by a Zn-ionomer and come to 60° gloss levels of (not lower than) 40 %, without loss of physical properties and without orange peel. Their results show, that gloss reduction is unaffected by either the acid number or the degree of neutralization of the ionomer. Best results, concerning mechanical properties and surface appearance, are obtained by low melt viscosity and small particle size (100 μm instead of granules) ionomers.

Further studies demonstrate that gloss reduction, without loss of mechanical properties, can also be obtained with systems using TGIC instead of HAA.

The matting principle is again based on the double curing mechanism.

Alcan Chemicals [36] recommends a solid and a liquid aluminum organic compound (AOC's) for matting (and fine texturing) of powder coatings, as has been described in Chapter 3.5.10.

DSM Resins [37] uses melamine as matting agent, which is only suitable for their "Uranox" aliphatic oxirane binder system. Gloss values of about 20 % at 60° angle can be obtained with melamine in "Uranox" systems.

In 2011 Evonik Industries introduced a novel HAA with the tradename "Vestagon EP-HA 368" [38]. This **unique matt hardener** is suitable for one shot matt applications with 10 to 45 gloss units at 60°. It can be used with a variety of different polyester resins, preferably with a single binder partner. No influence of filler, film thickness, or stoving temperature can be seen. In addition "HA 368" seems to be color stable at overbake conditions and weather resistant.

A number of different technologies have been given above – mainly based on the principle of chemically induced incompatibility – for producing a wide range of gloss reductions in powder coatings. Nearly all of them have their own advantages and limitations. Recognizing these facts and using this knowledge to select the appropriate technology for a certain application, will minimize problems during the production and use of the reduced gloss coatings. *Table 3.25* shows a comparison of different matting methods.

Table 3.25: Comparison of different matting methods [38]

Matting Method (Gloss level at 60°)	Fillers waxes	Dry blend	Dry Blend dissimilar	One Shot PES-blend	Matte hardener	Novel HAA "HA 368"*
Flat matte (0-10)	no	no	O	++	O/no	no
Matte (10-25)	no	O	+	++	++/O	+
Semi matte (25-45)	no	+	+	++	++/O	++
Half gloss (45-70)	O	1), 2)	1)	1)	1)	1), 3)
State of the art for	All	HAA/TGIC	TGIC/PUR	PUR	HY/TGIC	next generation HAA
Cost raw material	++	O	O	--	-/-	-
Cost handling	++	--	--	+	++/++	++
Reproducibility	+	O	O	+ to O	++/-	+
Surface	O to -	+ to -	+ to O	++ to +	++ / O	++
Handling	++	-	-	++	++ / ++	++

1) Fillers in use for price reason, 2) no superdurables, 3) superdurables possible
**"VESTAGON EP HA 368" (Evonik Industries)*

3.6.1 References

[1] *Lange, P.G. de,* Proc. XIX FATIPEC Congress, Aachen, (1988) Vol. II, p. 399
[2] Wagemakers' Lakfabrieken, NL Pat. 138009 (1973)
[3] Shell Oil Co., US Pat. 3.824.035 (1974)
[4] UCB (incl. EMS), Techn. Brochure "Crylcoat Powder Coating Resins"
[5] DSM Resins, Techn. Brochure "Powder Coating Resins"
[6] Vianova, Techn. Data Sheet "Alftalat 03402" for dry-mix matt systems"
[7] *Asami, Y.,* Techno-Cosmos, 18, 8-17, 2005
[8] Nippon Paint Co., WO 2004083320, 2004
[9] Chemische Werke Hüls, EU Pat. 0044030 (1980)
[10] Chemische Werke Hüls, DE Pat. 3.311.404 (1983)
[11] *Schülde, F.,* Proc. XIV. FATIPEC Congress, Budapest (1978) p. 605 or Defazet, 33, (2), 46 (1979)
[12] Degussa AG, WO 2007028661, 2007
[13] *Cock, C., and Godts, M.,* Proc. XXI FATIPEC Congress, Amsterdam (1992) Vol. 4, 215
[14] SKW Trostberg, Techn. Info "Dyhard PC 45"
[15] Th. Swan & Co., Techn. Info "Casamid" curing agents for epoxy resins
[16] Ciba-Geigy, EU Pat. 0165 207 and Data Sheet "Matting Agent DT 3329-1"
[17] *Chetcuti, P., Dreher, B., and Gotis, P.,* Proc. 21st Int. Waterborne, High Solids and Powder Coatings Symp. New Orleans, Feb. 1994, or Modern Paint & Coatings, July1995, p. 28
[18] *Schmidthauser, J.,* Powder Coating, Sep. 2002, p. 51
[19] *Skora, A., and Green, C.D.,* Congress Papers PCE 2002, Nuremberg (2002) p. 175
[20] *Freudenberg, U., Meier-Westhues, U., and Laas, H.,* European Coatings Journal, 9 (1987) p. 804
[21] *Wenning, A., Weiss, J-V., and Grenda, W.,* Congress Papers PCE 98, Amsterdam (1998) p. 147
[22] *Freudenberg, U., Thometzek, P., Meier-Westhues, U.,* Proc. 26th Int. Waterborne, High Solids & Powder Coatings Symp., New Orleans, Feb. (1999) or Proc. 5th Nuremberg Congress (1999) or FARBE&LACK 105, (1999), 9, 46
[23] DuPont Powder Coatings Iberica, EP 1942124, 2008
[24] *Skora, A., Green, C.D.,* PCE 2002, Nuremberg, 175, 2002
[25] *Dumain, E.,* Coatings World, Vol. 1, Oct. 1996, Part 1, p. 27

[26] *Iwamura, G.*, Proc. 5th Nuremberg Congress (1999) Paper 20
[27] *Maetens, D., Baudour, M., Loutz, J., and Van Kerkhove. M.*, Proc. XXIII. FATIPEC Brussels (1996) Vol. B, p. 389
[28] UCB, US Pat. 5 346311
[29] a) Cytec, WO 2008125417, 2008
b) *de Becker, B.; Jacquin, J.; Minesso, A.; Fink, D.*, European Coatings Journal, (12), 90-93, 2009
[30] EMS, WO 2009083121, 2009
[31] Ciba, Data Sheet "Matting Agent DT 3357"
[32] Th. Swan & Co., Techn. Info. "Casamid 2228" f. Lowgloss Exter. Durable Powder Coat.
[33] Degussa AG, EP 1426423, 2004
[34] *Loar, D.*, Proc.20th Int. Waterborne, High Solids & Powder Coatings Symp., New Orleans, Feb. (1993) p. 536
[35] *Loar, D., and Verlé, P.*, Polym. Paint Col. J., Vol. 184, (4359) (1994), p. 504
[36] Alcan Chemicals, Techn. Data "AOC's" and Harris, S., Powder Coating Bulletin 14, (6), p. 4 (Oct. 1991)
[37] DSM Coating Resins, Brochure "Powder Coating Resins", Jan. 2002
[38] a) *Grenda, W., Spyrou, E.*, Congress Papers European Coatings Congress (ECS), Nuremberg, 113, 2011
b) Evonik Industries, WO 2011110624, WO 2011110640, 2011

4 Parameters influencing powder coating properties

Powder coatings have a specific place in the paint industry, being different in many aspects from the liquid coating systems. Compared to the conventional organic coatings similarities can be drawn with respect to the ultimate exploitation properties of cured films, where the same general rules are valid for both systems. However, all similarities end there. The analysis of the **differences between powder and conventional solventborne coatings** should also include a comparison between the production technologies, ways of application and the process of film formation. With respect to the manufacturing process, powder coatings should be classified in the group of plastics rather than paint technologies. The application techniques for powder coatings are unique and not applicable to the other organic surface coatings used commonly by the industry. Finally, despite the same crosslinking chemistry, the physics of the film formation of powder coatings differs completely from that of the solventborne paints.

The problems with powder coatings begin when the processing and application characteristics have to be combined with the final coating performances. In most cases the requirements attributed to these properties are contradictory. The application characteristics of the solventborne systems can be adjusted by choice of a proper solvent or solvent blend with little influence on the ultimate properties of the cured film. In the case of powder coatings this possibility does not exist, simply because the solvent is absent. Although it seems to be a simple system considering the number of constituents in the coating formulation, powder coatings are rather complex with regard to the combination of processing, storage, film forming and final service properties of the coating.

The **absence of solvent**, which is desirable from an ecological point of view, **decreases the degree of freedom** of the coating formulator. Most of the properties of a powder coating are determined exclusively by the binder. Therefore a good understanding of the resin parameters influencing the coating properties is a prerequisite for the resin and paint chemist in order to produce powder coatings with well-balanced overall properties.

A basic understanding of the correlations between different resin and paint parameters and their influence upon the final coating properties can help the paint chemist more than the most extensive list of powder coating formulations. Therefore this chapter addresses the important resin parameters such as molecular weight, functionality, glass transition temperature and viscosity, with the aim of correlating them with the coating properties. The description of the elements of the paint formulation, such as the binder/crosslinker ratio, pigment volume concentration, catalyst level, surface tension of the molten powder paint follows the discussion of the resin parameters. The last two sections consider the influence of powder particle geometry and the characteristics of the stoving line to the application and film forming properties of the powder coating during the curing stage.

Emmanouil Spyrou: Powder Coatings: Chemistry and Technology

ISBN: 978-3-86630-884-8

4.1 Molecular weight of binder systems

Like all polymeric materials, the resins used in powder coatings are blends of molecules of different molecular weights. Therefore averaging of the molecular weights is necessary. From the several different ways of averaging the molecular weights two are most important for the properties of the powder coatings: the number of the average molecular weight (M_n) and the weight of the average molecular weight (M_W), expressed by the well-known equations:

Equation 4.1

$$M_n = \frac{\sum M_i N_i}{\sum N_i}$$

Equation 4.2

$$M_w = \frac{\sum N_i M_i^2}{\sum N_i M_i}$$

Mechanical properties of powder coatings such as tensile strength and impact resistance are mainly dependent on the **number average molecular weight**, while the **weight average molecular weight** governs mainly the **melt viscosity** of the resin. For polymers, mechanical properties can often be expressed by the following formula [1]:

Equation 4.3 $X = X_\infty - A/M_n$

where X is the property considered, X_∞ is its value at very high molecular weight and A is an empirical constant.

Based on a wide range of collected data it has been established that molecular weight should be between 20,000 and 200,000 for a commercial polymer in order to ensure that they exhibit good impact resistance and tensile strength [2].

The lower the molecular weight of the polyester the closer the polyester/epoxy ratios to the stoichiometric ratio must be. In other words, **the system is more sensitive to the deviations** that are quite common – i.e. in weighing raw materials during the coating manufacture, determination of acid, hydroxyl or epoxy equivalent values, degree of hydrolysis of the oxirane rings during the storage of the coating or the epoxy crosslinker itself and the extent of the reaction during the curing cycle.

The possible manufacture of polyesters with higher molecular masses has the unfortunate consequence of increasing the viscosity that causes problems concerning processability and flow of the coating.

4.2 Functionality of the coating composition

The problem of sensitivity of the coating formulation to the correct ratio between the functional groups can be overcome by **increasing the functionality of the crosslinker** or the functionality of the resin. In this way, the system will be less sensitive to the necessary stoichiometry for obtaining an infinite network. Using *Gordon's* theory of branching processes with cascade substitution [3-6], calculations have been made for a system, based on carboxyl functional polyester with a number average molecular weight of 3800 and functionality between 2 and 3.25 and bisphenol A epoxy resin with a number average molecular weight of 1500 and epoxy functionality of 2.

Figure 4.1 represents the dependence of the number average molecular weight of the system at different degrees of conversion of the epoxy groups during the curing process.

As expected, the build-up of the molecular weight is much faster with increasing functionality. A number average molecular weight of 20,000 is obtained at a conversion of 86 % in the case of a functionality of 2, and 62, 24 and 8 % when the functionality of the polyester resin is 2.5, 3 and 3.25, respectively.

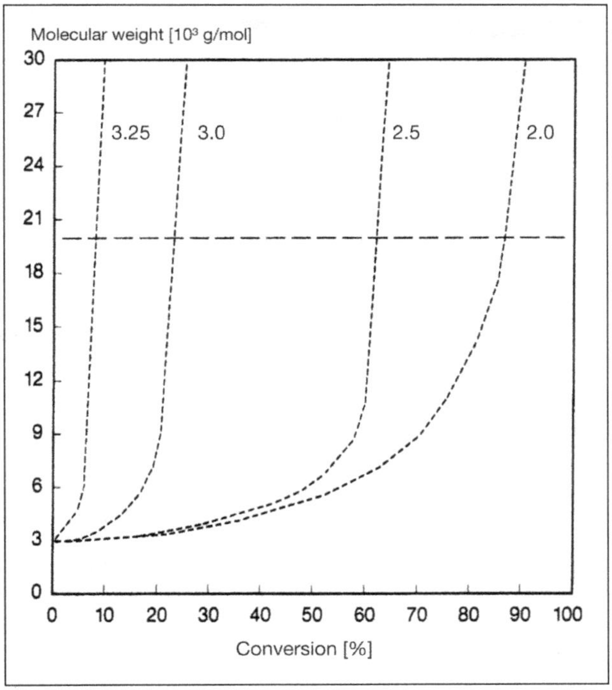

Figure 4.1: M_n as a function of conversion during the curing of polyesters with different functionalities with bisphenol A epoxy resins Source: Oil and Colour Chemist's Association

As usual, the formulator is once again faced with several difficulties when taking this option as a solution to the problem. High functionality will have as a consequence a **fast increase of viscosity**, thus shortening the time available for flowing of the coating during film formation. The result will be poor flow with an emphasized orange peel effect. *Scholtens, Van der Linde* and *Tiemersma-Thoone* [7], using thermo-viscoelastic analysis (TVA), have determined the influence of the variations of the molecular mass and the functionality of the polyester resins on the rheological behavior of a molten polyester/epoxy unpigmented hybrid powder coating. An increase in functionality increases the crosslinking density whereas an increase of molecular mass decreases the crosslinking density.

Functionality (F_n), the number average molecular weight (M_n) and the acid value (AV) or the hydroxyl value are not independent variables of the polyester resins. In the case of acid terminated polyester resins which are to be crosslinked with epoxies, this dependence can be expressed by the following equation:

Equation 4.4

$$F_n = \frac{AV * M_n}{56100}$$

This means that for the same acid value of the resin (in order to keep constant the weight ratio of polyester/epoxy) if one wants to increase the functionality, a higher molecular weight has to be chosen. The other option is to keep the same molecular weight but then to increase the acid value, which has a direct influence on the desired polyester/epoxy ratio.

The increase in the functionality also leads to a higher mole fraction of branching components in the resin formulation. In the case of a random esterification process the number average molecular weight increases linearly with the functionality at a constant acid value, but the **increase in the weight average molecular weight is very rapid**.

Using the *Flory-Stockmayer* theory for molecular weight distribution in non-linear polymerization [8, 9] a calculation has been made for a series of resins based on neopentyl glycol, isophthalic acid and trimethylolpropane as a branching component with an acid value

Table 4.1: Dependence between F_n, M_n and M_w in the case of nonlinear polyesterification

F_n	M_n	M_w
2.0	3205	6550
2.5	4000	12,830
3.0	4800	384,000
3.5	5600	infinite
4.0	6400	infinite

of 35 and functionality between 2 and 4. The results are presented in *Table 4.1*. At a functionality of 3.5 the resin is already beyond the gel point. Because of the assumptions that have been made in deriving the *Flory-Stockmayer* theory (especially neglecting the intramolecular reactions that lead to cyclization), there is a certain discrepancy between the theoretical predictions and the practical results [10].

Therefore, it is possible to make a resin with a functionality of 4, but the viscosity will be too high for practical application. In addition, the gel point distance defined as the ratio between the extent of the reaction at the gel point and the extent of the reaction necessary to obtain a resin with certain characteristics will be too small to produce a safe production process [11].

4.3 Glass transition temperature

Although a wide range of crystalline polymers are used in the production of powder coatings, their use is almost exclusively restricted to the thermoplastic coatings. Thermosetting powder coatings have become more important because of the reasons discussed in Chapter 3.

While the crystalline melting temperature is one of the decisive factors in choosing a suitable thermoplast as binder for a powder coating, in the case of thermosetting powder coatings that use mainly amorphous polymers the glass transition temperature of the coating composition is a parameter to which the resin and paint chemist must pay great attention. It influences directly or indirectly the **physical and chemical stability** of the coating composition during storage, the rheological behavior during the production and film formation, and finally the development of internal stress in the cured film during its service life.

Although the glass transition temperature (T_g) as an indication of a phase change resembling a thermodynamic second-order transition is exhibited by many low molecular weight compounds, it is a parameter that is considered typical for polymers. Very often, those approaching polymer science for the first time are not familiar with this term [12]. In the case of a second-order transition, the plot of a primary quantity (specific volume, for example) shows a change in slope, while the plot of a secondary quantity (such as the expansion coefficient) shows a sudden jump. In contrast, in the case of a first-order transition the primary quantities (specific volume, for example) show an abrupt change in their values.

Thermodynamically speaking, the T_g is not in fact a true second-order transition, since it is kinetically controlled. The value of the glass transition temperature is dependent on the rate of cooling or heating. Although from an exact thermodynamic point of view the classification of T_g as a second-order transition cannot be fully defended, it is very suitable for the paint chemist. It is directly related to the specific and the free volume of the polymers helping to explain many phenomena in the paint field. At least, the definition of the T_g as the temperature above which there is an increase in the temperature coefficient of expansion of the polymer, is much better than that which defines T_g as the temperature below which the polymers are brittle and above which they are flexible [13]. It is not necessary to go far into the powder coating world to find evidence that the latter definition is wrong. As will be seen, all powder coatings in cured states have T_g much higher than their exploitation temperature, but they still exhibit a certain flexibility. On the other hand, the **definition of the T_g by means of a brittle and flexible state is short and simple**, and possibly because of this it is widely used.

There are several different theories of the glass transition temperature that can be classified into three main groups [14]:

- the free volume theory,
- the kinetic theory and
- the thermodynamic theory.

The concept of the free volume and the free volume theory of the glass transition is very convenient for elucidating many phenomena in the paint world, and is probably the most acceptable for the paint chemist.

The term free volume, defined as the volume not occupied by the molecules making up the material, was introduced in the 1930's by *Eyring* [15] who assumed that the molecular motion in the bulk state depends on the presence of holes into which the molecules can jump. Moving into these holes, the molecules and the holes exchange their places which results in a flow of the material. This concept was applied to the polymers in the 1950's by *Fox* and *Flory* [16, 17]. They found that the value of the specific free volume above T_g, in the case of polystyrene, is proportional to the temperature only for a polymer with an infinite molecular weight. However, below T_g, this relation is valid independently of the molecular weight. A very obvious conclusion is that below T_g the configurational arrangement of the polymer segments is the factor determining the free volume and that it is not dependent on the molecular weight and temperature. In other words, below T_g the free volume reaches a constant value and does not decrease as the material is cooled further. The contribution of the chain ends to the creation of a free volume differs from that of the other polymer segments. When the molecular weight is lowered, the contribution of the chain ends to the free volume is more emphasized, and, in fact, the dependence of the free volume on molecular weight is valid only for high molecular weight polymers.

Polymers exhibit two types of motion: a localized mobility of the segments called segmental motion and total mobility of the molecule called molecular motion. In a glassy state the motions in the polymeric materials are restricted to vibrations on a short range and rotational movements. Below T_g only one to four chain atoms are involved in the motions. Reaching T_g with an input of thermal energy, vibration of segments on a long range in a coordinated manner begins. From a plot of T_g versus the molecular weight between the crosslinking points (M_c) it has been calculated that this segmental motion in the polymers involves 10 to 50 chain atoms [18-20]. Above T_g the vibrations of the segments are strong enough to free the neighboring segments, thus creating new empty holes, which contributes to an increase of the free volume of the polymer. The overall result is a change in the slope of the curve representing the dependence of the specific volume of the polymer on temperature. This is the so-called **rubbery or viscoelastic state**. If the system is not crosslinked, the molecules jump into the holes in a random Brownian motion exhibiting mobility on a molecular scale. When stress is applied, the jumps are in preferred directions relieving the stress, and the polymeric material flows.

4.3.1 T_g and powder stability

Assume now a layer of powder coating particles exposed to a certain pressure derived from the weight of the powder particles above. If the T_g of the powder is higher than the storage temperature, because of the absence of segmental mobility there is no diffusion of the material on a segmental or molecular scale between different particles. In the case of having T_g lower than the storage temperature, the segmental mobility is high enough to produce, in the long term, a considerable degree of interpenetration of the molecular chains between different powder particles, resulting in blocking of the same. This phenomenon is known as physical instability of the powder coating. Therefore a high T_g value of the powder coating is a prerequisite for good **physical stability**. It is difficult to judge what should be the minimum T_g that guarantees good powder stability. It is dependent, of course, on the storage temperature of the coating. For practical reasons it is widely accepted that the **T_g of the powder coatings should not be lower than 40 °C**.

In the case of thermosetting powder coatings, the glass transition temperature also influences the **chemical stability** of the system. Since both reaction partners, the resin and the crosslinker, are present in the coating, it could be expected that a certain reaction will take place during storage. From a thermodynamic point of view, a reaction takes place only when the reactants that meet each other, possess enough energy, at least equal to the activation energy. When the powder coating has a T_g higher than the storage temperature, the probability that two functional groups belonging to the resin and the crosslinker meet each other is very low because of the restricted mobility of the polymer segments. Paints, which must be formulated as two component systems, because of limited storage stability in the case of liquid coatings, can exist as one component systems in the form of a powder. Good examples are polyester/epoxy powder coatings based on bisphenol A epoxy resins and polyesters with carboxyl functional groups. While they can be formulated into powder coatings with practically unlimited storage stability, the solventbased counterparts have a storage stability of only a few days at room temperature or a few months when not catalyzed [21, 22].

4.3.2 *T_g and melt viscosity*

According to the free volume theory, the glass transition temperature is related to the unoccupied volume by the polymer segments or molecules. The latter is directly proportional to the difference between the glass transition temperature and the temperature at which the viscosity of the polymer is monitored. The unoccupied volume facilitates the flow of the material or, with viscosity of the polymer defined as a resistance to flow, should be directly dependent on T_g. Although many aspects of the rheological behavior of polymers in melts still have to be elucidated, the dependence of the melt viscosity on T_g is a fact that was recognized, and was the reason for many attempts to find expressions giving a quantitative description of this dependence. Among several empirical, semi-empirical and theoretical equations, the relation suggested by *Williams*, *Landel* and *Ferry*, or the well-known WLF equation [23], best fits the experimental data.

Equation 4.6

$$\log \eta_T = \log \eta_{Tg} - \frac{C_1(T - T_g)}{C_2 + (T - T_g)}$$

where

η_T = viscosity of the polymer with the glass transition temperature T_g at temperature T
η_{Tg} = viscosity of the polymer at a temperature equal to its glass transition temperature
C_1, C_2 = constants

Although it has been suggested that C_1 and C_2 have universal values, they differ in fact for different polymers. When data are not available, with good approximation, values for C_1 = 17.4 and C_2 = 51.6 can be used.

The validity of the WLF equation is restricted to the temperature interval $(T-T_g) < 100$ °C [24, 25]. It has been presumed that for $(T-T_g) > 100$ °C, the melt viscosity is governed more by the thermal activation of the bond rotation and secondary segment interactions than by the free volume availability, which makes an *Arrhenius* type of equation more suitable for such a case [26]:

Equation 4.7

$$\log \eta = \log K + \frac{E_a}{RT}$$

where E_a is the activation energy of viscous flow and K is a material specific constant.

The WLF equation cannot be directly used to calculate the absolute values for the viscosity of the powder coatings, because of the complicated interdependence of the elements influencing the rheology of the melt, but it can be very **useful for comparing the expected values for resins with different glass transition temperatures**. For example, a comparison of viscosities of two resins with T_g = 60 °C and T_g = 30 °C obtained by using the WLF equation for a temperature of 150 °C gives the following results [27]:

Equation 4.8

$$\eta_{Tg-60} = \eta_{Tg} \cdot 10^{-11}$$

Equation 4.9

$$\eta_{Tg-30} = \eta_{Tg} \cdot 10^{-12}$$

Since the viscosities of different polymers at T_g are of the same order ($1 \cdot 10^{-12}$ Pa · s) the comparison shows that the resin with a T_g of 30 °C will have at 150 °C a ten times lower viscosity than the other one.

The T_g of the powder coating resins are much higher than the T_g of the resins for solventborne coatings, for reasons related to the physical and chemical stability of the coatings, as discussed in the previous section. Consequently, the **melt viscosities of the powder coatings are several hundred times higher than those of the solventborne coating** at the curing temperature. The fact that there is no solvent present to reduce the viscosity during the film formation emphasizes this difference even more. This has an adverse effect on the flow and the leveling characteristics of powder coatings.

4.3.3 T_g and thermal stress development

The internal stress development in a cured coating is not a characteristic typical only of powder coatings. This phenomenon exists in almost all types of coatings, but it is certain that powder coatings, together with the radiation curing materials, suffer the most.

The reasons for internal stress development in the coatings during curing and its determination are described in many papers [28-34]. The dimensional changes in the coating during film formation are the main reason for the stress formation. The shrinkage of the coating during crosslinking is very noticeable in the case of the radiation curing coatings, where the stress can be a reason for development of micro cracks in the cured film or in extreme cases can cause even a detachment of the coating from the surface. Shrinkage of the other coating materials, except the high solids coatings, is not of such a magnitude.

Powder coatings, especially the carbon/epoxy curing types, do not undergo major shrinkage due to crosslinking. The ring opening reaction of crosslinking in some cases is even followed by expansion [35, 36]. Stress in this case is mainly of a thermal character, and is related to the high curing temperature combined with a high glass transition temperature of the cured films.

The expansion coefficients of the coating and the substrate are in general different. Therefore, during cooling of the thermally cured coatings the dimensional changes in the substrate and the coating are different. As a result of this difference an **internal stress will develop as the film cools**. When the exploitation temperature of the coating is above T_g, the internal stress is completely relieved due to the fast relaxation processes. Below T_g the relaxation processes are very slow, and the coating layer cannot follow the dimensional changes of the substrate. Therefore, a certain stress will build up, proportional to the difference

Table 4.2: Stress values at 21 °C and 0 % relative humidity

Type of coating	T_g [°C]	S [mPa]
Polyurethane thermoplast	<0	0.24
Air drying alkyd	<0	0.60
Conventional polyester/melamine	25	0.80
High solids polyester/melamine	33	2.10
Powder coating	62	3.60

between the T_g of the cured film and the temperature to which the coating layer is cooled down after the curing; the thermal stress S_m can be calculated if the thermal coefficient of the coating and of the substrate at a temperature T are known [34].

Values of internal stresses for different types of coatings are listed in *Table 4.2* [37]. It is clear that powder coating has the highest value for the stress, but is still comparable with one of the high solids coatings.

Although the stress development in principle is not a typical problem related to powder coatings, as it is with radiation curing coatings, for example, it should be considered especially in cases where problems concerning adhesion of the coating to substrates are encountered.

4.3.4 Molecular weight and T_g

For a given molecular structure, the T_g of the resins depends up to a certain value on the molecular weight. The influence of molecular weight can be expressed by the equation of *Fox* and *Loshaek* [38] conveniently for practical use by

Equation 4.10

$$T_g = T_{g\infty} - \frac{K}{M_n}$$

where $T_{g\infty}$ is the glass transition temperature for a polymer of infinite molecular weight and K is a constant for a polymer with a given molecular structure [39, 40]. The original form of this equation comes as the result of the theoretical analysis of *Fox* and *Flory* [16]. It shows that the glass transition temperature rises with an increase of the molecular weight, leveling off asymptotically to a certain value typical for a specific polymer. This can be qualitatively explained by the free volume theory of glass transition, assuming that the **chain ends contribute more free volume to the system** than similar segments along the chains.

Since the resins used for manufacturing thermosetting powder coatings are of relatively low molecular weight, in most cases around 3000, they are in principle in the region where the slope of the curve representing the T_g-molecular weight dependence is steep. This means that a small increase or decrease in molecular weight could have a considerable influence upon the T_g of the resin.

The molecular weight distribution will influence the T_g of the resins. This is to be expected and is easily explained. The low molecular fractions will decrease considerably the T_g of the resin acting as plasticizers. *Kumler, Keinath* and *Boyer* [41] found for polystyrenes of different polydispersities that the **T_g values are a function of M_n** rather than M_w. For example, polystyrene with M_n of 2740 and M_w of 25,100 has a T_g of 43 °C, while the blend with M_n of 650 and M_w of 23,300 has a T_g of -25 °C. *Holda* [42] considers this phenomenon as an explanation for the lower T_g of the polyesters for powder coatings based on phthalic anhydride, in comparison with those where terephthalic and isophthalic acids are used as diacids.

The glass transition temperature in principle should be influenced by the chain entanglements which restrict the mobility of the system. The molecular weights of most polymers for thermosetting powder coatings fall below the critical value at which there is a change in slope of the curve representing the viscosity versus log M_w. This change in slope is an indication of the contribution of the entanglements to the rheological behavior of the polymer. Therefore, for most thermosetting powder coating resins, the existence of some critical molecular weight, concerning molecular mass-T_g dependence that may be attributed to entanglements should not be expected. In fact, there is no extensive evidence of such behavior of the polymers, although hints of this character have been made in the literature [43, 44].

According to the free volume theory, branching of the polyesters should contribute to a higher free volume and decrease the glass transition temperature. There is no broad evidence for this phenomenon in the literature. In a paper dealing with saturated polyesters for powder coatings, *Holda* [42] gives a comparison between different polyesters with a different degree of branching, concluding that the branched types have a lower T_g than the linear ones. The presented results are easily discussed, because the polyesters in question differ in molecular weights, and moreover their molecular masses are in the region of 2000, when small differences between them result in large differences in T_g. It seems that the influence of the branching upon the T_g in the case of powder coating resins is too small to play an important role. On the other hand, the branching of the resin will have distinctly more influence on the T_g of the cured powder coating, in particular if high crosslink densities are reached.

4.3.5 Chemical structure and T_g

The influence of the chemical structure on T_g is not fully elucidated, although some general rules are well established. Since the T_g is related to the co-operative motion of the groups or atoms making up the polymer chain, any molecular parameter affecting the chain mobility, will influence the glass transition temperature. More precisely, any group or structure that reduces the chain mobility will increase T_g.

Polymers with perfectly repeating units prone to crystallization usually show a higher T_g than should be expected. This is the result of limited chain mobility in the crystalline state. Some of these polymers crystallize very fast and cannot be quenched easily in an amorphous form. Very often their T_g is a matter of disagreement. For example, in the case of polyethylene, the published values for T_g vary from -30 to -108 °C [45]. Polymers with carbon-carbon bonds in the main chain, without side groups, have a very low T_g, as mentioned for polyethylene. The rigid bulky substituents in the chain decrease the flexibility and affect the steric hindrance to rotation, increasing T_g. Thus, for example, in contrast to polyethylene, polypropylene, polystyrene and poly(α-methylstyrene) have T_g values of -12, 100 and 180 °C respectively.

The **nature of the side chain also plays quite a large role**. Long side chains in the case of acrylic polymers, which act as internal plasticizers, have little effect on the freedom for rotation, since they are nearly flexible. On the other hand, they keep the main chains further apart, contributing more free volume. According to the free volume theory the result will be a decrease of the glass transition temperature in the series of different polyacrylic or polymethacrylic esters, which is the case in practice. Polymethyl methacrylate for example, has a T_g of 105 °C, and the higher polymethacrylic esters like polyethyl, polybutyl and polyhexyl methacrylate have T_g values of 65, 22 and -5 °C, respectively.

The chain stiffness and the nature of the constituents in the heterochain polymers, like polyesters for example, play an enormous role concerning the T_g of the polymer. Polytetramethylene oxalate has a T_g of -5 °C, while the corresponding polytetramethylene terephthalate has a 50 °C higher glass transition temperature. Again, the presence of side bulky groups has the same influence as in the case of homochain polymers. Polyhexamethylene terephthalate (T_g = 8 °C) is a good example to be compared with polyneopentyl glycol terephthalate (T_g = 63 °C).

Intermolecular attractions introduced by the presence of **polar groups or hydrogen bonding increase the T_g** of polymers, as one would expect. Atactic polystyrene with a T_g of -19 °C should be compared with polyacrylonitrile with a T_g of 104 °C. It has been suggested that the influence of the polarity of the polymers (in terms of their cohesive energy density, CED) to the T_g should be expressed as

Equation 4.11 $$T_g = K_1 (CED)^{1/2}$$

where K_1 depends on the structure of the polymer [46].

The hydrogen bonding will have the same effect as the polarity, increasing the T_g. Polyhexamethylene adipate has a glass transition temperature of -63 °C, while polyhexamethylene adipamide has a T_g of 80 °C. The **effect of polar groups** on the T_g plays an important role, since in most cases the thermosetting powder coating resins contain functional groups of a polar nature, or groups introducing hydrogen bonding forces.

In systems without solvents, like the powder coatings, one could expect that the intermolecular attraction forces between the binders, and the pigments and fillers should increase the T_g. In an extensive review on the effect of the fillers on T_g, *Toussaint* [47] noted that it is difficult to find a predictable dependence. The **fillers can increase or decrease the glass transition temperature**, or have no remarkable effect. The increasing effect on T_g resulting from the build-up of reversible internal structures between the pigments and fillers on one side and the resin on the other, can be compensated by looser packing, which is an explanation for the noticed decrease of density of the filled polymers [48-50]. In general, the influence of the pigments and fillers on the T_g of the powder coatings, in concentrations in which they are used, is too low to be considered from a practical point of view.

4.3.6 *T_g of polymer blends*

The phenomenon of decreasing the glass transition temperature of powder coatings by incorporating **crosslinkers** is well known by coating chemists. Very often this is the reason for having unstable systems during storage. The possibility of predicting the T_g of a blend resin/crosslinker should be very helpful, especially in developing new powder coating formulations or examining new crosslinkers.

The hardeners that are used for powder coatings are solid materials at room temperature. Many of them are low molecular weight compounds, also having a low glass transition temperature. The fact that they are solids at room temperature is related to the molecular symmetry which allows crystallization. In a blend with the resin after melting in the extruder and cooling down to the storage temperature (which is always lower than the T_g of the coating), the **crosslinker does not recrystallize**, because of the restricted mobility in the system below its glass transition temperature. Since there is no phase separation, the resin and the crosslinker in the powder coating are in a form of homogeneous blend exhibiting only one T_g. Because the T_g of the crosslinker is generally low, the resulting glass transition temperature of the blend is always lower than that of the resin.

Many attempts have been made to relate the T_g of the blends to the blend composition. The results have shown that, in principle, equations that are used for predicting the T_g of the amorphous random co-polymers are valid for polymer blends or blends of polymers with low molecular compounds.

A well-known formula for predicting the glass transition temperatures of polymer blends or amorphous random copolymers is derived by *Gordon* and *Taylor* [51] but a more simple and useful equation is that of *Fox* [52] involving only the glass transition temperatures of the blend constituents (T_{g1} and T_{g2}) and their weight fractions (w_1 and w_2):

Equation 4.12 $$\frac{1}{T_g}=\frac{w_1}{T_{g1}}+\frac{w_2}{T_{g2}}$$

An empirical equation describing the glass transition temperature of one phase system involves the parameter K, which, once determined for one ratio of the components in the blend, can be used for the whole concentration range [53]:

Equation 4.13 $$T_g = w_1 T_{g1} + (1-w_1)\, T_{g2} + w_1\,(1-w_1)\, K$$

The value of K can also be determined from the initial slope:

Equation 4.14 $$\frac{dT_{g(0)}}{dw_1}=T_{g1}-T_{g2}+K_{g(0)}$$

Couchman [54-57] has developed, on the basis of the thermodynamic theory an equation for predicting the glass transition temperature of a blend, knowing the heat capacities of the components at constant pressure. If the heat capacities of the blend constituents are equal, or close to each other, then the following formula, which is also used semi-empirically, results from the *Couchman* equation:

Equation 4.15 $$T_g = w_1\, T_{g1} + w_2\, T_{g2}$$

Another relatively simple expression gives a relation between the $\ln T_g$ of the blend and the blend composition [58].

4.4 Viscosity

The discussion about molecular weight and glass transition temperature of the binder, related to the desired properties of the powder coating, leads to the conclusion that both parameters should have as high values as possible. This is in favor with better powder stability and a powder coating formulation with mechanical properties which are not too sensitive to the possible deviations from the best coating composition. However, high molecular weight combined with high glass transition temperature will lead to an increase of the melt viscosity of the polymer, adversely affecting the processing performances of the powder coating during its manufacture and the flow properties during film formation.

The viscosity dependence on T_g was discussed in the previous chapter, and can be quantitatively described by the WLF equation.

Melt viscosity of the polymers will increase with increasing molecular weight, as could be expected. The plot of the logarithm of the melt viscosity versus the logarithm of the molecular weight in arbitrary units (*Figure 4.2*) exhibits **two distinct regions**: region A in which the melt viscosity in terms of molecular weight can be expressed as

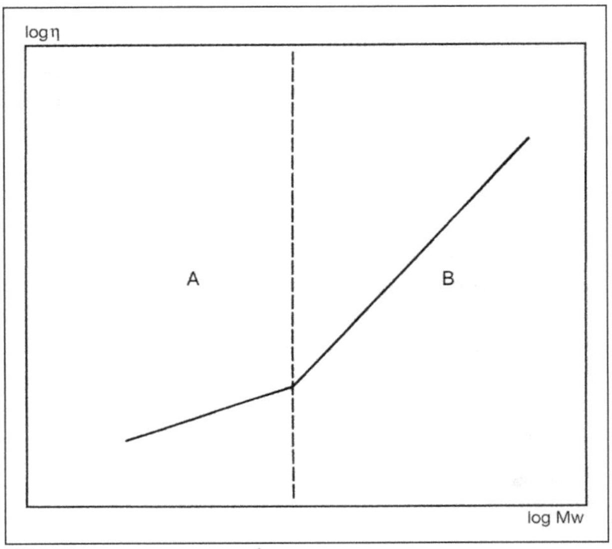

Figure 4.2: Melt viscosity as a function of molecular mass

Equation 4.16 $\log \eta = A \log M_W + K_l$

and region B for which the following expression is valid:

Equation 4.17 $\log \eta = B \log M_W + K_h$

where M_W is the weight average molecular weight of the polymer, η is viscosity and K_l and K_h are constants for low and high degrees of polymerization.

Fox [59] has shown that below a certain critical value of the weight average molecular weight, melt viscosity is proportional to M_W (A = 1). Above this value, melt viscosity of the high molecular weight polymers depends on M_W to the 3.4 power (B = 3.4).

The **critical value for M_W**, above which, due to the chain entanglements, a viscosity increase follows Equation 4.17, differs for different polymers. However, it has been found that, in general, the critical weight average molecular weight corresponds to a weight average degree of polymerization of 600 or a **number average molecular weight of 30,000** [14, p. 389].

It is interesting that this number average molecular weight is close to the low end of molecular weights at which most of the polymers exhibit good mechanical properties, as mentioned in Chapter 4.1.

The number average molecular weights of powder coating resins for thermo-setting powder coatings are far from this region. Therefore, Equation 4.16 can be considered as the correct way to describe melt viscosity dependence on molecular weight.

In contrast to Equation 4.3, which indicates that the mechanical properties of the polymers level off after a certain molecular mass has been reached, Equation 4.16 indicates that with increasing molecular mass the viscosity increases exponentially to infinity.

In an article dealing with the melt viscosity of saturated polyester resins for powder coatings, *Holda* [42] has proposed that the viscosity-molecular weight relationship obeys the following equation:

Equation 4.18 $\log \eta = A \log M_n + B$

Holda relates the failure of the general Equation 4.16 involving M_W to the choice of technique used to determine the M_W in the case of the powder coating polyesters in question. He also points out the contribution that polarity and hydrogen bonding make to the melt viscosity of the resins with relatively low molecular weights.

4.4.1 *Viscosity and processing performances*

Viscosity of the resins has a direct influence on the processing characteristics of the powder coatings during hot melt compounding. The probability that a pigment cluster will be de-agglomerated during extrusion is directly proportional to the energy density present to overcome the forces holding the primary particles together [60]: The energy supplied by the motor in the case of smearing types of machines like the extruder is proportional to the viscosity of the media in which the pigments are to be dispersed. Also in these types of machines, where the pigments are dispersed in a gradient of shear, the shear stress will be the determining quantity with respect to the efficiency of the de-agglomeration process. The equation that describes the mutual dependence between the shear stress and the viscosity is as follows:

Equation 4.19 $\tau = \eta D$

where

τ = shear stress
η = dynamic viscosity
D = shear rate

In other words, high resin viscosity is favorable considering the efficiency of the extruder as a dispersing machine. On the other hand, the **wetting of the pigment**, which is in fact displacement of the air or other contaminants (water, for example) adsorbed on the surface of the pigment particles by the resin, is controlled by the velocity U with which the resin penetrates the pores on the pigment surface given by [61].

Equation 4.20

$$U = \frac{K \cdot r}{\eta}$$

where

r = pore radius,
η = viscosity of the resin
K = constant of surface tension coefficient of the resin

It is quite obvious that the viscous resin will facilitate the wetting of the pigments. From these two equations it can be concluded that there is an **optimum viscosity at a given temperature** (which is again restricted by the reactivity of the system) that provides the best dispersing efficiency of the extruder. This optimum depends on the type of extruder, the coating formulation and the types of pigments used.

4.4.2 Viscosity and film forming properties

The layer of coating powder, when applied on the surface to be coated, cannot be regarded as a continuous system. This is, among others, one of the principal differences from the solventborne coatings, where the film formation comes as the result of evaporation of the solvent from a continuous low viscous layer of the applied coating, followed by subsequent crosslinking in the case of thermosetting coatings.

Conversely, film formation of powder coatings results from melting powder particles and their coalescence to form a continuous film. This is followed by flow of the formed film, to obtain a smooth surface from the irregular one produced after the melting.

Different workers considering **the coalescence of particles during film formation** have derived equations which describe this phenomenon with considerable success [62-66]. The derivations of these equations are beyond the scope of this book. Only two of them will be presented to illustrate the influence of viscosity on the speed of sintering during film formation of powder coatings. In both equations the gravitational forces are neglected, and it is assumed that for films thinner than 100 µm the single driving force for coalescence is the surface tension of the coating [67].

One of the first equations derived by *Frenkel* [62] relates the time (t) necessary for two spherical particles to coalescence, radius of the particle (R), radius of the contact area between the particles (x), viscosity of the material (η) and its surface tension (τ):

Equation 4.21

$$(x/R)^2 - \frac{3\,\tau\,t}{2\,R\,\eta}$$

Nix and *Dodge* [63] have proposed something different, assuming that the surface tension is the only reason for coalesce and flow and viscosity of the molten particle the only resistance to flow. In such a case they propose that the pressure causing two particles to flow together will be proportional to the surface tension divided by the average radius of the curvature (R_c). The latter is in a first approximation equal to the mean particle radius. According to them, the time necessary for coalescence will be given by the following equation:

Equation 4.22 $$t - \frac{\eta R_c}{\tau}$$

Using an instrumental technique to compare the surface profiles of coatings with different viscosities, they have confirmed in principle the validity of the above equation, admitting that some other factors concerning the pigment concentration and dispersion have to be included for a better quantitative description of the sintering process.

Although these equations refer to a simplified model of a single spherical particle, which is not the exact description of a real powder coating composed of irregular particles of different size, they both show that the higher the **melt viscosity**, the longer the time necessary for particles to coalesce.

Melt viscosity of the powder coating also influences the possibility of the coating flowing during the second stage of the film forming process, after the sintering has been accomplished. The equation of *Orchard* [68], which was originally derived for liquid coating systems applied by brush, can in principle be used here to illustrate the influence of the viscosity on the flow. This equation connects the time (t) necessary for a coating layer of a thickness (h), melt viscosity (η) surface tension (τ) and a wavelength of striation (λ), to decrease the initial amplitude of striation (a°) to a value (a_t), as presented in *Figure 4.3*.

Equation 4.23 $$\ln (a_t/a^\circ) = -(16\,\pi^4 h^3/3\,\lambda^4) \int_0^t (\tau/\eta)\,dt$$

The integrated form of the term dt/η in the Orchard equation known as integrated fluidity, often designated by F, is

Equation 4.24 $$F = \int_0^t (dt/\eta)$$

Spitz [69] has shown in the case of thermoplastic powder coatings that the experimental results fit Orchard's equation. On a doubly logarithmic plot against $h^3\,t/\eta$, the wavelength increased linearly with a slope of 0.25.

Hannon, Rhum and *Wissbrum* [70] confirmed the validity of the equation of *Orchard* for thermosetting powder coatings performing rheological measurements during curing using a thermo-mechanical spectrometer and the obtained data for calculation of the integrated fluidity F.

The model was successfully tested on a commercial white TiO_2-pigmented epoxy powder coating, using a thermo-mechanical spectrometer to measure the time and temperature dependence of the viscosity, showing good agreement with experimental results.

An improved model describing the non-isothermal viscosity changes in unpigmented acrylic systems has been developed by *Kuwano*, as reported by *Nakamichi* [71]. He has taken into consideration the influence of an increase in molecular weight during curing on the activation energy of viscous flow. This effect was disregarded in the model of *Hannon*.

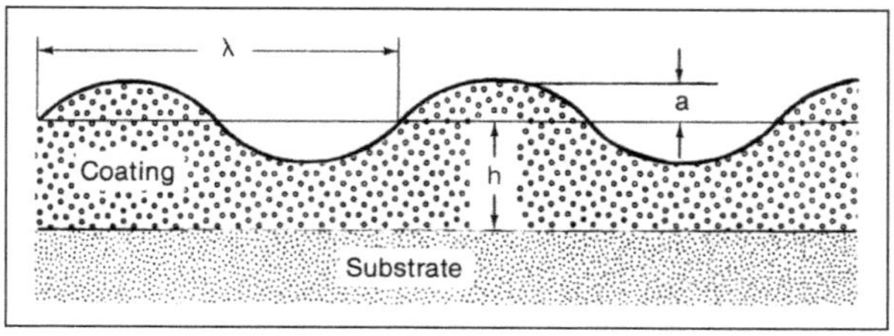

Figure 4.3: Illustrative explanation of the terms involved in the equations of Orchard

The equation of *Orchard* shows that for good leveling properties of the powder coating, a low melt viscosity is desirable, as was also the case for fast coalescence of the powder particles in the first stage of the film formation process. However, too low a viscosity leads to bad edge coverage and even sagging on vertical surfaces. *Walz* and *Kraft* [72] have suggested that a **good compromise should be within a viscosity range of 6000 and 10,000 mPa s**.

The flow of powder coatings cannot be completely described by integrated fluidity only, because it is a more complex phenomenon. However, for comparison, integrated fluidity can be quite a useful parameter. **Thermo-viscoelastic analysis** (TVA), described in Chapter 3.2.3, is a very suitable experimental method for determination of integrated fluidity of powder coatings. The temperature profile can easily be adjusted to simulate the normal stoving conditions. The area under the reciprocal viscosity-time curve gives the value for the integrated fluidity.

Nakamichi [71] and others [73-75] have used the rolling ball technique to determine the viscosity changes during the non-isothermal curing of acrylic resins with different functionalities and melt flow index as an indication of the viscosity. Although *Nakamichi* addresses some remarks to the reproducibility of the technique, it seems to be a very simple and practical method for determination of the non-isothermal viscosity change during the curing process, with even better simulation of the real stoving conditions than with thermo-viscoelastic analysis.

Recently *Bosma* et al. [101] developed a quantitative method to predict the sag and leveling behavior of powder coating formulations. For this model they took into account the film thickness (**paint flow scales with the cube of the thickness**), the temperature, the heating rate (the higher, the better), the reactivity (measured by DSC) and the rheological behavior.

4.5 Resin/crosslinker ratio

The ratio between the resin and the crosslinker has already been discussed in the section dealing with the molecular mass of the resin. It was stated that the proper ratio between the resin and the hardener is very important for obtaining a certain molecular weight of the cured material, and indirectly for obtaining good mechanical properties of the coating. Since the mechanical properties are directly proportional to the molecular weight up to a certain point, in principle the higher the molecular weight of the cured coating, the better. The highest molecular weights of the cured coating can be obtained when the resin and the crosslinker are in **stoichiometric proportion**.

Because powder coating resins and crosslinkers are usually of a much lower functionality compared to the solventbased systems, it is very important to have a proper amount of crosslinker in the coating composition. The self-condensation reactions which are present in the polyester/melamine or acrylic/melamine systems leading to harder films do not proceed in the powder coating systems based on epoxy or polyisocyanate crosslinkers. Therefore more hardener in a powder coating does not mean a harder film. The result can even be the opposite - a softer film without flexibility and impact resistance.

The resin producer usually gives the parameters of the resin important for determination of the stoichiometric amount of the crosslinker in the coating formulation. These parameters are the acid value of the resin in the case of carboxyl functional binders and the hydroxyl value in the case of resins that are to be crosslinked by polyisocyanates. The following formulas can be useful for calculating the amount of crosslinker in the coating formulation,

Equation 4.25
$$EC = \frac{R * AV * EEW}{56100}$$

Equation 4.26

$$IC = \frac{0.0749 * R * HV}{NCO}$$

where

EC = amount of epoxy crosslinker [g]
IC = amount of isocyanate crosslinker [g]
R = amount of carboxyl or hydroxyl functional resin [g]
AV = acid value of the resin [mg KOH/g]
HV = hydroxyl value of the resin [mg KOH/g]
EEW = epoxy equivalent weight of epoxy crosslinker [g]
NCO = content of isocyanate groups in the isocyanate crosslinker [%]

Powder coatings cured by epoxy/carboxyl curing reactions, like TGIC powders for example, are extremely sensitive to the ratio between the reactive groups [77]. A combination of incomplete curing and non-stoichiometrical ratio between the reactive groups in the coating composition can be disastrous for mechanical properties. Even when the curing reaction proceeds completely, **the ratio between the reactants is of extreme importance**.

In the case of polyester/TGIC powder coatings, a shortage of TGIC in the coating composition has a more dramatic effect on the mechanical properties than an excess of the same [78]. Therefore, a **small excess of TGIC** in polyester powder coatings for outdoor application is desirable. *Greidanus* [77] explains this phenomenon by a higher consumption of the epoxy groups than that calculated with respect to the amount of available carboxyl groups, assuming that part of the oxirane rings undergoes esterification with the formed hydroxyl groups during curing. Although this could be a reason for such a behavior, the difference in crosslinking density between compositions with an excess and shortage of three-functional crosslinkers like TGIC should also be considered.

4.6 Catalyst level

The **kinetics of the curing process** in thermosetting powder coatings is greatly influenced by the choice and level of the catalyst. The normal tendency towards lower curing temperatures and shorter curing times in order to reduce the energy consumption during curing stimulates the paint chemist to search for systems which offer a combination of high curing rate and acceptable storage stability. On the other hand, in the case of powder coatings, premature curing during the processing in the extruder should be minimized in order to avoid undesirable viscosity increase leading to inferior flow and leveling properties. The influence of the catalyst on the integrated fluidity as an indication of the flow properties of certain coating compositions has already been mentioned (Chapter 4.4.2).

The graph of a dynamic DSC run of thermosetting powder coating exhibiting good flow (Chapter 3) is characterized by a horizontal plateau between the temperature onset flow (T_{of}) and temperature onset cure (T_{oc}). This means that the **ideal catalyst** should not be active at room and processing temperatures, and will manifest its activity at the stoving temperature, permitting shorter cure times.

In contrast to the solventborne coatings where the activity of the catalyst at room temperature can cause considerable problems with respect to the storage stability, chemical stability of powder coatings during storage is significantly less influenced by the reaction kinetics. The storage temperature of powder coatings is always lower than the glass transitions of the coating composition in order to avoid blocking of the powder particles. In such conditions, the mobility of the system is restricted to vibrations on short range and rotational movements only, which are not favorable conditions for reaction, although in particular smaller molecules might still show prereactions. Therefore, curing systems such as epoxy/carboxyl, which cannot be used in one component solventborne coatings because of storage instability, can be employed without incurring large problems in powder coatings. However, catalysis of the curing has a considerable influence on the processing stability of the powder coating and its flow and leveling during film formation.

Since in the early stage of the reaction at relatively low viscosities the curing rate is determined predominantly by the reaction kinetics, a logarithmic form of the following well-known *Arrhenius* equation, which describes the **temperature dependence of the rate constant**, can be quite helpful for providing general guidelines in choosing the correct catalyst:

Equation 4.27

$$\ln K = \ln A - \frac{E_a}{RT}$$

Let us suppose a powder coating has, at the stoving temperature T_S, a rate constant K_{S1} and an unacceptably long curing time. The curing process could be accelerated by increasing the concentration of the catalyst, reaching the rate constant $K_{S2} > K_{S1}$. As a consequence it will have an increasing rate constant at the onset flow temperature to a value K_{of2}. In other words, more pre-reaction will take place during the extrusion of the powder coating, and viscosity build-up will be faster in the period until the stoving temperature is reached. This will worsen the flow properties of the powder coating composition, resulting in a film with a more emphasized **orange peel effect**.

The kinetics of the curing process of powder coatings have been studied by means of differential scanning calorimetry [79-81]. It has been found that the overall curing reaction could be considered to be of an order between 1 and 2 [79, 81].

Supposing, for simplicity, that the curing reaction is of second order; then the rate constant can easily be calculated by the following equation, which is derived from basic kinetics principles:

Equation 4.28

$$K = \frac{1/C_a - 1/C_{a0}}{t}$$

where C_{ao} and C_a are the concentrations of the reactive groups at the beginning of the reaction and after time t.

For an analysis of the system let us choose a polyester/epoxy powder coating with 70 % polyester resin and 30 % epoxy crosslinker, or the so-called 70/30 hybrid. Polyester resin with an acid value of 37 mg KOH/g and epoxy resin with an epoxy equivalent value of

Table 4.3: Rate constants, activation energies and pre-exponential parameters for polyester/epoxy powder coating at extrusion (K_e) and stoving (K_s) temperatures supposing different degrees of pre-reaction during the processing time

Pre-reaction [%]	K_e [l/eq s]	K_s [l/eq s]	E_a [kJ/mol]	A [l/eq s]
0.1	$2.78 \cdot 10^{-8}$	0.0324	103.6	$8.8 \cdot 10^{9}$
0.5	$1.81 \cdot 10^{-4}$	0.0324	76.1	$8.1 \cdot 10^{6}$
1.0	$3.65 \cdot 10^{-4}$	0.0324	65.8	$5.9 \cdot 10^{5}$
5.0	$1.90 \cdot 10^{-3}$	0.0324	41.6	$1.3 \cdot 10^{3}$
10.0	$4.01 \cdot 10^{-2}$	0.0324	30.6	$0.8 \cdot 10^{2}$

ca. 700 is a typical binder/crosslinker combination for such a system and will correspond to an initial concentration of the carboxyl groups of 0.46 equi./l. In most cases powder coatings are formulated in such a way that **90 % conversion during curing** should be enough to obtain good mechanical properties. In order to reach 90 % conversion at a typical stoving schedule of 10 min at 200 °C, the rate constant calculated from Equation 4.28 should be $3.24 \cdot 10^{-2}$ l/equi.

Five different rate constants at extrusion conditions at 100 °C are calculated supposing that 0.1, 0.5, 1.0, 5.0 and 10.0 % conversion has taken place within the **residence time in the extruder of 1 min**, the first one being the most favorable and the last one the worst concerning the flow and stability of the system. Using a pair of values for K_e and K_s (rate constants at extrusion and stoving temperatures), by means of the equation of Arrhenius the values for the activation energy E_a and pre-exponential parameter A are calculated. The results are presented in *Table 4.3*. Experimental results in the same range as those calculated and presented in *Table 4.3* are reported by *Franiau* for the catalyzed polyester/hybrid system [81]. He has found by DSC measurements a value for E_a of 104 kJ/mol and for the pre-exponential parameter of $3.6 \cdot 10^9$ l/equi.

It is obvious that in searching for a good catalyst in a proper concentration at a given stoving schedule, one should look for a combination that gives an increase in the activation energy of the system and the pre-exponential parameter, resulting in a series of Arrhenius plots passing the same point and representing the necessary rate constant at the stoving temperature.

The same approach of choosing a catalyst by which the reaction kinetics will be characterized by **a large Arrhenius activation energy (E_a) and large pre-exponential parameter (A)** to compensate for the small values of K has been suggested by *Pappas* and *Hill* for one component liquid coatings [82].

For an **optimal combination of a desired curing rate and good storage** stability in the case of first-order curing reactions, A values of 10^{12} s^{-1} are recommended [77]. Whereas bimolecular reactions have in general A values not exceeding 10^{11} l/mol s, unimolecular first-order reactions are characterized by pre-exponential parameters of up to 10^{16} s^{-1}. Most of the thermosetting coatings employ bimolecular curing processes. For good storage and, in the case of powder coatings, processing and leveling characteristics, these curing reactions have to be preceded by rate-controlling unimolecular reactions, utilizing latent catalysts or latent reactants. A typical example of latent reactants are blocked isocyanates, where the first-order deblocking reaction precedes the formation of the urethane linkage by a second-order condensation between the isocyanate and hydroxyl groups.

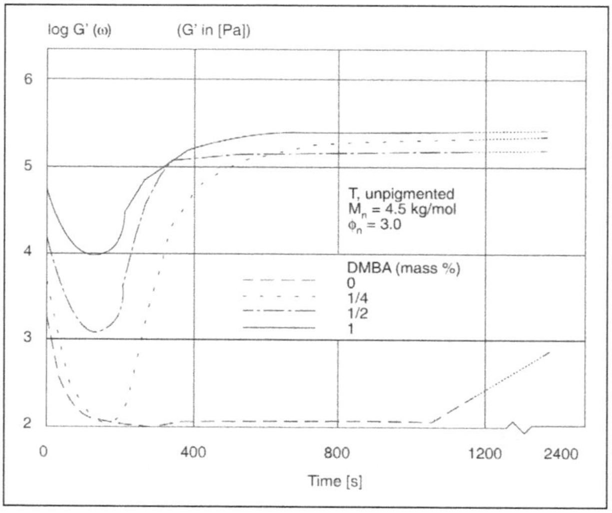

Figure 4.4: TVA isofrequent heating runs at ω = 2π rad/s for an unpigmented polyester/epoxy hybrid system at four concentrations of DMBA catalyst at T = 110 to 200 °C (heating rate 20 °C/min) [78] *source: DSM Resins BV*

The type and concentration of the catalyst depends on the binder/crosslinker system, the stoving schedule and the requested flow and leveling of the film. Even for the same type of curing reaction the effect of the catalyst on the flow properties of the system can be very different. *Figures 4.4* and *4.5* present results obtained by thermo-viscoelastic measurements on mixtures of polyester resins with TGIC and bisphenol A epoxy resin respectively, using N,N-dimethyl benzyl amine as catalyst [78]. The storage modulus is plotted semi-logarithmically versus time. As the heating proceeds at a rate of 20 °C/min there is an initial decrease of the storage modulus (and viscosity), and after 200 s an abrupt increase can be observed which is caused by the crosslinking process. The **minimum modulus** is almost **independent of the catalyst concentration in the TGIC systems**, but it varies strongly in the epoxy systems. The subsequent increase in the modulus due to network formation begins at an earlier stage with increasing catalyst concentration. An additional observation is that the **reaction in the uncatalyzed epoxy formulation is extremely slow**, in contrast to the TGIC formulation.

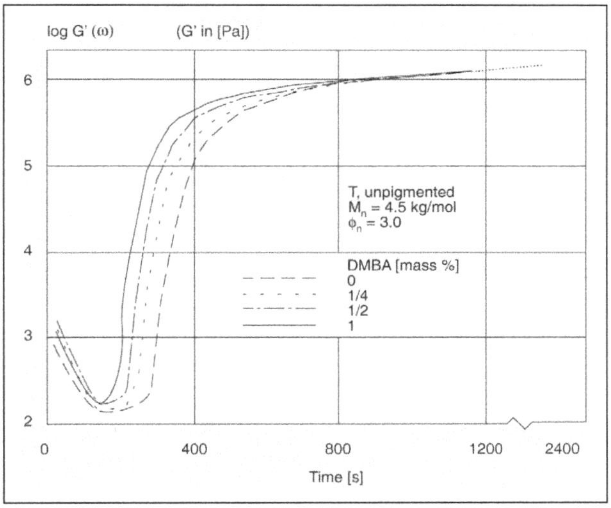

Figure 4.5: TVA isofrequent heating runs at ω = 2π rad/s for an unpigmented polyester/TGIC system at four concentrations of DMBA catalyst at T = 110 to 200 °C (heating rate 20 °C/min) [78] *source: DSM Resins BV*

4.7 Surface tension

The Orchard equation (Equation 4.30), referring to the flow behavior of the powder coating after melting, indicates that **high surface tension will have a positive influence on the flow** properties. On the other hand, wetting of the pigment particles and of the substrate is dependent also on the surface tension of the resin and the coating. It will be shown later that good wetting of the pigment particles is essential for a low yield value and melt viscosity of the powder coating, both having a considerable influence on the leveling of the film. Good wetting of the substrate is a prerequisite if surface defects are to be avoided, a phenomenon known as cratering of powder coatings. All of this shows that the influence of surface tension on the film forming properties of powder coatings is very complex. It is the intention in this section to elucidate the most important aspects of surface tension relevant to the leveling, wetting and cratering of powder coatings.

4.7.1 Wetting properties and surface tension

A molecule present in a drop of liquid in contact with a solid surface, depending on its position, can undergo attraction from the liquid phase only (position A), the liquid and vapor phase (position B), the liquid and solid phase (position C) or the liquid, solid and vapor phase (position D), as depicted in *Figure 4.6*.

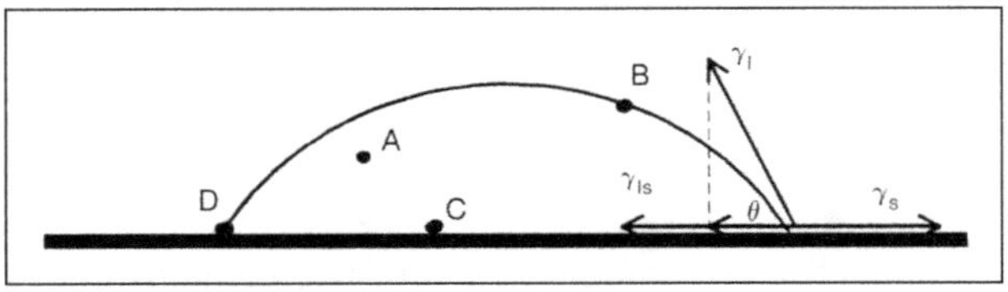

Figure 4.6: Drop of liquid in contact with the surface

In position A the molecule is in equilibrium in a uniform field. However, in positions B, C and D the molecule undergoes an attraction that is not uniform in all directions, since the cohesive and adhesive forces at the liquid/vapor, liquid/solid and solid/vapor interfaces are of different magnitudes. In order to reach equilibrium, the system will tend to **minimize its free energy** per unit area, which results in a contraction of the surface to a minimum, thus obtaining a typical form of drop. According to the well-known Young equation, the three existing interfacial forces in the equilibrium state are connected by the following dependence:

Equation 4.29

$$\gamma_{sv} = \gamma_{lv} \cos\theta + \gamma_{sl}$$

where

γ_{sv} = surface tension of the substrate in contact with the vapor from the powder coating
γ_{lv} = surface tension of the powder coating in contact with its vapor
γ_{sl} = interfacial tension between the powder coating and the substrate
θ = contact angle, representing the angle between the tangent of the liquid surface drawn at the three phase boundaries and the plane of the substrate surface

A spontaneous spreading will occur only when $\gamma_{sv} > \gamma_{lv} \cos\theta + \gamma_{sl}$,

Since it is very difficult to measure γ_{sv}, Equation 4.29 is of limited practical use.

On the other hand, the **contact angle θ can be easily measured** in a relatively simple way by a contact angle goniometer. Since it is related to the partial wetting of the substrate, it is a useful parameter for estimating the wetting properties of the powder coatings.

From *Figure 4.6* the contact angle θ can be defined by the following equation:

Equation 4.30

$$\cos\theta = \frac{\gamma_{sv} - \gamma_{sl}}{\gamma_{lv}}$$

Since cos θ can never be greater than 1, it follows that no spontaneous spreading can occur; it can happen only in an extreme case when θ = 0. On the other hand, **a lower contact angle promotes better wetting.** Equation 4.30 shows that for lower values of θ the surface tension of the powder coating γ_{lv} should be as low as possible and that the surface tension of the substrate should be maximized, having no contaminants with low surface tension.

4.7.2 Surface tension and cratering

The cratering phenomenon of the coatings is a very common problem facing the paint chemist and has been a matter of interest to many scientists [83-91]. In solventborne coatings, the viscosity of the film forming material is relatively low and the cratering is caused mainly by surface tension gradients, which are responsible for the flow of the material from one point to another (from low surface tension areas, to areas with higher surface tension). This is the so-called *Maragony* effect. Mathematically it can be described by the equation of *Fink-Jensen* [83] which gives the dependence of the flow of the material (q) as a function of the film thickness (h), viscosity (η) and the surface tension gradient across the surface (γ):

Equation 4.31

$$q = \frac{h^2\, \gamma}{2\, \eta}$$

It seems that, for **good control of the cratering**, thinner films with higher viscosities and equalized surface tension gradients across the surface should be applied.

Scholtens, Van der Linde and *Tiemersma-Thoone* [78] believe that the *Maragony* effects do not play a role in cratering of the powder coatings and that the **crater formation is caused by an insufficient wetting** of the substrate by the molten coating.

Feyt and *Bauwin* [84] have found that the surface tension of the substrate decreases linearly with an increase in temperature from 10 to 185 °C. The same dependence was found by *Gabriel* [87] for bisphenol A type epoxy resins. It is interesting that the slopes of the straight lines representing the values of the surface tension against temperature are almost the same in both cases. This means that the contribution of the temperature in an eventual decrease of the contact angle due to the change of surface tension of the powder coating and the substrate is almost negligible.

However, *Gabriel* has shown that the wetting of the substrate is improved by increasing the temperature, and has attributed this to the reduction of the interfacial tension between the coating and the substrate (τsl) which according to Equation 4.30 will increase the degree of the spreading.

The results obtained by *de Lange* [67], in experiments where the surface tension of an epoxy powder and polyester powder coatings was varied by addition of acrylic additive, support at first sight the idea that the cratering is caused by bad wetting of the substrate; namely, in both cases, the powder coatings without acrylic additive having a contact angle with the substrate of 60° (epoxy powder) and 35° (polyester powder) showed obvious cratering. The addition of 0.25 % of acrylic additive decreases the contact angles and reduces the cratering tendency to a certain extent, but does not prevent the cratering completely. **The cratering disappears completely at a concentration of 0.5 % of acrylic additive**. Further addition has a positive effect on the flow, reducing the orange peel effect.

Figure 4.7 [67] represents the dependence of the contact angles of the polyester and epoxy powder coating with the substrate as a function of the concentration of the acrylic additive. In both cases there is an obvious minimum at a concentration of 0.25 %, after which the contact angles increase again and level off at a concentration higher than 0.5 %. Very similar results were obtained by *Feyt* and *Bauwin* for epoxy powder coating and linear acrylic polymer additive of medium molecular weight bearing unbranched side chains. The presence of craters at the minimum contact angle and their absence at higher concentrations when the contact angle is increased does not support the hypothesis that the cratering is caused only by poor wetting of the substrate. If this is true, then the best results concerning the cratering should be obtained with 0.25 % of acrylic additive. Obviously the **cratering is a more complex phenomenon**.

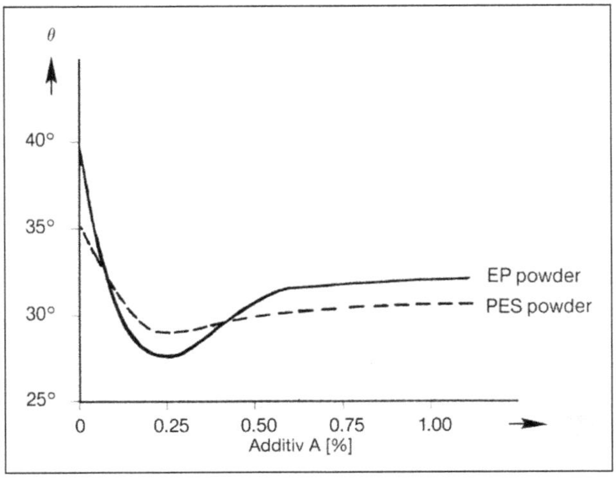

Figure 4.7: Effect of percentage of acrylic additive on contact angle for unpigmented epoxy and polyester powder coatings [67]
source: Federation of Societies for Coating Technology FSCT

Feyt and *Bauwin*, and *de Lange* give an identical explanation of the action of the acrylic additive. They suppose that the lowering of the interfacial tension between the substrate and the coating, manifested by the lower contact angle, has a positive effect on the wetting of the substrate, but it is not the only element necessary to prevent the crater formation.

The acrylic additive, which is a linear acrylic polymer with unbranched side chains, has a lower surface tension than the main resin in the coating formulation. Consequently, it will migrate to the surface with the polar backbone dissolved in the polymer, while the alkyl chains will tend to escape into the surrounding environment, arriving at the surface and thus equalizing its chemical potential [67, 92]. A certain minimum concentration of the linear side chains of the acrylic additive per unit surface area is necessary for equalization of the surface tension across the film surface, thus minimizing the surface tension gradient. In this way, the flow of the material on the surface is reduced (*Fink-Jensen* equation), resulting in a decreased cratering tendency. The further addition of acrylic additive has no effect upon the cratering.

The difference in surface tension is supposed to be **reason for contamination effects** with, e.g. acrylic impurities in polyester formulations. According to *Kerssen* [93] this difference can be adjusted by appropriate additives, thus enhancing the compatibility of acrylics with other powder coating systems.

4.7.3 Surface tension and film leveling

According to the equation of *Nix* and *Dodge* (Equation 4.22), which gives a quantitative description of the coalescence of powder particles during their fusion into a continuous layer, and the equation of *Orchard* (Equation 4.23), describing the leveling of the film in the next stage, **high surface tension promotes better flow** of the powder coatings during the film forming process. On the other hand, it was shown that **low surface tension facilitates the wetting properties**. This lowers the yield value and viscosity of the coating, thus improving the flow of the material and preventing the formation of craters during curing. It is obvious that in order to obtain a powder coating with good flow and without surface defects, a compromise must be found to combine these two contradictory requirements.

Gabriel [87] and *de Lange* [67] have suggested that a good balance between the two opposite influences of the surface tension can be obtained by a combination of two different types of additives - one decreasing and the other increasing the surface tension of the coating. It has been found that additives being copolymers of acrylic esters (ethyl acrylate/2-ethyl hexyl acrylate, for example) with a number average molecular weight of ca. 8000 reduce the contact angle, thus improving the wetting. The other types of additives which increase the contact angle are epoxidized soybean fatty acids or abietyl alcohol. These should improve the flow by raising the surface tension [67]. In this way it seems that no advantage can be obtained because one additive nullifies the effect of the other, unless a higher freedom is expected when only the surface tension is adjusted. In fact, this is only the case for certain fixed temperatures. The curing of the powder coating is not a completely isothermal process. The object temperature varies between the normal ambient, when the article enters the stove, and the stoving temperature at the end of the curing process. The **surface tension of the coating decreases linearly with increasing temperature**. The addition of an acrylic based additive will improve the wetting of the pigments during the dispersion and the wetting of the surface, thus avoiding crater formation. However, since the surface tension, once lowered, undergoes additional decrease during curing due to the temperature increase, the leveling properties of the coating are badly affected. An ideal powder coating should have a low enough surface tension to facilitate pigment dispersion and substrate wetting, which will remain constant during the curing, thus not affecting the leveling of the film.

Gabriel [87] has found that the slopes of the straight lines representing the change of the surface tension with temperature for pure bisphenol A epoxy resin, additive which decreases the surface tension (additive O), additive increasing the surface tension (additive P), epoxy resin with additive O and epoxy resin with additive P are not the same. *Figure 4.8* represents the plots of the surface tension for the pure resin and the additives against temperature, as well as blends of the resin with 0.4 % additives.

It can be seen that neither the resin alone, nor the resin with additive O only, or additive P only, behave in a desirable way. The resin without additives has a high surface tension affecting the pigment and surface wetting, which decreases rapidly with increasing temperature. The addition of additive O improves the wetting properties, but affects the leveling at a late stage of curing. Addition of additive P only has as a consequence an unacceptably high surface tension over the whole temperature range, promoting cratering problems. Fortunately, **the combination of both additives in equal concentrations gives a good compromise**, lowering the slope of the straight line and bringing it closer to an ideal horizontal position.

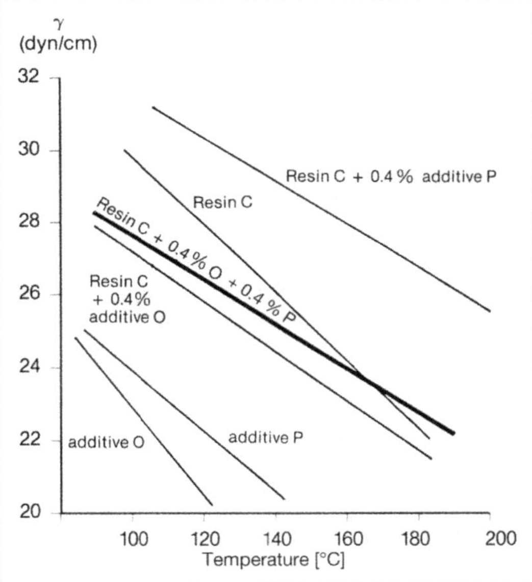

Figure 4.8: Surface tension versus temperature for bisphenol A epoxy resin, the same resin with 0.4 % additive O, 0.4 % additive P, 0.4 % additive O + 0.4 % additive P, and the pure additives [87]

source: Oil and Colour Chemist's Association

De Lange comes to the same conclusions, summarizing them in the following four statements:

- Epoxy resin without additives leads to cratering and the orange peel effect
- Epoxy resin + type P additive (increasing surface tension) leads to cratering but no orange peel effect
- Epoxy resin + type O additive (decreasing surface tension) leads to the orange peel effect but no cratering
- Epoxy resin + both types of additives leads to good flow without craters and the orange peel effect.

He suggests that the term 'flow control agent', which is commonly used in practice, is misleading and should be replaced by 'flow promoters' for additives increasing surface tension and 'wetting promoters' for those decreasing surface tension.

The effect of different additives on flow and leveling properties of powder coatings **depends very much on the type of binder**. The results obtained with epoxy resins should not be automatically transferred to other systems. In the case of polyester powder coatings, the addition of type P additives has completely the opposite effect, promoting even more orange peel effect. Therefore, in this coating system it is common practice to use acrylic types of additives only in order to prevent cratering.

Grundke [94] from the Institute of Polymer Research in Dresden used recently an axis-symmetrical drop-shape technique to determine at the same time both the temperature dependence of the density and the surface tension of polymeric melts from a single sessile drop. This way the effect of different leveling additives on the surface tension were evaluated.

4.8 Pigment volume concentration and pigment dispersion

The ideal liquid has a constant viscosity for a given temperature for any shear rate. This is the definition of a Newtonian liquid or Newtonian flow behavior. A linear plot of the shear stress against the shear rate is a straight line passing through the origin with the slope representing viscosity.

Coatings usually exhibit a different behavior than Newtonian liquids. For many of them, and especially the pigmented coatings, **a certain minimum shear stress must be exceeded before the material begins to flow**. This non-Newtonian flow behavior is known as plastic flow, and the liquids exhibiting it are commonly called *Bingham* liquids.

A linear plot of the shear stress against shear rate for plastic flow results in a straight line which does not pass through the origin. The slope of the line represents the plastic viscosity, and the intercept on the Y axis is the minimum shear stress that has to be exceeded to initiate flow. This minimum shear stress (Y) is called the **"yield value" and is a typical characteristic for powder coatings**.

The reason for the non-Newtonian behavior of powder coatings exhibiting plastic flow is the interaction forces between the pigment particles. The yield value results from the work necessary to break the reversible physical network structure built up by the interparticle pigment interaction. This interaction increases with increasing pigment volume concentration, resulting in higher yield values and higher viscosities. Consequently, it will affect the flow and leveling properties of the powder coatings.

The role of the yield value in determining the extent of flow of pigmented epoxy powder coatings has been examined by *Stachowiak* [95], who has found that for excellent flow **the yield value of the powder coatings should not exceed 3 Pa**. A rough classification made by him could serve as a guideline for the expected flow behavior of the powder coatings dependent on the yield values:

Yield value	**Flow properties**
0 to 3 Pa	Very good to excellent
3 to 7 Pa	Moderate
7 to 50 Pa	Poor
>50 Pa	Very poor

Figure 4.9 represents results reported by *de Lange* [67], who measured the yield values and **viscosities for different powder coatings** as a function of the pigment volume concentration (titanium dioxide) at various temperatures, for three different powder coatings: epoxy/DICY, epoxy/polyester (50/50) and polyester/TGIC (93/7). A dramatic increase of yield value (and viscosity!) is quite obvious at pvc levels of approximately 20 %.

It has already been mentioned that powder coatings with excellent flow should not have yield values that exceed 3 Pa. Since the usual pvc levels in powder coatings are about 15 %, from the intercept between the curves representing the yield values as a function of pvc and the line Y = 3 Pa (*Figure 4.9*), *de Lange* draws the conclusion that for good flow epoxy powders should be cured at 160 °C, polyester/epoxy hybrids at 180 °C or higher and TGIC/polyester powder coatings at a minimum temperature of 200 °C [67].

The interaction between the pigment particles and the binder in powder coatings has been extensively studied by *Ghiljadow* [96] and others. Three acrylic resins with carboxyl functionality (RCOOH), hydroxyl functionality (ROH) and without functional groups (RH) were used in combination with a rutile type of titanium dioxide treated with aluminum, zinc and silicon salts. The polymer/pigment interaction was estimated by IR spectroscopy.

The spectrum of the rutile pigment taken after mixing in melt with RCOOH acrylic resin shows an absorption band at 1550 to 1580 cm^{-1} typical for the COO^--ion, indicative of the formation of a strong adsorption band (chemisorption) between the polymer and the pigment surface. This absorption band does not exist in the spectrum of the pigment itself, but it is still present, and even more pronounced, after separating the pigments from the binder by washing the pigment particles with a solvent in a Soxhlet apparatus. The same results were obtained with a blend containing 75 % non-functional acrylic resin and 25 % carboxyl functional acrylic resin. This indicates the preferential interaction between the pigment particles and the carboxyl functional polymer.

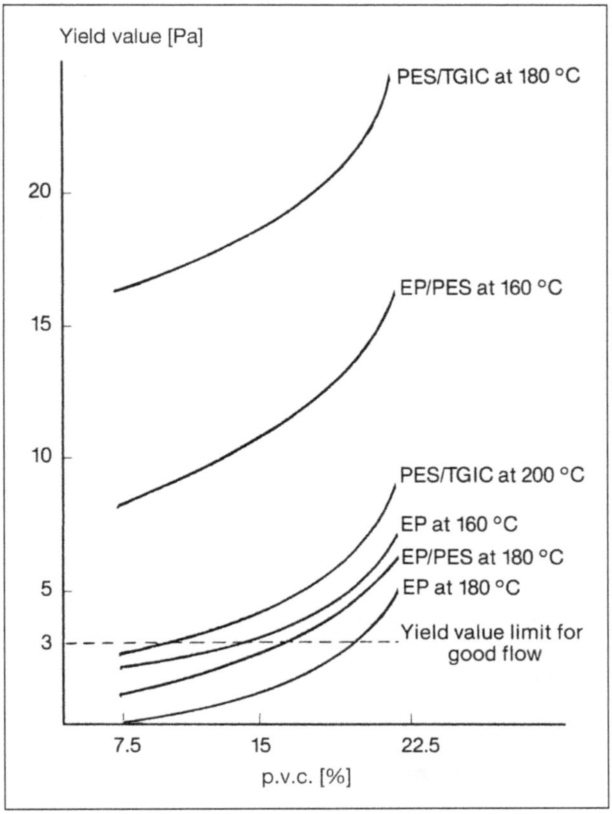

Figure 4.9: Influence of the pigment volume concentration and stoving temperature on yield values for three powder coatings [67] *source: Federation of Societies for Coatings Technology*

The analogous experiments with ROH acrylic resin also showed that the resin cannot be washed out from the pigment surface with a solvent, due to the strong hydrogen bond formation between the hydroxyl groups from the polymer and the surface hydroxyl groups of the pigment. Although the absorption band typical for the C=0 group remains after washing with a solvent in the case of the ROH acrylic resin, it almost disappears when the RH non-functional acrylic resin is treated in the same way. These results show that a considerable part of the polymers containing functional groups capable of interaction with the pigment forming salt-type compounds or hydrogen bonds is irreversibly adsorbed on the pigment surface.

The strong pigment/polymer interaction will result in the formation of a thin polymer layer around the particles, reducing in this way the interparticle interaction. This should lead to a reduced melt viscosity of the powder.

The influence of the addition of the RCOOH polymer modifier is quite remarkable. The activation energy of viscous flow is reduced even below the original value for the unpigmented non-functional acrylic resin, due to the formation of an adsorption layer on the pigment. This indicates the **importance of good wetting of the particles during dispersion of the pigments** in the extruder. *Nix* and *Dodge* [63] have shown that the degree of dispersion of the pigment in an acrylic powder coating has a considerable influence on the orange peel effect typical for these coatings. The worse flow of the coating with **poorly dispersed pigment is accompanied by much higher viscosities**, which in extreme cases for the same coating formulation can vary by a factor of 10. This could be expected, since in the case when the pigments are not completely wetted by the binder, the intermolecular attraction forces between the pigment particles will build up reversible internal structures immobilizing the system.

Schellenberger, Thometzek et al. [97] investigated the optical effects of pearlescent pigments in β-hydroxyalkylamide based polyester powder coatings. For the optical effects the pigments close to the surface play a decisive role. The hardener had a positive effect on the desired horizontally orientation of these pigments. This effect might be attributed to the differences of the wetting properties between the monomeric hardener and the polymeric resin.

In a recent article [98] *Dullaert* et al. explored the influence of the pigment loading on the gloss of powder coatings. They found that the combination of short wave and long way patterns determine the optical properties of the coating. The pigment loading has an effect on the long wave roughness, in particular over 45 wt-%, whereas the short wave roughness is predominantly influenced by the pigment particle size.

4.9 Particle size

Particle size is an important parameter of the powder coatings and is often expressed by two different terms: micrometres (μm) and mesh screen number. On the other hand, the measures for the film thickness are reported in micrometres, Hegman gauge and mils. For example, 1 mil film corresponds to a thickness of 25.4 μm and 150 mesh screen allows spherical particles of up to 100 μm to pass during the sieving operation.

Because of the specific nature of the film forming process, **the particle size of powder coatings plays an important role in determining the leveling properties** of the coating. It is difficult to expect good leveling of a coating consisting of coarse large particles which have to be sintered in a uniform continuous film during stoving. Higher particle packing densities will promote the formation of films with less voids, pinholes, orange peel and shrinkage during curing. Supposing spherical particles to be of uniform size, the voids amount to 48 % when particles are packed in a cubical array or 26 % when the package has a tetrahedral or pyramidal structure. Even when it is possible to freely choose the best possible packing of particles with a particular particle size, the voids will still amount to 15 % [99].

During the application by electrostatic spraying, smaller particles will be charged higher with respect to the unit weight than larger ones. Because of the Faraday cage effect which is specially emphasized in corona spraying guns, the smaller particles will penetrate less into the openings on the article to be coated, thus causing non-uniform distribution of the particles across the coated surface. On the other hand, due to the larger mass and lower velocity, the larger particles are more likely to fall out of the spray cloud. In the longer term, in continuous operation with recirculation of the powder coating this means **a permanent change of the particle size distribution with time**.

Crowley et al. [99] report that a probe introduced into freshly applied powder causes particle movement over a hundred particles distant from the point of the probe, concluding that the charged particles behave as individual dipoles attracted to each other in chains, circles and loose agglomerates.

Two stages can be distinguished during film formation of the powder coating. In the first stage a continuous film that flows is formed through coalescence of the molten particles, transforming the irregular film surface to a smooth one in the second stage. Below a thickness of 100 μm the gravitational forces play no significant role in the film formation [67]. It has already been pointed out that the major driving force that causes flow is the surface tension. The time required for two particles to coalesce can be expressed by the already mentioned equation of *Nix* and *Dodge* [63]:

Equation 4.32

$$t = \frac{k\eta R}{\gamma}$$

where

t = time necessary for two particles to coalesce
k = constant
η = viscosity of the powder coating
R = radius of the powder particle
γ = surface tension of the coating

Equation 4.32 indicates that **finer powder particles will improve the coalescence speed**, which means more time is available for flow of the formed continuous film in the second stage. The end result will be a coating with **improved flow properties**.

To obtain a powder coating with good flow and gloss when applied at low film thicknesses down to 25 µm, the particles with a radius larger than 50 µm should be removed by proper particle classification.

Deane [100] describes AkzoNobel's patented "**Particle Management Technology**", a manufacturing method, which allows supply of powders that have significantly different and improved handling and application characteristics. Higher first-pass transfer efficiency, reduction of film thickness from 45 to 50 µm to a true thin film of 30 µm with good flow, and more uniform coating of complex shapes can bring the user **savings of at least 16 % on applied coating costs**. In addition it provides better penetration, increased line speed, lower maintenance costs and less clean-up/changeover time.

4.10 Stoving temperature profile

Melt viscosity of the resin as a function of temperature can be expressed very well by the WLF equation. This equation is valid only for thermoplasts. In the case of thermoreactive systems, which most powders are, the situation is more complicated. Immediately after melting the coating viscosity decreases to a certain value with an increase in temperature, but at the same time the network formation that follows causes a dramatic increase in the viscosity.

It is very difficult to derive a mathematical expression that will quantitatively describe the viscosity profile during curing. **Viscosity build-up depends on the kinetics of the curing reaction**. The *Arrhenius* equation can be used to calculate the degree of conversion with time in a dynamic temperature profile of the coated surface, but it will be very difficult to correlate these results with their implication to the viscosity of the system. The situation is even more complicated because the viscosity decreases with an increase in temperature. The viscosity profile for a certain temperature pattern during curing will be the result of several factors combined in a rather complicated manner for mathematical interpretation.

The approach of *Hannon, Rhum* and *Wissbrun* [70] (Chapter 4.4.2) which takes into consideration the viscosity change in non-isothermal curing of the powder coating due to the change of temperature with time and the network formation during curing can be used to simulate the melt viscosity change with time, performing numerical integration with a computer.

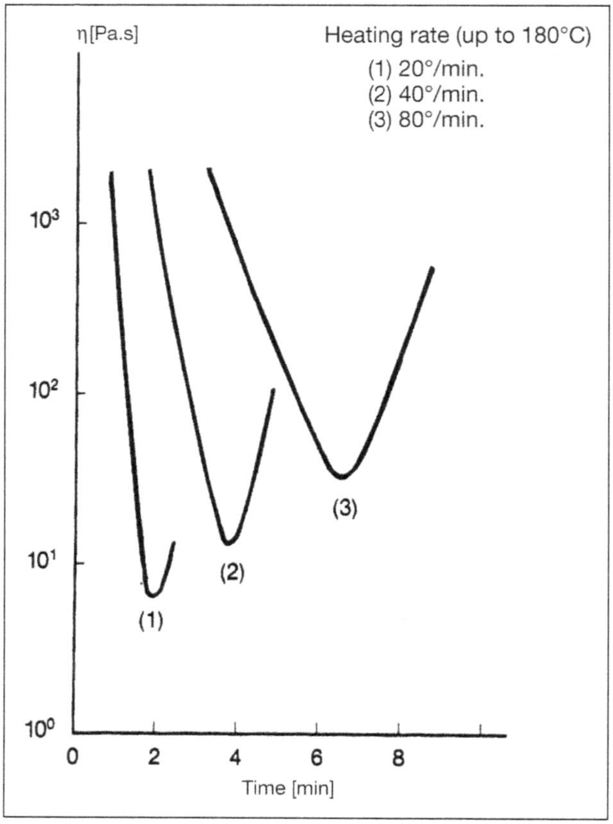

Figure 4.10: Viscosity profiles of epoxy powder coatings cured at different heating rates [67]

source: Federation of Societies for Coatings Technology

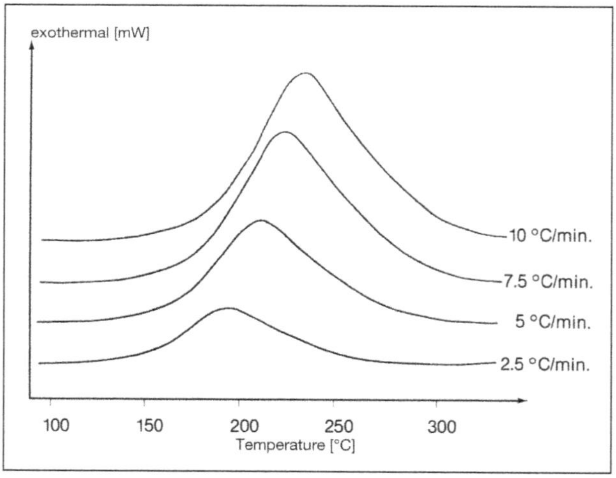

Figure 4.11: Effect of heating rate on the cure behavior of the DGEBA/DDS system [101]

source: Society of Plastics Engineers

The experimental results obtained by *de Lange* [67], and *Ghijsels* and *Klaren* [101] referring to the viscosity profile of epoxy powder coatings cured with different heating rates showed that they attained lower minimum viscosities and consequently **better flow by the use of higher heating rates**. Viscosity profiles obtained at three different heating rates are presented in *Figure 4.10*.

Better flow is obtained when powder coatings are applied on preheated substrates or on non-massive articles which have a low heat capacity.

Figure 4.11 shows the DSC curing exotherms at four different heating rates (2.5, 5.0, 7.5 and 10.0 °C/min) for a composition containing stoichiometrical amounts of diglycidyl ether of bisphenol A (DGEBA) and diamino diphenyl sulphone (DDS) reported by *Galy*, *Sabra* and *Pascault* [102]. It can be seen that the peak of the exotherms shifts to higher temperatures with an increase in the heating rate. This results in lower viscosity of the coating obtained during the stoving process.

Walz and *Kraft* [72] have established that for good flow the time necessary to reach a **viscosity of 50,000 mPa · s** at a certain curing temperature should be **longer than 100 s**. The graphs in *Figure 4.10* show that even at heating rates higher than those normally used in practice (10 to 30 °C/min), this time of 100 s is not enough to reach the minimum viscosity during curing. In other words, high heating rates are favorable for obtaining good leveling of the powder coating during film formation.

4.11 References

[1] *Krevelen, D.W. van,* Properties of Polymers, Elsevier Sci. Publ. Co., Amsterdam 1976, p. 19

[2] *Govarkier, V.R.,* Polymer Science, John Wiley & Sons, New York, 1976, p. 98. 3. Gordon, M., Proc. Roy. Soc. London, A268, 240, 1962

[4] *Gordon, M., and Malcolm, G.N.,* Proc. Roy. Soc. London, A295, 29, 1966

[5] *Dusek, K.,* Adv. Polym. Sci., 78, 1, 1986

[6] *Tiemersma-Thoone, G.P.J.M., Scholtens, B.J.R., and Dusek, K.,* 1st International Conference on Industrial and Applied Mathematics (ICIAM'87), CWI track 36, 1987, p. 295

[7] *Scholtens, B.J.R., Linde, R. van der, and Tiemersma-Thoone, P.J.M.,* DSM Resins-Powder Coating Resins Symposium, Noordwijk, May 1988

[8] *Flory, P.J.,* Principles of Polymer Chemistry, Cornell University Press, Ithaca, N.Y., 1953

[9] *Stockmayer, W.H.,* J. Polym. Sei., IX (1) 69, 1952

[10] *Munari, A.,* La Chimica e L'Industria, V67 (N12) 694, 1985

[11] *Misev, T., and Belder, E.,* DSM Resins-Powder Coating Resins Symposium, Noordwijk, May 1988

[12] *Collins, A.E., Bares, J., and Billmeyer Jr, F.W.,* Experiments in Polymer Science, John Wiley & Sons, New York, 1973, p. 216

[13] *Wicks Jr, Z.W.,* J. Coat. Technol., 58, 743, 23, 1986

[14] *Sperling, L.H.,* Introduction to Physical Polymer Science, John Wiley & Sons, New York, 1986, p. 254

[15] *Eyring, H.,* J. Chem. Phys., 4, 283, 1936

[16] *Fox, T.G., and Flory, P.J.,* J. Appl. Phys., 21, 581, 1950

[17] *Fox, T.G., and Flory, P.J.,* J. Polym. Sci., 14, 315, 1954

[18] *Boyer, R.F.,* in Encyclopedia of Polymer Science and Technology, Suppl., Vol. 2, Ed. N. M. Bikales, Interscience, New York, 1977, p. 822

[19] *Katz, D., and Zervi, I.G.,* J. Polym. Sci., 46C, 139, 1974

[20] *Brandruys, J., and Immergut, E.G.* (Eds.) Polymer Handbook, Vol. III, 2nd ed., John Wiley & Sons, New York, 1975, p. 139

[21] Union Carbide Corp., US Pat. 3 397 254, 1968

[22] *Kurka, A.K., and Herbert, M.B.,* US Pat. 3 027 279, 1959

[23] *Williams, M.L., Landel, L.F., and Ferry, J.D.,* J. Am. Chem. Soc., 77, 3701, 1952

[24] *Ferry, J.D.,* in Viscoelastic Properties of Polymers, 3rd Ed., John Wiley, New York, 1980, pp. 280-8

[25] *Nielsen, L.E.,* in Polymer Rheology, Marsel Dekker, New York, 1977, p. 33

[26] *Paul, S.,* in Surface Coatings-Science and Technology, John Wiley & Sons, Chichester, 1985, p. 711

[27] *Misev, T., and Belder, E.,* J. Oil Colour Chem. Assoc., 72 (9) 363, 1989

[28] *Corcoran, E.M.,* J. Paint Technol., 41 (538) 635, 1969

[29] *Shimbo, M., Ochi, M., and Arai, K.,* J. Coat. Technol., 56 (713) 45, 1984

[30] *Shimbo, M., Ochi, M., and Arai, K.,* J. Coat. Technol., 57 (728) 93, 1.985

[31] *Perera, D., and Eynde, D.V.,* J. Coat. Technol., 53 (677) 39, 1981

[32] *Aronson, P.D.,* J. Oil and Colour Chemists' Assoc., 57, 66, 1974

[33] *Sato, K.,* Prog. Org. Coat, 8, 143, 1980

[34] *Perera, D., and Eynde, D.V.,* J. Coat. Technol., 59 (748) 55, 1987

[35] *Tagoshi, H., and Endo, T.,* J. Polym. Sci., Polym. Lett., C26, 77, 1988

[36] *Bailey, W.J., Sun, R.L., Katsuki, H., Endo, T., Iwama, H., Tsutshima, R., Saigou, K., and Bitritto, M.M.,* ACS, Symp. Ser., 59, 38, 1977

[37] *Perera, D., Greidanus, P., and Misev, T.,* unpublished results

[38] *Fox, G.T., and Loshaek, S.,* J. Polym. Sci., 15, 371, 1955

[39] *Cowie, J.M.G.,* Eur. Poly. J., 11, 297, 1975

[40] *Boyer, R.F.,* Macromolecules, 7, 142, 1974

[41] *Kumler, P.L., Keinath, S.E. and Boyer, R.F.,* J. Macromot. Sci., Phys., B13, 631, 1977

[42] *Holda, E.M.,* J. Paint Technol., 44 (570) 75, 1972

[43] *Bueche, F.,* Physical Properties of Polymers, Interscience Publishers, New York, 1962

[44] *Boyer, R.F.,* Rubber Chem. Technol., 36, 303, 1963

[45] *Davis, G.T., and Eby, R.K.,* J. Appl. Phys., 44, 4274, 1973

[46] *Schalaby, W.S.,* in Thermal Characterisation of Polymeric Materials, Ed. *E. A. Turi,* Academic Press, New York, 1981, p. 261

[47] *Toussaint, A.,* Prog. Org. Coat., 2, 273, 1973-4

[48] *Kwei, T.K.,* J. Polym. Sci., A3, 3299, 1964

[49] *Kumins, C.A., Roteman, J., and Ralle, C.J.,* J. Polym. Sci., Al, 541, 1963

[50] *Lipatov, Yu.S., and Geller, T.E.,* Vysokomol. Soedin., A9, 222, 1967

[51] *Gordon, M., and Taylor, J.S.,* J. Appl. Chem., 2, 493, 1952

[52] *Fox, T. G.,* Bull. Am. Phys. Soc., 1, 123, 1956

[53] *Jenckel, E., and Heusch, R.*, Kolloid-Z, 130, 89, 1953
[54] *Couchman, P.R.*, Macromolecules, 11, 1156, 1978
[55] *Couchman, P.R.*, Polym. Eng. Sci., 21, 377, 1981
[56] *Couchman, P.R.*, J. Mater. Sci., 15, 1680, 1980
[57] *Couchman, P.R., and Karasz, F.E.*, Macromolecules, 11, 117, 1978
[58] *Pochan, J.M., Beatty, C.L., and Hinman, D.F.*, Macromolecules, 11, 1156, 1977
[59] *Fox, T.G.*, J. Polym. Sci., C9, 35, 1965
[60] *Winkler, J., Klinke, E., and Dulog, L.*, J. Coat. Technol., 59, (754) 35, 1987
[61] *Patton, T.C.*, in Paint Flow and Pigment Dispersion, 2nd ed., John Wiley & Sons, New York, 1979, p. 377
[62] *Frenkel, J.*, J. Phys., 9 (5) 385, 1945
[63] *Nix, V.G., and Dodge, J.S.*, J. Paint. Technol., 45 (586) 59, 1973
[64] *Kuczynski, G.C., and Neuville, B.*, Conference on Sintering and Related Phenomena, Paris, 1960
[65] *Geguzin, J.E.*, Physik des Sinterns, Leipzig, 1973
[66] *Hann, H., and Jonach, B.*, Plaste and Kautschuk, 24 (2) 135, 1977
[67] *de Lange, P.G.*, J. Coat. Technol., 56 (717) 23, 1984
[68] *Orchard, S.E.*, Appl. Sci. Res., A 11, 451, 1962
[69] *Spitz, G.T.*, Am. Chem. Soc., Div. Org. Coat. Plast. Chem., 33, 502, 1973
[70] *Hannon, M.J., Rhum, D., and Wissbrun, K.F.*, J. Coat. Technol., 48 (621) 42, 1976
[71] *Nakamichi, T.*, Prog. Org. Coat., 8, 19, 1980
[72] *Walz, G., and Kraft, K.*, Proceedings of 4th International Conference on Organic Coatings Science and Technology, Athens, 1978, p. 56
[73] *Taylor, J.R., and Foster, H.*, J. Oil Colour Chem. Assoc., 54, 1030, 1971
[74] *Quach, A., and Hansen, C.M.*, J. Paint. Technol., 46 (592) 40, 1974
[75] *Güring, W.*, FARBE&LACK, 83, 270, 1977
[76] *Bosma, M.; Brinkhuis, R.; Rensen, E.; Watson, R.*, Progress in Organic Coatings, 72(1-2), 26-33, 2011
[77] *Greidanus, P.J.*, 12th Congress of the Federation of Scandinavian Paint and Yarnish Technologists, Helsinki, 8-11 May, 1988
[78] *Scholtens, B.J.R., Linde, R. van der, and Tiemersma-Thoone, P.L.M.*, DSM Resin-Powder Coating Resins Symposium, Noordwijk, May 1988
[79] *Henig, A., Jath, M., and Mohler, H.*, FARBE&LACK, 86 (4) 313, 1980
[80] *Linde, R. van der, and Belder, E.G.*, 7th International Conference on Organic Coatings Science and Technology, Athens, 1981, p. 359
[81] *Franiau, R.P.*, Paintindia, September 1987, p. 33
[82] *Pappas, S.P., and Hill, L.W.*, J. Coat. Technol., 53 (675) 43, 1981
[83] *Fink-Jensen, P.*, FARBE&LACK, 68, 155, 1962
[84] *Feyt, L., and Bauwin, H.*, J. Coat. Technol., 52 (664) 87, 1980
[85] *Nowak, E., Deutch, J.M., and Berker, A.N.*, J. Chem. Phys., 78 (1) 529, 1983
[86] *Gusman, S.*, Off Dig., May 1952, p. 296
[87] *Gabriel, S.*, J. Oil Col. Chem. Assoc., 58, 52, 1975
[88] *Hahn, F.J., and Steinhauer, S.*, J. Paint Technol., 47 (606) 54, 1974
[89] *Pesach, D., and Marmur, A.*, Langmuir, 3, 519, 1987
[90] *Kitayama, M., Azami, T., Miura, N., and Ogasawara, T.*, Trans. ISU, 24, 743, 1984
[91] *Hansen, C.M., and Pierce, P.E.*, Ind. Eng. Chem., Prod. Res. Develop., 13 (4) 218,1974
[92] *Schwartz, A.*, J. Polym. Sci., 12, 1195, 1974
[93] *Kerssen, G. W.*, JOT, Journal fuer Oberflaechentechnik, 44(2), 18-21, 2004
[94] *Grundke, K.*, Proceedings of the Annual Meeting of the Adhesion Society, 27th 4-6, 2004
[95] *Stachowiak, S.A.*, 7th International Conference on Organic Coatings Science and Technology, Athens, 1981, p. 469
[96] *Ghiljadow, S.K.*, XII FATIPEC-Congress Book, 1974, Garmisch-Partenkirchen, p. 617
[97] *Schellenberger, S.; Entenmann M.; Hennemann, A.; Thometzek, P.*, Farbe + Lack, 113(4), 130-135, 2007
[98] *Dullaert, K.; Steeman, P.; Bolks, J.*, Progress in Organic Coatings, 70(4), 205-212, 2011
[99] *Crowley, L.D., Teague, G.S., Curtis, L.G., Foulk, RG., and Ball, F.M.*, J. Paint. Technol., 44 (571) 56, 1972
[100] *Deane, R.*, Congress Papers PCE 2000, Amsterdam, Jan. 2000, p. 331, or FARBE&LACK, 107, (7) 46, 2001
[101] *Ghijsels, A., and Klaren, C.H.J.*, XII FATIPEC Congress Book, Garmisch-Partenkirchen, 12-18 May, p. 269, 1974
[102] *Galy, J., Sabra, A., and Pascault, J.P.*, Polym. Eng. Sci., 26 (21) 1514, 1986

5 Technology of production of powder coatings

With respect to the manufacturing process, the production of powder coatings is much closer to the plastic rather than paint technologies. The manufacturing techniques are based on **equipment widely used in the plastic and fine powder industries**. Therefore, the design of the powder paint plant differs substantially from one producing wet coatings. The manufacture of powder coatings consists of different processes connected in a logical sequence as presented in *Figure 5.1*.

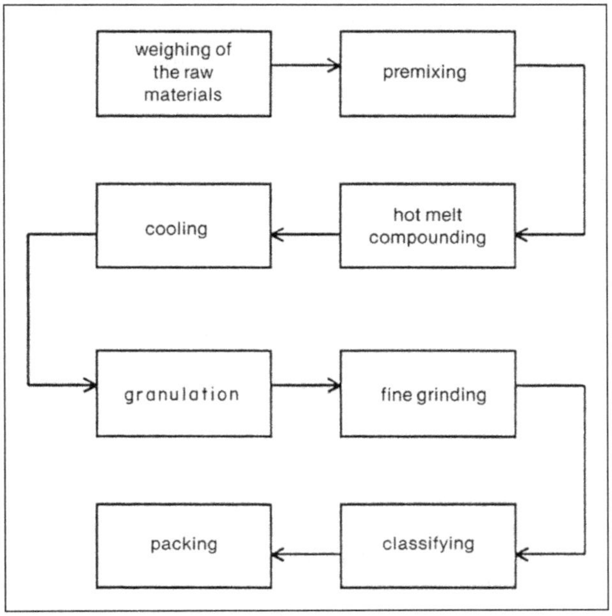

Figure 5.1: Schematic illustration of powder coating production procedures

Factors which influence the plant effectiveness are analyzed by *Scott* by means of a mathematical model whose algorithms incorporate data readily obtainable from production records [1]. The model includes personnel absence loss factors, plant unavailability loss factors, a premix preparation step, main plant preparation time and time for quality control.

The process of production can be discontinuous, which is very often the case with small plants. However, for plants with a high capacity, the individual process steps are usually integrated in a continuous form of production.

Figure 5.2 represents schematically the design of a modern plant for **continuous production of powder coatings** [2]. Separately weighed-out components are thoroughly mixed in the premixer. The resulting premix is then uniformly metered into the extruder. Before entering the grinding mill, the material leaving the extruder has to be cooled and granulated. The last step of the production process includes particle classification with recirculation of the oversized product back to the grinding mill. This lay-out of the plant for powder coating production can be modified depending on the capacity of the production line and the types of products. Plants with a small capacity operate in a completely discontinuous way. Regardless of the type of process (continuous or discontinuous) every production step involves specific equipment which will be described in details in the following sections.

Emmanouil Spyrou: Powder Coatings: Chemistry and Technology

ISBN: 978-3-86630-884-8

5.1 Premixing

Premixing of the ingredients of the powder coating prior to hot melt compounding is an operation whose importance must not be ignored. This step of the powder coating manufacture plays a decisive role in determining the performances of the coating. Insufficient premixing of the ingredients, especially of those which are employed in small concentrations such as additives and tinting pigments cannot be compensated for later on in the second stage by the hot melt extrusion process. The hot melt compounding is a continuous process and a uniform feed of homogeneous material through the entire extrusion step is a prerequisite for good quality products. Insufficient premixing in the first step may lead to inhomogeneous composition of the coating which is later manifested by inferior flow, bad mechanical properties, non-even color of the coated surface, reduced gloss, surface defects, etc.

Therefore, if a powder coating expert is asked how to make good powders, he will answer, that there are three important factors, namely:

- first premixing,
- second premixing and
- third premixing.

And one thing is quite certain: it is virtually impossible to correct poor premixing by good extrusion!

In principle the ingredients of the powder coating composition are solid materials. The best results in the premixing step are obtained when the particles of the constituents of the blend are of the same size. However, it is obvious that it can never be the case, since the difference in the particle size between the pigments and the resins is of several orders of magnitude. *Anand* [3] used the Monte Carlo method to simulate the mixing of particles. Segregation is observed in particular in cohesionless systems. Therefore enhancing the

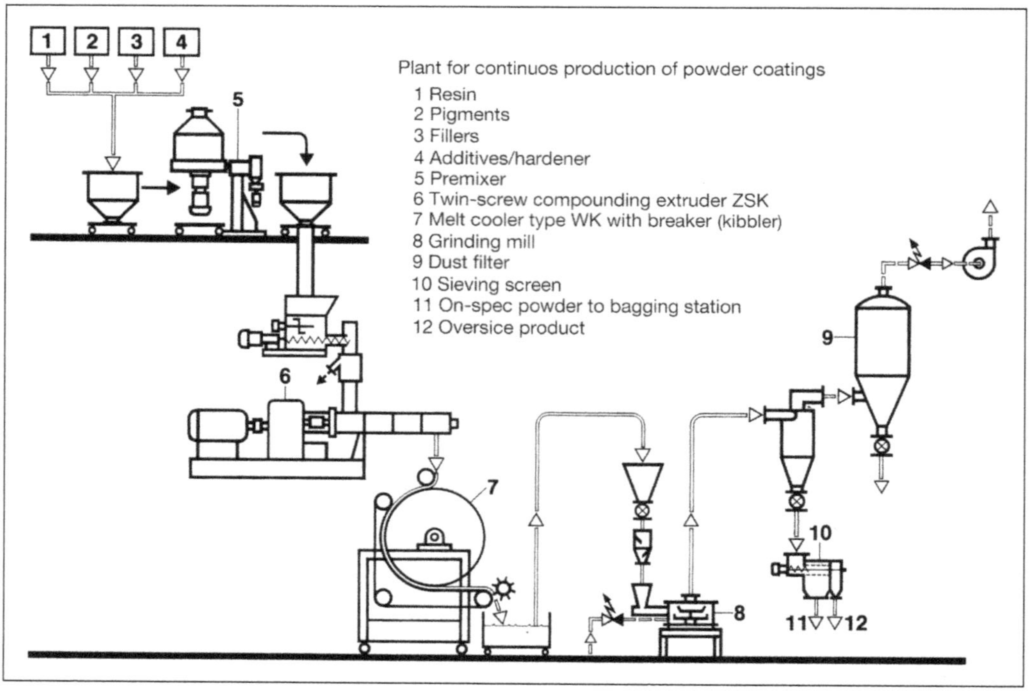

Figure 5.2: Plant for continuous production of powder coatings [2] *source: Coperion GmbH*

cohesive forces decreases the trend to segregation. The problem of incorporation of small quantities of additives such as flow agents or catalyst is successfully overcome by using the **master batch technique**.

It is important to run the mixing cycle so that the binder is not broken down excessively. Ideally, the binder should have a particle size of 2 to 4 mm when mixing is completed. If the particles are too small the premix will tend to fluidize during feeding, and if they are too large excessive power will be needed to flux the premix in the compounder [4].

The choice of premixing equipment can be made on the basis of mixing intensity, color change frequency and whether the process is continuous or a batch process. The variety of formulations in the production plant influences the type of premixer selected, because changeover and cleaning times become crucial in the production operation. For lines with a **regular color change high-speed mixers** or container mixers are preferred, due to their easy cleaning and short mixing time of 2 to 3 minutes. For lines with very **little color changes, ribbon blenders are more advantageous** due to lower investment and larger batch size [5].

Apart from the process requirements, to which a great amount of attention should naturally be paid, it is necessary to have an adequate mechanical design in order to prevent inconsistent performance or even breakdown of the mixing equipment. This is especially important when the mixer has to be stopped in mid process either accidentally or intentionally in order to check the progress of the operation. It is essential that the drive in such cases must be able to withstand repetitive starting under full load. The most preferable driving construction is with direct online electric motors which provide two to three times additional torque than normal on starting. When this cannot be tolerated, mixers with different direct driving arrangements such as slip rings, fluid couplings and centrifugal clutches or indirect arrangements in the form of V-belts are available. Different driving arrangements together with design details which provide safe loading of the mixer, vessel and agitator designs, construction of the mixer outlets and cleaning methods are discussed by *Craddock* [6]. Subsequent sections describe different types of mixing equipment typically used in the production of powder coatings.

5.1.1 Tumbler mixers

Two types of tumbler mixers are used in the production of powder coatings. The **simplest and cheapest example is the drum hoop mixer** which consists of a frame with a strapping that is placed over the top of a previously charged drum, and the whole is turned sideways and placed over rollers close to the floor. The time of mixing is normally 10 to 15 minutes. The size of the drum used in this type of mixer does not usually exceed 200 liters.

Drum hoop mixers are cheap machines suitable for small size discontinuous production characterized by a frequent color change. They are easy to operate machines with many advantages. There is no delay between charging and mixing separate batches since several drums may be used and prepared for mixing. There are no delays due to color change. The same drums may be used for storing the premix and then transporting it to the feed hopper of the extruder. After use the drums can be cleaned easily. However, they are **not very effective in mixing small amounts of tinting pigments**. Another disadvantage is the accumulation of fine pigments and extenders on the walls of the drums in layers with a thickness of several millimeters [7].

5.1.2 Double cone blenders

Much higher capacities of mixing space up to 4000 liters are available at the cone blenders which are very often obliquely angled with respect to the rotating shaft. During tumble blending of the ingredients, the spiral movement of the material due to the offset angle of the cones ensures a completely uniform distribution of the particles with minimum product breakdown in a totally closed environment. The blender can be supplied for operation under vacuum conditions or with a jacket for heating and/or cooling.

As in the case with drum hoop mixers these machines are **very easy to operate and maintain**. Because they do not contain baffles in the inside of the cones, cleaning the machine when the color is changed is not difficult. Easy discharging of the mixed material is provided by a simple diaphragm valve. The same problems of **low efficiency of mixing** when small amounts of tinting pigments are present and a build-up of finely divided pigments on the walls of the cones are not successfully solved.

5.1.3 Horizontal mixers

Horizontal mixers consist of two main parts: the horizontal body, which can be semi (so called "U" body) or full cylindrical in shape, and the horizontal rotating shaft carrying mixing blades. The design of the blades, the geometry and specific weight of the materials to be mixed determines the rotational speed of the shaft.

The **traditional "U" trough mixers** have a long narrow shape, and as the agitator rotates, mixing is achieved by moving the product from both ends to the opposite ends at the same time in a countercurrent manner. The longer the mixer the longer the agitator ribbons or the greater the number of elements where the agitator is of the interrupted ribbon design. The new so-called "HE" series of ribbon mixers produced by Gardner are characterized by a small length/width ratio, providing some special benefits. To complement the short aspect ratio of the mixer (length to width), a technically advanced double interrupted ribbon agitator has been developed which has only four helix segments, and two scraper elements, irrespective of mixer capacity. The combination of the agitator design and short mixing trough ensures substantial reductions in mixing cycles. The minimal number of helix elements, together with reduced mixing times result in less work being done on the product to achieve the desired mix, thus minimizing the product degradation in terms of particle size reduction. As will be seen, this is a rather important condition for good compounding efficiency of the single screw extruders. Having an easily accessible short wide trough and only four helix elements on the agitator, these mixers are easy to clean.

Standard HE series mixers are available with working capacities of 500 to 10,000 liters. The so-called "Turbulent Rapid Mixer" of Draiswerke GmbH (*Figure 5.3*) offers the possibility that the important units of the machine are exchangeable, thus facilitating subsequent conversion to suit a change of production. The machine is designed for operation in various speed ranges to permit optimum utilization for a given application.

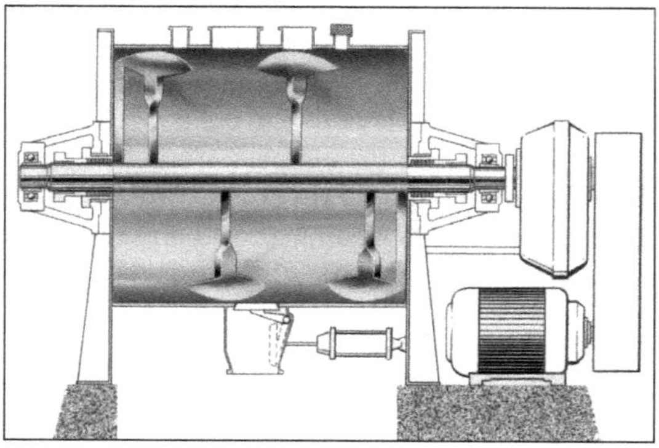

Figure 5.3: Schematic presentation of "Drais turbulent rapid mixer" *source: Draiswerke GmbH*

In the low or subcritical speed range the machine functions as a blender, with mixing elements pushing the material in the axial and radial directions, according to the countercurrent principle, to ensure intimate mixing of all the ingredients. The advantage of this system is that practically the entire internal space of the tank can be utilized and power requirements are small in relation to the

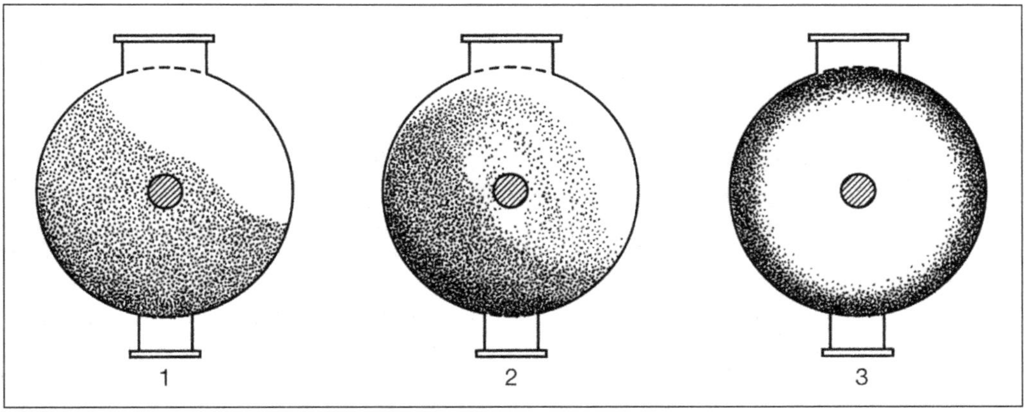

Figure 5.4: Schematics of mix patterns at different mixer speeds *source: Draiswerke GmbH*
1 Low-speed mixing with blades constantly pushing the material forward; 2 High-speed turbulent mixing; 3 Turbulent peripheral layer produced by mixing at extremely high speed

volume of feed. On the other hand, the mechanical stressing of the material is minimized, so the braking of the particles of the binder during the mixing operation is not excessive, ensuring in this way trouble-free feeding of the extruder in the next step of the powder coating production.

If the mixing elements rotate at high speed close to the so-called supercritical speed range, then the mixing pattern differs from that produced at low speed rotation of the shaft, as presented in *Figure 5.4*. The work performed by the mixing elements in circulating the ingredients within the mix is of secondary importance. Mixing efficiency is primarily dependent on the effect of particle impact. The individual rotating “T” elements plough through the mix alternatively, with varying angles of attack. Large chunks of material are thereby torn out of the mix and thrown into series of crossing paths, which in turn cause the material to be uniformly scattered over the entire length of the mixing chamber before falling back into the bulk of the mix. Although in **this way the mixing efficiency is much higher**, the machine should operate in this centrifugal mixing mode only when the particle disintegration does not cause problems with respect to the feeding of the extruder.

At very high speeds, the centrifugal action impels the mix towards the inner wall of the cylindrical mixing chamber, where it forms a turbulent annular layer of increasing density in the course of the outward travel. Interaction with the braking effect of friction at the wall of the mixing chamber results in large differences in particle velocity within the annular layer and thus intensive mixing turbulence. Consequently **uniform disintegration and blending of the constituents is achieved in a matter of seconds**.

Depending on the mode of operation, mixing elements with different design are employed. Drais turbulent mixers are produced in various sizes ranging from 2.5 to 50 liters for research applications and from 100 to 40,000 liters for production purposes.

An advantage of the horizontal mixing machines is the high efficiency of mixing. Dispersion is adequate for most powder blending purposes, including the incorporation of small amounts of pigments. The easy removal of the shaft from the mixing chamber allows good cleaning access to the cylinder and the blades.

5.1.4 High-speed blenders

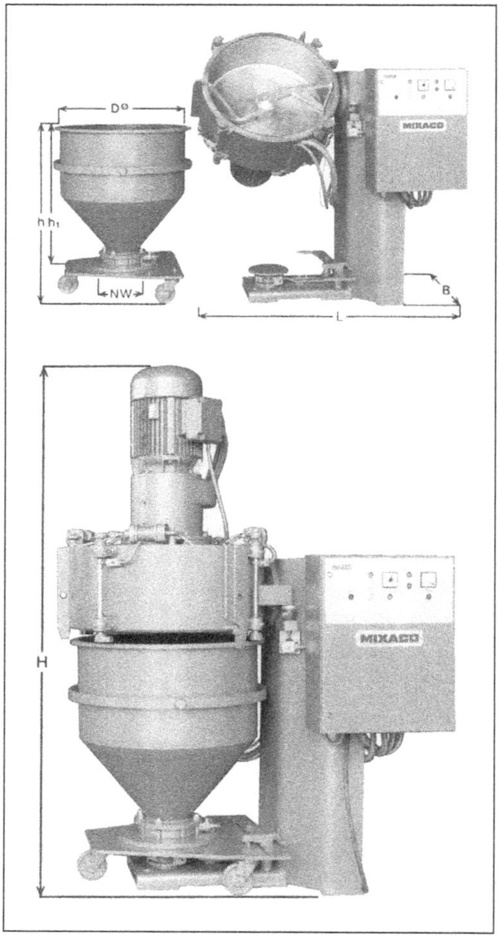

Figure 5.5: "Mixaco" container mixer

source: Dr. Herfeld GmbH & Co KG

High speed blenders are **very effective mixing equipment** for preparation of dry powder blends. A typical item of this type is the widely used "Mixaco" container mixer, presented in *Figure 5.5*. The mixer is designed for quick mixing, normally between 10 seconds and 10 minutes, of powders, pellets and viscous liquids with fillers, pigments or other additives. Its design allows the easily accessible mixing head to serve a number of batch containers, one at a time. The head of the mixing equipment can be easily and quickly cleaned when frequent color change is required. The mixing rotor operates at 5 m/s circumferential speed. The mixer is also produced in two-speed version with 5 and 10 m/s circumferential rotor speed.

The operational sequences of the premixing step are as follows: After adding the ingredients of the batch in the container it is pushed under the mixing head. The container is then pneumatically lifted and automatically locked to the mixer head by manual or pneumatically actuated clamps dependent upon the container size. By means of a gear motor the head/container assembly is inverted into the mixing position. In this position the head serves as a mixing chamber. The mixing rotor design, in its inverted operating position, centrifugally flings up the material against the mixing chamber and the container wall. As the material rises and reaches the conical area of the inverted container it falls back on to the mixing rotor. This flow pattern takes place in seconds and continues through the timed mixing cycle, thus creating a uniform mix. After the timed mixing cycle is completed the mixer is tilted down and the container is released. The **mixer is now available for another container**.

Several advantages of the high-speed mixers compared to the slow-motion blenders make them very attractive mixing equipment for powder coating manufacture. A distinguished feature of these machines is the **very short mixing time accompanied by extremely high mixing efficiency**. Even small amounts of tinting pigments can be easily incorporated into the premix, being homogeneously dispersed through the premix. Considerable savings can be achieved in this way, reducing the amount of the pigments used to obtain pastel shades, since the rapid mixing action helps the development of full color. Small amounts of liquids can be homogeneously dispersed, which eliminates the expensive master batching technique for incorporation of liquid additives. Finally, quick cleaning combined with the possibility of using several mixing containers with one mixing head and a short mixing time eliminates the problems of frequent color change, at least in the pre-blending step.

The high workload of these machines can cause overheating, although in most cases the time taken to attain a uniform premix is no longer than a few minutes. The excess heat generated by high-speed mixing is an operating parameter of the "High Intensity Mixers", produced by Thyssen Henschel. They are available in explosion proof versions equipped with cooled jacket and cooled mixing tools for the "bonding" of metallic pigments into the basic powder coating mix. The proportion added varies from 1 to 10 %, depending on the intended application. Prior to the bonding process, the basic powder coating mix is heated by friction to a temperature slightly lower than its softening point. Nitrogen is added to generate a protective inert gas atmosphere. The mixing tool speed is then reduced and the metallic pigment is added. During the bonding process interaction of the mixing tools and an adjustable deflector ensures optimum dispersion of the components, and bonding of the metal to the powder particle surface is completed in several minutes. The temperature in the mix is kept constant by the feed back control of both, the temperature of the container and the mixing tools and the motor speed. A programmable logic control, which responds to a temperature sensor at the deflector, monitors the entire mixing process and ensures its exact reproducibility. In order to prevent lumping, the finished metallic coating powder is discharged into a cooling mixer "KM", which is equipped with a special cooling tool. The gentle mixing action of the cooling mixer prevents separation of the metal powder from the basic powder coating mix.

5.1.5 Conical mixers

Conical mixers offer **low energy consumption to a high product output ratio** and make optimum use of fractional batch sizes without detracting from the mixing ability. They usually comprise a single or multicone static body with an orbital moving screw around the wall of the container. The main characteristic of the conical blenders is the three-dimensional mixing action as a combination of the rotational movement of the screw, causing horizontal rotation of the material, upward movement of the ingredients due to the helical shape of the screw, and vertical downward movement due to the gravitational forces which begin when the material reaches the top wide part of the conical container.

Conical mixers require a minimum floor area. As they are rather tall machines, in many cases they need a two-story building to be installed. They provide efficient mixing which is practically independent of the batch size. The cleaning of the conical mixers is not a very easy operation. Therefore, they are more suitable for a large uniform production rather than a production characterized by frequent color changes.

Hosokawa Mikron[8], the supplier of the "Vrieco-Nauta" conical batch mixer, offers a special conical mixer, the "Cyclomix". With "Nauta" mixer being particularly suited for free flowing powders and granules, "Cyclomix" mixer has been primarily developed for the mixing of (extremely) cohesive materials. For such types of products it is important to create an intensive mixing action by means of high shear forces.

The mixing principle of "Cyclomix" (presented in *Figure 5.6*) differs markedly from existing mixing techniques, owing to the specific interaction between mixing element and vessel wall.

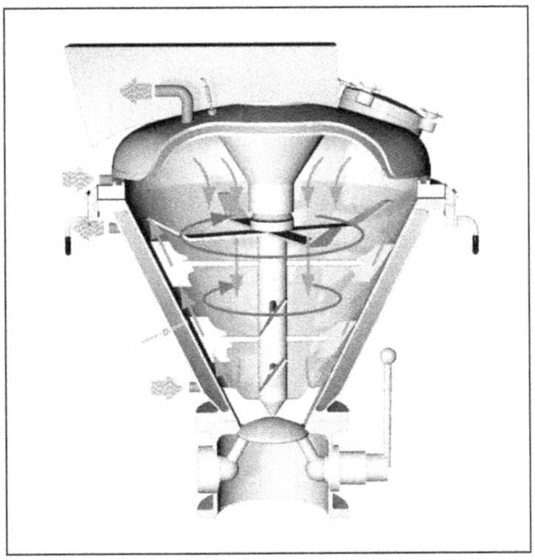

Figure 5.6: Mixing principle of "Cyclomix" conical mixer[8] source: Hosokawa Micron BV

The product is loaded into the conical mixing vessel from the top. The degree of filling can range between 30 and 100 % (dependent on the nature of the mixture). A centrally located high-speed rotating shaft is driven from the mixer cover, thus eliminating seals and bearings from the product zone. The shaft is fitted with paddle-shaped mixing elements, which rotate close to the inner vessel wall. Due to the high speed rotation of the paddles and the conical shape of the vessel, the product is forced from the bottom to the upper zone of the vessel. Upon reaching the top, the product flows downwards into the center of the vessel. This flow pattern results in fast "macro"-mixing. During the up-ward motion, the product particles are accelerated by the paddles and intensively mixed ("micro"-mixing) by friction with the wall of the vessel. The product temperature can be efficiently controlled by the excellent heat transfer between product and vessel wall. Special features of the "Cyclomix" mixer are:

- very intensive and fast mixing (mixing times from 30 seconds to 5 minutes),
- fully self-emptying vessel,
- no bearings or seals in the product zone,
- good control of the product temperature,
- wide range of applications.

The "Cyclomix" can also be used for "bonding" of metallic pigments.

5.2 Hot melt compounding of powder coatings

The **compounding of powder coatings involves several processes** which find a place in one single piece of equipment. During this step of the production the binder and crosslinker (not always) are molten and mixed on a molecular scale, the pigment agglomerates are broken to primary particles and homogeneously dispersed together with the additives in the molten binder/crosslinker mix and the surface of the pigment particle is wetted by the molten binder.

Three types of machines can be used in principle for performing the hot melt compounding. The main equipment for carrying out the compounding is the extruder, although simple machines such as Z-blades and heated twin-roll mils may be used. The energy necessary for melting the binder is supplied partly by heat transfer between the walls of the extruder and the material to be compounded and partly by the heat which is generated due to the high shear gradients in the machine.

The **dispersive mixing** which breaks down the pigment agglomerates is a process which requires high energy input. A relatively small amount of energy is required for the distributive mixing by which the primary pigment particles are homogeneously distributed throughout the molten binder.

The theory of deagglomeration of the pigment clusters during extrusion has been discussed by *Winkler, Klinke* and *Dulog* [9]. They point out that the efficiency of the dispersion process depends on the energy density present at the so-called effective extruder volume where the dispersion process takes place. *Franz and Bolt* have already indicated that for a given volume flow of the material through the extruder, the overall dissipated energy is directly proportional to the shear rate and the residence time [10].

The technical problems created by the new generation of fast curing powder coatings have almost completely excluded the Z-blade mixers from the machines used to produce contemporary powder coatings. The alternative spray drying technique starting from a liquid paint failed because of economical reasons. However, Z-blade mixers are still in use for **batchwise production of thermoplastic powder coatings** where the relatively long residence time at high temperature during production does not affect the rheological behavior of the product.

5.2.1 Batch compounding by Z-blade mixers

Z-blade mixers are steel heat-jacketed mixers which are charged with a solid or liquid component from the top and later discharge the homogeneous molten mass from the bottom. They are equipped with strong internal agitators that are shaped like the letter "Z".

Usually the solid resin is charged into the mixer and melted within several hours. Then the blades can be put into operation and the pigments, fillers and additives are charged without the curing agent. The components are processed at temperatures ranging from 110 to 130 °C for one to four hours depending on the dispersion efficiency that has to be obtained. In the next step the temperature is lowered as much as possible and the hardener is incorporated in as short a period as possible. Before discharge the temperature is increased quickly to 130 °C in order to reduce the viscosity for rapid run-out into cooling trays.

Although these machines are rather old and not widely used for production of powder coatings, they still possess some advantages. The pre-blending operation is completely avoided since the residence time in the machine is rather long. Liquid materials can be easily introduced without master batching techniques. It is easy to match the color of the coating by addition of tinting pigments. In principle larger amounts of pigments and fillers can be incorporated thus reducing the raw material costs. However, decisive disadvantages have contributed to the demise of this machine, which has now almost disappeared from powder coating plants. Power consumption of the machine per weight unit produced of powder coating is very high. Compared to the extruders, **Z-blade mixers have a much higher heat history** which has certainly had a bad impact in the case of thermosetting powder coatings. It is very difficult to produce high gloss powder coatings by this technique. In this respect the extruder compounding technique is superior. As with any batch process, lot-to-lot uniformity is a constant quality control problem. An additional increase of the production costs is caused by low production rates, difficult cleaning, which in many cases employs solvents and higher working power requirements.

5.2.2 Continuous compounding

Three types of machines are used in the production of powder coatings; **single screw extruders** with a rotating and oscillating screw (Buss "co-kneader"), self-cleaning co-rotating **twinscrew extruders** (Coperion, APV Baker Perkins, OMC, Prism, Theysohn and TSA extruders), and **planetary roller extruders** (Schalke and Entex types). All three extruder types operate at pressure-free conditions by having a number of openings in the cylinder wall through which the materials can be added to the mix. The system is also characterized by a great variety in screw design. Application of transporting, neutral and contra-transporting knead elements in the twin screw extruder and transport, half transport or knead elements (varying from each other in the number of elements, gaps in the elements and pins through the walls) in the Buss "co-kneader", or application of the continuous rolling process in the planetary roller extruder influences the mixing and kneading operations [11].

5.2.2.1 Buss "co-kneader"

Figure 5.7 represents schematically the inside of the kneading section of the Buss "co-kneader". In contrast to the other single-screw and twin-screw extruders, the screw of the Buss "co-kneader" executes **not only a radial movement but also an axial movement**. In this way the kneading flights positioned on the shaft knead the mixture axially against the three rows of stationary kneading pins in the barrel casing. The movement of the screw flights relative to the kneading pins produces a shearing gap in which the product is subjected to shearing with pronounced reorientation. In this way very high surface regeneration

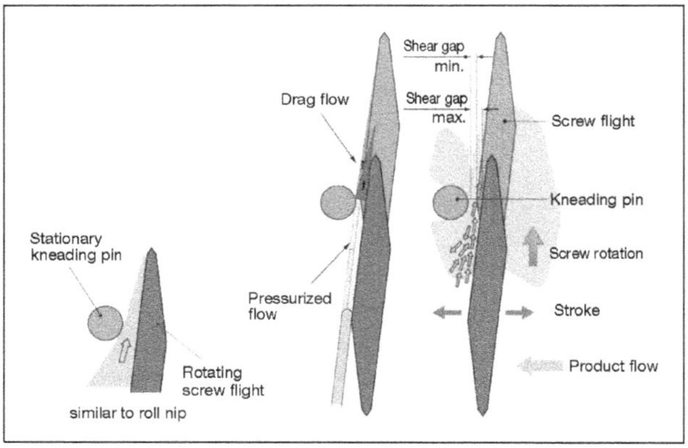

Figure 5.7: Schematic representation of Buss "co-kneader"

source: Buss AG

is achieved providing **homogeneous mixing despite extremely short processing length**. At the same time the process conditions are rather gentle due to balanced shearing and controllable temperature.

The process of producing powder coatings with the Buss "co-kneader" is schematically presented in *Figure 5.8* [12]. The container with the premix is placed on the metering unit and serves as a feed hopper for the metering unit. The metering unit ensures uniform, continuous, volumetric metering of the premix via the inlet screw into the extruder. To protect the downstream plant components, the premix is lead through a metal detector after the metering unit. For all kneader sizes, the barrel zones and the screw are temperature controlled with water. Thus, start-up of the plant can be accelerated by preheating the extruder to temperatures of 10 to 20 °C above the melting point of the resin to be processed. As soon as the molten material emerges from the die, the heat control is switched to direct cooling for removal of the heat. As a result, some energy introduced mechanically by dispersive mixing is

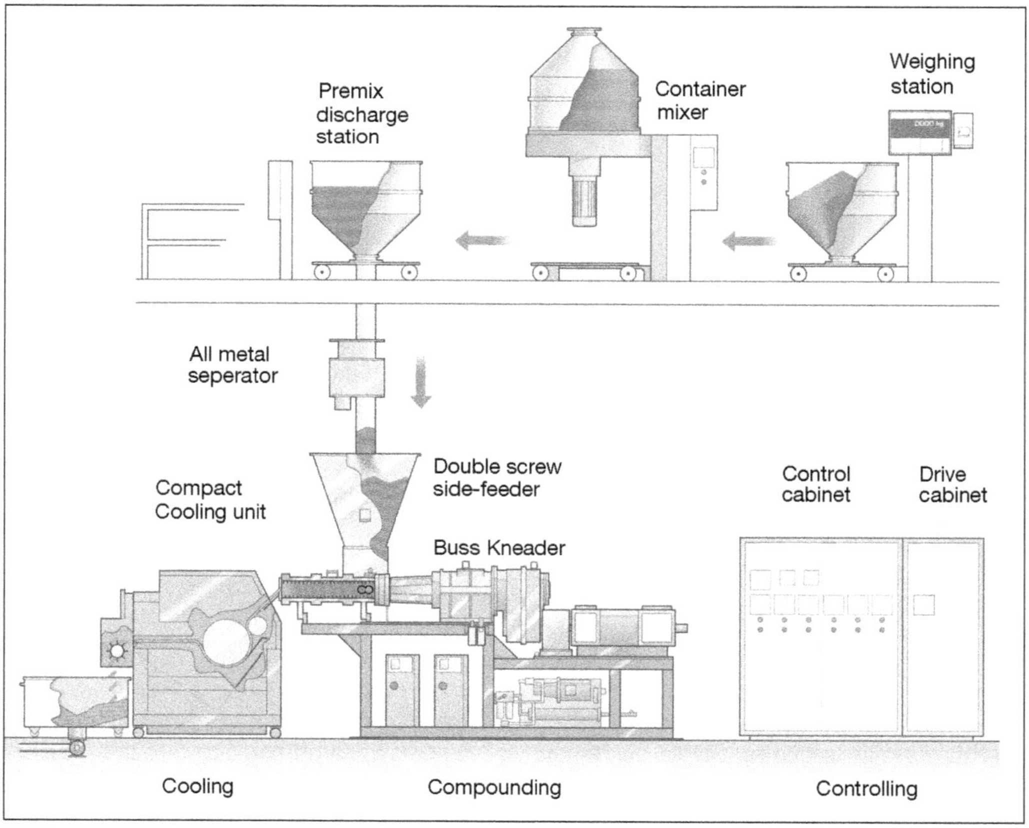

Figure 5.8: Buss concept for production of powder coatings [12]

source: Buss AG

removed. This in turn accelerates dispersive mixing and ensures a mild heat history typical for this type of machine. Temperature peaks, which can have an undesirable effect on the product quality, particularly in the case of rapidly curing powder coatings, are excluded by virtue of the principle of operation. Electronically controlled heating/cooling units keep the preselected temperature for each exactly constant.

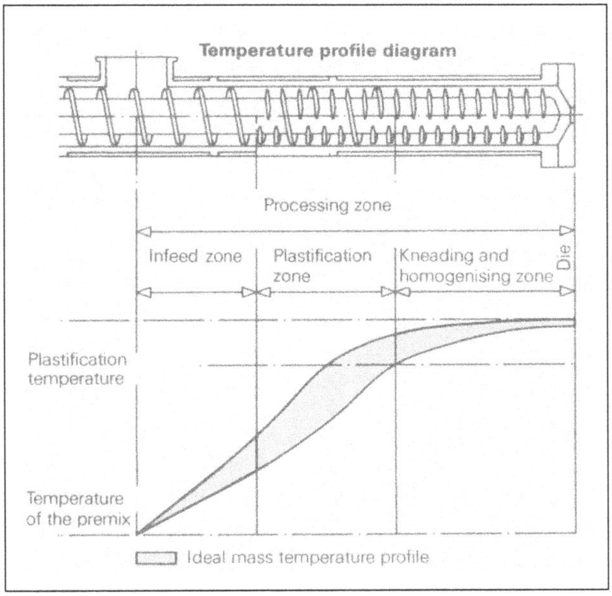

Figure 5.9: Temperature profile diagram of Buss "co-kneader"
source: Buss AG

In *Figure 5.9* the temperature profile diagram of the Buss "co-kneader" is presented. The barrel length can be divided into three zones; infeed zone, plastification zone and kneading and homogenizing zone. The graph indicates the ideal mass temperature profile.

The **self-purging principle** which is one of the distinguishing features of this machine enables very easy cleaning of the barrel by flushing the kneader with resin or cleaning compound when colors are changed. Inspection of the machine can be carried out easily and quickly, since the two halves of the barrel are convenient to open and the mixing and kneading shaft is freely accessible without any dismantling.

Figure 5.10: Interior of a Buss "co-kneader" *source: Buss AG*

The product emerging from the kneader is fed to a pair of cooling rolls. The water-cooled rolls spread out the cooled extrudate on the cooling belt. Additional cooling on the belt is provided by spraying with water underneath until the product becomes brittle before entering the crusher where it is cut in the form of flakes. The chips are collected in a container and conveyed to a sifter mill in the next production step. *Figure 5.10* shows the interior a Buss "co-kneader".

Buss "co-kneader" are produced in different sizes and capacities starting from laboratory equipment with a capacity of 5 kg/h up to machines with throughput up to 2500 kg/hour.

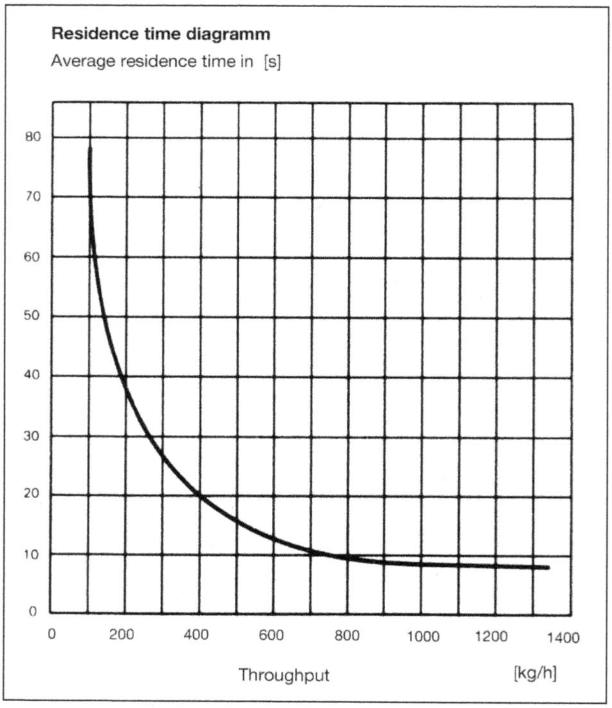

Figure 5.11: Residence time diagram of Buss "co-kneader"

Source: Buss AG

Figure 5.11 represents the residence time diagram of the material in the extruder casing. The type of the machine in the diagram (*Figure 5.11*) is the PLK, which is still used by many powder coating manufactures.

Harris [7] lists several **advantages and disadvantages** of the Buss "co-kneader". He counts the low power requirements, good temperature control, good compounding efficiency, short dwell time in the barrel, self-wiping action and easy way of cleaning among the strong points of this machine. The unsuitability for reprocessing finely divided powder and the reduced throughput when the premix entering the extruder is finely divided can be classed as disadvantages of the Buss "co-kneader".

These disadvantages could be eliminated in 1995 by the introduction of a new generation of Buss "PCS extruders", based on the same Buss "co-kneader" principle, but offering optimum feeding of all materials (including fines) from a double screw side feeder.

As the technology developed during the past ten years, the design of the Buss "co-kneader" has been steadily improved, to meet the requirements of the new powder coating chemistries. So it may happen that new formulations make it necessary to replace the "standard" screw (already suitable for a wide range of powders) by new screw geometry. If increased shear and dispersion are required, more kneading elements have to be used. If, on the other hand, a longer residence time is needed, elements must be provided, which move the product only a little bit or even in the opposite direction. Buss adapted to these new(er) chemistries and came with an extended version of their single screw extruder "PCS 55-11 L/D" at the PCE '98 exhibition in Amsterdam, followed by the complete series of "Extended PCS" extruders in 1999. The proposed design and features of this new range of Buss "PCS" extruders are shown clearly in the prototype, and are described by *Franz* and *Maier* [13]. Giving an average 40 % higher throughput for 20 % higher costs, this new design contains many improved features, such as a wider range of shear rates (provided by an extended process section and increased screw speed), further improved side feeder design, increased shear surface for improved dispersion, integrated and compact machine design, and minimum cleaning and maintenance down time by split barrel.

The "Extended PCS 55-11 L/D" has a maximum throughput rate of 480 kg/h. A further extension was obtained with the Buss "Modular Kneader System" (MKS), using a reciprocating screw with 3 screw flights and only intended for extrusion of thermoplasts.

Franz and *Maier* [13] describe some details of the newer design innovations, which include e.g. segmented barrel liner elements and segmented screw elements. They can be easily removed and replaced, when the elements become worn, without the need to dismantle the barrel or screw shaft.

Two methods may be used to heat the screw and barrel, depending on the type of product being extruded. One system consists of water, circulating in-side the screw, while the temperature is controlled by a separate unit. The other consists of a combined system with electrical heating and cooling water, the temperature being controlled directly inside the barrel. Accurate measurement of the internal working temperature of the newer generation Buss extruders can be assured by replacing one or more of the kneading pins by a thermocouple, which has the same profile as the kneading pins.

During extrusion energy is required to break down agglomerates of finely divided pigments and fillers. However, the energy applied by high impact forces may lead to localized overheating and to the formation of gel particles in the cured powder coating. To prevent this, high impact energy is here replaced by the application of "fatigue" forces, which apply moderate portions of energy, resulting in safe processing, even for rapid and low temperature curing powder coatings. Such a so-called "micro-mixing" efficiency provides a certain inter-molecular homogeneity, which is essential for uniform curing and thin film application of powder coatings.

Heat transfer efficiency within the extruder barrel is also a point of attention by *Franz* and *Maier*. The formation of heat-insulating layers at the wall can greatly reduce heat transfer efficiency, but the self-wiping and close clearance tolerances of the new generation Buss extruders reduce these layers very effectively. In addition, wall thicknesses of the barrel liners and the cooled screw core are kept to a minimum, which reduces the resistance to heat transfer.

Control of the reactivity of thermosetting powder coatings during extrusion is very important. Therefore, it is essential to ensure short residence times, a narrow residence time distribution, and a high self-cleaning efficiency. These aspects of the new-generation Buss extruder designing have been confirmed by the use of radioactive tracers.

At the ECS 2011 exhibition in Nuremberg, Buss launched their single-screw "PCS 70-11L/D", which is a further improved extended version of the proven Buss machines within the powder coating industry. One feature on the new machine generation has been the ability to run the machine at **higher speeds** (up to 650 rpm). This allows a **higher throughput** as well as better product quality due to higher sheer rates. In addition the design of the machine has been optimized to the needs for powder coating customers. E.g. the lubrication system for the gear box has been eliminated hence the maintenance of these machines is much easier.

5.2.2.2 Twin screw extruders

Most widely used are the Coperion (formerly known as Werner&Pfleiderer) "ZSK" extruders and the APV Baker extruders. The Coperion "ZSK" extruder (*Figure 5.12*) is based on a **co-rotating and closely intermeshing twin screw system**. The complete extruder consists of the drive and the processing part, all assembled on a common base frame to form a compact unit. The drive is made up of a variable speed a.c. motor, a mechanical safety clutch and a speed reduction and distributing gear with two co-rotating output shafts.

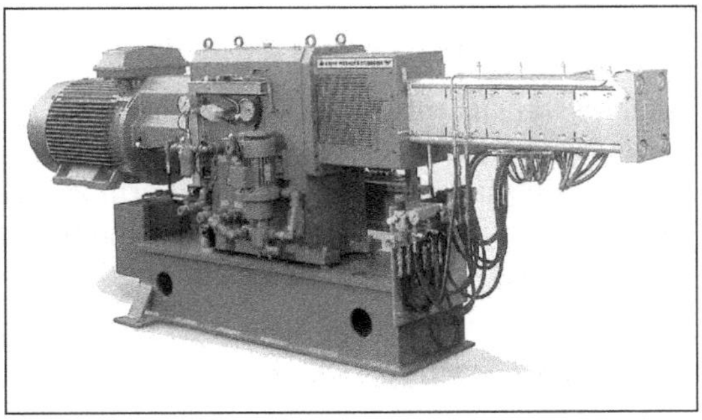

Figure 5.12: Twin screw extruder "ZSK 50 Mc" for powder coatings, with throughputs up to 600 kg/h, built around 1998 *source: Coperion GmbH*

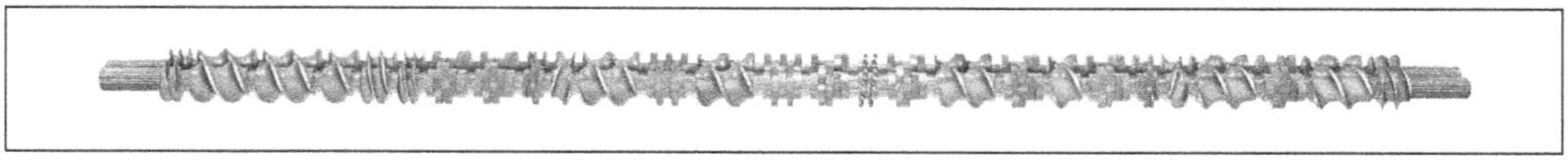

Figure 5.13: Building block principle of "ZSK" screws source: Coperion GmbH

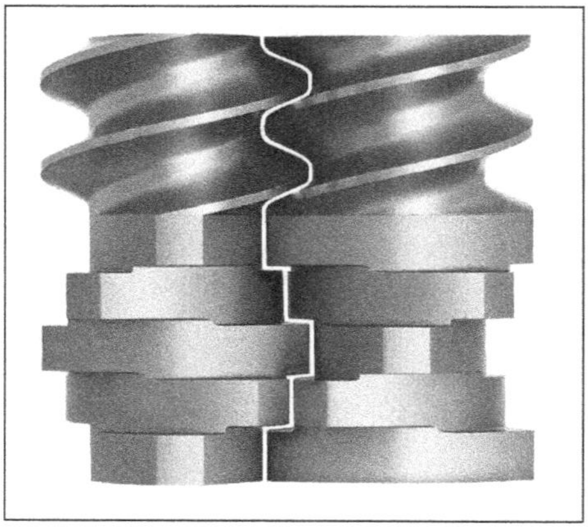

Figure 5.14: Sealing profile of the self-cleaning "ZSK" conveying and kneading elements source: Coperion GmbH

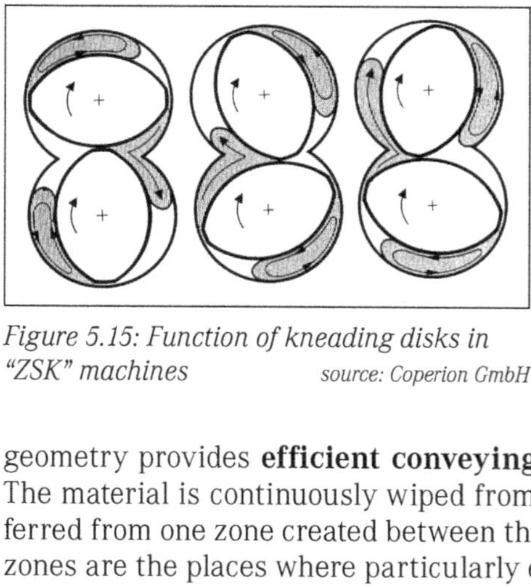

Figure 5.15: Function of kneading disks in "ZSK" machines source: Coperion GmbH

The processing section, i.e. the barrel and the two co-rotating screws is designed on the building block principle. The barrel consists of separate sections bolted together to provide the required processing length.

The processing section for machines up to 62 mm screw diameter is assembled by tie rods which allow easy maintenance by the option to exchange single barrels from the top without a complete dismantling of the process section. Another advantage is the use of heating cartridges instead of heater shells which saves energy and which can be used in combination with the tie rod design. The screws can be easily withdrawn from the barrel. With two sets of screws, the production can continue with the second set while the first is taken away from the extruder to be cleaned. In this way utilization of the whole plant significantly increases.

The screws as presented in *Figure 5.13* and *5.14* consist of a pair of one-piece screw shafts onto which a suitable sequence of deepflighted screw and kneading elements, each for specific effect, is arranged.

The cross-cut operational scheme (*Figure 5.15*) of the closed twin cylindrical barrel containing two screws with transverse lens-shaped kneading elements set at 90° shows how the screw geometry provides **efficient conveying, pressure build-up and self-cleaning action**. The material is continuously wiped from the agitator screw and the barrel wall and transferred from one zone created between the wall and the kneading element to another. These zones are the places where particularly effective dispersing and homogenizing take place.

An important feature of "ZSK" screws is their profile. Irrespective of the shape of the elements, they always intermesh and in operation the crest of one constantly wipes the root of the other while maintaining a very small uniform gap between them. This achieves a high degree of self-cleaning. The kneading elements and the conveying screws have identical cross-sections so that the self-cleaning effect is maintained along the full length of the screws.

The "ZSK" extruder is characterized by a **high torque capability and a large free volume in the screws channels**. Power and volume are matched to optimize processing parameters as well as the economics of the compounding operations.

The building block principle of the machine with interchangeable screw elements and barrel sections allows the user to vary and accurately control conditions anywhere along the process length, which can range for powder coating applications from 16 to 48 L/D. The individual barrels (*Figure 5.16*) having an L/D ratio of 4 are produced in different designs, enabling multiple feeding of ingredients or, if necessary, venting of moisture or removal of other volatiles. In this way solid, liquid or even gaseous components can be introduced anywhere within the processing section. Both the barrel elements and screw components in the feed and melt zone are equipped with a special wear-protection. The barrels have removable wear-protected liners (*Figure 5.16*), the screw elements are completely manufactured out of the wear resistant material, besides the soft core which enables high torque transmission. The rest of the machine is made of nitrided steel. This makes it possible for only defined sections of the total length of the processing part that need to be wear-protected to be replaced after long use.

Figure 5.16: "ZSK" barrel with removable wear-protected liner source: Coperion GmbH

The temperature control of the barrel can be achieved by electric heating combined with water cooling. The cooling channels are placed as close as possible to the product and are surrounding the whole barrel. Depending on the temperature profile, the cooling is achieved by full water flow or by water evaporation. In addition, on several types of "ZSK" machines, there is a possibility for internal cooling of the screw shafts.

The production of powder coatings on "ZSK" twin screw extruders is characterized by an **extremely short residence time** which on average throughput rates ranges from 10 to 20 seconds and at the same time a narrow residence time distribution, caused by the self-cleaning effect. Considering that the product is in the molten state in only approximately 65 % of the barrel length, this means that the material is exposed to a correspondingly shorter thermal stress. This allows highest shear rates without pre-reaction of standard powder coatings or a very gentle treatment of low curing or shear sensitive powder coatings.

The molten compound emerging from the extruder falls between chilled rolls and is squeezed into a wide thin strip. The additional cooling takes place on an endless conveyor band before entering the breaking roll. If the available floor space is limited, a special design of cooling roll device with a breaker is recommended as presented in *Figure 5.2*. This compact cooler design helps in addition to cool down the product very fast, which is important for low viscose powder coatings like acrylics, and to avoid dust contamination by the closed and covered design.

Since the appearance of the first extruders, manufacturers of twin screw extruders, suitable for modern thermosetting powder coatings, have made a number of modifications, designed to improve the compounding efficiency of these machines.

One of the problems of producing TGIC-free polyester powder coatings, using hardeners such as HAA or "Araldite PT 910" glycidyl esters, has been the **great difference in melt viscosity** between the resin and the curing agent. To avoid the necessity of double extrusion of these products, **much higher shear rates are required** to process these materials.

The first step to meet the new requirements was the development of the "ZSK MEGAcompounder" in the 1990ies. This machine still retain their standard features with the building block principles for barrel and screw elements in the conveying and mixing sections, and their specific design features, such as intensively cooled feed barrels, cooled screw shafts and optimal adjusted screw mixing section. The distinguished features of the "MEGA" technology are [14]:

- Increased maximum permissible screw speed of 1200 rpm (instead of the standard 300 rpm). Such a high screw speed creates a high shear rate in the melting section.
- Increased length/diameter ratio (from L/D = 16 to L/D = 20).
- No risk of "overshear" and too high temperature, due to the extremely short residence time in the compounding section.
- New screw configuration (matched to the higher speeds and the new L/D ratio).
- Improved wear protection in the process section. The barrels are equipped with wear resistant oval liners and the screw elements incorporate finely dispersed carbides to protect against abrasive fillers or extenders.
- No repeated passes through the machine with difficult products.
- Improved feed intake with low bulk density powders and fines.
- Twice the throughput rate with the same extruder size. Alternative: smaller-size extruder for the same throughput and thus also shorter cleaning times.

The "ZSK MEGAcompounder" has been further developed in the past years into the "ZSK MEGAcompounderPLUS" and, presented on the K 2007, into the "ZSK Mc18". Both developments have set the focus in a **higher transferable torque** which can be used especially in applications to produce engineering plastic, but which also shows the high mechanical safety of gear boxes and screw shafts.

All types of "ZSK" are available in a second design series called "ZSK MEGAcompact". The machine, drive and control system form then a compact unit that can be started up in one day.

The "MEGAcompact" features the same performance characteristics of the "MEGAcompounder", along with the same high torque/high screw speed capabilities and resulting high throughput rates.

These compact twin-screw machines, with screw diameters of 32 to 70 mm, employ the same engineering principle as the "MEGAcompounder".

What is new for the "MEGAcompact" is the pre-engineered, pre-wired design. They are built from standardized components, assembled ready for use and, together with the electric and electronic devices for drive and control, form a ready-tested independent unit.

Because the "MEGAcompact" is pre-engineered, it is a cost-effective, economical alternative to the "MEGAcompounder". Installation costs are greatly reduced, thanks to system pre-testing and the lack of separate cabinets to be field-wired. As the control system whereby the once separate switchgear cabinets for the drive and control are now integrated into the machine, less space is required.

At the PCE 2002 exhibition in Nuremberg Coperion introduced their new "ZSK MEGAvolume" twin screw compounder, to enhance the profitability of coating powder production, which is available in the meantime as "ZSK MEGAvolumePLUS" with 30 % higher available torque and which has become in its 10 years of production the market leading equipment for powder coating production worldwide.

Featuring screws with an increased channel depth, its free volume is 40 % larger than with the "ZSK MEGAcompounder" which had been the state of the art until 2002. The increased screw speed, now reaching up to 1800 rpm, also permits a further increase in throughput. The longer process zone, extended by one barrel to 24 D, guarantees a constantly high level of product homogeneity. The new developed intake zone, including a redesigned intake barrel and a new screw configuration, allows now a constant throughput for almost all kind of premixes, including a premix with 20 % fines. The powder manufacturer is simultaneously assured of greater flexibility in the selection of the speed, residence time and throughput processing parameters.

The "ZSK MEGAvolume" series consists of various machine sizes with screw diameters ranging from 34 to 125 mm. For powder coatings, the sizes up to 62 mm are successful established in the market, covering the throughput range from 50 to 1,800 kg/h.

At APV Baker [15] high speed extruders have been introduced in 1997. Their "MP-PC" range of twin screw extruders have a high intake for difficult to feed premixes, a screw speed of 600 rpm, with an increase in motor power. A 750 rpm option can further increase power to provide high speed extrusion. Claimed to be one of the most advanced in the market, the "MP-PC" series contains many new features to reduce operating costs. The "clamshell" barrel opening, pioneered by APV, has been improved with a simple clamping mechanism to give faster opening.

This also allows a new design of liner insert to be used, so that each of the liner segments can be **easily replaced** with the barrel opened. Installation has been simplified by a new base design, which eliminates the need for optical alignment on site and gives easy access for maintenance. The problem of powder leakage at the screw shafts has been tackled by "active gland packing", which eliminates the need for frequent manual tightening of the packing and so reduces maintenance. Overhang of the extruder barrel allows direct discharge into the nip-rolls of the stainless steel cooling band, used for product flaking. "MP-PC" production machines range now from 40 to 100 mm screw diameter, with outputs from 50 to 3000 kg/h and with powers ranging from 21 to 225 kW. Different screw elements and configurations are now available for all powder formulations from rebar to clear acrylic automotive topcoats.

5.2.2.3 Planetary roller extruder

The invention of the roller extruder by Schalke was the first impulse to a rapid development of extrusion technology. Entex [16a] made the decisive step into a new dimension during the eighties, by further development and perfectioning of this system. Therefore, one can say, that the roller extruder represents the principal item of compounding technology, controlled by the latest process technique.

The planetary extruder (schematically presented in *Figure 5.17*) is based on a toothed planetary roller system. It utilizes a helically grooved center (main) shaft, surrounded by smaller intermeshing (planetary) shafts, similar to a gearbox in design. On rotation of the central spindle the small planetary spindles are rolling on the central spindle and the outer cylinder assembly (which also carries helical grooves) and rotate like planets. During the extrusion process the premix will be caught and by the plunge of the spindle teeth into the corresponding tooth space the material will

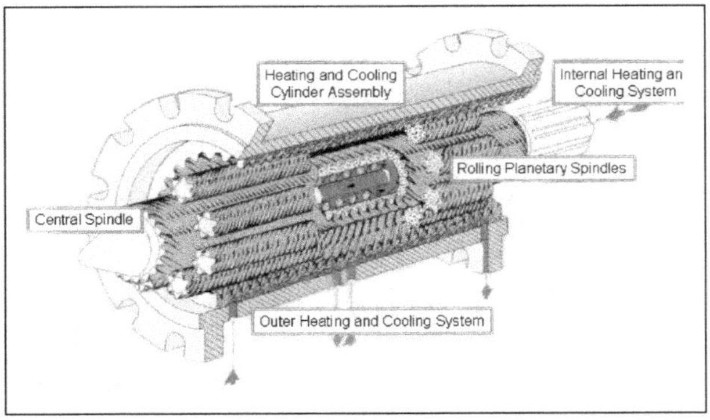

Figure 5.17: Schematic illustration of a planetary roller extruder *source: Entex Rust & Mitschke GmbH*

be drawn out into thin layers and transported by means of the 45° helix. With other words: the mixing action takes place at the interfaces of all of the shafts (between the intermeshing faces).

All steps of development by Entex are based on their new production process of the heatable thin wall cylinder by means of "spark (or depth) erosion". This electro-carbon machining can be used with any grade of steel, whatever the hardness and it is possible to produce liners of 1500 mm length in one piece, which cannot be achieved by any other machining process.

The following **advantages** of this patented process will be mentioned:

- Modular system with extremely thin wall thickness with possibilities of pressure measurement and exact control of temperature. This means that thermally delicate premixes can now be compounded very well.
  Lüchtefeld [16b] claims, that there is no other existing compounding system, which allows not nearly such an excellent temperature control in the compounding process.
- Injection of viscous components and degassing between the modules allows a compounding technology in new dimensions.

Harris [7] warns that it is necessary, during operation, to secure the die-head backing plate into position, to prevent any linear movement of the planetary screws, which would allow these to be ejected from the extruder. The planetary roller extruder is also operating as an open, pressure-free system.

The main compounding characteristics of an extruder include free volume, distributive and dispersive mixing, kneading, heat transfer and temperature control, cleanability, and flexibility (which is the suitability of the machine to extrude the full range of powder coating types). These characteristics, as they apply to each type of extruder, are described and compared by *Harris* [7] and about twenty years later by *Lovatt* [17], *Peitz* [18] and *Bodnar* [19]. *Harris* comes to the conclusion that compared to the single screw extruders, twin screw extruders consume somewhat more energy per unit of produced coating powder. However, they have several advantages. The extremely short dwell time allows safe production of powders with very high reactivity. The processing of finely powdered materials is less problematic and the dispersion is more efficient, so that smaller amounts of strong colors are required, due to the more efficient color development.

Concerning distributive mixing and temperature control, *Lovatt* and *Peitz* see no differences between the three types of extruder. They all score good to very good. *Bodnar* remarks, that the extruder manufacturers claim certain **economical and technological advantages** for their machines and he gives the following examples:

- for the single screw extruder better temperature control and high speed;
- for the twin screw extruder better dispersion and high speed;
- the planetary roller extruder for metallic powders and for recycling very fine powders.

However, due to the great variety in powder coating formulations, *Bodnar* always recommends to carry out preliminary tests in the suppliers' pilot plants, to be able to select the right extruder type.

5.3 Fine grinding

The fine grinding process follows the hot melt compounding. After leaving the extruder die, cooling and squeezing between the chill rolls, additional cooling on the cooling band to a brittle, easily breakable continuous strip and flaking by a kibbler unit, the material is usually collected in movable containers and transported to the grinding mills.

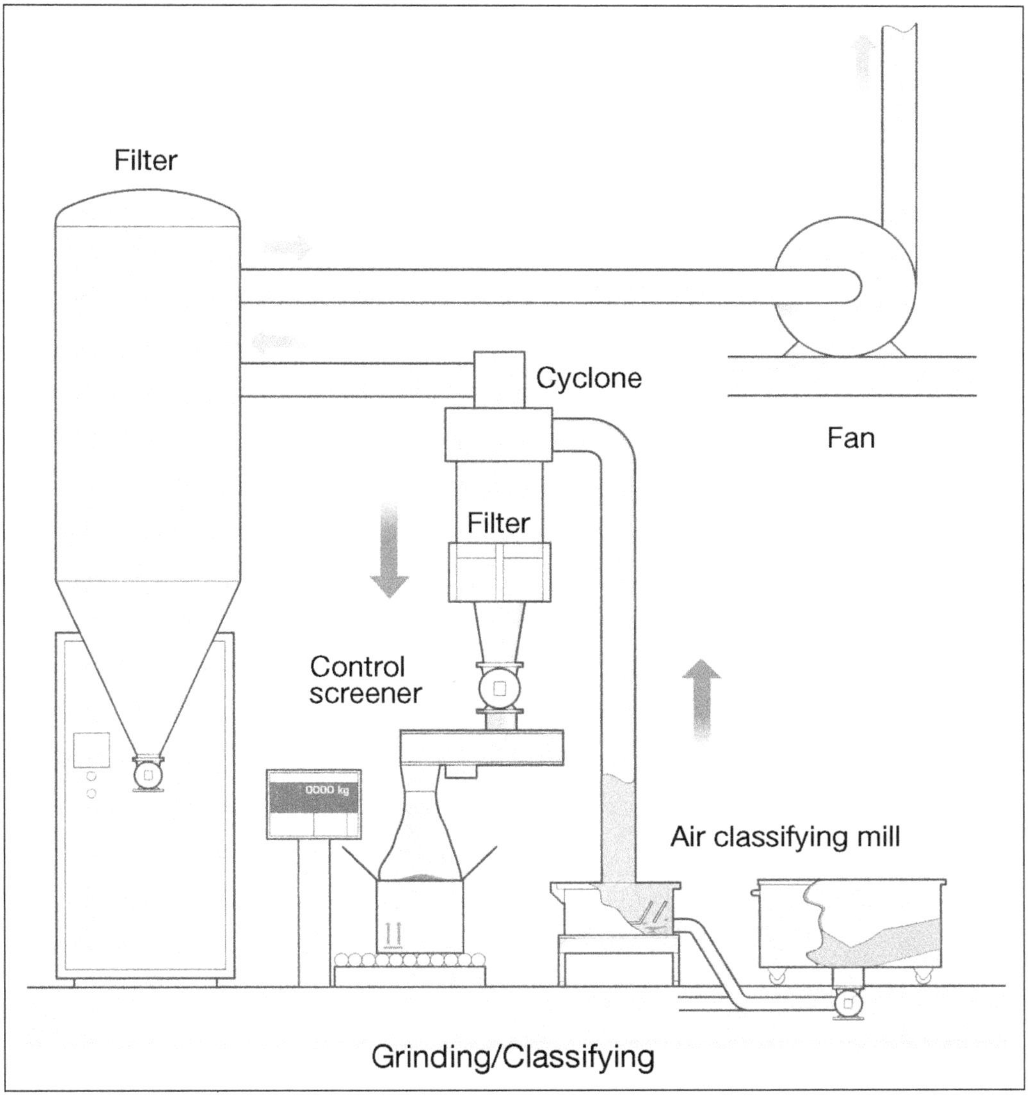

Figure 5.18: Scheme of a grinding, classifying and dedusting unit for powder coating production

source: Buss AG

Powder coating is formulated in a way that the **glass transition temperature of the system is at least 40 °C** to ensure storage stability of the coating. If the grinding of the material is performed at room temperature by appropriate removal of the heat developed during the grinding process, the material is subjected to the mechanical forces action below its glass transition temperature. Polyester, epoxy or acrylic resins used for the manufacture of thermosetting powder coatings usually have molecular weights between 3000 and 6000. With such a molecular weight, below their glass transition temperature, these binders are brittle materials with a low modulus which allows easy non-problematic grinding. High molecular weight thermoplastic binders, however, are extremely tough and resilient. **Cryogenic grinding** at low temperatures is the only method for reducing the size of the particles of the thermoplastic powder coatings.

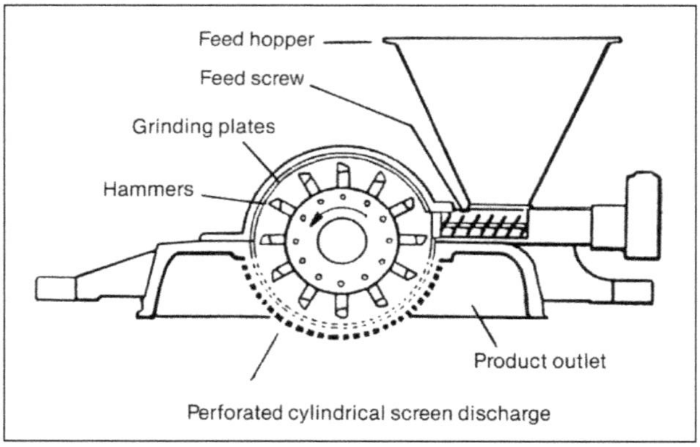

Figure 5.19: Scheme of a hammer mill

source: Federation of Societies for Coatings Technology, U.S.

Different types of grinding mills are used by the powder coating producers to bring the particle size of the coating within a desirable range. All of them utilize a process of attrition in which the size reduction is the result of impact of the solid granules or flakes with the grinding surface or with each other. Contemporary grinding systems are delivered in one set together with the particle size classifiers and dedusting devices, as presented schematically in *Figure 5.18*.

Since the 1960's efforts of the mill producers concentrated on developing a grinding machine that combines both, the grinding and classifying procedures in one device which will simplify the plant design. Multitude of patents referring to this system issued in the last 45 years indicate the permanent improvements leading to a complete grinding/classifying line that provides **clean and safe continuous grinding of powder coatings**.

5.3.1 Hammer mills

Hammer mills are among the first devices used for fine grinding of powder coatings. The **construction of the mill is rather simple** as presented on *Figure 5.19*. The granulate to be grinded enters the milling chamber through an appropriate feeding unit. The grinding takes place as a result of impact between the granulated material and the rotating hammers. Contemporary hammer mills operate with a stepless speed variation of the rotor between 1000 and 5000 rpm [20]. In this way it is possible to influence the output of the machine and the quality of the ground powder.

The outlet of the machine is separated from the milling chamber by an exchangeable screen. The type of screen also influences the output of the machine. For powder coatings production herringbone slotted or cross slotted screens are preferred because they give a finer particle size [7]. Hammer mills are produced with a water cooling option on request, so that if necessary the heat developed during the grinding process can be removed efficiently.

5.3.2 Pin disc mills

5.3.2.1 "Mikro ACM" air classifying mills

In 1962 "Mikro ACM" (Hosokawa) came on the market as the first air classifier mill and since then it has been continuously developed further and adapted to market needs. It is suitable for the size reduction of various products for all industries, including powder coatings. The combination of impact grinding with integral dynamic classification offers a solution for fine and ultrafine milling of soft and medium soft products. The principle of operation can be described as follows:

The product is fed into the mill pneumatically. Size reduction takes place by the **impact of material particles on the rotating grinding surfaces** and on the fixed liner. With a striking edge velocity up to 140 m/s, finenesses of d_{97} = 10 μm are achieved. The cooling, conveying and classifying air drawn through the mill by the downstream fan entrains the metered feed material which then passes the blades of the shroud ring. A characteristic feature is the integral dynamic classifier.

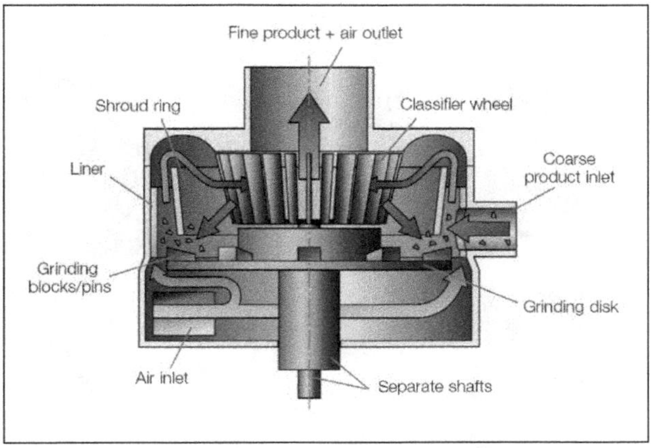

Figure 5.20: Scheme of "Mikro ACM CX" air classifying unit

source: Hosokawa Alpine GmbH

The air/product mixture is uniformly distributed by the blades to the rotating classifier. Due to two opposing forces and the different masses of the material particles the separation into coarse and fine fraction is achieved in the classifying section. Those of high mass are thrown away from the wheel by centrifugal force. The rejected particles return to the milling zone. The fine material is carried with the air through the classifier to the outlet connection. A scheme of "Mikro ACM" mill (type CX) is given in *Figure 5.20*. The cut point is adjusted by means of the classifier speed and is infinitely variable even during grinding operation. "Mikro ACM" is available in different designs. As a result of continued development and improvements during decades, important features have been optimized:

- grinding and classifying in one machine,
- compact space saving design,
- cool and smooth grinding,
- steep particle size distribution,
- sharp classifier cut point, instantly adjustable,
- easy cleaning and maintenance,
- high accessibility,
- low noise emission,
- low specific energy requirements,
- pressure shock and wear resistant.

For grinding abrasive products, the grinding surfaces can be protected by the use of wear resistant materials. Four different design variants plus a laboratory version of the "ACM" have been developed: "ACM CX" - The coaxial bearing arrangement of both classifier and grinding disc shafts makes a vertical product discharge possible, which in turn prevents product deposits and makes the cleaning of the mill quite simple. "ACM CL" - with direct classifier drive is ideal for all those applications where the preferred configuration is a product discharge duct that is horizontal with the shortest possible connection to downstream system components. Arrangement of mill is typically at the upper platform level. "ACM EC" (Easy Clean) - was developed to meet the demands for quick and easy cleaning. This is especially important for those branches of industry where the product is changed frequently or for batch production. *Figure 5.21* shows a "Mikro ACM EC" mill.

"ACM EC-CL" - combines the advantages of the CL model, namely compact configuration with short connections to downstream system sections, with those of the EC model, with the main focus on short cleaning times. "ACM 2" and "5" - for laboratory and small-batch production. With the newly developed easy-clean concept, the "Mikro ACM" machine sizes 2 and 5 offer a high degree of user-friendliness and a minimum of cleaning effort when changing products. A space requirement of just 15 m^2, a mains power socket and a compressed air connection is all this unit needs for operation.

Figure 5.21: "Mikro ACM 30" EC mill

source: Hosokawa Alpine GmbH

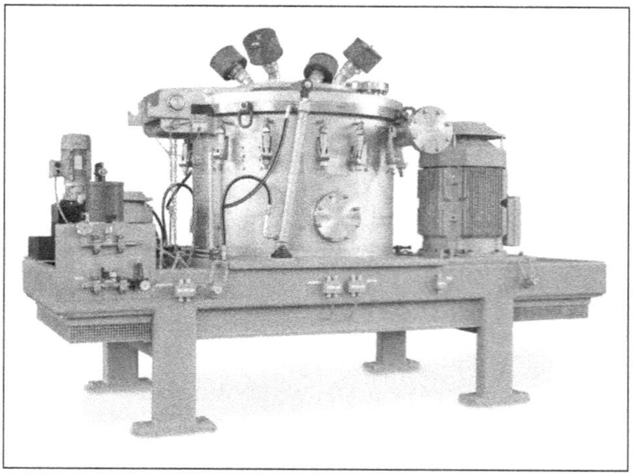

Figure 5.22: Dynamic air classifier "Condux CSM 560"

source: Netzsch

Netsch offers a dynamic air classifier (*Figure 5.22*) integrated into the grinding chamber housing of this fine impact mill, which manufactures fine particle sizes with a defined maximum particle size limitation. Internal classification of the coarse material results in a more stable and energy efficient process than a mill with an external classification circuit. Applicable for finenesses from 10 to 150 μm (d_{97}).

For **thin film powder coatings** it is necessary to reduce the particle size. A film thickness of e.g. 25 to 30 μm requires a maximum particle size of approx. 60 μm (instead of 90 to 100 μm) and the whole particle size distribution curve is shifted into the finer direction, with the mean particle size d_{50} in the range of 22 to 25 μm. In such finer powders the amount of "fines" below 10 μm will increase and often it will be necessary to reduce this amount to improve the fluidity (or free flow properties) of the powder and to prevent caking during storage and application. According to *Sanders* [21] a standard powder, ground with the "ACM" mill, will have 7 to 9 % by volume below 10 μm, while the thin film powder will have 14 to 18 % by volume of fines.

Sanders and *Galk* [21, 22] describe the technique for grinding and classifying these thin film coating powders. The amount of "fines" depends on different parameters. The main grinding parameters, governing the particle size distribution, can be listed as follows:

- particle size of feedstock (normally 2 to 3 mm thick flakes; excessive fine material can be too easily sucked into the mill, which will lead to a high level of fines);
- feed screw speed (should be set to give the maximum amperage of the mill);
- mill speed (should be set at the maximum);
- rotor design (round rather than square pins);

- liner design (the ridged multiple deflection type is recommended; they give a harder grind, with less need for recycling within the chamber; and hence less fines in the finished product);
- classifier speed (in rpm) (adjustment of this speed only moves the PSD curve to the finer or coarser end, without affecting the shape of the curve);
- classifier type (long-, medium- or short-bladed types are available, and the shorter the blade length, the less the amount of internal recycling, and the smaller the amount of fines will be);
- air flow rate through the mill (for an "ACM 60" an air flow rate of at least 8500 m^3/h is recommended and it may be advantageous to go still higher);
- outlet temperature (should be kept low).

The **amount of fines below 10 μm** can principally not be decreased to zero. But examination of the physical limits of grinding has shown that the amount of fine particles in the range of 10 μm, produced by "ideal" grinding, is only 1 % less than produced with "ACM" mills [22]. Nowadays the base polyester/epoxy industry of powder coatings produces about 5 % of "fines". *Grisa* et al. [23] described a way to recycle these fine particles in adhesives for carpet on cement slab and high density fiber. In a recent article [24] *Fu* et al. pointed out that fines are difficult to recycle in powder coatings due to the undesired effect on the flow properties.

Another way to optimize the process of coating powder production is to use air classification with a high sharpness of cut. The "MikroClassifier CC" (Hosokawa) is derived from the cyclone classifier "VME-CLC" and its main field of application is for coarse particles down to 10 μm. It can be used for offline and inline classifications. The essential advantage of inline classification is that less additional devices (such as pipe-work, feeding equipment, filter and rotary airlocks) are needed [22]. With the "MikroClassifier CC" the mass proportion of coarse material (also called "yield") can be increased by approx. 19 %, decreasing the loss by about 42 %, according to *Galk*.

An improved rotor design, secondary air flow and integrated central fittings create an optimum of material and air flow. As a result, agglomerates are significantly well dispersed. The "Mikro Classifier" operates in accordance with the well proven principle of a cyclone classifier. The feed material is fed tangentially to the air classifier along with the classifying air. While the fines are discharged through the classifying wheel entrained in the classifying air, the coarse product migrates to the classifier wall and is transported downwards.

A flow of secondary air can be introduced tangentially above the product discharge. This can be used on the one hand to control the residence time of the product in the classifier, and on the other hand to recirculate any unclassified material to the classifying wheel for renewed classification, whereby agglomerates are dispersed and the material is then distributed uniformly in the classifying zone. As a special feature, the "Mikro Classifier" is equipped with patented fittings which separate the down flow zone from the central up flow zone, thus preventing any mixing effects. In combination with the high-efficiency classifying wheel, the "Mikro Classifier CC" achieves a much higher precision of cut than conventional classifiers of this type. In addition, the **low pressure drop results in low energy consumption** and thus lower operating costs.

The "Mikro Classifier CC" is suitable for all types of fine to medium-fine products. It is especially suitable for dedusting powder coatings in the particle size range below 5 to 10 μm. It is recommended to apply classifiers for the production of well-defined particle size distributions in powder coating production and processing, especially in two cases:

- production of powder coating with optimum narrow particle size distribution
- stabilization of powder coating recycle circuit during the coating process

Galk [22a] also describes the “ICM Impact Classifier Mill” and “Cyclone Classifier” of Neumann & Esser, to be used for the production of thin film powder coatings.

The feeding of the extrudate chips is done by a dosing rotary airlock and the chips are transported to the mill by airflow. Chips are ground in the “ICM” by impact grinding, using hammers with a tip speed of up to 130 m/s and are comminuted at the exterior circumference of the grinding disk where they impact against the profiled surface of the internal wall of the mill housing.

An internal circuit transports the ground particles to an impeller wheel classifier, which allows particles of the desired size to pass. Coarse particles are rejected and returned to the grinding zone. The main task of this classifier is to determine the top size of the product. A high sharpness of cut guarantees a well-defined top size free of spattered particles and the internal classification ensures a short residence time in the mill, which avoids over-milling of the product. The final product with the desired particle size will not be returned to the grinding zone and therefore, a narrow particle size distribution can be obtained. The product leaving the classifier mill is either separated from the airflow by means of a cyclone, or if a reduction in fine particles is required, an additional classification step is introduced. This is realized by a cyclone classifier and in this way a product with good flow properties is achieved.

Different particle size distributions produced by the “ICM” are possible. Top size can range from 100 μm to below 40 μm with x_{50} values between 50 and 30 μm. To achieve a product with a low D_{50} value, it is necessary to increase the tip speed of the grinding elements to give the high intensity needed for a finer particle size distribution (PSD). This will shift the PSD to the finer side and increase the amount of fines.

The amount of fines produced, below 10 μm, varies between 5 and 30 %. It depends on the process conditions and the material properties of the chips. To reduce the problems of excessive fines, temperature rise and impact fusion of small particles, Hosokawa Micron Powder Systems have developed a new milling technique, which combines gravity-flow long gap milling and vertical axis air classification with an external circulation mechanism. It gives easy temperature control, low air flow rate and fewer fines than the conventional “ACM”. With this new “PCM” mill it is possible to feed materials up to one inch in dimensions and reduce the product to median sizes as low as 5 to 15 μm. A comparison between the “PCM” and the “ACM” learns that the “PCM” can produce similar milled products at a higher throughput rate with a lower airflow than with the “ACM”. Moreover in tests *Huang* of Hosakawa found approximately 1.4 % fewer fines below 10 μm in the ground material. A full description of the “PCM” mill is given by *Huang* [22b].

5.3.3 *Jet mills*

In the conventional grinding machines the main grinding mechanism involves contact between the particles whose size has to be reduced, the grinding media (hammers, rotating pins, rods or discs) and the mill lining. For obtaining superfine powder qualities, this type of mechanical grinding is associated with rapid rise in energy consumption which makes the grinding uneconomical.

The contemporary tendency towards thin layer powder coatings with ultra-fine particle sizes makes the use of the opposed jet-mills very attractive. In the case of jet milling, one takes advantage of the fact that as compressed gas exits a nozzle, it is accelerated to extremely high speeds. In expanding, the energy contained in the compressed gas in the form of heat is converted to kinetic energy. Compressed air of 20 °C and 6 bar overpressure is frequently used as the grinding gas, and delivers nozzle exit speeds of around 500 m/s. The grinding mechanism is based on **inter-particle collisions between opposed suspension jets** in the grinding chamber. With these mills particle sizes between 10 and 35 µm with 97 % of the material being below 12 µm can be easily obtained [25]. This is usually combined with a pneumatic micro classifier in closed circuit with the jet mill as presented in *Figure 5.23*.

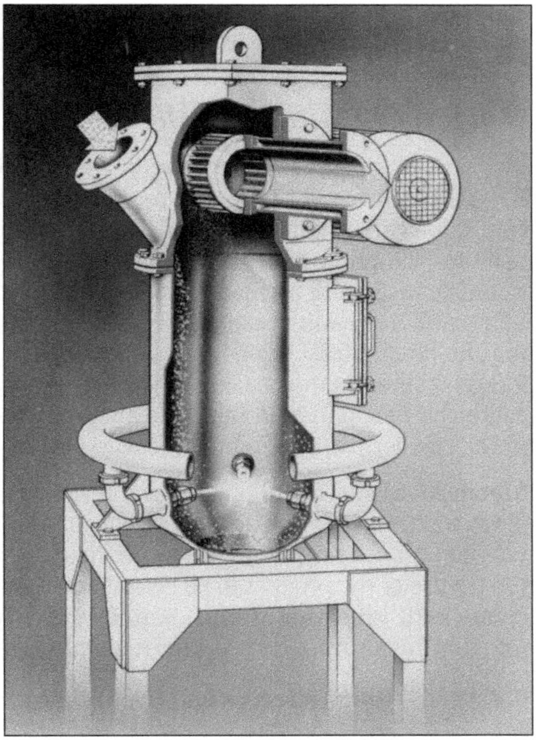

Figure 5.23: Operational principle of the opposed jet mill source: Hosokawa Alpine

Spiral jet mills have a static air classifier integrated into the mill housing, whereas fluidised bed opposed jet mills are equipped with a dynamic deflector-wheel classifier. The **fineness is** set as **a function of the classifying wheel speed**. All-important is a high product loading of the nozzle jets in order to achieve a high concentration of particles and thus high impact probabilities. The patented Megajet nozzles consist of four small nozzles which as a result of their close proximity generate an underpressure at their common centre, and thus draw particles from the fluidised bed direct into the centre of the nozzle jet. The product level in the machine is controlled by means of load cells or by monitoring the current loading of the classifier drive. The **mechanical jet power** not only increases as **a function of the gas flow rate** but also as **a function of the grinding gas pressure and temperature**. Different conditions of the grinding gas, e.g. pressures, temperatures, have to be compared on the basis of the mechanical jet power. During compression, the grinding gas heats up. However, the less it is cooled, the more energy is transported to the nozzles and is then available for the size reduction.

5.4 Particle size classification

The contemporary grinding equipment is usually delivered as a complete unit including a classifying device. The output of the mill is adjusted to the capacity of the classifier and maximum allowed particle size. However, the classifying mills are not the only equipment used by the powder coating producers and neither is the air stream classification the only way of separating the coarse particles from the fine powder. The hammer and some pin disc mills do not provide particle classification and this operation must be done in a separate step of the process. As seen in the previous section, the opposed jet mills have also to be accompanied by a classifier in order to obtain a desirable particle size range.

The screen is probably the best known type of classifier which is widely used by the industry for separation of powders according to size alone. For such an operation industrial screen classifiers are within a range of screen apertures between 10 mm and 75 µm. The **capacity of the screen classifiers decreases rapidly** when the particle size of the material to be classified is **lower than 70 μm**. For dry classification below 70 μm and particularly below 50 μm centrifugal air classifiers are most commonly used [26].

5.4.1 Classification by sieving

Mechanical sieving or screening is one of the oldest known process techniques for classification still used in current practice in the production of powder coatings. The same technique can be applied not only to particle size classification but also to scalping, a term used for the removal of small quantities of oversize from a feed which is predominantly fine. The screening method in principle is used in the dust removal filters in the final step in the powder coating production where the de-dusting filter has to retain all particles leaving the de-dusting cyclone in the exhausting air flow. Finally, the policing, or removal of foreign contaminants accompanied by the breaking down of powder agglomerates which can block the spray gun nozzle is another possible application of the screening technique in the powder paint overspray recovery.

Mechanical sieving allows separation of particles as small as 50 μm although separations down to 20 μm may be achieved with an **air-jet sieve** [7]. The industrial sifting machines range in capacity from a few kilos per hour throughput up to hundreds of tons depending on the nature of the product to be classified. They can be designed for single or multi-product splits. Both **vibrational and centrifugal sifting machines** for classifying the product into fine and coarse fraction are frequently used in the production of powder coatings.

5.4.1.1 Tumbler screening machines

The movement of the material to be classified across the screen is achieved by a **three-dimensional rotation of the screening drum**. This rotation is produced by a rigid crank drive which has an adjustable slope. A vertical movement is superimposed on the horizontal gyratory movement to result in a drum oscillation with the parameters speed, eccentricity of the crank, radial slope, and tangential inclination. The tumbler screening machines are designed in such a way that all mentioned parameters are infinitely variable. The material to be screened can move over the screen deck at any required speed or remain on the screen deck for any desired period. The interaction of tangential and radial inclination results in a spiral movement of the material across the screen.

The tumbler screening machines can be equipped as single, double, triple or quadruple screens to provide continuous screening up to five fractions. Assembly of the machine is rather simple as presented in *Figure 5.24*. Because of their circular construction these machines can be easily equipped with supplementary devices which provide continuous trou-

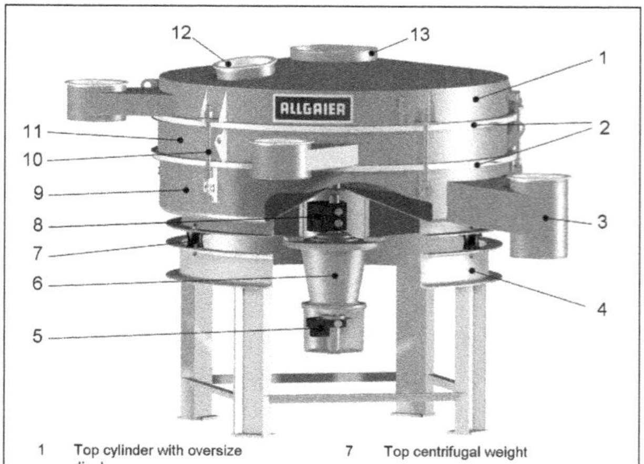

1	Top cylinder with oversize discharge	7	Top centrifugal weight
		8	Rubber elements
2	Screen frame	9	Bottom pan
3	Fine outlet	10	Clamp bolts
4	Support	11	Intermediate cylinder
5	Lower centrifugal weight	12	Inspection port with rubber cover
6	Vibratory motor	13	Material inlet

Figure 5.24: Five-unit construction system for a double screen tumbler sifter *source: Allgaier-Werke GmbH*

ble-free screening of the material by permanent cleaning of the screen bottom from the accumulated fines. In its simplest construction it can be done by bouncing balls made out of highly wear-resistant rubber which are fitted between a ball tray and the underside of the screen. They clean the screen continuously by their impact and wiping action. Another accessory for the same purpose is the roller brush cleaning device composed of two or more brush bearing arms carried on a bowl mounted drive unit driven from the main shaft. Cleaning of the screen can be achieved by an air cleaning device which comprises nozzle arms that rotate beneath the screen deck and blow air through the screen to clean it. A combination of air cleaning and rotational roller brushes is very affective for extremely difficult materials that stick to the underside of the screen.

5.4.1.2 Vibratory screening machines

These screening machines are especially suitable for **removal of the oversize from the fines**, the so-called "scalping". Oversized particles are discharged at side, while the discharge of the fines is done through the center of the hopper shaped basic cylinder. The continuous discharge of the fine particles allows relatively high capacity.

The vibrational motion of the screen is achieved by the action of two unbalanced motors, each with a set of flyweights, offset against each other in order to obtain a circular motion forcing the particles to travel across the screen from the center to the periphery. The intensity and direction of this oscillation are adjustable according to the characteristics of the material to be sieved. The vibratory screening machines are constructed in modules, to give a single or double deck unit providing classification up to three fractions. A combination of vibratory and tumbling action is realized in the vibratory tumbler screening machine which is presented in *Figure 5.25.*

The sifter is available in a single or multi-deck unit depending on the number of fractions to be collected. Material is fed centrally and travels in a spiral screening pattern across the screen cloth to the periphery. Drive is by a vibratory motor arranged in the machine center which generates **circular tumbling vibrations** by appropriate adjustment of the flyweights. The tumbling motion is easily adjustable in three dimensions in order to obtain an optimum machine setting taking into consideration the flowing and screening characteristics of the material.

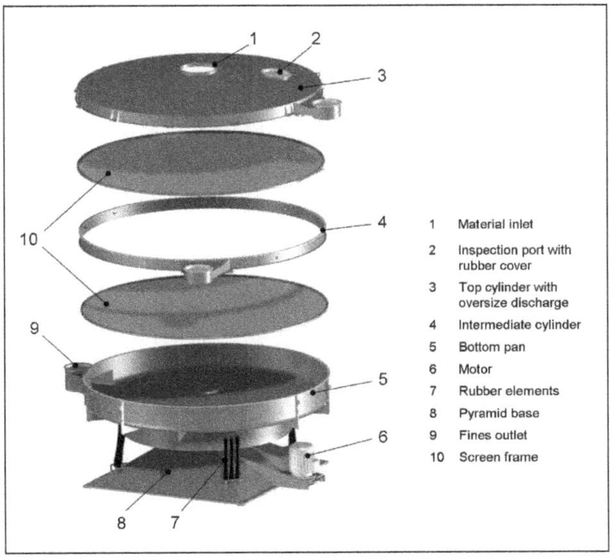

Figure 5.25: Vibratory tumbler screening machine
source: Allgaier-Werke GmbH

With the introduction of large(r) classification mills there was a need for bigger sieves and for the continuous removal of oversize material as well as contaminants. For this reason larger vibratory screens, up to 48 inch (1200 mm) in diameter, and rotary centrifugal sieves were introduced. With such sieves doubled or even trebled throughput rates could be realized. But "**mesh blinding**", due to static adherence, remained the major problem in high volume coating powder production. The solution to the problem of mesh blinding is to apply **ultrasonic**

vibration to the mesh. This device was first introduced by Russell Finex more than 30 years ago. Over this period there has been a series of progressively improved designs to their "Vibrasonic" system, the latest (2012) being still the "Vibrasonic 2000" [27].

The principle of this technique is the application of a secondary vibration to the mesh in the frequency range of 35 kHz, which prevents particles, both slightly larger and smaller than the aperture size, from blinding the mesh. This vibration is applied by attaching an ultrasonic transducer to the mesh, connected by a single cable to an ultrasonic generator. The transducer probe is attached to the sieving mesh by bolting it to a disc, which is bonded to the mesh. Dimensions of the disc and its position are critical to successful operation, as well as the use of a suitable adhesive to bond the mesh to the frame and the disc to the mesh.

The ultrasonic frequency has been designed to excite the wires of the mesh, rendering the mesh virtually friction free. By use of this system Russell Finex is also able to dissipate the static build-up and also to assist in deagglomeration of the powder. But the main advantage is that it keeps the mesh free from blinding.

In addition to the development of the "Vibrasonic" system, Russell paid attention to the sieving machine itself, and introduced the "Compact Vibratory Sieves", which fit neatly into production lines, where headroom is restricted.

In response to the demand for improved performance, higher throughput rates and more accurate sieving, recent developments to the "Vibrasonic 2000" (which is equipped with the "Compact Vibrasonic Sieve") include a powerful generator, which is able to handle up to 2 m diameter screens. Therefore the "Vibrasonic 2000" system provides:

- High capacities. Russell has examples of 1200 mm screens, sieving well in excess of 1 ton per hour.
- Accurate consistent particle specification.
- Performance matched to all available micronizer systems.
- Fine mesh sieving can now be realized. Thin film coating powders require sieving at production rates through meshes as fine as 75 μm.
- Mesh life is increased from weeks to months.
- A low headroom inline installation, which is an ideal system for fitting over automatic or semi-automatic bagging installations.
- Due to its simplistic design, the whole system is easy to effect clean-down between color changes as well as mesh changes.

A nearly identical sieving system is delivered by the Swiss firm Telsonic [28], with their "Sonoscreen" ultrasonic supported sieving technology. The "Sonoscreen" system consists of an ultrasonic generator, converter and sieving resonator. High frequency electrical oscillations (in the micron range) produced by the ultrasonic generator, are converted into mechanical oscillations, which are spread evenly through the sieve mesh surface by the resonator. The converter/resonator unit is "dust-ex-proof" certified and absolutely leak-tight designed, which enables wet cleaning, if required. "Sonoscreen" may be integrated and retrofitted to most types of sieving machines.

5.4.1.3 Pneumatic tumbler screening

Pneumatic tumbler screening installations are used for **separation of fine materials which are difficult to screen**, especially in cases when the fine fraction ranges down to 30 μm. Even with high acceleration, the inertia forces of the individual particles are not sufficient to overcome the surface tensions that arise through electrostatic charges, moisture or Van der Waals-forces. In such cases the air-jet screening is a good solution for overcoming these forces which block the efficiency of the sieving machine.

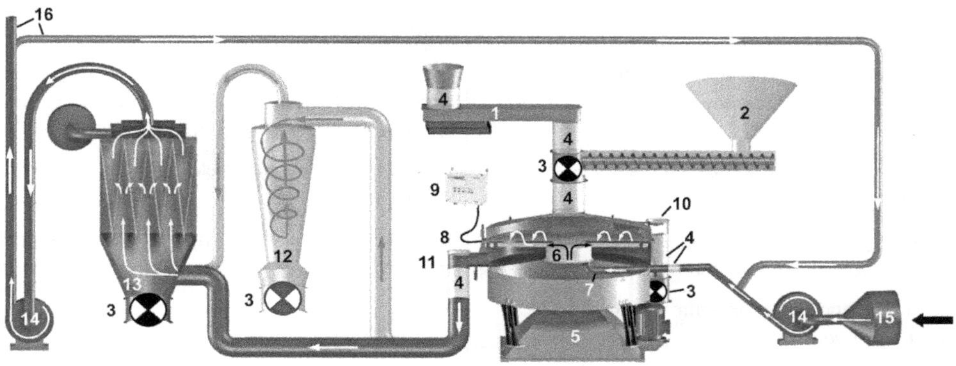

1	Vibratory feeder	9	Ultrasonic generator
2	Feed hooper with screw conveyor, alternative	10	Coarse grain outlet
3	Bucket wheel valve	11	Fine grains sucket out here
4	Flexible hose connection	12	Alternative: preliminary separation by cyclone
5	Pneumatic Tumbler Screening Machine AVTM	13	Filter
6	Air bell with nozzle arms	14	Fan
7	Air intake	15	Fresh air filter for through-air operation
8	Ultrasonic screen	16	Butterfly valve for circulating air operation

Figure 5.26: Layout of a pneumatic tumbler screening installation *source: Allgaier-Werke GmbH*

A pneumatic tumbler screening installation is presented in the diagram in *Figure 5.26*. The **screen deck is cleaned from below by jets of air** from a rotating nozzle system. This causes the material to be fluidized. The coarse material separates from the fine material and by means of the tumbling motion of the screen is continuously discharged. The fine material is drawn down through the screen and transported with the extracted air to an appropriate filter system. The pneumatic screening equipment can be fed either with fresh air or with recirculated air. The recirculation makes it possible to work with an inert gas which thus provides an additional safety with respect to the dust-explosion hazards.

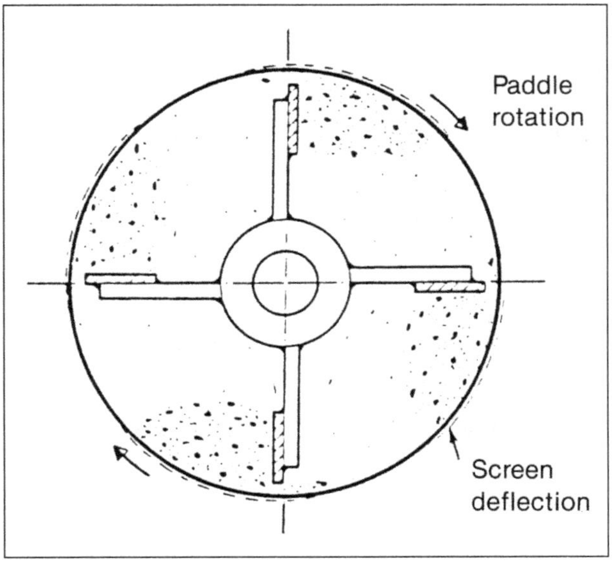

Figure 5.27: Uneven circumferential product loading in the centrifugal sifting machine[29] *source: Paint & Resin Turret Group plc*

5.4.1.4 *Centrifugal sifters*

Centrifugal sifting machines where the vibrating flatbed screen is replaced by a stationary cylindrical screen have gained **wide acceptance in powder coating manufacture**. The material to be screened is fed via the sifter inlet by means of a rotary

Table 5.1: Benefits and limitations of centrifugal sifting machines compared to flatbed sifters [17]

Benefits	Limitations
Higher throughput in relation to screen area for a given separation efficiency.	Design limited to single split type i.e. output of two product streams only. Cascade of machines necessary for multi-split operation
5 to 20 % higher separation efficiency compared to other mechanical screening machines.	Not suitable for very friable and very abrasive products.
Vibration free operation, thus less demanding with respect to supporting structures.	Limited to medium size capacity; however not relevant in case of powder coating manufacture.
Less noisy and dusty operation compared to vibrating sieves.	
Quite strong de-agglomerating action, less fines contaminating the coarse fraction.	

feeder, conveyor screw or vertical conveyor. The feed screw conveys the material into the cylindrical sifting chamber where it is picked up by the rotating fluidizing bars and thrown centrifugally against the screen. The helix configuration of the fluidizing bars ensures that the powder particles are traversed axially along the whole length of the machine. The fines penetrate the screen while the coarse particles after passing round the end baffle are discharged continuously.

The **rotating fluidizing bars** provide efficient means of distributing material over the screen periphery. However powdered material to be sifted is not uniformly distributed in either the circumferential or axial direction. The uneven circumferential product loading is presented schematically in *Figure 5.27* [29] Each fluidizing bar carries a wave of semi fluidized material in front of it leaving an empty space between the front of the wave and the next bar. This produces a **very useful secondary vibrating effect** which helps in the self-cleaning of the screen. The sack fibers are screened without a beard forming at the screen, which adversely affects the screening process. In order to avoid the appearance of electrostatic charges during the sifting of powder coatings, the use of carbon fabrics is recommended. The benefits and limitations of the centrifugal sifting machines as an alternative to flatbed sifting, discussed by *Tunnincliffe* [29] are summarized in *Table 5.1.*

5.4.2 Centrifugal air classifiers

Operating principles of the centrifugal air classifiers are based on the **action of a vertical flow of air in both rotary and cyclonic devices**. In the case of cyclones this is obtained by the tangential air inlet as presented in *Figure 5.28.* Rotary centrifugal air classifiers make use of a high/speed rotor located inside the classifier. The vortex formed in the cyclone causes an inward spiral movement of the air stream from the tangential inlet to the central outlet. Similar stream lines can be drawn for the air which enters at the rotor periphery in case of rotary centrifugal air classifiers.

The tangential velocity component (V_t) at any radial position (R) in the separating device [26] is given by

Equation 5.1 $$V_t = K \cdot R^n$$

where K and n are constants whose values depend on the geometry of the equipment. There is a difference in the tangential velocity component profiles between particles exposed to the action of free vortex formed by the tangential air inlet in the case of cyclones and forced vortex which exists in the rotary centrifugal air classifiers. The exponent n has value of -1 in the case of cyclonic classifiers and 1 in the ideal case for the rotary types. The actual tangential velocity profile for the rotary centrifugal air classifiers is a combination of these two extreme cases.

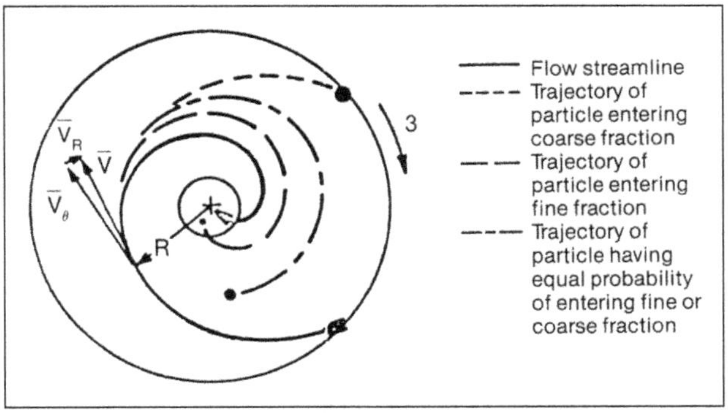

Figure 5.28: Principles of centrifugal air classifiers

Particles which enter the classifier together with the air stream are subjected to the action of two forces: centrifugal force, which is proportional to the cube of the particle diameter and the square of the particle tangential component velocity, and drag force proportional to the particle diameter and the ratio between the velocities of the air and the particle in the radial direction. Consequently, the trajectories of the **large particles are dominated by the centrifugal forces** and they will be flung out to the wall of the container. In the case of cyclones they will sink in a circular path through the bottom exit. In the case of rotary classifiers they will be collected as a coarse fraction at the periphery of the rotor. The **small particles** exposed to a large drag per unit mass **will follow the air stream** and will leave the cyclone or the rotary classifier through the central outlet.

Depending on the rate of the volumetric air flow (Q), the diameter (D_o) and the width (H) of the high speed rotor used to obtain the spiral-vortex air flow path, the rotational speed of the rotor (ω), viscosity of the air stream (η) and the particle density (δ), there is some critical particle diameter ($D_{50\%}$) known as the cut size where the drag and centrifugal forces are equal [30]:

Equation 5.2

$$D_{50\%} = \left(\frac{3k}{\pi D_o \omega}\right)\left(\frac{\eta Q}{\pi H \delta}\right)^{1/2}$$

In the Equation 5.2 k is a constant dependent on the flow pattern and has to be determined for a specific classifier design. The particles with this critical diameter have an equal chance either to enter the coarse or the fine fraction.

Even a superficial qualitative analysis of the equation defining the critical particle diameter shows that it will be **more difficult to remove the fine particles** from unpigmented powder coatings having a low particle density. Removal of the fine particles will also be adversely affected by increased air flow rates. On the other hand, increasing the rotor speed decreases the cut size.

In the ideal case when the operating conditions of the system completely comply with the mathematical description it will be a sharply defined boundary between the fine and the coarse fraction. In practice, however, such a **sharp classification can be never achieved**. Several reasons contribute to a deviation of the practical results from the ideal. Powder particles are not of the spherical shape that is assumed in the derivation of *Equation 5.2*. The interactions between the particles in the air stream are also neglected. The collision between the particles affects the particle velocity during its residence in the classifier which can, for example, decrease the velocity of the tangential component of the coarse particles, resulting in contamination of the fine fraction with particles having larger diameter than the critical. On the other hand, fine particles can enter the classifier in the form of agglomerates behaving as having a larger diameter and are carried to the coarse fraction. This phenomenon

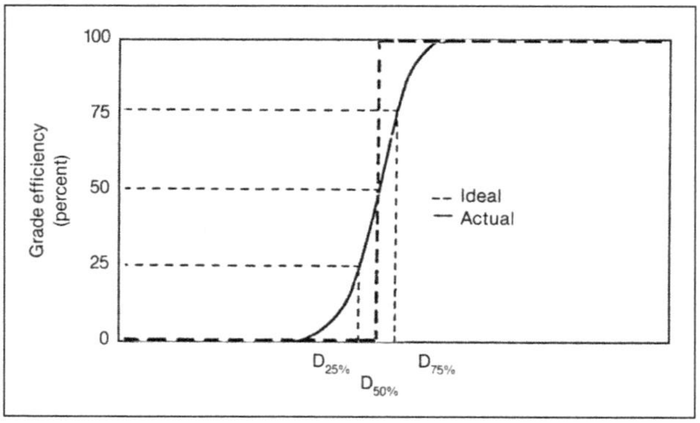

Figure 5.29: Representation of classification performance by the grade efficiency curve [26]

source: Federation of Societies for Coatings Technology

is especially emphasized when extremely fine powders have to be classified. Decreasing the particle size rapidly increases the surface/volume ratio, resulting in increasing interaction between the particles due to the surface forces such as Van der Waals-forces. For these reason it is **very difficult to obtain a particle size lower than 5 μm** by centrifugal air classifiers [26].

It is difficult to choose which criteria should be considered as relevant in the evaluation of the classifier efficiency since the cut size, the sharpness of the cut and the output of the machine as important elements in judging the equipment performances are mutually dependent. The geometrical method suggested by *Eder* [31] seems to be widely accepted [26, 32]. It uses the following equation:

Equation 5.3 $\phi = D_{75\%}/D_{25\%}$

in which $D_{75\%}$ is the particle size diameter corresponding to 75 % grade efficiency and $D_{25\%}$ is the particle size corresponding to 25 % efficiency (*Figure 5.29*).

The **grade efficiency** which is plotted against the particle diameter in *Figure 5.29* is defined as:

$$n_{(D)} = 100 \frac{\text{weight of size D entering coarse fraction}}{\text{sum of weights of size D in fine and coarse fraction}}$$

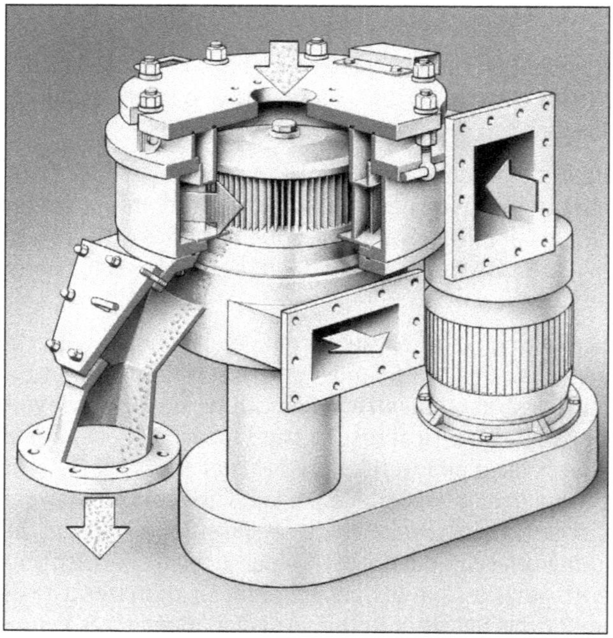

Figure 5.30: High-tech Classifier "TSP" *Source: Hosokawa Alpine AG*

The choice of the most suitable classifier depends on the **desirable cut size and the output of the machine**. High capacity units are usually used in closed circuit grinding systems and have a cut size range between 20 and 70 μm. This is the usual particle size range of powder coatings. For classifying below 20 μm units with moderate capacity are used. Laboratory centrifugal units with a capacity of a few grams per hour can achieve classifications even below 1 μm.

The principle construction of a widely used centrifugal air classifier ("TSP" from Hosokawa) is depicted in *Figure 5.30*.

The feed material is dispersed by the classifying wheel and trans-

Figure 5.31: High-tech Classifier "TTSP" *source: Hosokawa Alpine AG*

ported through the same by the classifying air. The dust entrained in the air is transported through the support wheel and discharged via the fines discharge. The coarse product is rejected by the classifying wheel and conveyed to the coarse material discharge. The product routing is designed such that **no back-mixing can occur and a high quality of dust removal is ensured**. The special configuration with the support wheel permits optimum access for quick and thorough cleaning at minimum downtime.

The tandem classifier "TTSP" (*Figure 5.31*) is a refinement of the classifier "TSP", primarily with the aim of improving the precision of cut even further by means of two consecutive classifying stages. At the same product loading of the classifying air, the tandem classifier achieves a much higher precision of cut than the "TSP", or at a constant coarse product quality, achieves a higher loading factor. This makes it possible to almost treble the throughput in comparison with the "TSP". Features are:

- central feeding
- uniform product concentration
- stable flow pattern
- low energy consumption
- low content of fines
- low pressure drop

5.5 Powder collection and dedusting

The collection of fine particles leaving the outlet of the air classifier is usually done by so-called side removal filters or more often by a combination of a cyclone for preliminary fine material separation and a side removal filter. Through-air operation without preliminary fine material separation via cyclone is the most economic design from the investment and maintenance standpoint, but certainly not the most economical for exploitation.

Secondly through-air operations with preliminary separation of the fine material via a cyclone are used. **Cyclones have become the essential component** in the modern grinding systems for powder coating. They reduce considerably the cleaning expense in comparison with filters and permit great availability of the system even in the case of frequent color changes. The collecting efficiency of the cyclone, however, decreases with increasing dimensions. This makes

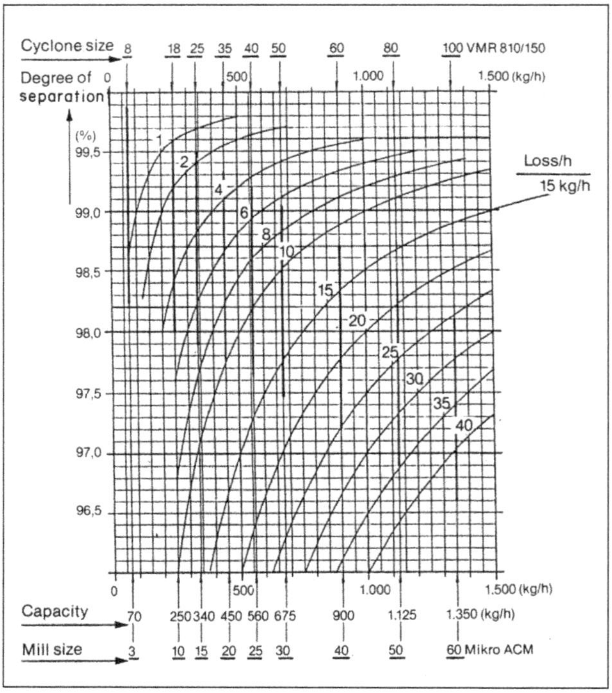

Figure 5.32: Cyclone collecting efficiencies in "Mikro ACM" systems for powder coatings *source: Hosokawa Micron GmbH*

the use of **cyclones for capacities higher than 1000 kg/h inefficient**. *Figure 5.32* represents the cyclone collecting efficiency in terms of product loss per hour as a function of the cyclone size and capacity [33].

The combination of cyclone with a side removal filter permits the lowest product loss regardless the capacity of the plant. In large batch production the side removal filter is more efficient than the cyclone. For medium and small batches the cyclone can almost completely replace the side removal filter. The choice between cyclones as powder coating collectors and side removal filters depends on the plant capacity and the diversity of powder coatings produced with respect to the types and colors. *Table 5.2* gives a comparison between these two systems.

Very often residual dust filters are used as safety devices for the side removal filters. In such cases the dust emission of the grinding system can be reduced to less than 5 mg/m^3. Dust removal filters also function as safety equipment in the case of malfunction of the side removal filter.

Residual dust removal filters require almost no maintenance work, increase the reliability of the grinding and classifying operation and improve the pollution control. Among several com-

Table 5.2: Comparison between cyclone and side removal filters

Cyclone advantages versus side removal filters	Cyclone disadvantages versus side removal filters
Quick access	Lower product recovery rate
Easier, faster and perfect cleaning	High loss of additives (fumed silica for example)
Less powder deposits	Decreasing efficiency when portion increases
No fiber contamination from the filter media	Dropping efficiency in case of air leakage
Stabile pressure drop and continuous powder flow	No performance control while in operation. Product losses are realized when the batch is finished
Constant granulometry	More sensitive to deviations in the operating conditions. Supervision by trained personnel necessary for correct cyclone use
Lower investments, no spare parts, no maintenance	
No frequent change of filter bags and no risk of malfunction	

mercial **residual dust removal filters** the so-called "MikroPulsaire" bag filters are very often in use for production of powder coatings. They consist of removable filter bags placed in a housing with different sizes and form, dependent upon the required filter capacity.

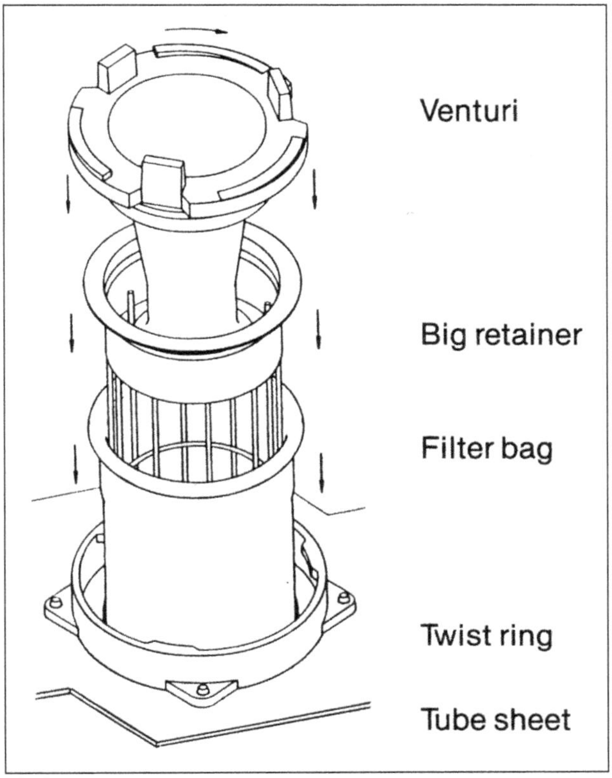

Figure 5.33: Principle of venturi/bag retainer/filter bag assembly Source: Hosokawa Micron GmbH

The air contaminated with the powder particles enters the air plenum and flows from the outside to the inside of the filter bags. The fastening of the bags, which are slipped over bag retainers, can be done either by means of bag clamps or by clampless twist venturi as shown in *Figure 5.33*. With both versions the blow tubes can be moved out in groups to allow quick access to the filter bags. The dust collects on the outside of the filter bag, while the clean air passes through the filter media into the inside of the bag and leaves the filter through the outlet nozzle. Cleaning of the filter bags is accomplished by very short compressed air jets coming from the nozzles of the blow tubes, which are mounted above the filter bags. During the injection of compressed air into the venturis, extra air is added using the jet pump principle. Relaxation occurring when the cleaning air leaves the venturi tube produces a pressure wave that continues over the whole filter bag area. This flexes the filter bags. The accumulated dust is thrown off into the hopper and from there is discharged through the discharging device, which is usually an airlock rotary valve.

The cleaning sequence is automatically controlled by an electronic control instrument and by the diaphragm and solenoid valves fitted to each blow tube.

Different designs of the cylindrical types of "MikroPulsaire" dust collectors allow bag changing on the row gas side, clean gas side by means of slewing or tilting top, or on the clean gas side having a high walk-in clean gas plenum. The raw gas inlet can be in the hopper which is a suitable construction for heavy dusts prone to agglomeration. For light non-agglomerating and very fine dusts as in the case of powder coating production, the raw gas inlet is below the tube sheet in the tangential or radial version. Constructions with the tangential raw gas inlet and cyclonic pre-separation and the tangential raw gas inlet and internal cyclonic pre-separation are also available.

The clampless bayonet-venturi technique permits quick and secure installation of filter bags and bag retainers. A metal ring is sewn into the collar of the filter bags, serving as a seal and preventing the bag from dropping through the tube sheet opening. If the bags need to be dropped into the hopper during the replacement, the metal ring is replaced by a flexible one. "MikroPulsaire" dust filters are highly effective devices for gas cleaning. They are produced in various models and sizes so that it is easy to choose for a model that best fits the plant capacity. **The cleaning operation is fully automatic**. The filter operates at a constant pressure drop between 8 and 15 mbar due to continuous efficient cleaning of the filter media. The only moving parts of the filter are the diaphragms of the externally mounted compressed air pulse control valves, which require a minimum of maintenance.

5.6 New production technologies

5.6.1 "VAMP" process

Almost twenty years ago Ferro developed a completely new powder coating manufacturing process [34]. In this process the mixing of raw materials is performed in **supercritical carbon dioxide as the solvent**. "VAMP" (VEDOC Advanced Manufacturing Process), as the process is called, eliminates the melt-mixing (extrusion) step and often the grinding step (dependent on the type of coating powder and on the application requirements) and significantly reduces the processing temperature. This is an important aspect, since it allows for the application of more reactive systems, without the risk of pre-reaction during processing. It may also be expected, that a better dispersion of the pigments can be obtained, using this technology.

The powder coating raw materials, most of which are dry solids, are charged into a pressure vessel with a stirrer. The vessel is then charged with carbon dioxide until a supercritical state is reached, in which the binder system is soluble. The system is stirred until the pigments are finely dispersed and a homogeneous fluid material is obtained. Then the batch is discharged from the vessel through nozzles, converting it to a fine powder (like a spray drying technique), requiring minimal further processing before packaging. The entire manufacturing process is controlled and monitored by a computerized control system.

For carbon dioxide (CO_2), which is low in cost, readily available and environmentally acceptable, the supercritical state is rather easily attained, when it is compressed and heated above the critical point. Its critical temperature is about 31 °C (very attractive) and its critical pressure is about 73 bar. In the supercritical state the carbon dioxide has both gaseous and liquid properties. The "VAMP" process conditions can range from 55 to 80 °C and 165 to 220 bar. Additional **benefits** of the new process are, according to Ferro:

- The lower processing temperature prevents not only gelling of the product, but also reduces rejects and improves powder quality and consistency.
- "VAMP" increases batch size significantly over the present system. This batch size can be as high as 9000 kg compared to the normal batch size of 700 to 900 kg for extruder premixes. By consequence "VAMP" allows for more efficient processing of raw materials and minimizes manufacturing variations. Realistic process parameters for such a 9000 kg batch are 4 h mixing at 220 bar and 70 °C.
- Corrections to formulations can be made during the process and there are facilities for adding pigments and additives.
- The homogeneous liquid material can be filtered before discharging, which is not possible with the extruder process.
- Quality control can be carried out at any stage of the processing step.
- There is a better control of particle size distribution, which is realized by adjustment of the pressure during discharging and by the design of the nozzles.
- "VAMP" also offers the possibility to achieve control over the nearly spherical particle shape – impossible with the current manufacturing technique.
- Powders, which may be capable of processing by "VAMP", include the crystalline resins that possess sharp melting points and very low melt viscosities, and cannot be processed

with extruders. This opens the possibility to produce very low-bake (= 100 °C) coating powders, even with excellent film flow.

According to a company spokesman, investments for a "VAMP" production facility for 9000 to 10,000 kg batches (which means a production capacity of 2500 kg/h) were $ 46 million some years ago (buildings not included). Investments for the same capacity with the extrusion technology (needed 4 extruders) would be at that time approximately $ 5 million.

Although smaller batches can also be produced on such a "VAMP" production model, it is felt, that for small batches the conventional extrusion and grinding process may provide the best cost/performance option. More details of the "VAMP" process are given by *Bush* [35], *Brand* and *Ortiz* [36], and *Mandel* [37].

With Ferro's patented invention a breakthrough in the thermosetting coating powder production, which is still based - for too long a time - on the thermoplastic compounding technique, could become a reality. Unfortunately, **the costs of this process were prohibitive**, due to the large high pressure vessel required for homogenization, the relatively long residence time of the reactive mixture in the autoclave (minutes to hours) and the discontinuous mode of operation. For these reasons the "VAMP" process seems to have vanished.

5.6.2 Continuous powder coating spray process (CPCSP)

The concept of using supercritical carbon dioxide for the production of powder coatings has created considerable interest by other powder coating manufacturers, e.g. Morton and DuPont.

Based on the PGSS (Particles from Gas Saturated Solutions) technology, a continuously operated process has been developed by the Ruhr University of Bochum, in cooperation with DuPont Performance Coatings. In this so-called CPCSP (**continuous powder coating spray process**), described by *Petermann, Weidner* and *Rekowski* [38], the main components of the powder coating mixture - binder and hardener - are melted in separate vessels (V1, V2, see *Figure 5.34*) to avoid premature reaction of the polymers. Using high pressure

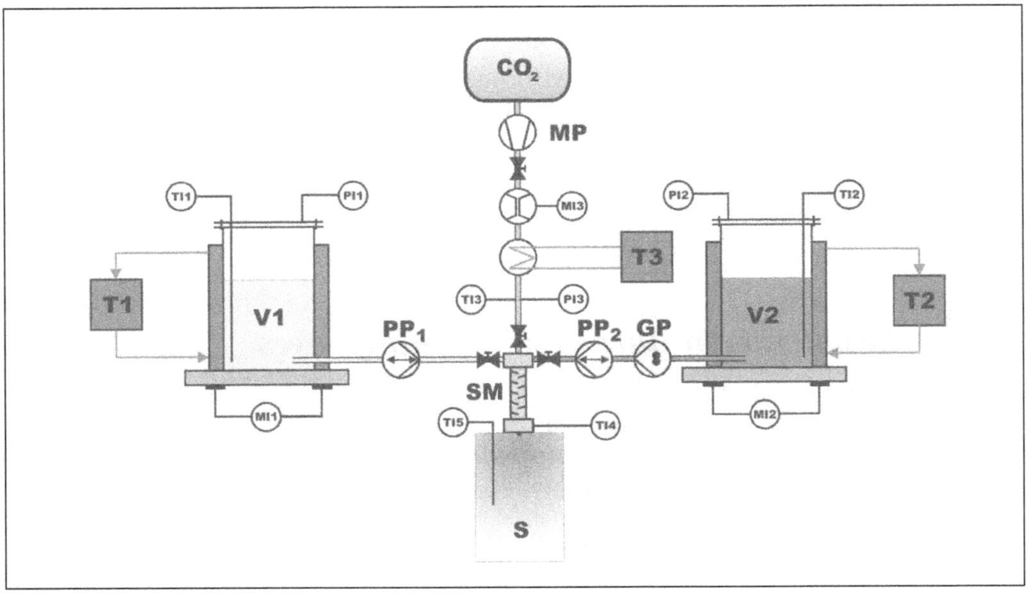

Figure 5.34: Flow sheet of the CPCSP process

pumps (PP_1, PP_2, GP), the melts are dosed to a static mixer (SM). Additionally carbon dioxide is dosed to the mixer with a membrane pump (MP). In the mixer the streams are homogenized and under pressures from 60 to 250 bar the gas is partially dissolved in the melts. The so formed partial solution is instantly released into a spray tower (S) through a nozzle. Thereby, the gas escapes explosion-like from the melt and small melt drops are formed. At the same time the droplets freeze by the Joule-Thomson effect of the expanding gas, and **a fine powder with adjustable particle size and morphology is formed**.

The new process has been tested with single powder coating components and with complete formulations in epoxy, polyester and acrylic systems. In these tests the effect on particle size and particle size distribution by varying the temperature, pressure and type of nozzle was examined. It could be demonstrated, that **tailor-made spherical particles** with particle size lower than 40 μm and a very narrow particle size distribution (PSD) could be obtained. Also the effect of increasing pressure in the static mixer on particle size and PSD, and the phase behavior of a polyester system in the presence of carbon dioxide have been investigated. With increasing pressure particle size is reduced, the PSD will be narrower and more CO_2 is dissolved in the melt. At a pressure of 177 bar a narrow PSD is formed and nearly 16 wt-% of CO_2 is dissolved in the melt. The sensitivity of the process is shown by further increasing the pressure to 190 bar, when only 1 wt-% extra of CO_2 is dissolved in the melt. Nevertheless, the particle size is further reduced and an excellent PSD is obtained.

With the CPCSP-technique another new process to produce powder coatings is offered, that can be seen as a "RE-VAMP". Apart from the already mentioned advantages, this new process has shown by experiments to provide the following benefits:

- Suitable for conventional powder coating formulations, as well as for new low-melting and high reactive systems.
- The dissolved CO_2 leads to a decrease of the melting temperature of the polymers, which means, that it is possible to set the temperature in the static mixer very low, to avoid reactions between the powder coating components.
- Very short residence time in the mixer (some seconds to some ten seconds), so that also at high temperatures a reaction between the components is prevented.
- Reduced mean particle sizes (<40 μm) with a (nearly) spherical shape and very narrow PSD's.

Experiments with this CPCSP project have been carried out with a lab scaled plant capable of producing 15 kg of powder per hour. This has already been scaled up to a 50 kg/h pilot plant and it is expected (or already realized) to increase capacity to the industrial size of 350 kg/h. And moreover, with a rate of consumption of CO_2 of only 0.5 to 2.0 kg per kg of powder coating, the continuous process is also promising from an economical point of view. This technology has not been commercialized up to 2012, but DuPont still holds the basic patents.

5.6.3 *Production by ultrasonic standing waves*

Mattern [39] describes a research program, in which the main points are, besides lowering of the curing temperature and improving the application technique, **reduction of the film thickness and simultaneously improving film flow**. Better flow and reduced film thickness not only improve optical quality, but also have cost advantages, as less material is needed. Normally the molten extrudate is cooled by using a cooling belt, which squeezes the extrudate into a thin sheet through two cooled rolls and then deposits the sheet onto a cooled steel belt. The material moves along the belt until it is cool enough to be flaked and ground. When the powder is fine enough, it passes through a classifying sieve.

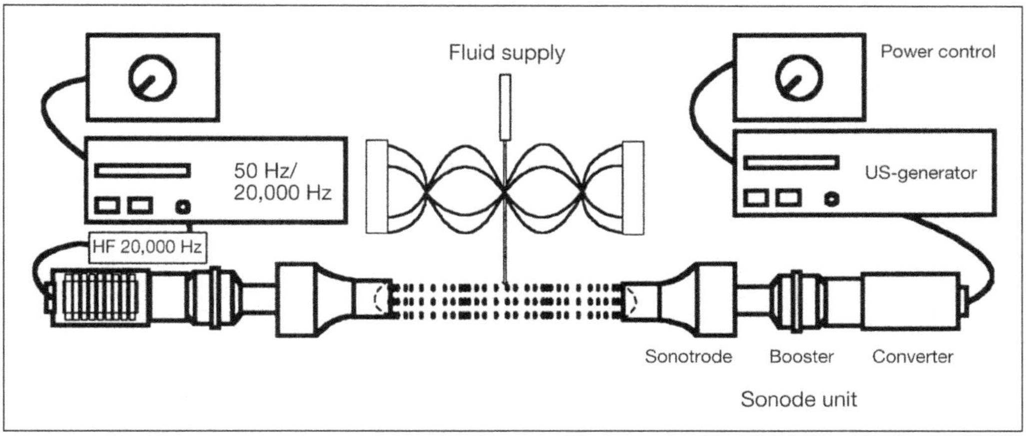

Figure 5.35: Test installation for the generation of ultrasonic standing waves[39]

Mattern eliminates all these steps and brings the molten extrudate immediately into a field of ultrasonic standing waves (illustrated in *Figure 5.35*), where the liquid mass disintegrates into fine droplets. The droplets cool within parts of a second and form fine spherical powder particles, of which a microscopic photo is given in *Figure 5.36*.

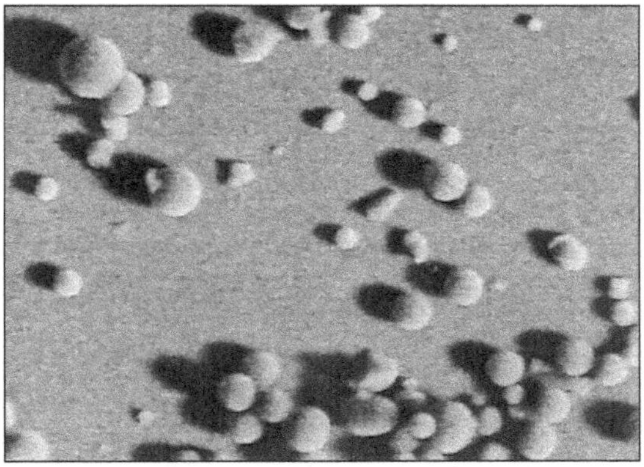

Figure 5.36: Microscopic photo of fine spherical particles, produced by USW[39]

Advantages of the ultrasonic standing waves atomization are:

- Less production steps (reduction of investments and production costs)
- Reduced energy costs (elimination of grinding and sieving)
- Narrow particle size distribution and less "fines" (saving of material)
- Production of spherical powder particles, which minimizes the contact surfaces between the particles and by consequence gives improved physical storage stability
- Spherical particles also provide improved transport through the tubes and in the spray guns
- The spherical shape leads to a more uniform distribution of the electrical charges on the powder particles and to a more homogeneous deposition of the powder on the substrate, which gives improved film flow (also with thin films)

5.6.4 Other new or modified production methods

Zweifel et al. [40] describe a process for the atomization of the molten raw materials, using a **high speed rotating hot-disk**: A premix of an epoxy resin, pigments, fillers and additives is brought into a hopper. The hardener is brought into a separate hopper. The premix and the hardener are melted and separately pumped to the rotating hot-disk, where premix and hardener are mixed. Due to the high centrifugal forces the liquid mixture leaves the disk as molten droplets, which are immediately cooled and form spherical powder particles. After classification the powder is ready for packaging.

Howell [41a] mentions, among spray drying and aqueous powder suspensions, a **suspension agglomeration method** for the preparation of **clear acrylic powder coatings**. This rather complicated process also produces spherical particles with a narrow particle size distribution.

Acrylic polymers, containing glycidyl groups, are prepared by free radical polymerization of glycidyl methacrylate and other (acrylic) monomers. This polymer, together with dodecanedioic acid and benzoin, are dissolved in xylene and form a clear solution. This solution is dispersed, at 25 °C, in an aqueous solution of a suspension stabilizer, to provide an average particle size of 4 to 5 µm. The suspension is then heated to 70 °C under reduced pressure. Now the particles agglomerate and give spherical particles of about 10 µm. The xylene is distilled off during this process. Then the suspension is cooled, centrifuged and water washed to remove the suspension stabilizer. The dried clear powder shows **better fluidization, improved storage stability**, as well as better transfer efficiency.

A patent by ICI [41b] describes a "new dispersion" process, in which a powder coating formulation is mixed in an extruder, whereby the mixture is very finely dispersed in water. A free flowing powder coating, with narrow particle size distribution and again spherical particles, is obtained after spray drying of the dispersion. Mean particle sizes between 1 and 7 µm have been obtained with this process. Mitsubishi Gas [42] recommended **using a solvent during the extrusion** step, although this technique would require an additional drying step. AkzoNobel [43] patented a similar approach by using an immiscible liquid during the extrusion process. This liquid could even be superheated pressurized water [44].

5.7 *Quality control*

This section refers to the methods used for quality assessment of powder coatings by tests performed either by the powder producers or by the end users. Since many of the test **procedures are standardized** there is no need to spend much time describing the details. In such cases the principle of the test method will be briefly described with references to the standardized test procedures. Somewhat more details will be given for the test methods typical for powder coatings only. The testing procedures for quality assessment of powder coatings have been a matter of official standardization [45, 46]. Since there are applications including a **warranty of quite a long time** (e.g. 30 years for the German Federal Railways) quality control is getting even more importance [47].

Acid number determination of resins

The acid number or the acid value [48] is defined as the number of milligrams of potassium hydroxide (KOH) required to neutralize the carboxyl groups of one gram of resin. Since most of the powder coating resins have a relatively low solubility, the toluene/ethanol solvent blend, normally used for acid value determination of most of the resins for conventional solventbased coatings, is commonly replaced by tetrahydrofuran/water blend in the ratio 10:1. This solvent blend is suitable for dissolving most of the powder coating resins. The following formula is used to **calculate the acid number**:

Equation 5.4 $\text{Acid Number} = 56.1\ AN/W$

where:

A = potassium hydroxide solution used for titration [ml]
N = normality of potassium hydroxide solution
W = weight of the resin [g]

Hydroxyl number determination

The hydroxyl number [49] is defined as the amount of potassium hydroxide (KOH), in mg, which is equivalent to the hydroxyl (OH) content of one gram of resin. The normal procedure for determination of the hydroxyl number of powder coating resins involves dissolution of the sample in a blend of cyclohexanone and pyridine (ratio 1:1), reaction with acetic anhydride, addition of water to react with the excess of acetic anhydride, addition of tetrahydrofuran to redissolve the sample and titration of the reaction product with 0.5 N potassium hydroxide. This procedure which slightly differs from the standardized methods has been specially developed for polyester powder resins and related products in the laboratories of DSM Resins BV [50]. The hydroxyl number can be calculated from the formula:

Equation 5.5 $\text{Hydroxyl number} = 56.1(B-A)\,N / W + \text{Acid number}$

where:

A = potassium hydroxide solution spent for titration of the sample [ml]
B = potassium hydroxide needed to titrate the blank [ml]
N = normality of the potassium hydroxide solution
W = weight of the resin sample in grams (recommended to be calculated as W = 280/expected hydroxyl value)

Epoxy equivalent weight determination

The perchloric acid method widely used for this purpose is based on the stoichiometrical reaction of 1,2-epoxy group with hydrogen bromide, which is generated during the reaction of perchloric acid ($HClO_4$) with tetraethylammonium bromide (NEt_4Br). The procedure involves titration of a solution of epoxy resin dissolved in methylene chloride containing tetraethylammonium bromide and crystal violet indicator with a standardized solution of perchloric acid in glacial acetic acid. When the epoxy groups are completely consumed, the free hydrogen bromide causes a color change in the crystal violet indicator.

The epoxy equivalent weight (EEW) [51] of the resin is calculated by the following formula:

Equation 5.6 $EEW = 100\,W/AN$

where:

W = epoxy resin sample [g]
A = perchloric acid solution used for titration [ml]
N = normality of the perchloric acid solution

Determination of the glass transition temperature

Differential scanning calorimetry is the most commonly used method for determination of T_g of powder coating resins and powder coatings. This technique was extensively described in Chapter 3.2.1 and will not be a matter of consideration in this section. Since T_g is dependent on the heating rate, the report on T_g should always be accompanied by the data relating to the heating rate used in the test. For powder coating resins it is common practice to use a heating rate of 5 °C/min [52].

Determination of the melting point and melting range

Polymers used for production of thermoplastic powder coatings exhibit a certain degree of crystallinity and a more or less well-defined melting temperature [53]. This temperature determines both the processing conditions during production of powder coatings and the stoving condition during their application. On the other hand polymers which are used for production of thermosetting powder coatings are mainly of an amorphous nature. They do not exhibit defined melting temperatures like the crystalline polymers, but rather a broad melting range in which the material begins to flow. This melting range should be considered as a minimum temperature interval for processing coating powders during extrusion.

Although the DSC technique can be used for determination of the melting point of the crystalline polymers, a much simpler and cheaper technique employs the so-called capillary method. This method can be used to determine the melting point, melting range, sintering point and fusion point of the resins.

The resin which has to be tested is powdered with a mortar and pestle and the powdered material is filled in a capillary tube to a dept. of 1.5 to 2.0 cm. The capillary is then placed in a bath with controllable speed of heating. The temperature is raised quickly to about 20 °C below the expected melting range. Thereafter the heating rate is slowed to 1 to 2 °C/min up to the melting point. The report may include the following:

- The melting range, defined as the two temperatures in between which the sintering and fusion of the material takes place.
- The melting point, defined as the temperature at which all the material just becomes molten.
- The sintering point, defined as the temperature at which the first signs of contraction of the powdered column in the capillary occurs and at which the material column frees itself from the capillary wall.
- The fusion point, defined as the temperature at which the material shows a fusion or translucency at any point in the capillary tube.

For practical reasons laboratories, which possess a gradient oven, can make use of this apparatus to evaluate the melting range.

Determination of the softening point

A method [54, 55] that has been well established for years is the so called Ring and Ball method. The procedure is rather simple and consists of melting and pouring the molten resin into a preheated metal ring with standardized dimensions. After cooling the excess resin is cut by a preheated knife and removed. The ring with the solid resin is then placed in a horizontal position in a bath with glycerol or other heat transfer liquid. A ball with a standardized dimension and weight is placed on the surface of the solid resin and the glycerol bath is then heated with a speed of 5 °C/min. The softening point of the resin is recorded as the temperature at which the material touches the lower horizontal plate.

Viscosity of powder coating resins

Two instruments are typically used for determination of the resin viscosity: the “Emila Rheometer” with a special measuring cup at elevated temperatures and the “ICI Cone and Plate” viscosimeter. Although different in construction, both instruments operate on the same principles [56, 57].

In the case of the “Emila rheometer” a hollow measuring cylinder is rotated by a motor at a constant speed in the measuring cup filled with the molten resin. The measuring cup is heated by a stream of oil coming from a thermostat or directly by electronically controlled electric heating elements. The rotation of the cylinder causes laminar flow in the liquid. The resistance against this rotation is proportional to the dynamic viscosity of the molten resin and can be read from the torque engine.

The operating principle of the “ICI cone and plate” viscosimeter is similar. It differs from the “Emila rheometer” in the construction of the measuring head which consists of a thermostatted plate and rotating cone instead of a cup and cylinder (*Figure 5.37*).

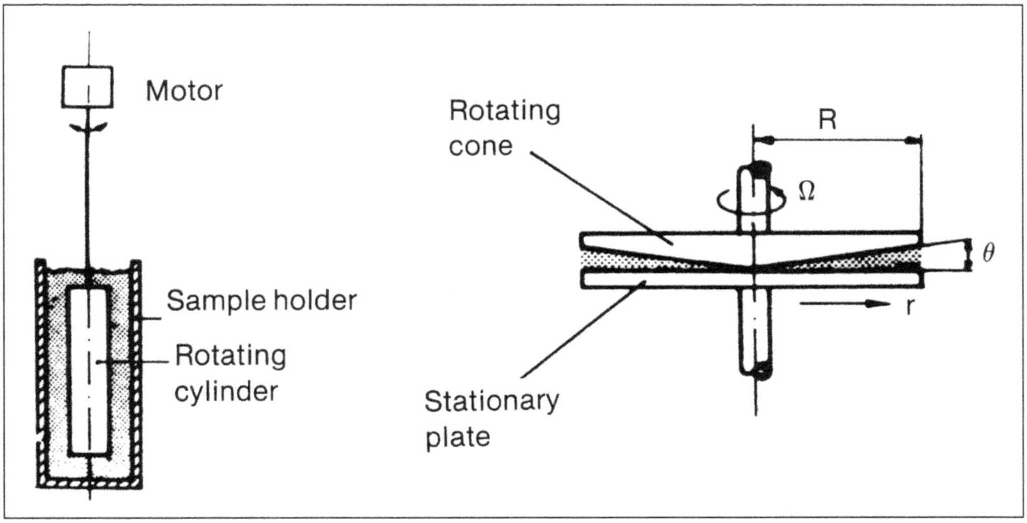

Figure 5.37: Principal construction of Emila rheometer (left) and ICI cone and plate viscosimeter (right)

The "ICI cone and plate" viscosimeter is a handler instrument, easily cleaned and requiring less material for testing. However, the temperature control is not as accurate as with the "Emila rheometer". Moreover, in the case of high viscosities, the resin creeps up and detaches from the plate, thus reducing the contact surface between the rotating cone and the stationary plate realized through the resin layer, which leads to inconsistent results in high viscosity regions. The "Emila rheometer" is widely used for melt viscosity determination of powder coating resins in Europe [57]. The "ICI cone and plate" and the procedure for measuring melt viscosity is recommended by the Powder Coating Institute in the United States [56] and has been standardized by ISO [58].

Color of powder coating resins

The most widely used method for determination of the color of the powder coating resins [59] is the "Gardner 1953" using the Gardner-Hellige varnish comparator. The resin is placed in a clean Gardner viscosity tube in a 50 % solution in dimethyl formamide (toxic!). Nowadays other solvents with lower toxicity are applied, too. Determination of the color is done visually by comparing the color of the sample with the reference color from the color disc that best matches the color of the sample.

Determination of gel time

The gel time of powder coating resins [60] mixed in a given proportion with the suitable crosslinker gives an indication of the reactivity of the system. It is the standard test for resins and for powder coatings. The method is simple, and although it seems to be very subjective it gives useful results.

By the very common manual method for gel time determination described in the ISO Standard 8130-6, ca. 0.25 ml of the material is placed by means of a measuring spoon in a convex shaped cup in a preheated thermostated plate. The stop watch is started immediately after all the material in the cup has melted. Stirring is commenced. The molten material is stirred gently in small circular movements over the total surface of the cup by a needle of very low heat capacity. When thickening starts, whilst maintaining the overall stirring action, the stirring device is lifted every 2 to 3 s to about 10 mm away from the molten powder. When the threads thus formed become brittle, such that they break and can no longer be drawn from the molten powder, the timer is stopped and the time is recorded. This is the gel time.

The value below which the absolute difference between two single test results, obtained on identical material by one operator in one laboratory within a short interval of time using the standardized test method, may be expected to lie, with a 95 % probability, is 5 %. Due to the nature of the material, reproducibility cannot be determined. A procedure, which differs in small detail, is recommended by the American Powder Coating Institute [61].

Automated versions of the gel time testers eliminate the errors due to differences in individual operators. *Weder* [62] reports a standard deviation of 2.1 s at an average gel time of 185 s with an automatic gel time tester. Compared to this the manual operation provides a standard deviation of 11.6 s at an average gel time of 193 s, according to *Weder*. The instrument operates on thermo-mechanical principles. A probe of (200 ± 10 mg) is heated to the desired temperature in a heating block. A precision measuring dolly made of glass or stainless steel dips into the specimen and is pulled out at programmable intervals The movement of the dolly which is directly dependent on the viscoelastic properties of the molten powder coating is plotted against the setting time to produce a penetration curve. All functions of the dolly are set on a central control unit which evaluates at the same time the measuring data and transmits them to either a graphical recorder or a small computer, dependent on the configuration. In this way the reciprocating movement of the dolly is recorded as a measure of the degree of hardening against time. Compared to the manual methods, the automatic methods for determination of the gelation time offer much better reproducibility.

Accelerated stability test

When the glass transition temperature of the coating composition is lower than the storage temperature, powder particles undergo agglomeration which in extreme cases can completely destroy the free flowing properties of the powder. Such a product is unsuitable for application by either fluidized bed or electrostatic spraying. In the case of rapid curing thermosetting powder coatings a premature reaction can take place during storage that has an adverse effect upon the film forming properties during curing.

A procedure for the accelerated stability test recommended by ISO allows rather quick evaluation of both the physical and chemical stability of powder coatings [63]. The method advises storage of a small quantity of powder coating at 30 and 40 °C under a standard load (to simulate the conditions at the bottom of a container) and periodical visual testing of free flowing properties of the material, followed by determination of the gel time. The test is discontinued when the powder is no longer free flowing and/or the formed lumps cannot be easily broken.

The number of days before physical properties of the powder coating are affected is reported as a measurement for physical stability of the product.

Chemical stability of powder coatings is usually expressed in terms of decrease in gel time compared to the original readings. It is very difficult to establish certain norms which will define chemical stability of the coating. Therefore, it is recommended to use these results in comparison against standard thermosetting powder coatings with a well-established chemical stability. To obtain reliable results care should be taken that the standard powder coating employs the same curing chemistry. A similar procedure is recommended by the Powder Coating Institute [64].

Density of powder coating materials

The liquid displacement pyknometer method is recommended by the Powder Coating Institute [65] and has been standardized by ISO [66]. It can be used for determination of the density of powder coatings, pigments, resins, crosslinkers and other solid ingredients of powder coatings. The method is based on a determination of mass and volume of the test portion.

The method uses a 50 ml pyknometer, according to ISO 787-10, in which part of the liquid (in most cases hexane) is displaced by the solid material. The density of the powder coating material can be calculated by the following Equation 5.7:

Equation 5.7

$$D_p = \frac{W_{fp} - W_f}{50 - (W_{fpl} - W_{fp}) / D_l}$$

where:

W_f = weight of the empty pyknometer
W_{fp} = weight of the pyknometer containing the powder sample
W_{fpl} = weight of the pyknometer filled with the powder sample and the displacement liquid up to the mark of 50 ml.
D_l = density of the displacement liquid used.

When the volume of the pyknometer (or other measuring flask) is not known, the above equation becomes:

$$D_p = D_l \frac{W_{fp} - W_f}{\left(W_l - W_f\right) - \left(W_{fpl} - W_{fp}\right)}$$

where:

W_l = weight of the flask or pyknometer, filled with the displacement liquid.

The precision and accuracy of the method have not been determined statistically, but a limited survey of testing performed by various laboratories indicates that results appear to differ by less than 0.05 g/ml.

The liquid displacement pyknometric method is relatively inexpensive, but can give erroneous results if the powder swells in contact with the liquid (in most cases hexane or heptane are suitable) or if the liquid does not totally displace the air between the powder particles. Therefore ISO connects the pyknometer with a vacuum pump. To prevent such shortcomings, the gas comparison method, which has also been standardized by ISO [67] can be used.

The **gas comparison pyknometer method** is based on the determination of the volume of a weighed powder sample, by measuring the volume of the gas, displaced within a receptacle, by the introduction of the powder sample. This is achieved by equalizing the pressure difference that arises due to displacement of a gas (or air). The powder density is calculated from the known weight and volume of the test portion.

A typical gas comparison pyknometer uses air as the medium. It consists of two cylinders (A and B) with pistons of exactly equal dimensions. The cylinders are connected by a valve and a pressure difference meter. The powder test sample, contained in a 50 ml beaker, is placed in cylinder B. Both pistons are moved by an equal amount, which gives a pressure difference between cylinders A and B. The measuring piston in cylinder B is shifted by an amount proportional to the volume of the powder sample, whereupon the pressure equalizes. The volume of the test sample is read on the scale. The density of the coating powder can be calculated, using the equation:

$$D_p = \frac{W_p}{V}$$

where:

W_p = weight of the test portion [g]
V = volume of the test portion [ml], read on the scale.

Precision data of the method have not yet been determined statistically, but if the volumes of two readings differ by more than 2 %, the results should be disregarded and the procedure should be repeated with a new sample.

Compatibility of powder coatings

It is known, that contamination or the (mutual) incompatibility of two coating powders is a serious problem, as even traces of a "foreign" contaminant can disturb the film quality of the main powder. The ISO standard specifies a method for the determination of the tendency for the mixing of two different coating powders to result in the deterioration of the surface quality of the final coating [68]. The results depend on the following characteristics of the coating powders:

- their chemical composition,
- their chemical reactivity,
- their melt properties.

The onset of the deterioration in appearance, its nature and its extent will greatly depend on the ratio in which the powders are mixed. This test can indicate or predict a possible incompatibility arising from mixing different powders both during the production process and during application of the coating powder.

The nature of the deterioration in surface appearance may manifest itself in various ways, such as change (mostly loss) of gloss, appearance or increase of orange peel (or textured finishes), presence of craters, (micro)pinholes, graininess and color contamination. For the test two coating powders are mixed together in a range of ratios and each mixture is applied to standard test panels, stoved and the surface inspected for evidence of incompatibility. The mixtures should be prepared in the following weight ratios, by weighing each component to an accuracy of 10 mg:

100/0; 99.9/0.1; 99/1; 90/10; 50/50; 10/90; 1/99; 0.1/99.9; 0/100.

After inspection of the test panels, a verbal description of the appearance of the coatings will be sufficient for the purpose of the standard method. Where appropriate, a quantitative measure of change in gloss may be obtained by measuring gloss values, according to ISO 2813.

As very small amounts of contaminants can greatly influence the gloss level, it may be useful to plot the gloss measurements graphically. In this case it is recommended [69] to employ a probability chart in which the gloss measurements are plotted as abscissa and the weight ratios of the powders in the mixture as a probability distribution on the ordinate scale. In this way it is possible to obtain a "compatibility curve", in which the influence of small concentrations of one powder on another are clearly demonstrated. Such a "compatibility curve" is illustrated in *Figure 5.38.*

Precision data are inappropriate for non-quantitative measurements. But, when designing compatibility curves in duplicate, they should overlap each other with a maximum deviation of ±10 gloss units.

Particle size analysis

Particle size is an important factor influencing powder coating application performances. Not only the average particle size, but also the particle size distribution becomes more important, especially in areas where powder coatings are expected to meet sophisticated requirements. *Pugh* et al. [70] emphasize the advantages of online as well as at-line particle size analyses.

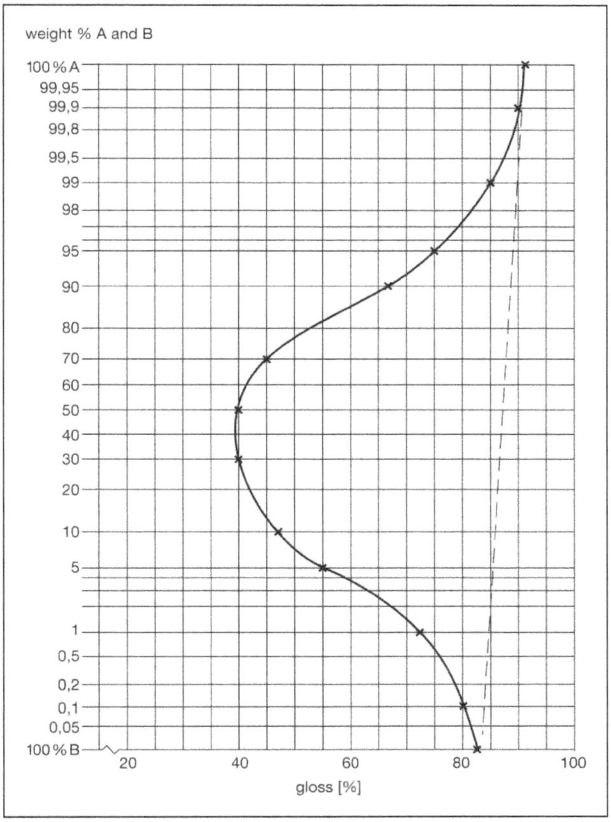

Figure 5.38: Typical compatibility curve of two different powders [69]

Many different methods for particle size determination [71] are used in practice. The list includes microscopic analysis, optical arrays, flow ultramicroscopy, light scattering, sieving, filtration, sedimentation methods, electrozone sensing, turbidimetry, optical diffraction, optical particle counters, velocimetry, neutron scattering, centrifugation etc. Excellent articles review the methods developed up to 1975 [72] and 1989 [73]. The reader interested in the methods typically used to determine the particle size of powder coatings can consult the book of *Harris* [74]. Because of space reasons, only the principles of the technique based on laser diffraction particle sizing, which has been available over two decade, will be described in this section. This technique is at the moment certainly the best method offering speed, performance, ease of use and reliable results, and is unmatched by other techniques.

The layout of the laser diffraction particle sizer is produced by Malvern Instruments Ltd [75]. A small helium-neon laser provides a parallel monochromatic laser beam which is expanded in diameter to ensure that a large number of particles will be in the measuring zone at any one time. The sample is placed in a special feeder attachment. It is dispersed in a fast air stream which carries the particles in a controlled flow through the measuring zone. As the sample passes through the beam, individual particles scatter light at angles related to their respective sizes; small particles scatter light at large angles and vice versa. The scattered light is collected by a Fourier transform lens and sent onto a silicon photodiode array detector. The detector measures the light scattered over a range of angles and produces a signal related to the composite light energy pattern of all particles in the sample. This signal is digitized and analyzed by the computer. The particle size distribution is then displayed or printed in tabular In 2002 the International Standards Organization ISO has published a standard for this method. See therefore: ISO 8130 - Part 13: Particle size analysis by laser diffraction.

Measurement of electrical properties of powder coatings

Resistivity of powder coatings is an indicator for the chargeability of the particles and their behavior during application by electrostatic guns. *Hughes* [76] describes one of the simplest methods for measuring resistivity of coating powders which makes use of the so-called resistivity cell. A sample of coating powder is subjected to a steady direct current potential while the steady state conduction current is measured. Knowing the cell geometry, the resistivity of the powder coating (r) can be easily calculated by the following equation:

Equation 5.8 $r = V \cdot A/I \cdot d$ $[\Omega m]$

where:

A = electrode surface
d = distance between the electrodes
V = cell potential
I = steady state electrical current

Contrary to what can be expected, the evaluation of different resistivity cells indicates that resistivity is minimally affected by powder compression. Therefore, prior to commencing measurements, the powder is gently tapped, so that particle packing gives reproducible results [76].

As will be seen in the following section, the relative **permittivity of the powder** (ε_r) might be an important parameter influencing the charge acceptance of the powder particles. Measurement of the permittivity can be done with the same resistivity cell using a standard capacitance measuring instrument. The relative permittivity can be determined as a ratio between the capacitance of the resistivity cell filled with coating powder (C_r) and the air capacitance of the empty cell (C_o).

Equation 5.9 $C_o = \varepsilon_o \cdot A/d$

Equation 5.10 $C_r = A \cdot \varepsilon_o \cdot \varepsilon_r/d$

where:

ε_o = permittivity of the free space = $8.854 \cdot 10^{-12}$ F/m
ε_r = relative permittivity of the powder particle

From Equation 5.9 and 5.10 follows:

Equation 5.11 $\varepsilon_r = C_r/C_o$

Charge-to-mass ratio (q/m)

Measurement of this very important charging characteristic of powder particles is relatively simple, although certain precautions have to be taken in practice. The so-called Faraday cup arranged as presented in *Figure 5.36* is the most commonly used device for determining the charge-to-mass ratio [76]. A small metallic cup is enclosed by a larger grounded metallic cup. The inner cup is connected to an electrometer instrument so that any charge associated with the collected sample sprayed by the electrostatic gun onto the test cell can be recorded. Dividing the recorded charge by the difference in weight of the cup before and after the test leads to the charge-to-mass ratio for the powder in question.

This simple technique is quite suitable for determining q/m values of powder coatings when applied by triboelectric guns. However, the technique is vulnerable to errors when the efficiency of high-voltage corona guns has to be assessed. In such a case the flying powder particles are always accompanied by a stream of free ions emanating from the gun nozzle. The charge of the free ions entering the inner cup will be registered by the instrument although they do not contribute to the charge-to-mass ratio. As a result much higher measurements than the true values are obtained. An improvement of the measuring technique involves an attached grid screen across the inlet of the Faraday cup which traps the free ions before they enter the inner cell. In practice this technique does not perform satisfactorily.

The grid itself is very soon coated with powder particles which initiate back-ionization thus discharging further oncoming particles and adversely affecting the accuracy and validity of the obtained results. An improvement on the technique using a water irrigated grid has been proposed by *Moyle* and *Hughes* [77]. In this method powder particles accumulating on the grid trap are permanently washed out by a stream of water, allowing an accurate measurement of the charge-to-mass ratio of a corona-charged powder cloud. Details about the construction and practical use of a commercially available electrostatic powder coating diagnostic kit are given by *Hearn* [78].

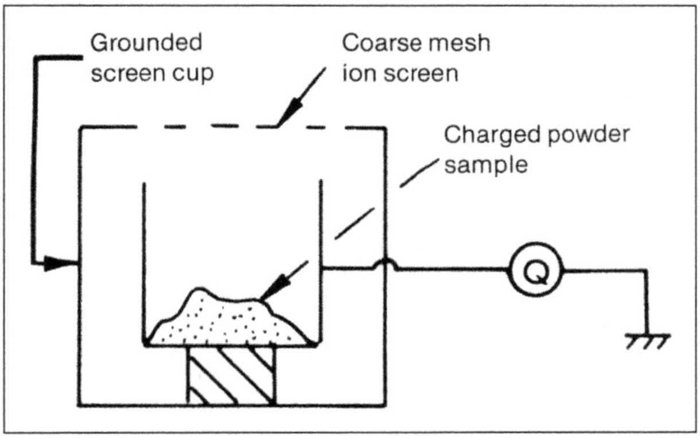

Figure 5.39: Schematic illustration of a Faraday cup test cell
source: Oil and Colour Chemists' Association

Inclined plane flow test

The flow of powder coatings is too complex a phenomenon to be determined by a single test. However, the inclined-plane flow test can be considered as a useful indicator of the degree of flow that may occur during the curing cycle contributing considerably to the surface appearance of the cured film. The test procedure is rather simple. A pellet of coating powder, prepared by means of a mold press, is placed horizontal to the glass plate that has been preheated to the test temperature. The glass plate stands in an oven on a metal plate rack assembly which is capable of being maintained in both horizontal and inclined positions by means of a lever without opening the oven door. When the pellet is slightly molten in a flat position for a time of 30 s (including the time, needed to close the oven door), the plate is tilted to 65° and kept in this position for 15 min. The amount of flow is measured as the length of the trace that is made by the molten powder. Since the oven design, angle of inclination and pill variations affect the results, making interlab reproducibility difficult to correlate, a standardized procedure has been developed by ISO [79], which prescribes a test temperature of 180 °C (or the recommended stoving temperature, if known), pill thickness of 6.5 mm, pill weight one half of the specific gravity of the test powder, and angle of inclination of 65°.

The precision of this standard method, as obtained by statistical examination of interlaboratory test results, is described as follows: The value below which the absolute difference between two single test results, each being the mean of duplicates, obtained on identical material by one operator in one laboratory within a short interval of time, using the standardized test method, may be expected to be with a 95 % probability is 5 %.

Gradient oven test

The gradient oven laboratory apparatus manufactured by Byk-Labotron is a very useful instrument for evaluation of the stoving characteristics of powder coatings. The essential part of the oven is the heating bank which consists of 45 heating elements embedded in aluminum blocks. The elements are arranged side by side, and are separated by insulation material so that each element may be programmed with a different temperature. The microprocessor through which the control of the heating bank is affected allows a temperature rate control between 2 and 30 °C/min and simulation of real time-temperature profiles of industrial ovens. The oven can produce a linear gradient with a maximum span of 100 °C within the operating range of the instrument between 40 and 250 °C and a stepwise gradient with two, three or four steps. In all cases the maximum difference between the steps can be 50 °C.

The coating to be tested is applied on a panel which is placed on the heating bank. An electrically driven pressing device ensures an efficient contact between the panel and the heated surface, so that the time taken for the sample to reach the programmed surface temperature varies between 20 and 120 s dependent on the thickness of the test panel.

The gradient oven can be used to determine the melting point or melting range of the resins used in the manufacture of powder coatings or the coatings themselves. However, the most important use of the gradient oven is in the evaluation of the curing characteristics of the thermosetting coating. The possibility to set up different curing temperatures on a single test panel results in considerable savings in experimental time since several tests may be conducted simultaneously. For example in one single experiment with a fixed curing time one can easily determine the minimum curing temperature needed to reach good mechanical properties. At the same time an impression of the coating behavior at under-cure and over-cure can be obtained with respect to the mechanical properties, together with the temperature dependence of the yellowing tendency. The use of the stepped gradient facility allows larger test areas of a single temperature to be produced in cases where the optimum curing temperature is more closely defined. Finally, the instrument allows a simulation of the temperature profile of industrial ovens. In such a case the temperature of the heating bank is uniform along the whole length at any given time following the profile of the real industrial oven [80].

Measuring the thickness of powder coatings

The simplest method for measuring the thickness of powder coatings involves causing mechanical damage to the film in order to reach the substrate and measurement of the depth of the niche by micrometer [81, 82].

Modern methods for measuring the film thickness of powder coatings are based upon high-frequency eddy current and electromagnetic induction [83]. The eddy current method is used to determine the thickness of coatings applied on non-ferrous metals such as copper and copper alloys, zinc and zinc alloys, aluminum and aluminum alloys and other magnetic alloys including some stainless steels. For coatings on magnetic steel such as iron and iron alloys including some stainless steels, the electromagnetic induction method is virtually always used since it has been proved to be the most reliable [84]. Instruments that are commercially available are produced in both hand-held and bench located versions. Contemporary hand-held instruments automatically calculate the mean result from either a single series, or a multiple of series or batches. They provide the range of thickness measured and often a whole series of statistical data. Many of them can store the measurements data for later off-loading into a printer or computer.

Non-destructive, hand-held coating thickness gauges are available to measure on all kinds of metals with just one gauge and one probe. Bench located versions which are more suitable for laboratory use have all the facilities of the hand-held instruments. However, the larger size permits addition of options that cannot be incorporated in the hand-held version, such as continuous monitoring of the film thickness when large numbers of simply shaped components are to be controlled.

Instruments for the contactless measurement of the powder layer thickness (before stoving) belong to the latest developments in the field of thickness measurements.

To correlate the film thickness to other properties like optical appearance or chemical resistance statistical methods are recommended. According to *Koban* et al. [85] the general linear model proves to be superior to either correlation analysis or regression analysis.

Solvent cure test

The methods for monitoring the curing of powder coatings described in Chapter 3 are strong tools in the hands of the paint chemist, that are necessary in order to develop a proper powder coating formulation. However, they are too complicated, time consuming and therefore unsuitable for quick evaluation of the state of cure that is very often necessary for the end user to carry out.

A solvent cure test method has been developed by which one can differentiate between cured, partially cured and uncured thermosetting powder coating [86]. The method resembles the well-known acetone resistance test used in solventborne coatings. A 3 · 3 inch soft cloth pad is attached to the ball end of a 2 lb ball peen hammer. The hammer is used to provide a constant force on the test panel and to eliminate operator dependence. The pad is saturated with the testing solvent which is methyl ethyl ketone (MEK) for pure epoxy powders or a blend of 10 % MEK and 90 % xylene for hybrids, polyester/TGIC (and alternatives) or polyurethane coatings. The pad is slid on the test panel looking for signs of coating failure. Although it is stated that less than 25 double rubs indicate insufficient curing, 25 to 100 rubs partial curing and more than 100 rubs full curing, results obtained by this test should always be compared with known cured panels of the same type of coating.

Hardness of cured coatings

The common test methods used to determine the hardness of organic surface coatings are also used to measure the hardness of cured powder coatings [87–92]. These are: indentation hardness, pencil hardness and pendulum hardness. The test methods have been standardized and the procedures are described in detail in the corresponding standards.

The **Tukon indentation** tester is the device used to determine the hardness of the films with an indenter that is a pyramidal diamond of prescribed dimensions known under the name Knoop indenter [87]. After placing a fixed weight on the indenter for a fixed time period the indentation of the diamond is determined by a microscope. Dividing the applied weight by the projected area of the indentation the so-called **Knoop Hardness Number** is obtained. Test results are rather inconsistent for rubbers and materials with a T_g lower than the ambient testing temperature, because of partial recovery of the deformations. Fortunately, this is not the case with powder coatings where rather reproducible results can be expected. A similar indentation hardness method uses the so-called **Pfund indenter** which is hemispherical quartz or sapphire indenter of prescribed dimensions. The result of the test is reported in terms of Pfund Hardness Number. Another standardized method for determination of the hardness of powder coatings is the **Buchholz indentation test** [88].

Determination of the hardness by means of a square shaped graphite pencil is a very popular test method because of its simplicity. A set of calibrated wood pencils having the following scale of hardness: 6B, 5B, 4B, 3B, 2B, B, HB, F, H, 2H, 3H, 4H, 5H, 6H which increase from left to right, is used as a tool for determination of the hardness of the film. The hardness is reported as the grade of the hardest pencil that does not cause any marring of the surface [89].

Pendulum hardness test methods make use of the **Persoz pendulum**, the **König pendulum** and the **Sward Rocker** [90–92]. The principles of the tests are similar although the constructions of the testers differ considerably. In the pendulum testing devices, the pendulum is in contact with the coated surface through two steel balls. In the case of the Sward Rocker the contact between the tester and the film surface is assured through two flat metal rings. In both cases the testing element (the pendulum or the ring) is replaced from its equilibrium position and oscillates freely, swinging back and forth through a small angle. The oscillation amplitude decreases with time due to absorption of mechanical energy by the coating. Harder coatings give longer dumping times while soft coatings are characterized by short dumping times. In the case of the Sward Rocker the dumping is the result of both absorption of mechanical energy by the coating and energy loss due to the rolling friction between the rings and the surface. Test results in the case of pendulum hardness are reported as the time in seconds required for reduction of the oscillation amplitude from a specified starting angle to a specified ending angle. Similarly, the test results with the Sward Rocker are reported as twice the number of swings required for a fixed amount of dumping.

Flexibility tests

Mandrel bend tests being a matter of standardization [93-95] consist of bending the coated metal panels over metal supports of cylindrical or conical shape. In the case of cylindrical mandrels the test results are reported as the smallest diameter of bending at which a crack of the coated surface is not detected by the unaided eye. When a conical mandrel is used as a support the test results are reported as the distance from the small end of the cone to the end of the crack in the panel. Both methods provide formula according to which the elongation at break of the coating can be easily calculated and reported when necessary.

Impact resistance of cured coatings

Impact resistance of powder coatings is determined in a simple way by dropping a weight down a guide tube onto a hemispherical indenter which rests on a coated panel. Due to the hole opposite the indenter, the panel together with the coating undergoes very rapid deformation during which cracking of the coating may occur. The "Gardner Impact Tester" [96-97] having 2 lbs, an 4 lbs weights and a 40 inch guide tube with a maximum impact of 160 in·lbs (= ca. 16 Nm) is a widely used instrument for performing the tests. The results are reported as the maximum in·lbs (or Newtonmeter) value obtained at which there is no cracking of the film. The report must contain an indication of a direct or reverse impact depending on whether the coated surface is in an up position in contact with the indenter or down. As a rule, the reverse impact test is more severe than the direct impact. Impact resistance depends very much on the thickness of the film. Therefore for comparison purposes impact tests should be carried out at the same coating thickness. Adhesion to the substrate also plays an important role. Insufficient adhesion can lead to a misinterpretation of the results.

Adhesion of cured films

The so-called **crosscut adhesion** test is the most common test method for assessment of the adhesion of a powder coating to the substrate. A series of parallel cuts are made with a special knife in the film deep enough to touch the substrate surface. A second series of parallel cuts is then made at right angles to the first. By means of pressure sensitive adhesive tape the squares which are detached from the surface are pulled off. The result is reported as the number of squares remaining on the surface after this treatment or with a number rating from 5B to 0B depending on the percentage of the affected area of the lattice. The method as such, although not very accurate, has been standardized [98, 99] and is widely used in practice.

More reliable results which can be directly correlated to the adhesion properties of the film provides the pull-off test method [100]. The method is based on determination of the greatest perpendicular force that a coated surface area can bear before a plug of material is detached. However, since this method makes use of tensile stress as compared to shear stress applied by scratch or knife adhesion, the results obtained by both methods may not be comparable.

Rosenthal [101] recommends the so called notched coating adhesion (NCA) test. The coating is notched with a razor blade and a tension is applied in a tensile machine perpendicular to the notch. The critical strain at which the delamination starts to propagate is recorded and analyzed with mathematical methods to determine the adhesive energy.

Gloss

The gloss of powder coatings is determined as specular gloss, which is defined as the relative luminous reflectance factor of a specimen in the mirror direction. The relative luminous reflectance factor on the other hand represents the ratio of the luminous flux reflected from a specimen to the luminous flux reflected from a standard surface under the same geometric conditions. According to ASTM [102] and ISO [103] the standard surface is a polished glass plate. Measurements are performed by means of commercially available optical instruments. Commercially available gloss meters consist of a light source furnishing an incident beam, means for locating the surface of the specimen, and a receptor of the reflected light, which is a photosensitive device responding to visible radiation. Measurements are made usually with 60° and 20°, and sometimes with 85° geometry. The 60° geometry is normally used for determination of the gloss. However, if the gloss rates are higher than 70 at an incident angle of 60°, then 20° geometry is more suitable. In contrast, when the gloss rates are lower than 10 at 60°, then 85° geometry gives more consistent and suitable results for comparison.

Voelkel et al. [104] recommend an automatic measurement of gloss, leveling and hiding power ensuring a higher quality of the coating.

Salt spray resistance

The salt spray resistance test is very often performed in order to evaluate the performance of the coated ferrous and non-ferrous metals under heavy corrosion conditions. The conditions in salt spray testing, the apparatus required, the preparation of the test specimens and the reporting of the results have been a matter of standardization [105–107]. The test method consists of spraying 5 % water solution of sodium chloride, with pH between 6.5 and 7.2, over the coated panel which is placed in the salt spray chamber at a temperature of 35 °C. When it is desired to determine the development of corrosion from an abraded area in the coating, a scratch is made with a sharp instrument so as to expose the underlying metal before testing. The period of test is usually a matter of mutual agreement between the supplier and end user of the coating or it complies with already existing standards depending on the end use of the material. Regardless of the testing time, the test is continuous for the duration of the entire test period except for short interruptions necessary to inspect, rearrange or remove the test specimens.

The salt spray test is useful in rating the relative resistance of the protective coatings in atmosphere containing salt. However, the reproducibility of the test is highly dependent on the type on the substrate. It should also be noted that the obtained results should not be related to the corrosion resistance in other media.

Alternative techniques with similar results like the electrochemical impedance spectroscopy are even more time consuming (up to ten months). However, *Heinze* et al. [108] reported that measurement of the ion permeability is a reliable method and can be performed in 22 hours for an epoxy powder coating. For aluminum metallic pigmented coatings *Maardalen* et al. [109] found that dipping 32 h in 10 % HCl can replace easily a 1000 h salt spray test.

Transport and spraying characteristics of coating powders

It is well-known, that the transport and spraying characteristics of powders [110] are highly dependent on their flow properties in bulk and in air.

ISO 8130 specifies a method for determining the flow properties of a mixture of coating powder and air, which reflects commercial practice in powder spraying. The results obtained are influenced by the chemical composition of the coating powder, its density, particle size distribution and particle shape together with the tendency of the particles to agglomerate and to accept a triboelectric charge [111].

The procedure, described in this method, is considered to be more meaningful than the "angle of repose" or "flow angle", sometimes used to evaluate bulk flow properties. In the latter, the angle of the cone, formed when a powder is allowed to flow through a vertical funnel onto a horizontal surface, is measured. A certain mass of powder with good free flow properties forms a shallower cone than an equal mass of a powder with poorer free flow. The objections to the angle method are, that it is basically qualitative and that the powder is used alone, whereas during application it is mixed (and fluidized) with air [112].

The ISO procedure is that a specified quantity of powder is placed in a vessel and is fluidized under standard conditions of atmospheric pressure and temperature. The height of the powder during and after fluidization is measured and the rate at which the fluidized powder flows through a specified orifice is then measured. These measurements are used to calculate the fluidization factor ϕ and the powder flow rate (or "flow factor") R. Together they define the transport and spraying characteristics of the powder. The fluidization factor ϕ is calculated, using the Equation 5.12:

Equation5.12 $$\phi = h_1/h_0$$

where:

h_1 = height (in mm) of the fluidized powder bed
h_0 = height (in mm) of the powder at rest.

The powder flow rate (or "flow factor") R is calculated, using the Equation 5.13:

Equation 5.13 $$R = m \cdot \phi$$

where:

m = weight of powder (in g), collected after flowing through the orifice during 30 s.

It may be expected, that two results, each being the mean of duplicates, obtained on identical material by one operator, differ not more than 5 %.

Deposition efficiency

Deposition efficiency (also called transfer efficiency) is defined as the proportion of the weight of a powder deposited on a target, compared to the weight of powder sprayed, expressed as a percentage.

Standard test methods to determine the deposition efficiency have been developed by ASTM [113] and later-on by ISO [114]. As the ISO-method is less complicated, only this one will be described here.

The method is applicable to powders applied by corona and tribo charging, and may be used to compare the deposition efficiency of different powders with the same gun, or of different guns with the same powder. Deposition efficiency is also a measure for the so-called "overspray", which should be kept as low as possible, for economical and technological reasons. The results, obtained with this method, are dependent on numerous properties of the powder and the test conditions (see also Chapter 6.2.1.4), such as: chemical composition, density, particle size distribution and particle shape, moisture content, and flow properties of the powder in its mixture with air; but also gun voltage, gun polarity, gun air pressure, spray pattern produced by the gun, and air humidity.

The ISO method consists of spraying charged powder (at a known flow rate and under known conditions of atmospheric temperature and humidity) at the central one of a series of five similar steel tubes, each wrapped in aluminum foil. The mass of powder, deposited on the central tube, is determined (by weighing the aluminum foil), from which the deposition efficiency can be calculated. The deposition efficiency E, expressed as a percentage, is calculated using the equation:

Equation 5.14

$$E = \frac{m_p \cdot 60 \cdot 100}{P_f \cdot t}$$

where:

m_p = weight of powder, deposited on foil (in g)
t = spray time (in s)
P_f = powder flow rate (in g/min.)

Experience shows, that the powder flow rate may be determined with a precision of ±3 %. The deposition efficiency with both corona and tribo charged guns may be determined with a precision of ±5 %.

Calculation of lower explosion limit

The **lower explosion limit** (LEL) [115] is that concentration of a coating powder (in g/m^3), in a mixture of powder and air, below which self-propagation of flames is not possible. With other words, it is the minimum concentration of a coating powder in air, which could form an explosive mixture.

Reliable methods for the measurement of this quantity require the use of special apparatuses, which may not be readily available. As the lower explosion limit is an important quantity, which may be asked for safety and legal reasons, the following calculation procedure, developed by ISO, is considered to be a suitable alternative. It is based on the knowledge of the gross calorific value (H_0) of the product, as determined by the method described in ISO 1928, or on calculation by summation of the gross calorific values of the combustible constituents of the product. When calculating the gross calorific value (H_0) of a powder from its combustible constituents, the procedure is: Multiply the gross calorific value of each combustible constituent by its weight fraction present in 1 g of powder, and sum together. The quantity H_0 will then be noted as

$H_0 = c_1 H_1 + c_2 H_2 + \ldots,$

where c_1, c_2 etc. are the weight fractions of the constituents $_1$, $_2$, etc. and H_1, H_2, etc. are their individual calorific values. The lower explosion limit C can be calculated, using the Equation 5.15:

Equation 5.15 $C = A + B/H_0$

where:

A = 2.5 g/m^3
B = $1.24 \cdot 10^6$ J/m^3
C = lower explosion limit (g/m^3) of the powder
H_0 = gross calorific value (J/g) of the powder

A Japanese research group [116] investigated the ignitability of polyester powder coatings in a concentration of 0.6 kg/m^3 at a particle size of 32 μm. They found a decrease of the ignitibility in an electrical field created by corona discharge.

Abrasion resistance

Abrasion resistance tests involve usually mounting a specimen (approximately 100 mm square or round and typically less than 12.5 mm thickness) to a turntable platform that rotates at a fixed speed (*Figure 5.40*). Two abrasive wheels, which are applied at a spe-

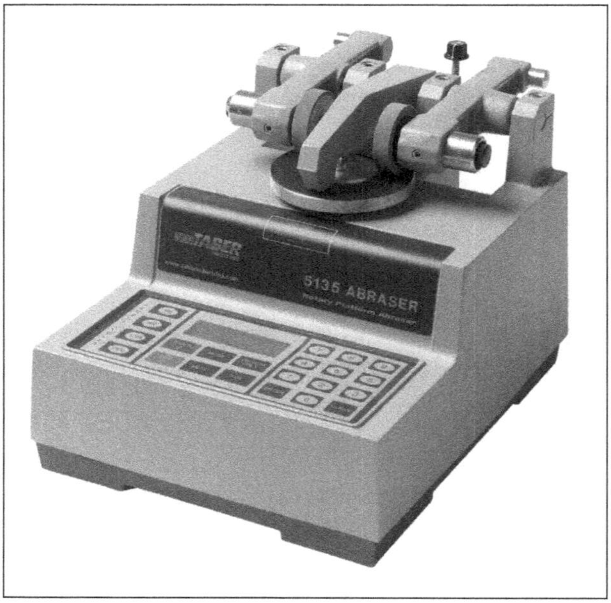

Figure 5.40: "Model 5135 Taber Abraser" source: Taber Industries

cific pressure, are lowered onto the specimen surface. As the turntable rotates, the wheels are driven by the sample in opposite directions about a horizontal axis displaced tangentially from the axis of the sample. One abrading wheel rubs the specimen outward toward the periphery and the other, inward toward the center while a vacuum system removes loose debris during testing. A characteristic rub-wear action (sliding rotation) is produced on the surface of the test piece and the resulting abrasion marks form a pattern of crossed arcs in a circular band that cover an area of 30 cm^2. The method of evaluation depends on the type of material being tested but commonly includes the number of cycles required to reach a predetermined end point or how much material has been removed by abrasion.

Taber Industries offers a selection of abrading wheels to accommodate both flexible and rigid specimens. The wheels are comprised of silicon carbide or aluminum oxide abrasive particles embedded in a resilient or vitrified (clay) binder, and designed so that the binder material breaks down during use - thus exposing and creating a new abrading surface. For powder coatings and other organic coatings, the "CS-10" and "CS-17" are normally specified.

Review of tests for pipeline coatings

Powder coatings are widely used for protection of pipelines. The requirements with respect to the coating properties in this application area differ considerably from those which are common for very standard powder coatings. A number of test procedures have been developed specifically for testing coatings systems intended for use as pipeline coatings. Those standard methods which are applicable to the powder coatings are listed in the references [117-127] without summarizing the testing principles within this section.

5.8 References

[1] *Scott, J.A.*, Polymers Paint Colour Journal, 179, 4239, 407, 1989
[2] Werner&Pfleiderer GmbH, Brochure WER 05 066/3-1.5-VIII. 86
[3] *Anand, C.; Bellehumeur, C. T.; Azaiez, J.*, Plastics, Rubber and Composites, 34(10), 443-449, 2005
[4] *Franz, P.*, "Buss Powder Coating Technology", Buss brochure MK-T3P, October 1988
[5] *Walty, O.*, Production Process for Thermoset Powder Coatings, DSM Resins Symposium, Singapore, 1986
[6] *Craddock, I.*, Manufacturing Chemist, Jan. 1989, p. 24
[7] *Harris, S.*, The Technology of Powder Coatings, Portcullis Press Ltd, Redhill 1976
[8] Hosokawa Micron, Product Presentation "Cyclomix" at PCE '98, Amsterdam, Jan. 1998; or Brochure "Cyclomix"
[9] *Winkler, J., Klinke, E., Dulog, L.*, J. Coat. Technol., 59, 754, 35, 1987
[10] *Franz, P., Bolt, A.*, Powder Coatings, 4, 1, 3, 1981
[11] *Meijer. H.E.H.*, The Modeling of Continuous Mixers, DSM Resins Powder Coating Symposium, Noordwijk, May 1988

[12] Buss AG Brochure MK/832/03/1088/001/E, p. 4
[13] *Franz, P. and Meier, E.*, Proc. Powder Coating - Formulating and Production Conf. Indianapolis, Sept. 1998, or European Coatings Journal 1-2 (1999) p. 62
[14] Coperion Werner&Pfleiderer GmbH, Brochures "ZSK MEGAcompounders"
[15] a) *Swanborough, A.*, PPCJ, 188 4408, (Sept. 1998) p. 11
b) APV Baker, Brochure "MP-PC extruders"
[16] a) Entex, Product Presentation "Planetary Roller Extruders" at PCE '98, Amsterdam Jan. 1998, or Brochure
b) *Lüchtefeld, S.*, Congress Papers PCE 2002, Nuremberg, Jan. 2002, p. 299
[17] *Lovatt, J.*, APV-Brochure "Extruder Selection", 1995
[18] *Peitz, Th.*, Proc. Powder Coating - Formulation and Production Conf. Indianapolis, Sept. 1996, p. 55
[19] *Bodnar, E.*, European Coatings Journal, 7-8 (1996) p. 532
[20] Frewitt MGH, Hammermills Brochure, October 1987
[21] *Sanders, G.*, Proc. Powder Coatings Symp. - Solutions for Today - Goals for Tomorrow Birmingham (UK), 11-12 Mar. 1997
[22] a) *Galk, J.*, Congress Papers PCE 2002, Nuremberg, Jan. 2002, p. 291
b) *Huang, C.C.*, Powder Coating 2000 Conf. Proc., Indianapolis, Sept. 2002
c) *Galk, J.*, Product Pressent. at PCE '98, Amsterdam, Jan. 1998
[23] *Grisa, A. M. C.; Freire, E.; Aguzzoli, C.; Zeni, M.*, Progress in Rubber, Plastics and Recycling Technology, 20(1), 47-50, 2004
[24] *Fu, J.; Krantz, M.; Zhang, H.; Zhu, J.; Kuo, H.; Wang, Y. M.; Lis, K.*, Powder Technology, 211(1), 38-45, 2011
[25] Alpine AG, Brochure 017/9e, p. 12 "Process, Engineering and Systems for Mechanical Processing Technology"
[26] *Schaller, R.E., Lapple, C.E.*, J. Paint Technol., 44, 571, 87, 1972
[27] Russell Finex, Product Presentation "Vibrasonic 2000" at PCE '98, Amsterdam Jan. 1998
[28] Telsonic, Product Presentation at PCE '98, Amsterdam, Jan. 1998
[29] *Tunnicliffe, G.*, Paint & Resin, December 1987, p. 8
[30] *Newton, H.W., Newton, W.H.*, "A Study of Classification Calculation", Rock Prod., 35, 1932
[31] *Eder, Th.*, Aufbereitungstechnik, 4, 140, 1961
[32] *Rumpf, H., and Leschonski, K.*, Chemie Ing. Technik, 39, 1231, 1967
[33] *Schwamborn, K.H.*, Tendencies of Development in Grinding Systems for Powder Coatings, Buss Powder Coatings Training Course, Lucerne, November 1988
[34] Powder Coatings Bulletin (Sid Harris), Nov. 1995, p. 3
[35] *Bush, W.*, Powder Coating, March 1996, 7 (2), p. 58
[36] *Brand, D. and Ortiz, C.*, Proc. Pulverlack Kongress, Hamburg, 23-24 Sept. 1996
[37] *Mandel, F.*, Proc. Symp. Powder Coatings - Solutions for Today, Goals for Tomorrow, Birmingham (UK), 11-12 Mar. 1997
[38] *Petermann, M., Weidner, E. and Rekowski, V.*, Congress Papers PCE 2002, Nuremberg, Jan. 2002, p. 285
[39] *Mattern, R.*, FARBE & LACK, 105, 11 (1999) p. 52
[40] *Zweifel, Y., Franssen, P., Senuma, Y. Hilborn, J., Cotting, J. and Rickert, C.*, 213th ACS National Meeting (San Francisco), pp. PMSE (1997)
[41] a) *Howell, D.*, Conf. Paper "Powder Coatings - What's Next?" Birmingham (UK), 10 Mar. 1999
b) ICI, WO 97/45476
[42] Mitsubishi Gas Chemical Co., US 20050113490, 2005
[43] AkzoNobel Coatings, WO 2009034112, 2009
[44] AkzoNobel Coatings, WO 2010103019, 2010
[45] ASTM D 3451, Standard Practices for Testing Polymeric Powders and Powder Coatings.
[46] ISO 8130, Coating Powders Part 1-14
[47] *Schmeller, N.*, JOT, Journal fuer Oberflaechentechnik, 47(8), 38-40, 2007
[48] ISO 2114: 2000, Plastics (polyesters) and paints and varnishes (binders). Determination of partial and total acid value
[49] ISO 4629: 1996 (under review in 2001), Binders for paints and varnishes: Determination of hydroxyl value by titrimetric method
[50] DSM Resins, Powder Coating Resins Brochure, page 6.2, July 1988
[51] ASTM D 1652, Standard test Method for Epoxy Content of Epoxy Resins; or: ISO 7142: 1996 (under review in 2001), Binders for paints and varnishes: Epoxy resins - General methods
[52] DSM Resins, Powder Coating Resins Brochure, page 6.3, July 1988
[53] DIN 53181, Binders for paints and varnishes; Determination of the melting range of resins by the capillary method
[54] ISO 4625-1, Binders for paints and varnishes; Determination of softening point - Ring-and-ball method
[55] ASTM E 28, Standard Test Method for Softening Point of Resins by Ring-and-Ball apparatus
[56] Recommended Procedure for Measurement of Melt Viscosity of Powder Coating Materials, No. 5, Powder Coating Institute, Alexandria, VA 22314, USA

[57] DIN 53214, Testing of Paints and Varnishes; Determination of Rheograms and Viscosities by Rotational Viscosimeters
[58] ISO 2884-1: 1999, Paints and varnishes - Determination of viscosity, using rotational viscometer - Part 1: Cone-and-plate viscometer operated at a high rate of shear
[59] ISO 4630: 2002, Binders for paints and varnishes - Estimation of color of clear liquids by the Gardner color scale
[60] ISO 8130-6: 1998, Coating Powders - Part 6: Determination of gel time of thermosetting coating powders at a given temperature
[61] Recommended Procedure for Gel Time Reactivity, No. 6, Powder Coating Institute, Alexandria, VA 22314, USA
[62] *Weder, J.A.*, Finishing, 13, 4, 41, 1989
[63] ISO 8130-8: 1999, Coating Powders - Part 8: Assessment of the storage stability of thermosetting powders
[64] Recommended Procedure for Assessing the Storage Stability of Coating Powders, No. 1, Powder Coatings Institute, Alexandria, VA 22314, USA
[65] Recommended Procedure for Density of Powder Coating Materials, No. 4 Powder Coating Institute, Alexandria, VA 22314, USA
[66] ISO 8130-3: 1997, Coating powders - Part 3: Determination of density by liquid displacement pyknometer
[67] ISO 8130-2: 1997, Coating powders - Part 2: Determination of density by gas comparison pyknometer (referee method)
[68] ISO 8130-12: 1998, Coating powders - Part 12: Determination of compatibility
[69] *Lange, P.G. de*, XIXth FATIPEC Cngressbook, Vol. II, p. 399, Aachen 1998
[70] *Pugh, D.; Kippax, P.; Ward-Smith, S*, Schuettgut, 11(4), 254-258, 2005
[71] ISO 8130-1: 1997, Coating powders - Part 1: Determination of particle size distribution by sieving
[72] *Collins, E.A., Davidson, J.A., Daniels, C.A.*, J. Coat. Techol., 47, 604, 35, 1975
[73] *Barth, G.H., Sun, S.T.*, Anal. Chem., 61, 143, 1989
[74] *Harris, S.T.*, Technology of Powder Coatings, p. 143-153 Portcullis Press Ltd, London, 1976
[75] *Ward, J.* (Ed), Thermoset Powder Coatings, FMJ International Publications Ltd., Redhill, Surrey 1989, p. 39
[76] *Hughes, J.F.*, J. Oil Col. Chem. Ass., 4, 145, 1989
[77] *Moyle, B.D., Hughes, J.F.*, IEEE Transactions on Industry Applications, IA-20, 6, 1631, 1984
[78] *Hearn, G.L.*, Thermoset Powder Coatings, Ed. J. Ward, FMJ International Publications Ltd., Redhill, Surrey 1989, p.116
[79] ISO 8130-11: 1997, Coating powders - Part 11: Inclined-plane flow test
[80] *Cauchois, G.*, Thermoset Powder Coatings, Ed. J. Ward, FMJ International Publications Ltd, Redhill, Surrey 1989, p.128
[81] DIN 50986, Measurement of Coating Thickness; Wedge Cut Method for Measuring the Thickness of Paints and Related Coatings
[82] ASTM D 4138, Standard Method of Measurement of Dry Film Thickness of Protective Coating Systems by Destructive Means
[83] ISO 2808: 1997, Paints and varnishes - Determination of film thickness
[84] *Latter, T.D.T.*, in Thermoset Powder Coatings, Ed. J. Ward, FMJ International Publications Ltd, Redhill, 1989, p.124
[85] *Koban, M. E.; Delmotte, A.*, European Coatings Journal, (6), 38-43, 2005
[86] Recommended Procedure for Solvent Cure Test, No. 8 The Powder Coating Institute, Alexandria VA 22341, USA, 1998
[87] ISO 6441-2: 1999, Paints and varnishes - Determination of micro-indentation hardness - Part 2: Knoop hardness by measurement of indentation depth under load
[88] ISO 2815: 2003, Paints and varnishes - Buchholz indentation test
[89] ISO 15184:1998, Paints and varnishes - Determination of film hardness by pencil test
[90] ISO 1522: 1998, Paints and varnishes - Pendulum damping test
[91] DIN 53157, Testing of Paints, varnishes and Similar Coating Materials; Assessment of the mechanical damping behavior of coatings using König pendulum apparatus
[92] ASTM D 4366, Standard Test methods for Hardness of Organic Coatings by Pendulum Damping Tests
[93] ASTM D 1737, Standard Test Method for Elongation of Attached Organic Coatings with Cylindrical Mandrel Apparatus
[94] ISO 6860: 1996 (under review in 2001), Paints and varnishes - Bend test (conical mandrel)
[95] ISO 1519: 2002 (?), Paints and varnishes - Bend test (cylindrical mandrel)
[96] ASTM D 2794, Test for Resistance of Organic Coatings to the Effects of Rapid Deformation (impact)
[97] ISO 6272: 2002, Paints and varnishes Rapid deformation tests, Part 1: Falling weight tests, large area; Part 2: Falling weight tests, small area
[98] ASTM D 3359, Standard Methods for Measuring Adhesion by Tape Test
[99] ISO 2409: 1997, Paint and varnishes - Cross-cut test

[100] ISO 4624: 2003, Paints and varnishes - Pull-off test for adhesion
[101] *Rosenthal, C.; Kisin, S.; Trottier, E.*, European Coatings Journal, (9), 36,-43, 2008
[102] ASTM D 523, Standard Test Method for Specular Gloss
[103] ISO 2813: 1999, Paints and varnishes - Determination of specular gloss of non-metallic paint films at 20°, 60° and 85°
[104] *Voelkel, T.; Tragor, F.; Rademacher, J.; Hilger; C.;* Farbe und Lack, 110, 8, 65, 2004
[105] ASTM B 117, Standard method of Salt Spray (Fog) Testing
[106] ISO 7253: 1996 (under review), Paints and varnishes - Determination of resistance to neutral salt spray (fog)
[107] ASTM D 1654, Standard method for Evaluation of Painted or Coated Specimens Subjected to Corrosive Environments
[108] *Heinze, P., Keiner, A.*, Farbe + Lack, 114, 11, 20-24, 2008
[109] *Maardalen, J.; Lein, J. E.; Bolm, H.; Hallenstvet, M.; Rekowski, V.*, Progress in Organic Coatings, 63(1), 49-54, 2008
[110] ISO 8130-5: 1997, Coating powders - Part 5: Determination of flow properties of a powder/air mixture
[111] *Lange, P.G. de*, I-Lack, 42, (9), 339, 1973
[112] *Douma, G. and Selier, P.*, I-Lack, 41, (9), 361, 1973
[113] ASTM D 3451-19, Determining the Relative Deposition Efficiency of Organic Polymeric Powders on Moving Targets
[114] ISO 8130-10: 1998, Coating powders - Part 10: Determination of deposition efficiency
[115] ISO 8130-4: 1997, Coating powders - Part 4 : Calculation of lower explosion limit
[116] *Choi, K.-S.; Yamaguma, M.; Moon, K.-T.; Joung, J.-H.*, Japanese Journal of Applied Physics, 45(6A), 5207-5209, 2006
[117] ASTM G 6, Standard Test Method for Abrasion Resistance of Pipeline Coatings
[118] ASTM G 8, Standard Test Method for Cathodic Disbonding of Pipeline Coatings
[119] ASTM G 9, Standard Test Method for Water Penetration into Pipeline Coatings
[120] ASTM G 10, Standard Test Method for Specific Bendability of Pipeline Coatings
[121] ASTM G 11, Standard Test Method for Effect of Outdoor Weathering on Pipeline Coatings
[122] ASTM G 13 and G 14, Standard Test Method for Impact Resistance of Pipeline Coatings (Limestone Drop Test and Falling Weight Test)
[123] ASTM G 17, Standard Test Method for Penetration Resistance of Pipeline Coatings (Blunt Rod)
[124] ASTM G 18, Standard Test Method for Joints, Fittings, and Patches in Coated Pipelines
[125] ASTM G 19, Standard Test Method for Disbonding Characteristics of Pipeline Coatings by Direct Soil Burial
[126] ASTM G 20, Standard Test Method for Chemical Resistance of Pipeline Coatings
[127] ASTM G 42, Standard Test Method for Cathodic Disbonding of Pipeline Coatings Subjected to Elevated Temperatures

6 Powder coatings application techniques

6.1 Surface preparation

The substrate to which the powder coating is applied is in most cases metallic. Metallic articles are easily coated in one spray operation with a film thickens of 50 to 75 μm. This is quite an **advantage in comparison with the multi-coat wet paint system** when thickness of this range has to be achieved. With powder coating there is no flash-off period, no running away from the sharp edges, and no sagging due to excessive film thickness. However, a powder coating system does not avoid the necessity of surface preparation. The **importance of the surface treatment** prior to powder coating has been stressed in numerous papers [1-12].

A special pretreatment is necessary in particular for non-conductive coatings. In a recent article published by *Barletta* [13] an abrasive fluidized bed with copper powder is successfully applied on polyamide sheets prior to a hybrid epoxy powder coating. For MDF preheating is recommended, which frees charge-bearing or transmitting entities in the fiber enhancing conductivity. *Lin* [14] reported that special organic-inorganic hybrid emulsions containing conductive nano-materials are especially suited to improve the conductivity for MDF and wood.

The treatment of the metal substrate prior to coating includes cleaning of the surface followed by application of an inorganic layer of a so-called conversion coating. Chemical conversion coatings are applied to the metal substrate because of three reasons:

a) they provide temporary in-process corrosion protection to the metal prior to application of powder coating,
b) they promote good adhesion of the powder coating to the substrate during film formation and
c) they afford additional corrosion protection to the metal during the lifetime of the product.

The **cleaning of the surface cannot be avoided** even in cases when a special chemical pretreatment of the surface is not considered. The methods used in the cleaning of the surface can be classified into two groups: mechanical and chemical methods. Since chemical methods of cleaning depend very much on the type of the surface they will be discussed together with the surface pretreatment specifically for different substrates.

In his excellent review [15] *Kuhn* gives an overview on over 20 cleanliness test procedures. He distinguishes between direct observation and methods in which a liquid (water or solvent) is used as a mediator. The German VDA 19, based on ISO 1632-10.2, is a standard for particulate contamination.

6.1.1 Mechanical methods of cleaning

Mechanical methods of cleaning the surface before coating are widely used in practice. The principle is the same, using an abrasive action to remove the dirt in the form of tightly

Emmanouil Spyrou: Powder Coatings: Chemistry and Technology

ISBN: 978-3-86630-884-8

adhering contaminants on the steel surface. At the same time certain etching of the surface takes place, improving in this way the adhesion of the coating. The following forms of mechanical treatment are used: scratch-brushing by means of a rapidly rotating wire disk-brush, sanding by means of abrasive disks, wheels, cloths or papers, and dry or wet blasting with abrasive materials.

The most important method for mechanical preparation is **air blasting**. This is often called sandblasting because of the use of quartz sand as an abrasive medium, which in the form of sand jet is directed at the surface of the article through a special nozzle by compressed air.

A method that also works by throwing the abrasive material on the surface to be treated, without use of air stream, is the **airless centrifugal wheel blast**. The airless centrifugal wheel blasting is a more economical method from energy point of view. In the air blast systems high volumes of clean, dry compressed air are used to accelerate the abrasive particles. Therefore, the energy which is consumed by the airless systems is only 10 % of that, used in the air blast operation [16]. On the other hand by air blasting it is very easy to position the blowing nozzles to direct the blasts into nooks which are shadowed from the blast pattern of the centrifugal wheel. *Neise* et al. [17] explain criteria and give recommendation for the purchase of a sandblasting plant.

The air blasting or airless blasting techniques are the most appropriate for removal of mill scale from the mild steel surfaces, removal of oxides and corrosion products, removal of other dry soils such as mould sand residues, residues of paints, etc. It also removes the scratches and other surface irregularities, providing uniform roughness, which ensures good coating adhesion. Coating must take place immediately after the blasting because the cleaned surface is in a highly reactive state and corrosion occurs very soon.

6.1.2 Hygro-cleaning and pretreatment

The "Hygro-Cleaning System" (HCS) has been introduced by Dürr Ecoclean and is described by *Alvarez* [18]. In this method the substrate is impacted by a combination of air and cleaning liquid, based on the "air knife" principle, using a high velocity blast at 800 km/h. HCS gives deep pore cleaning, ensuring that neither streaks nor any other residues remain on the workpiece surface. Depending on the cleaning requirements, a cleaning medium or steam may be used to support the blowing air in its function as a carrier medium. Steam shows a conductance of less than 10 μS/cm, thus creating an affinity between the steam and the deposits on the workpiece surface.

During operation an enormous mechanical effect is created by the high kinetic energy, and due to the high temperature (100 to 130 °C) the **surface contaminant (oil)** is reduced. The cleaning medium is permanently renewed for each individual component, so the cleaning result is kept constant. Compared to classical pretreatment, HCS is shown to reduce particle contamination by about 50 % and in many cases (depending on the shape of the workpiece) it will strongly reduce oil/emulsion residues. **Advantages** of HCS are:

- low space requirement and low cycle times,
- reduced handling of components,
- air is re-circulated and not exhausted,
- waste water disposal is optimized.

6.1.3 Chemical cleaning and pretreatment

Surfaces on which oil or grease are present cannot be blasted directly, because these substances cannot be removed completely by blasting. Moreover, they contaminate the abrasive itself. Therefore, the oil or grease deposits must be removed before sandblasting. A chemical method that can be used for all types of surfaces is one employing solvents as degreasing agents.

For degreasing the surface a simple solvent wiping process can be used. Vapor degreasing by halogenated hydrocarbons like trichloroethylene is also used. It should be kept in mind that the vapor cleaning does not remove the solid dirt from the surface, and therefore, if the subsequent blasting does not take place, than wiping or liquid/vapor degreasing has to be used. **Trichloroethylene is still widely in use** despite the health and safety concerns and the necessary heating installations in order to provide adequate cleaning action at elevated temperatures.

Safety, health and environmental considerations work in favor of the aqueous cleaners. Nowadays the most common cleaning method, which is also the most economical, uses **emulsified cleaners**. These are cleaning agents such as organic solvents which have been incorporated in the water phase by means of emulsifiers. Since the surface on which the powder coating is applied can be of different nature, cleaning agents especially suitable for cold or hot-rolled steel, aluminum or galvanized steel have been developed.

Stainless steel is a usual substrate for powder coatings. Since phosphating doesn't work on stainless steel *Nunes* et al. [19] successfully applied a silane based primer ("Oxsilan MM-0705" from Chemetall) prior to powder coating. *Fedel* et al. were able to prove that silanes not only act as coupling agents but also provide a barrier effect against water and oxygen [20]. The next operation which follows the cleaning step is the application of the so-called conversion coatings, which also differ for different substrates. In the following sections the substrate preparation will be discussed separately for different types of material.

6.1.3.1 Cleaning and pretreatment of cold- and hot-rolled steel

Cold-rolled or the so-called mild steel is free from heat scale, and, when properly packed, even from rust.

Hot-rolled steel has a surface covered with a blue/black mill scale which is very brittle in nature and will flake off upon bending. Scale is a product of gaseous corrosion at high temperature and consists of the anhydrous oxides FeO, Fe_3O_4 and Fe_2O_3. Whereas ferric oxide, Fe_2O_3, is chemically stable, ferrous oxide FeO is less stable and readily changes to hydrated ferric oxide under the influence of water and oxygen. The cracked mill scale is cathodic to the underlying steel and promotes corrosion.

Hot-rolled pickled and oiled steel is also available on the market. This steel has been pickled to remove the hot scale, and oiled to prevent rusting during storage. All of these types of steel have to be cleaned before powder application. Several different types of chemical cleaners can be used for this purpose.

Alkaline cleaners and acid cleaners

A wide variety of different alkaline cleaners is available on the market formulated for each specific application. In general they are combinations of sodium phosphates, silicates and carbonates. Often they contain chelating agents, solvents and surfactants.

The **resistance of steel to alkaline attack** gives a great degree of freedom for adjusting the properties of the alkaline cleaners (e.g. 3 to 20 % of NaOH in water) to the type of the material undergoing the cleaning operation. A contemporary tendency in the development of alkaline cleaners is to lower the operating temperature by means of a proper choice of surfactants and chelate systems providing benefits of lower operating costs. The cleaned substrate must undergo subsequent rinsing operation in order to remove the meta-silicates present in the cleaning agent. Significant progress has been made by use of the so-called inhibited cleaners containing certain titanium salts [5]. These salts act as crystal size regulators for the conversion coatings which are applied in the following pretreatment step.

New surfactant systems also enable efficient cleaning action at less alkaline conditions than was the case with the aqueous cleaning agent of the first generation. This is especially important for cases when **zinc phosphate conversion** coatings are used in the next pretreatment step. Zinc phosphate conversion coatings applied on a steel surface being treated with strong alkaline cleaning solutions in general have a larger size of crystal deposited on the surface. This decreases the mechanical properties of the conversion coating and the gloss of the powder coating applied as a top layer because of the rough nature of the surface. Research efforts in this area are also in the direction of development of cleaners/conversion coatings blends, or more specifically cleaners/iron phosphates [5].

The cleaning process may vary from hand wiping cold cleaning to immersion or spraying process. All processes have to end up with a rinsing procedure in order to remove the cleaning agent from the surface. Dependent on the amount and type of dirt, the cleaning procedure may be repeated several times. The best results are obtained at automatic hot-spraying cleaning lines where the cleaning stage at a temperature of 60 to 70 °C within 45 to 60 s is followed by rinsing within 30 to 45 s. In case where the immersion cleaning method is used, residence times between two and five minutes are necessary for efficient cleaning [16, p.49].

Rust can be very effectively removed by using mineral acids like hydrochloric or sulfuric acid. Also the heat scale can be removed in the same manner. Strong **mineral acids for rust removal** are used in principle only in cold operations, and except in highly specialized operations almost never in spray equipment. The reactions between iron and its oxides and sulfuric acid generate $FeSO_4$ and $Fe_2(SO_4)_3$.

Moderate or **mild acid cleaners** based on phosphoric acid or organic acids are used in spray applications. In a recent Chinese publication [21] it is recommended to use an alkaline and not an acid based cleaning step before pre-phosphating electro-galvanized steel due to superior corrosion resistance. On the market there are various acid cleaners convenient for hand wiping. Usually they contain detergents and thickening agents making them suitable also for brush removing applications.

Conversion coating of steel

The chemical pretreatment of the metal surface is the following operation which prepares of the surface after the dirt, metal oxides, oil and grease have been removed. Conversion coatings on the metal surface consist of insoluble metal oxides or metal salts. **Three main types of conversion** coatings are used for a steel surface: iron phosphate, zinc phosphate and chromate conversion coatings. Due to pollution control problems connected with the disposal of chromium and the additional equipment requirements, the latter coating has never become commercially important. Recently Nippon Paint [22] patented a conversion of hot rolled steel based on a combination of zirconium salt and an aminoalkyl alkoxysilane in particular suitable for vehicle chassis. The characteristics of the main types of conversion coatings for steel are listed in *Table 6.1* [23].

Iron phosphate

The compositions which are used to produce iron phosphate conversion coatings on steel are principally based on monosodium phosphate, although even the simplest treatment with phosphoric acid as such gives acceptable results. These mildly acidic compositions are available on the market in powdered or liquid form. The rest of the compositions are various accelerating agents in the form of nitrites, nitrates or chlorates. When at the same time,

Table 6.1: Characteristics of different types of conversion coatings for steel[23]

source: Spruyt, van Mantgem & de Does N.V.

Type of phosphate coating	Color	Deposition [g/m²]	Thickness [µm]	Porosity [%]	Pencil hardness
Iron phosphate $Fe_3(PO_4)_2 \cdot 8H_2O$	blue	0.1 to 0.5	0.1 to 0.5	0.5 to 1	H
Zinc–iron phosphate $Zn_2Fe(PO_4)_2 \cdot 4H_2O$	medium gray	10 to 30	5 to 15	0.05 to 0.4	HB
Zinc phosphate $Zn_3(PO_4)_2 \cdot 4H_2O$	gray	2 to 10	1 to 5	0.05 to 0.5	HB to >H
Zink calcium phosphate $Zn_2Ca(PO_4)_2 \cdot 2H_2O$	light gray	1.5 to 6	1 to 3	0.05 to 0.4	HB to >H
Manganese phosphate $(MnFe)_5 \cdot H_2(PO_4)_4 \cdot 4H_2O$	dark gray	8 to 40	3 to 25	0.5 to 3	B to >HB

next to the coating, a cleaning action is considered, surface active agents might be included. Depending on the acidity of the composition, rust removal can also take place. In this case the compositions have moderate or strong acidity.

The iron phosphate conversion coating can be obtained by simple treatment of the steel surface with solutions containing 0.2 to 6 vol.-% of phosphoric acid. Usually the composition of the conversion coating is designed in such a way that the rinsing operation after the treatment is not necessary. Good results are obtained by air drying the treated material, but even better results are obtained by drying the surface in a stream of warm air. During the drying process the steel surface reacts with the phosphoric acid forming a layer of iron phosphate ($FeHPO_4$). The layer of $FeHPO_4$ is rather thin, having a thickness between 0.2 and 0.5 µm.

If the concentration of phosphoric acid in the conversion coating solution is too high then a sticky layer of unreacted phosphoric acid remains on the surface. This hygroscopic layer decreases the corrosion resistance of the coated article and the adhesion of the powder coating on the substrate. Conversely, too low a concentration of phosphoric acid produces as a consequence the formation of a thin layer of porous powder such as iron phosphate which has a rather poor corrosion resistance.

Depending on the bath temperature, concentrations between 4 and 6 % by volume of phosphoric acid are used for operation at room temperature, and 0.2 to 2 % at 70 to 90 °C. For good phosphatization it is necessary to **control the concentration of iron ions** in the bath. Usually in continuous operation the concentration of iron in the bath increases and should be kept below 5 g/l.

Monosodium phosphate conversion coatings solution has a lower acidity than phosphoric acid solutions. The pH is usually in the range between 4.5 and 5.8. Several different reactions simultaneously take place during the treatment of the surface with monosodium phosphate based conversion coatings. In the first reaction a ferrophosphate is formed ($Fe(H_2PO_4)_2$). The formed ferrophosphate reacts with the sodium hydroxide and oxygen, producing insoluble iron phosphate ($FePO_4$) or iron hydroxide ($Fe(OH)_3$). During the drying of the surface insoluble iron oxide (Fe_2O_3) is formed.

The oxygen necessary for the oxidative action during the phosphating process can be supplied by the air, especially in the case where phosphatization takes place by the spraying method. However, contemporary phosphating preparations contain accelerators which are oxidizing compounds e.g. nitrites, nitrates, chlorates, bromates or borates. The accelerating action of the oxidizers consists in oxidizing the hydrogen formed during dissolution of the iron. Additionally, the accelerators oxidize divalent iron ions in the solution, promoting formation of insoluble iron phosphate which precipitates in the form of a sludge.

Due to formation of disodium phosphate, the pH of the bath in continuous operations increases permanently. Where a pH of 6.0 is reached, the phosphating efficiency of the bath is practically zero. Therefore, **permanent control and correction of the pH** of the phosphating solution in the bath with sodium phosphate or phosphoric acid is necessary to ensure high quality of the conversion coating.

The weight of the formed conversion coating in the case of monosodium based solutions is between 0.3 and 0.7 g/m^2, with thickness not greater than 1 μm. Depending on the ratio between the formed iron phosphate and iron oxide the color of the coating ranges from gray to deep blue.

The advantage of monosodium phosphate conversion coatings over the simple phosphoric acid based solution is that phosphatization and cleaning of the surface take place at the same time. Usually the conversion coating solution contains surface active agents, mostly of a non-ionogenic type.

The conversion coatings solutions can be applied in a very simple way using a rug or mop. The solution can be reapplied several times until the surface is clean and free of rust. Since this is a mainly hand applied technique the solution is in principle not preheated. To speed-up the phosphating action of the solutions they are very often applied by a steam gun. The solution is mixed with wet steam and applied by the gun when preheated to ca. 70 °C. Rinsing is usually done by the same gun using only fresh steam.

Iron phosphate conversion coatings can be applied by immersion application methods. In this case care should be taken that the substrate is extremely clean before coating, since during the immersion coating technique the cleaning action of the composition is not very good; this is also the case for spray or hand application. The cleaning action during the coating application can be improved by agitation. In this case substrates with a light soil can be treated in a one stage process. The temperature of the bath can vary from 30 to 70 °C dependent on the concentration of the active material in the immersion solution, the desired coating thickness, and capacity and design of the coating line.

The **most widely used method** for applying iron phosphate conversion coating is **spray washing**. Depending on the nature of the soil and the quality level requirements, spray systems operate in at least two stages (clean/coat in one stage and rinse with water in the second) up to even nine, including cleaning, rinsing, coating, post rinsing and drying steps subsequently distributed in a logical manner. Typical time schedules for an automatic line for complete pretreatment is 45 to 60 s for the cleaning steps, 45 to 60 s for the treatment and 15 to 30 s for the rinsing step. The temperatures in different zones of the line vary between 30 and 70 °C [16, p.52].

Iron phosphate conversion coatings are used for steel products destined for normal indoor use, since they do not provide outstanding corrosion protection. However, they are highly appreciated for their **superior paint adhesion characteristics**. The control of the phosphating operation is simplified by using liquid coating formulations which permit the use of automatic metering systems. For limited cleaning performances one-pack phosphating/cleaning solutions are available on the market. Twin pack formulations, on the other hand, allow the cleaner and the coater to be dosed separately to the same tank. This system provides greater flexibility concerning the treatment of materials with a different degree of contamination by oil, grease and other impurities [4].

Zinc phosphate

Compositions for obtaining zinc phosphate conversion coatings on the steel surface are in principle acid solutions containing zinc dihydrogen phosphate $Zn(H_2PO_4)_2$, accelerating agents like zinc nitrate and additives for grain control and coating weight control. In water solution zinc dihydrogen phosphate exists in several forms [24].

The equilibrium of the different species depends on the temperature. A higher temperatures causes precipitation of the less soluble zinc hydrogen phosphate $ZnHPO_4$ in the form of slurry. This can be avoided by addition of small quantities of phosphoric acid which shifts the reaction equilibrium towards formation of zinc dihydrogen phosphate [23, p.62].

In practice the concentration of tertiary zinc phosphate in the bath is almost close to saturation. The metal surface, when in contact with such a bath is attacked by the phosphoric acid giving as a product ferrodihydrogen phosphate according to the already mentioned reaction. In this way, by consumption of phosphoric acid, the pH at the metal-solution interface is increased and the solubility products of zinc phosphate and iron hydrogen phosphate are exceeded. They precipitate and adhere well to the metal surface at the very reaction sites, building up insoluble layer of considerable thickness. In the solution, on the other hand, the pH remains quite low, thus preventing the formation of insoluble sludge of zinc phosphate salts. The phosphating action can be accelerated by increasing the bath temperature. Reasonably high phosphating rates can be achieved at 90 to 100 °C. This of course is connected with relatively high energy consumption, especially in the case where massive articles with high heat capacity have to be treated.

A much cheaper and more efficient method is one that employs **accelerating agents**. In this way acceptable phosphating rates can be reached at relatively low temperatures. There are many different types of accelerators described in the patent literature, such as oxidizing agents (nitric acid, nitrates, nitrites, chlorates, borates), reduction substances (hyposulphites, phosphites, benzaldehyde, hydroxylamine), organic nitrogen containing compounds, copper salts, etc. The most efficient and the most widely used accelerators are those of oxidizing types. The accelerating action of the oxidizing agents will be outlined in general terms since the detailed mechanism is beyond the scope of this book. These materials oxidize the hydrogen present in a layer of molecular size thickness over the metal surface. At the same time the ferro (Fe^{2+}) ions are oxidized to ferri (Fe^{3+}) ions which precipitate because of their low solubility. Both reactions promote further solubilization of the iron.

The soluble iron salts in the solution have a very bad influence on the bath performance. In the first instance the phosphating rate decreases, but what is even worse, the obtained conversion coating has a lower corrosion resistance because of its porous nature. Therefore, the **control of the iron content in the bath is essential** if good results will be obtained. In the case of too high a concentration of ferro ions, the addition of a certain amount of oxidizing agents such as sodium nitrate, sodium nitrite, hydrogen peroxide or chlorates can help by oxidizing the ferro ions to insoluble ferri ions which are then removed from the bath in the form of sludge of ferri phosphate. Where the sludge is heavy, normal operation of the phosphating process becomes quite difficult unless the whole bath is thoroughly cleaned.

During phosphatization in the presence of iron, nitrate is reduced to nitrite or to a nitroso compound which can only then react with the hydrogen liberated by dissolution of iron by phosphoric acid [24, p.138]. Therefore direct addition of nitrites in the form of sodium nitrite is more effective in accelerating the phosphating action of the bath. Sodium nitrite has a double action in accelerating the process speed. It helps the formation of zinc phosphate by transforming the zinc dihydrogen phosphate, and at the same time oxidizes ferro into ferri ions [23, p.65].

Sodium nitrite is a very strong oxidizing agent. Therefore, the phosphating rate in a nitrite containing composition based on zinc phosphate is very high. At bath temperatures of 40 to 80 °C the residence time in the immersion operating baths is only 3 to 10 min depending on the bath temperature and the desired layer thickness. Using spray phosphatization the phosphating time is only 1 to 2 min [23, p.66]. Sodium nitrite is not stable in acidic media. This instability which is manifested by the formation of nitrogen oxides is even more emphasized at higher temperatures. A low operating bath temperature is desirable to achieve good bath stability. **Temperatures higher than 80 °C** should be avoided when working with nitrite accelerated zinc phosphate conversion coatings.

If the concentration of the nitrite in the bath decreases because of its consumption during oxidizing the ferro salts in the solution or because of its decomposition, the concentration of ferro ions in the bath increases lowering the efficiency of the bath operation. In continuously operating phosphating baths, the concentration of the nitrite in the bath solution must be kept on a certain level by subsequent additions of fresh sodium nitrite. On the other hand, too high a nitrite concentration will cause a heavy slurry of ferri phosphate to form. In extreme cases, when the addition of sodium nitrite is much too high, the bath becomes alkaline which causes precipitation of zinc hydrogen phosphate, $ZnHPO_4$. In such cases the slurry formation is even heavier. The other negative influence of an excess of sodium nitrite is the formation of a very thick irregular zinc phosphate layer on the metal surface. The **optimal concentration of sodium nitrite** in the bath providing good operation is between 0.1 and 1.0 g $NaNO_2$ per liter. At good controlled operation conditions, the use of the nitrite accelerated bath is practically unlimited. It goes without saying that the slurry formed during the phosphating operation must be removed periodically from the bath.

Sodium chlorate is another oxidizing agent that has wide use in accelerating the phosphating process. The advantage of sodium chlorate over sodium nitrite is that it is possible to add this material to the concentrated zinc phosphate/phosphoric acid solution, since it is stable in acidic conditions. This allows ready-made one-component zinc phosphating solutions to be prepared, which is especially convenient when the concentration of the bath has to be corrected by addition of fresh phosphating agent. In this way the proportion zinc phosphate/accelerator is kept constant, thus providing the deposition of a layer of the same quality by time. For continuous operating lines this is of great importance. A disadvantage of sodium chlorate accelerator is the formation of sodium chloride during the bath operation, which remains in the bath permanently increasing its concentration.

Metal salts in combination with oxidizing agents are often used as accelerators for phosphating of steel. In principle copper, nickel, cobalt, silver, gold and platinum can be used as accelerators, but because of obvious commercial reasons the practical use is limited to the first two metal salts. During the phosphating process these metals precipitate on the steel surface and serve as a cathode, accelerating in this way the dissolution of the ferro phosphate. At the same time they catalyze nitrate decomposition and promote oxidation.

The **protective function of the phosphate coatings** depends very much on the crystalline character of the film. In general, protection increases with decreasing grain size. The size of the crystals of the zinc phosphate conversion coatings is dependent to a great extent on the type of cleaning agent used in the previous step. Especially when strong alkaline cleaning solutions are used, which is very often the case in immersion cleaning processes, the crystals of deposited zinc phosphate are of rather an undesirable size.

Refinement of the crystalline structure can be obtained by increasing the bath temperature. However, by using suitable additives, the temperature can be reduced with no adverse effect on the grain size. This can be achieved in two ways: by adding regulators of the crystal growth in the bath before phosphatization takes place or by adding **crystal growth regulators** directly in the phosphating solution.

Addition of titanium salts in concentrations of 1 to 10 mg/l in the alkaline cleaning solution decreases the size of the crystals of the deposited zinc phosphate. Good results can be obtained with a special rinsing operation. For example rinsing by immersing the article in a bath containing oxalic acid helps considerably to obtain a fine crystalline structure in the subsequent pretreatment with conversion coating. The action of the oxalic acid is a simple neutralization of the alkalies which remain after the cleaning and degreasing step.

A very fine crystalline structure can be obtained by modifying the phosphating solution by the addition of so-called regulators of crystal growth. These are calcium or barium phosphates and organic complexing agents like citric acid or tartaric acid. The best results are obtained by using calcium ions. It is supposed that the composition of the deposited fine crystals on the steel surface is zinc calcium phosphate expressed by the formula $CaZn_2(PO_4)_2 \cdot 2H_2O$. Because of the fine crystalline structure, the deposited conversion coating is more compact and less porous than the common zinc phosphate layer. Therefore, it provides superior corrosion protection. The thickness of the layer is of the same range as zinc phosphate.

In principle the application methods used for iron phosphate can also be used in this case. The **brush-on treatment** by a cold phosphating solution produces an acceptable coating thickens within 2 to 5 min. The **steam-gun application method** has not found commercial importance because of the poor quality of the coating and short life time of the operating equipment.

The **immersion application method** can be successfully used for zinc phosphate conversion coatings. The typical immersion technique involves five stages: clean, rinse, treatment, rinse and post-rinse. The time required to develop the zinc phosphate immersion coating is 2 to 5 min. The working temperature varies in the same range as iron phosphate between 30 and 70 °C.

The spray-washer application accounts for the largest proportion of the zinc phosphate treatment processing. Depending on the surface and quality requirements the treatment can involve five to nine steps which include cleaning, rinsing, treatment with the coating solution and drying, subsequently arranged in the most suitable way. Time requirements for each step are the same as those for the iron phosphate coatings.

The zinc phosphate pretreatment produces films of greater thickness than the iron phosphate one (1 to 5 g/m^2 in comparison with 0.1 to 0.5 g/m^2). Therefore, it is favored for articles which will be subjected to the most severe corrosion conditions. On the other hand, because of its crystalline structure, the conversion coating absorbs the powder coating which comes on top of it during the film formation and improves the adhesion of the powder coating to the substrate. The **best corrosion protection** of the surface is obtained **by thick conversion coatings**. On the other hand, too thick zinc phosphate can be disadvantageous for the flexibility of the complete system because of the brittle structure of the crystals.

A conversion coating can reduce the gloss of the organic coating applied on top of it, since the crystalline surface has an uneven micro structure. This is not very important for powder coatings, because in most cases they are applied in relatively thick layers, diminishing in this way the undesirable influence of the pretreated surface on the final gloss of the coating.

According to a recent study of *Tepe* et al. [25] zinc phosphate, a zirconium-based nano-scaled pretreatment and iron phosphate was applied on hot rolled steel. The corrosion resistance decreased in the aforementioned order. However, a recent US-patent [26] claimed, that zirconium phosphate ("X-Bond 4200DS") pretreatment is superior to zinc phosphate. After powder coating the first one showed a scribe loss value of 2.8 mm after 400 h of neutral salt spray test, the latter a value of 7.0 mm.

6.1.3.2 Cleaning and pretreatment of aluminum

Aluminum is one of the metals which is characterized by a certain degree of **self-protection towards corrosion** by building up a thin layer of aluminum oxide on its surface. However, this protection is not good enough to provide a long service life to aluminum articles exposed to normal weathering conditions. Moreover, aluminum oxide itself is not a very good base on which to apply powder or any other coating. The adhesion of organic coatings is in general very bad on aluminum unless the surface is specially treated in the ways described below.

The treatment of the aluminum surface before applying powder coatings includes phosphatization, chromium phosphatization, chromating and, because of the toxicity of chromates, a non-chrome pretreatment of the surface. The other method is electrochemical oxidation of the surface, which in most cases is used as the sole method, providing anticorrosion and decorative characteristics of the treated surface. It is almost never used as a pretreatment step in covering aluminum articles with powder coatings.

As in the case with steel, the chemical pretreatment of the aluminum is preceded by a necessary cleaning step in order to remove dirt, oil, grease and the corrosion products from the surface. Alkaline, acid and neutral types of cleaners are used for this purpose.

Alkaline cleaners

The **aluminum surface is sensitive to alkaline attack**. Alkaline cleaners for aluminum are milder than those used for cleaning steel. Alkaline salts such as sodium carbonate, sodium phosphate, potassium carbonate and sodium silicate are suitable for preparation of cleaning solutions. Only in extreme cases when heavy dirt has to be removed, alkaline cleaners containing small to moderate amounts of caustic soda can be used very carefully.

Depending on the pH of the cleaners and their compositions, they can be classified into two groups: weak alkaline or almost neutral cleaners without any or with a very weak etching action on the surface, and strong alkaline cleaners which dissolve the oxide layer on the surface which results in noticeable loss of weight. No correlation has been found between the removal rate and the corrosion behavior of coated aluminum [27].

The first type of cleaners include solutions of sodium silicate, blends of sodium silicate, carbonate and pyrophosphate, trisodium phosphate, sodium pyrophosphate, and blends of sodium chromate, carbonate and pyrophosphate. Very often wetting agents are also included in order to improve the cleaning action.

Sodium silicate and sodium chromate have no etching action on the surface, and the weight loss is zero or almost negligible. The reason is that sodium silicate and sodium chromate do not react with aluminum as a metal. On the other hand, they undergo chemical reaction with the aluminum oxide layer, but the result is a transformation of the aluminum oxide into aluminum silicate (Al_2SiO_5) or aluminum chromate ($Al_2(CrO_4)_3$), thus having almost no effect on the total weight of the treated material [23, p.87].

The weight loss of the aluminum surface treated with sodium silicate or chromate solutions is in the range from 0 to 3 mg/m^2. A weak etching effect is produced by treatment with sodium pyrophosphate generating $NaAlP_2O_7$ and trisodium phosphate forming $AlPO_4$, where the weight loss lies between 40 and 1700 mg/m^2.

Sodium carbonate or its combination with sodium triphosphate which belongs to the etching types of alkaline cleaners removes the aluminum oxide layer, but at the same time these solutions attack the metal surface too, causing a **weight loss** in the range between 8 and 12 g/m^2 by generating $NaAlO_2$.

Two methods are generally used for cleaning aluminum articles with alkaline cleaners: the power spray method, in which the articles pass through a tunnel, while the cleaning solution is pumped from an outside storage tank under high pressure onto the parts, and the simple immersion method. The temperature of the cleaning bath varies from 40 to 70 °C depending on the concentration of the cleaning agents, type and amount of dirt and the desired residence time of the articles in the cleaning station.

A special version of alkaline cleaning is the **electro-cleaning**. The articles are immersed in a bath in which a direct electrical current is passed through the solution. The parts to be cleaned serve as an anode, while other hanging electrodes are cathodes. The cleaning is improved by the scrubbing action of the oxygen bubbles which develop on the anode as a result of the electrolysis of water.

The simplest **hand-wiping method** can also be used, deriving benefit from the physical act of removing the soil from the surface by means of a cloth or sponge. This method is restricted to small discontinuous operating units or to cases where heavy massive articles have to be cleaned. For practical reasons it is not convenient in such cases to carry on a continuous cleaning operation with the article hanging on a conveyor chain or belt.

From an efficiency point of view **electro-cleaning is the best method** followed by spray, immersion and hand-wipe methods. On the other hand, the electro-cleaning method is the most expensive due to the higher concentrations of active components and the cost of the electricity. The next most expensive is the immersion method, followed by the hand-wiping method, while the cheapest is the spray-wash cleaning method.

Acid cleaners

These cleaners are composed of phosphoric acid or mild acidic salts of phosphoric acid alone or in combination with surfactants, or surfactants and solvents. The acid cleaners will remove the oxide film on the aluminum surface. On the other hand they are not as effective as the alkaline cleaners in removing common soils from the surface.

In some cases acidic salts of sulfuric acid in low concentrations are also used. So-called activators, which are salts of hydrofluoric acid are often used in combination with phosphoric or oxalic acid as pH regulators. Depending on the composition of the cleaner, $AlPO_4$, $Al(SO_4)_3$ and AlF_3 are formed during cleaning of the aluminum surface with acid.

In the case when aluminum alloys containing copper or silica have to be cleaned, the alkaline cleaners produce a black colored sludge which is very difficult to be rinsed out. With the help of nitric acid, sulfuric acid, fluorides or chromates these materials are transformed into a soluble form and can be easily removed from the aluminum surface.

The same methods used for alkaline cleaners, with the exception of the electro-cleaning method, can also be used for acid cleaners. The less expensive method again is the spray method, followed by the hand-wiping and the immersion methods. The **spray method is also the most effective** concerning the cleaning action.

Conversion coatings of aluminum

An **iron phosphate** solution can produce a conversion coating on aluminum which is not iron phosphate but aluminum phosphate. This is not an ideal system for aluminum but it is very suitable for cases where articles made out of steel and aluminum have to be coated on the same line.

Aluminum phosphate coating produced out of iron phosphate is **a very good base for powder coating**. The process is very easy to control and operate, but it can never be used exclusively for aluminum because the effective life of the bath is then very short. It is economical only when aluminum is processed together with steel.

A processing solution containing **zinc dihydrogen phosphate** $Zn(H_2PO_4)_2$ and fluoride produces a very good conversion coating on aluminum. The chemical reactions involved are similar to those for the chemical treatment of steel.

Since during the phosphatization of aluminum articles, the aluminum from the surface, like the iron in the case of steel, is dissolved and transferred in a soluble form into the bath solution, its concentration with time increases, which hinders the process. Addition of silicofluorides or fluoroborates helps the precipitation of solubilized aluminum in the form of an insoluble sludge. In continuous operation the zinc phosphate processing solutions will produce a large amount of insoluble sludge over a period of time, which may deposit on the heat transfer elements in the bath, decreasing their efficiency, or plug up the nozzles in the spray operation equipment. To maintain good operation conditions the process line must be cleaned very often. This is a rather weak part of this process, which otherwise produces a **zinc phosphate layer that is a very good base for powder coatings**. It has to be pointed out that this process can be used for lines which operate exclusively with aluminum articles.

As in all other cases, the process is accelerated by temperature, and on an industrial scale the operating line temperature is in a range between 40 and 70 °C. The coating produced in this way has a gray color with a thickness between 1 and 5 μm and a deposition weight of 2 to 6 g/m^2. It provides very good corrosion resistance and greatly improves the adhesion of the powder coating applied at the final stage of the coating process.

Chromium phosphate conversion coatings are obtained from solutions containing chromates, phosphates and fluorides. Chromate pretreatment of aluminum with solutions of the above compounds results in deposition of yellow, colorless or green layers depending upon the choice of chemicals.

During treatment of the surface, aluminum reacts with the hydrogen ions reducing them to molecular hydrogen which exists in the bath in a gaseous form. However, a portion of the hydrogen atoms is consumed in reaction with the six-valent chromium, reducing it to a valency of three. The latter reacts with phosphoric acid to form an insoluble chromium phosphate which deposits on the aluminum surface. The presence of fluorides accelerates these reactions by preventing formation of a passive phosphate or chromate layer. At low fluoride concentrations the deposition rate is slowed down because of passivation of the surface, while at high fluoride concentration undesirable high etching of the surface takes place. The **optimum ratio fluorides/chromates** depends on the concentration of phosphoric acid in the bath. For a very wide range of concentrations of phosphoric acid from 50 to 300 g/l the optimum mole ratio hydrofluoric acid/chromic acid falls in a rather narrow interval between 0.22 and 0.28. Concentrations of phosphoric acid higher than 300 g/l lead to too strong an etching of the aluminum surface and so they are avoided in practice. The same holds for the ratio hydrofluoric/chromic acid. For a given concentration of phosphoric acid there is a hydrofluoric/chromic acid mole ratio above which there is no formation of conversion coating due to strong etching of the surface. For 300 g/l of phosphoric acid the critical hydrofluoric/chromic acid mole ratio is ca. 0.2, increasing in a linear manner to 0.4 for a bath with 50 g/l of phosphoric acid content.

Too low a fluoric/chromic acid ratio, on the other hand, slows down the deposition rate. Below some critical value there is no formation of conversion coating within the processing times used in practice. This critical ratio is between 0.19 and 0.13 for a concentration of the phosphoric acid in the bath between 50 and 300 g/l.

In the process of precipitation of amorphous chromium phosphate conversion coating [23, p.90] $AlPO_4$, $CrPO_4$ and H_3AlF_6 is formed. The reaction is accelerated considerably by temperature. Depending on the desired deposition rate the bath temperature varies between 35 and 50 °C with a residence time between 5 s and 3 min. After drying at room temperature, a coating layer with an approximate composition of 50 to 55 % $CrPO_4$, 17 to 23 % $AlPO_4$ and 22 to 23 % water incorporated in the coating structure with traces of chromium fluoride is obtained [23, p.91].

In continuous operation, the aluminum content in the bath solution increases permanently. There are two possible ways to **control and keep the level of aluminum ions in the bath**. The use of ion exchangers is the best solution, while a relatively simpler way is to add sodium or potassium fluoride to the bath, which causes precipitation of the aluminum in the form of insoluble K_2NaAlF_6. A disadvantage of the latter method is the formation of a slurry of sodium potassium aluminum fluoride which has to be removed by filtration.

The chromium phosphate conversion coating is the best conversion coating base that can be applied on aluminum. It is more effective compared to the phosphate pretreatment where aluminum, zinc and steel articles can be treated on the same line. It exhibits better salt spray resistance and gives improved adhesion. The process can be easily controlled and is less expensive to maintain. However, it must be kept in mind that its use is restricted exclusively to aluminum surfaces.

Chromate conversion coatings for aluminum are solutions containing chromic acid or chromates and fluorides. Hydrochloric acid is often used to regulate the pH of the solution. Composed in this way, they deposit on the treated aluminum surface colorless to brown colored conversion coatings, which have a very complex structure that can be described by the general formula $xCr_2(CrO_4)_3 \cdot yAl_2O_3 \cdot zH_2O$. The composition of the layer by weight is the following: 24 to 28 % Cr^{3+}, 0.4 % Cr^{6+}, 1.5 to 7 % Al^{3+} and 0.3 to 4 % F^- [23, p.88].

Depending on the application method, a layer with an average weight between 200 and 300 mg/m^2 can be obtained within 5 to 120 s at room temperature. The thickness of the layer depends on the temperature and the method of application. The most effective method is the **spray application at temperatures not higher than 60 °C**. The colorless to light brown coating obtained is of an amorphous nature, providing outstanding corrosion protection. The colorless amorphous chromate layers obtained from diluted chromate solutions at low pH between 3 and 4 have a lower corrosion resistance than the colored coating obtained from more concentrated chromate solutions. The process is often called "yellow chromating".

Non-chromate conversion

Chromating and phospho-chromating have, until recently, proved to be the only acceptable systems for coating aluminum profiles and panels, used in building constructions. However, the **toxicity of chrome-based conversions** (hexavalent chromium is classified as carcinogenic!) has caused the industry to seek more suitable, non-toxic alternatives, in which neither chromates nor phosphates are used. A number of alternatives have been proposed, but the most reliable materials, to ensure **good anti-corrosive properties, are based on zirconium- and titanium compounds**. The obtained conversion coatings are equivalent to the yellow and green chromate coatings. The conversion solution consists of hydrofluoric acid and hexafluoric zirconium acid H_2ZrF_6 (or hexafluoric titanium acid). When very high corrosion resistance is required, a combination with an organic (film forming) polymer is needed. The solution deposits on the cleaned aluminum surface an extremely thin (about 0.01 μm) and nearly colorless conversion coating, consisting of complex zirconium-aluminum compounds, respectively complex titanium-aluminum compounds.

The Zr- (or Ti-) containing conversion layers have been investigated by *Droniou* et al. [24], using ESCA spectroscopy, and rather stoichiometric Zr- (or Ti-) oxide to Al-oxide ratios have been found, associated with fluorinated compounds.

Droniou gives the following chemical explanation: tetrahydric Zr ion is surrounded in water by four hydroxyl groups, and once in the coating, it is able to react, as a cationic oxide/hydroxide, with the functional groups of the paint, especially with carboxyl and hydroxyl groups, when there is control of temperature, time and pH conditions.

Highest qualities are achieved by systems with an **additional organic polymer**. The incorporated organic polymers have double functions and will react on one side with the substrate through polar inorganic moieties, and on the other side with the organic chains of the paint film, ensuring the coherence of the latter. According to *Droniou*, the organic polymers give coherence with the associated inorganic element (Zr or Ti mainly) through donor/acceptor bonding, providing some adherence to the substrate, again through donor/acceptor bonding. The organic polymers anchor the applied powder layer through reactions with the first molecular layers and these reactions are accelerated by the curing conditions of the powder coating. Recently this kind of pretreatment system has been introduced into the market under the name “Iradit CF” (chrome free) [28]. This product has been approved by GSB and Qualicoat.

Jüptner [29] suggests that the Cr-free conversion layer consists of $ZrOF_2$ (or $TiOF_2$), AlOOH, AlF_3 and Polymer. The process sequences for pretreatment of aluminum prior to high gloss powder coating are [29]:

alkaline cleaning - alkaline etch cleaning - rinse - deoxidizing - rinse - conversion - rinse - drying at 140 °C.

Advantages of chrome-free conversion coatings are [30]:

- lower safety precautions;
- less sludge formation and therefore less disposal costs;
- chrome reduction in waste water is not required;
- better working conditions (safer handling of chemicals: yearly health control of workers not required);
- cheaper storage of chemicals;
- reduced liability.

Disadvantages of the described Cr-free process are:

- colorless conversion layer;
- unpainted conversion layer is less corrosion resistant;
- less experience than with chromating.

Two other non-chrome conversion processes for aluminum should be mentioned here: the “Cerate” process and the “SAM” process.

The **Cerate process** [31] is based on cerium chemistry and has the advantage, that it provides a yellow surface, comparable to the color of yellow chromating. Test results show that cerium treatment has comparable performance to chromate coatings under paints, but bare corrosion protection (of the unpainted conversion layer) is somewhat less than with the chromate coating. Therefore the Cerate process is intended to be a pretreatment immediately prior to painting.

With the so-called **SAM-technology**, described by *Heitbaum* [31], *Milczarek* [32], *Adler* et al. [33] and *Harm* et al. [34], the conversion coating consists of a mono-molecular layer of **self assembling molecules**. The term self assembling molecules refers to a property of especially phosphonic acid derivatives and phosphonates, which makes, that these molecules (if they have a defined structure) can be chemisorbed with their "head" groups to the hydrated metal oxide layer (formed by a necessary etching step), whereas the "tail" groups are bound to the paint. During this process a continuous homogeneous film is formed, covering the whole of the surface, thereby providing excellent adhesion and corrosion protection. The reason for the corrosion protection of the conversion layer is its densely packed layer of molecules, which consist (apart from their phosphonic "head" groups) of alkane chains, which cause the layer itself to be hydrophobic.

The magic of the SAM system lies in the fact, that the process is self-limiting. That means, there is only one layer of SAM possible on the metal surface. The applied SAM's have a thickness of some nanometers and are colorless. For identification of these very thin layers a color test is available from Chemetall [31, 32], who can also give more details of the SAM-process.

6.1.3.3 Cleaning and pretreatment of galvanized steel

In a similar way to aluminum, the surface of the galvanized steel undergoes self-passivation when exposed to atmospheric conditions by formation of a thin layer of zinc oxide or zinc carbonate. However, this layer of so-called "white rust" is not a good protection in the long term, especially not in an environment of aggressive industrial atmosphere. The corrosion proceeds slowly forming a visible white/gray powder-like layer.

Depending on the method of production, the thickness of the zinc layer in the galvanized steel varies from 3 to 200 μm. Even in the case where the coating is applied directly on a freshly zinced steel surface not containing zinc oxide or zinc carbonate, a reaction between the zinc and the carboxyl groups of the binder takes place forming zinc soaps which reduce the adhesion of the coating to the surface. As usually, the cleaning step precedes the conversion coating deposition.

Alkaline and acid cleaners

Alkaline cleaners for galvanized steel differ from those used for steel in the strength of the alkali. These are in principle mild alkaline salts which prohibit attack of the zinc surface. Only when heavy dirt has to be removed from the zinc surface, small or moderate amounts of caustic soda can be added to improve the cleaning action. This addition will also cause certain etching of the surface which results in improved adhesion of the powder coating later on. The same cleaning technique used for aluminum, including electro-cleaning, can also be used for zinc, with the same remarks concerning the efficiency and the costs of the separate methods being appropriate.

The acid cleaners for galvanized steel are not as effective as the alkaline cleaning agents. Moreover the zinc surface is very sensitive to acidic conditions. These are the reasons why the acid cleaners are not widely used for cleaning galvanized steel. However, there are some special types of acid cleaners which are used to remove the white corrosion product from galvanized surfaces.

As with alkaline cleaners, an electro-cleaning operation can also be performed with the acid cleaners. The parts to be cleaned serve as the anode. The scrubbing action of the oxygen bubbles developing on the anode as a result of electrolysis of the water increases the cleaning efficiency.

Conversion coatings for galvanized steel

The **zinc phosphate** coating is the most important **conversion coating** applied on galvanized steel. The common conversion coating process known as iron phosphate does not produce a coating on galvanized steel. On the other hand the chromium phosphate process is considered to provide a conversion coating of inferior quality compared to that one obtained by zinc phosphate, especially with respect to the adhesion of the powder coatings which will be applied in the next step [16, p.43]. Chemical reactions which take place during the deposition of insoluble zinc phosphate on the galvanized steel surface are very similar to those in the treatment of a steel surface with zinc phosphate. The main reaction is between the elementary zinc with zinc dihydrogen phosphate generating zinc phosphate in a solution that contains free phosphoric acid as pH regulator.

As in the case of phosphatization of steel, oxidizing agents such as nitrates and nitrites are widely used in order to accelerate the process and/or to reduce the bath temperature. Salts of nickel and copper are also **effective accelerators** and **regulators of the growth of crystals**. To obtain specially fine crystals, the phosphating step is preceded by treatment of the surface with solutions containing titanium salts. These titanium salts are usually added to alkaline types of rinsing solutions after cleaning of the surface has been completed.

Zinc-aluminum alloys are very difficult surfaces to phosphate. In principle dissolved aluminum is a hindrance for the phosphating process. Even amounts as low as 0.5 g/l may completely prevent the formation of a phosphate coating [35, p.200]. In such cases addition of sodium fluoride or sodium fluor-silicate precipitates aluminum from the bath and improves the efficiency of the process. An alternative process is preliminary treatment of the surface with solutions of sodium or potassium hydroxide. As aluminum is more soluble, the surface layer consists almost entirely of zinc.

Due to the higher reactivity of zinc, the application of the conversion coatings can be done even at room temperature. This can improve the process efficiency in the treatment of a zinc surface containing aluminum. At room temperature the rate of aluminum dissolution is slowed down to a level which provides in many cases satisfactory results. The reaction rate rapidly increases with an increase in the bath temperature for pure zinc layers. However, because of obvious reasons this is not valid for zinc-aluminum alloys.

Application methods such as spray-wash, immersion or hand-wiping can be used for phosphatization of galvanized steel. The line operation steps include cleaning, rinsing, treatment with the conversion coating solution, rinsing, eventual post treatment, rinse with deionized water and drying. Line operating temperatures are the same as in the previous cases ranging from 30 to 70 °C. The advantage of the zinc phosphate conversion coating is the possibility of running articles of galvanized and cold rolled or heat rolled steel on the same line.

Chromium phosphate conversion coatings are usually blends of sodium bichromate and mineral or organic acid which form a colorless, yellow, brown or green layer on the zinc surface. The amount of the deposited material ranges from 500 to 700 mg/m^2. Zinc phosphate is generated in the first step of the chromating process. The hydrogen formed as a result of this reaction partly reduces the six-valent chromium ($Na_2Cr_2O_7$) to a three-valent chromium ($Cr(OH)_3$). The final product ($Cr(OH)_3 \cdot Cr(OH)CrO_4$) is formed on the metal surface [36, 37].

After drying the inorganic conversion coating with a film thickness of about 0.5 μm is obtained. Its composition can be described as $Cr_2O_3 \cdot CrO_3 \cdot 2H_2O$ [38, 39]. When in contact with water the chromate film hydrolyzes and free hexavalent soluble chromium ion is formed. The normal color of the coating is yellow. However, when the hexavalent chromium compounds are leached out, the yellow coloration disappears and the film takes on a greenish tint. In practice, chromate coatings on galvanized steel show colorations that range from yellows and golds to olive greens.

The residence time of the article in the chromating solution depends on the concentration of the active materials in the bath, the bath temperature and desired thickness of the deposited layer. In principle it is relatively short and in a bath temperature range between 15 and 40 °C the residence time varies between 5 and 60 s. Because of the speed of the process, the usual practice is to perform **passivation of the articles at room temperature**.

Non-chrome conversion method

Though zinc phosphate is already a non-chrome conversion method, that guarantees a high quality, it produces sludge and heavy metal containing waste water. Due to the steadily increasing use of galvanized steel, the customer requires a chemical pre-treatment for this metal, providing **high quality, improved ecological properties and lower overall process costs** compared to zinc phosphate. Therefore, for galvanized steel and zinc two new non-chrome conversion coatings have been developed:

The heavy metal free (HMF) phosphating process, introduced by *Henkel* [30], which is based on a novel combination of accelerators. It produces a dense layer of poly-amorphous phosphates. In combination with a chromium-free final rinse (e.g. based on complex zirconium fluorides) adhesion and corrosion protection are very good after coating. The process sequence is alkaline cleaning, rinsing, HMF conversion at pH 5 and 90 s/55 °C, rinsing, passivation, rinsing and drying. A comparison between the HMF process and a nickel-free zinc phosphating process, learns that the line of stages for HMF phosphating is shorter, the use of chemicals is reduced, there is much less sludge and the bath control is easier to perform than with zinc phosphate. The **cost saving potential** for the overall HMF process is estimated by Henkel at about 30 %.

- The Cr-free pre-treatment with "Zincon 1355" [29] an adhesion promoting special polymer mixture, also introduced by Henkel. Since some years "Zincon 1355" is successfully running in different lines for outdoor architectural use.
  A Dutch job-coater has tested this product in a so-called Duplex-system (= powder coated, pre-treated galvanized steel) at a location 25 m from the North Sea-coast. After 3 years exposure to the heavy marine conditions, no corrosion damages could be observed.
  The process includes alkaline cleaning, rinsing, deoxidizing, rinsing, conversion by immersion in "Zincon 1355" at room temperature, DI-water mist and drying.
  The Dutch Society of Industrial Spraying and Stoving Industries ("VISEM") recommends "Zincon" for the Cr-free pre-treatment of galvanized steel.

Table 6.2: Comparison between different application methods in surface treatment [16, p.54]

source: Society of Manufacturing Engineers

Method	Advantages	Disadvantages
Brush-on, wipe-on	+ No capital investments + Parts do not have to be moved + Large heavy parts can be handled + Can be started/stopped quickly + Suitable for intermittent work schedules	– Labor intensive – Low production rates
Steam gun	+ Low capital investments + Parts do not have to be moved + Large heavy parts can be handled + Can be started/stopped quickly + Suitable for intermittent work schedules	– Labor intensive – Low production rates
Immersion	+ Can be automated + Moderate production rates	– Moderate capital investments – Must be able to lift parts – Requires start-up time
Power Spray Washer	+ Normally automated + High production rate	– High capital investment – Requires start-up time

- Recently a duplex system containing of a pretreated hot dip galvanized steel and a conventional epoxy, polyester or polyurethane powder coating was tested for corrosion resistance. *Schütz* et al. [40] couldn't find any influence of the composition of the zinc layer but the pretreatment was crucial. Best results have been observed with chromium phosphate, close followed by trikation phosphating and titane-zirkone-fluorid-polymer conversion. Interestingly a mechanical pretreatment procedure ("sweeping") is claimed to show a similar influence on the corrosion resistance of the duplex system as the latter two chemical methods.

Regardless of the type of cleaning agent or conversion coating, four application methods are in common use in treatment of the surface. *Table 6.2* summarizes the advantages and disadvantages of the particular methods [16, p.54].

6.2 Application of powder coatings

Because of their different nature compared to the liquid coatings, powder coatings are applied on the objects to be coated by techniques that have little in common with the well-known methods of application of conventional solventborne coatings.

At first sight it seems that application of powder coatings should be more difficult than application of wet paints. However, this is only a prejudice emanating from the fact that people (even those who are not professionally involved in painting) are quite familiar with the use of the liquid coatings. Difficulties with powder coatings generally arise only in geometrically complicated objects, where among other things electrostatically screening effects may occur. These difficulties should not, however, give an impression that application of powder coatings is always problematic. Compared to the liquid coatings, **powder coatings even have several quite essential advantages**.

One of the greatest limitations associated with wet paint spraying is that the particles which miss the object are irretrievably lost. Only those hitting the workpiece are used for their purpose. The ratio between these two quantities, the so-called overspray percentage, can sometimes be so high, that depending on the object geometry, it is this factor alone that determines whether the coating process is economic or not. Very often this is the reason for switching to another wet application technique such as curtain or dip coating. Although the electrostatic paint spraying decreases the overspray rate, it can be never neglected as in the case of powder coatings where the powder is recycled in the spraying booth.

The **100 % utilization of powder coatings** together with the elimination of the solvents are among the elements which have enormously influenced the economy as well as the ecology of the painting process. These advantages far outweigh the technical difficulties encountered in the early stages of development of powder paint application techniques.

Four different powder coating processes have been developed during the last forty years: electrostatic spraying, fluidized bed, electrostatic fluidized bed and flame spray, the last one being developed in 1985 and not yet widely used [41]. These application techniques which will be described in the following sections offer the following advantages over the wet coating systems: self-regulating thickness of the applied powder coat, absence of sagging problems, good edge covering, low working power costs, low overall operating costs, low costs for training of the personnel, 100 % of utilization of the coating material, meeting the most stringent environmental compliances [42].

Some years ago a fifth coating process, comparable with the high-speed laser printer technique, has been developed by DSM Resins and Huber [43], namely the electromagnetic brush (EMB)-method, which will be described in Chapter 6.2.6.

6.2.1 Electrostatic spraying technique

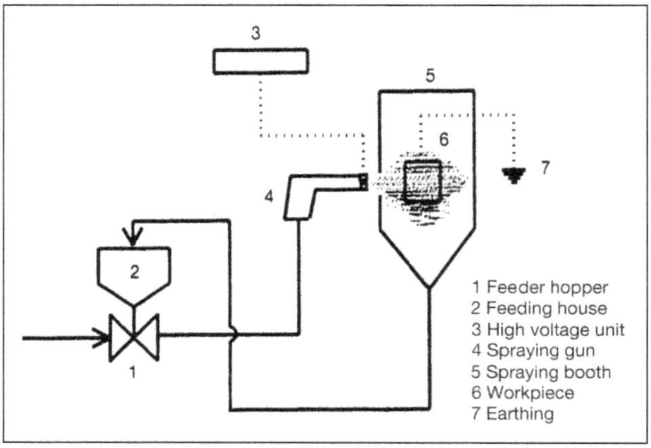

Figure 6.1: A schematic diagram of an electrostatic powder coating system

Electrostatic spraying is **the most common process used for application of powder coatings** in metal finishing. For low conductive substrates like engineered wood conductive fillers (e.g. silver particles coated with copper) [44] can be used. A diagram of the conventional powder coating spray system is shown in *Figure 6.1.* The basic principle of the process is propulsion of the dry powder by means of compressed air through a spray gun, in which it becomes electrically charged. The movement of the particles between the charged gun and the substrate to which the powder is applied is governed by a combination of electrical and mechanical forces. The electrical forces are the result of interaction between the charged powder particles and the electric field between the substrate and the gun, while the mechanical forces are derived from the air that blows the powder through the gun. Powder particles as electrically insulating material retain their charge - although this charge retention is reported to be influenced significantly by the storage temperature [45] - and adhere to the workpiece. While the workpiece is in front of the gun the particles are held on the surface by the electric field from the gun. When the field is removed, the charged particles are still held to the surface, attracted by image charges in the substrate. The charged particle and its image form a positive and negative pair and can be considered to be opposite plates of a capacitor. The oversprayed powder can be collected and reused. More than 98 % overall material usage is claimed in the industrial powder coating installations. Usually the recovery system consists of a cyclone, followed by a filter to remove the last traces of dust particles from the air.

6.2.1.1 Corona charging guns

These are the most widely used application devices for charging the powder particles. **Charging of the powder particles is imparted by high voltage** applied to at least one electrode located inside the sprayer. As well as charging the particles, the corona electrostatic spraying gun functions to shape and direct the flow of the powder supplied by the feeder, to control the pattern size, shape and density of the powder cloud emitted by the gun, and to control the deposition of the powder by electrostatic voltage levels.

In its simplest design, the corona gun may be a tube with a charging electrode in the form of a sharp pointed needle-like electrode or a fine gauge wire [46]. The electrode is connected to a high-voltage generator. Depending on the electrode geometry and the potential, a local electric field of about 3 MV/m can be created with field lines. This is the field strength at which electrical breakdown or discharge of the air itself will occur in the vicinity of the pointed needle. This is done in a controlled low energy continuous process which manifests itself as a bluish glow. As a result of the local discharge, the air will be ionized producing negatively charged ions, as is commonly the case with corona electrostatic spray commercial installation. The gun is constructed in such a way that **powder particles carried by compressed air are directed to pass through this charged space** picking up the negative ions on their way to the workpiece (*Figure 6.2*).

As can be expected, not all powder particles leaving the gun nozzle are charged, and not all negative air ions are picked up by the powder particles. Typically only 0.5 % of the ions produced by a corona are associated with powder particle charging. The remaining 99.5 % exist as free ions in the powder cloud [46]. In other words, the material leaving the electrostatic powder spray gun is composed of negatively charged powder particles, uncharged particles and negative ions.

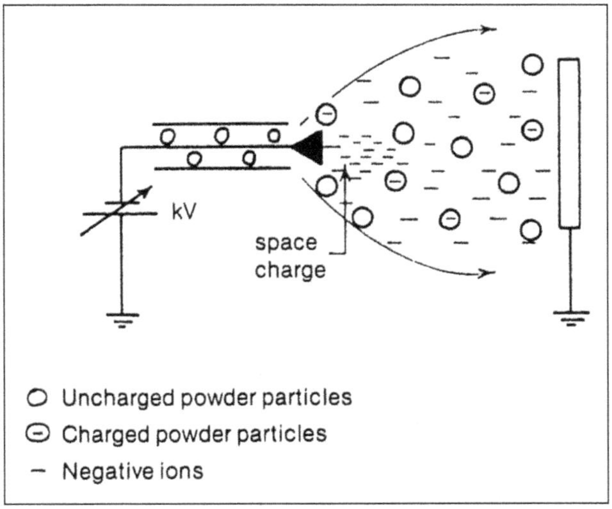

Figure 6.2: Corona charging of the powder particles [46]
source: John Wiley & Sons, Inc.

The mechanism of charging powder particles by negative ions is not well understood. The situation is especially complicated bearing in mind that the powder paints are materials with relatively high resistivity which can be considered as a constraint to efficient particle charging. The influence of resistivity and powder particle size on charging efficiency is investigated by *Corbett* [47, 48]. Although there is no sound fundamental clarification of the charging mechanism, it is believed that ion-trapping by the powder particles is purely a mechanical process as a result of direct collision between the flying powder particles and the ions when the latter are retained on the particle surface and "Locked into positions at the exact point of impaction" [46, p.5].

It is obvious that the powder particle in the ionic cloud could accumulate ions until its potential equals that of the surrounding field. This means that a certain maximum surface charge density cannot be exceeded. The theory of charging predicts that the saturation charge, Q, obtained by a spherical particle, in the size range of 1 to 200 μm, by air ions in a uniform electric field is given by:

Equation 6.1 $Q = 3kEa^2/(k + 2)$

where k is the dielectric constant of the particle, E is the charging field, which is in the case of corona charging proportional to the charging voltage, and a is the radius of the particle [49]. This **maximum surface charge** is known as the *Pauthenier* limit. For non-spherical particles, the charge is higher than that predicted from theory, but the difference is insignificant in practical applications [50].

The charge to mass ratio q/m is usually considered as a measurement for the **charging efficiency**. A somewhat modified version of Equation 6.1 which includes the permittivity of the free space can be used to calculate the maximum charge to mass ratio [51].

Equation 6.2 $(q/m)_{max} = 3\,\varepsilon_0\,E\,\{\,1 + 2\,(\,\varepsilon_r - 1\,)\,/\,(\,\varepsilon_r + 1\,)\,\}\,/\,(d_o)\,a$

where:

ε_0 = permittivity of free space
ε_r = relative permittivity of powder particles
a = particle radius
(d_o) = density of the particle
E = electric field to which the particles are subjected

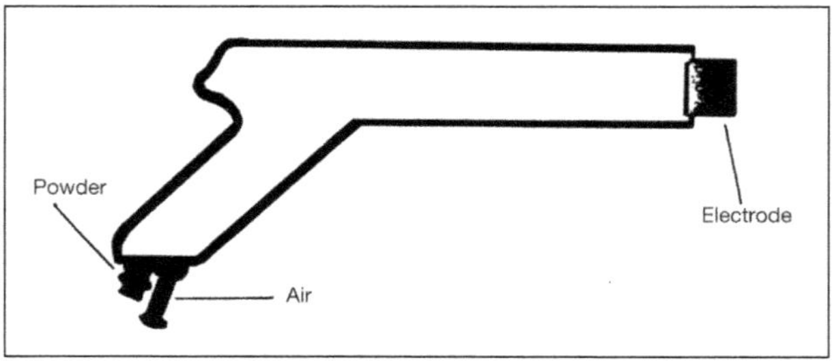

Figure 6.3: External corona gun

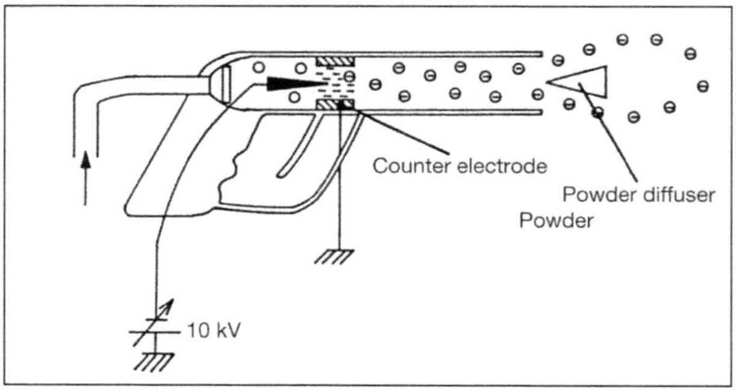

Figure 6.4: Internal charging corona gun [52]

source: Elsevier Science Publishers BV

Substituting corresponding values for the terms in Equation 6.2 *Hughes* has obtained a maximum charge to mass ratio of $8.8 \cdot 10^{-4}$ C/kg for an electrical field of 10^6 V/m [46, p.7]. The same Pauthenier limit is valid for positive charging by corona guns, although the mechanism of ion generation is different. In this case the electrons have to be stripped from the neutral air molecules creating positive ions. On their way to the earthed object these ions collide with the powder particles, imparting a positive charge in a manner identical to that of the negative ions.

The basic principle of charging in corona guns is the same, independently of the producer and type of gun. The differences are in construction details such as the design and location of the electrode in the gun, differences in the air diffuser at the gun nozzle, improved handling convenience, etc. These minor differences and improvements are very important to the user because they greatly influence the **deposition efficiency** of the equipment.

The most common type of corona gun incorporates a single short pointed electrode at the gun nozzle located at the center of the powder diffuser at the front of the spray gun. This is the so-called external charging gun (*Figure 6.3*) [16, p.78]. The electrode is usually fixed; the only adjustment of the powder transfer efficiency that is possible is by changing the potential applied to the electrode. These guns generate high voltage, low amperage electrostatic fields of 30 to 100 kV between the electrode and the workpiece. They provide **reasonably effective particle charging**, however, followed by an **excessive population of free ions**, which will be discussed later creates considerable problems during the spray operation. The electrostatic charge on the electrode is usually of a negative polarity, although the contemporary guns offer the possibility of choosing different polarities. This is especially important when powders that can be easily charged by friction with a positive charge have to be applied with corona. In such a case the negative corona charging gun is not the right choice. Switching to positive corona charging considerably increases the charging efficiency.

Developments in the equipment for a high-voltage supply by switching from a conventional oil-filled high-voltage generator to an epoxy encapsulated cascade type contributed very much to its **miniaturization**. This resulted in electrostatic corona gun with an **integrated high-voltage generator** [53]. In the internal charging guns the electrode is located inside the tube behind the gun nozzle with an earthed counter electrode (*Figure 6.4*). The charging is imparted by very thorough mixing in a turbulent flow in the region of the electrode. All powder emanating from the gun nozzle will have passed through the corona region, but few free ions are ejected from the nozzle. The result is **high specific charge with a fairly small voltage** on the electrode. Under typical operating conditions with a charging voltage of 10 kV, internal barrel potentials greater than 55 kV have been measured with this type

Figure 6.5: "Prodigy HDLV" automatic 'dense-phase' corona powder spray guns source: Nordson

of gun [52]. Due to the fact that the electrical field is built up inside the gun, the strong field developed by the corona gun between the electrode and the workpiece does not exist.

Although remarkably good in terms of high quality coatings, long term tests with the low-voltage internal charging corona guns have indicated deterioration in performance after long uninterrupted runs. This is associated with the growth of partially cured powder coating on the earthed counter electrode inside the gun. However, **back-ionization** at the counter electrode, or even nullification of the corona field, could account for this problem [52]. The only practical remedy for this problem is **frequent cleaning of the gun** at intervals of 1 to 2 h. However, this is a time consuming operation and must be performed with the utmost care due to the complex internal design of the gun. Another effect of back-ionization is claimed to be a size decreasing tendency of deposited particles [54].

Another improvement in the corona guns is claimed by the design of the so-called flat-jet nozzle with an air cleaned central electrode which operates at lower voltage than the standard gun and ensures optimum powder transport from the gun to the workpiece. This is achieved by the special nozzle geometry which produces a flat, well-directed spray jet which makes it easier to transport the powder into problematic zones such as recesses, internal edges and overlaps [53]. *Figure 6.5* shows a spray unit in a production process.

Problems associated with corona spray guns

The material leaving the gun nozzle consists of charged and neutral powder particles, and ionized air or so-called free ions. While the trajectories of the uncharged particles will depend predominantly on the air flow pattern, those of the charged particles and the free ions will result from the interaction between the mechanical forces emanating from the movement of the turbulent air flow, and the electrostatic field forces between the gun nozzle and the workpiece. Since the free ions are much smaller than the charged particles and have much higher mobility, their motion between the gun and the workpiece is very rapid. On the way to the workpiece they will collide with the neutral air molecules inducing an additional motion of the air itself in the same direction as the ions. This additional component of the air velocity to that created by the gun air supply is called **ion-wind** [55]. In some cases the ion-wind air velocity can be comparable with that of the gun air and cannot be disregarded as a component participating in an allover air dynamic pattern of the gun.

The trajectories of the charged particles, if not interrupted by the air movement, will tend to follow the lines of the electric field as presented in *Figure 6.6*. Because of this, by the electrostatic spray technique it is possible to coat shadow areas which are not directly exposed to the air stream leaving the gun as a result of wrap-around. Turbulent movement of the air in the environment of the workpiece is a prerequisite for achieving a wrap-around effect, but at the same time the linear air velocity near the object to be coated should not exceed the speed of the particle induced by the electrical attraction forces. Therefore, the air flow produced

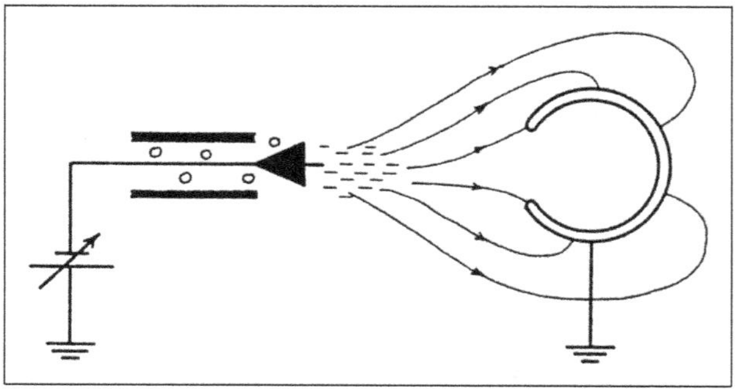

Figure 6.6: Typical flow pattern for corona charged guns[46]
source: John Wiley & Sons, Inc.

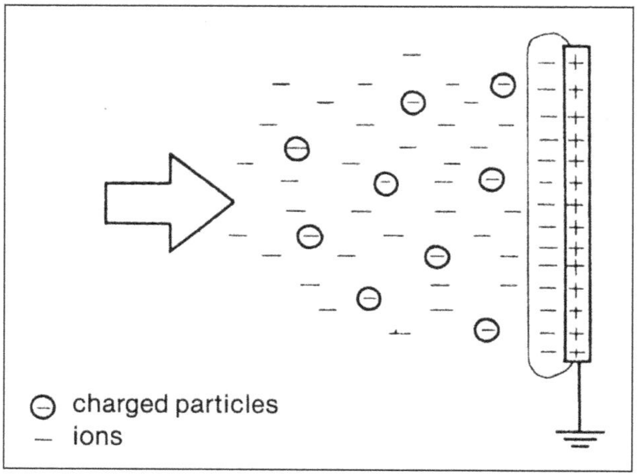

Figure 6.7: Back-ionization in the powder layer[46]
source: John Wiley & Sons, Inc.

by the gun can be even more important than the electrostatic field forces created between the gun and the workpiece. Confirmation of this is the **wrap-around effect** which is very effective in the case of tribo guns where the electrical field between the gun and the workpiece is essentially eliminated.

The air flow will play a predominant role for almost the whole distance between the gun nozzle and the workpiece, since the electrostatic attraction forces are shown to be significant only at distances shorter than 2 cm from the surface on which the particles deposit [56, 57].

The presence of the electric field between the gun nozzle and the workpiece creates enormous problems when articles with convex geometry have to be coated by corona spraying guns. As a direct result of the classical electrostatic law, no field lines can exist in or penetrate areas which are surrounded by the earthed metal boundary of the workpiece [46, p.18]. If the air velocity is low (and correspondingly the linear speed of the flying powder) the particles will follow the field line pattern that does not penetrate the inside surface of the workpiece *(Figure 6.6)*. As a rule, the electrostatic forces will deposit the material down into an opening to a depth equal to the minimum internal dimension of that opening. This is known as the **Faraday Cage effect**. The Faraday Cage effect is less pronounced in the internal charging guns, as can be expected due to the lower voltage applied to the electrode, but this effect certainly cannot completely be eliminated. According to *Trigwell* et al. [58] the generation of an almost ion-free powder cloud with high bipolar charge should provide a better coverage of recessed areas.

A phenomenon typically associated with electrostatic spraying of powder coatings with corona guns is the so-called back-ionization which was the concern of many studies [59-65]. *Figure 6.7* represents the development of a powder coating layer on the surface of the workpiece. As the powder layer grows, the potential across its thickness increases as a result of accumulation of charged particles and ions on the coated surface. At a certain moment the **breakdown potential is exceeded** and sparks will occur on the surface or within the layer. This breakdown potential is equivalent to that of the air, being 300 kV/m, since the coating layer is a loosely packed material in uncured form. The result of the discharge is the creation of bipolar ions. While the negative ions will be retained by the substrate, the positive ions will drift away towards the gun, creating the phenomenon known as back-ionization.

It has been found that with most commercial corona-charged gun systems, the back-ionization commences immediately after the first monolayer of powder coating has been deposited on the earthed surface. In terms of time it is approximately after one second spraying time [46, p.23]. The electrical discharging on the coated surface or within the coating layer is associated with a release of energy, which could lead to disruption of the deposited layer. This will certainly have a negative effect on the quality of the applied coating in terms of uniformity, although an explanation of the orange-peel effect by back ionization [46, p.24] is difficult to accept as the only contribution to this rather complex phenomenon.

Hughes and *Ting* [66] have found a considerable difference in the surface disruptions caused by the back-ionization effect between guns with different polarity. The spark associated with the back-ionization in the case of guns with negative charging polarity penetrates the depth of the deposited layer and creates surface defects such as pin-holes and craters. When a gun with positive charging polarity is used, the spark during the discharge is a surface phenomenon. Although in this case there is no pinholing, the surface defects tend to be more severe. *Hughes* explains these differences by the fact that the negative sparks are more energetic and can penetrate the bulk of the layer. He recommends the choice of **negative charging** corona guns in the case of coatings whose **aesthetic values** are of primary importance, and **positive charging** corona guns when the **surface protection** is the prime objective [46, p.27].

The **resistivity of the powder particles plays an important role** in the system behavior regardless the type of application equipment. At the gun head or in the vicinity of the charging electrode the powder resistivity should be as low as possible in order to maximize charge acquisition. However, the resistivity should be as high as possible in order to insure slow charge relaxation rates and good adhesion of the particles when they have reached the substrate. It has been stated that the lower acceptable limit of powder resistivity is about 10^{11} to 10^{12} Wm. Below this value charging will be good but adhesion will be poor [67]. Almost all powder coatings today usually have a resistivity higher than 10^{14} Wm and display excellent adhesive properties prior to curing. *Singh, O'Neill* and *Bright* [57] found that a minimum charge-to-mass ratio of around $2 \cdot 10^{-4}$ C/kg was necessary for adequate adhesion of powder to the workpiece. However, the fact of having this minimum charge-to-mass ratio does not necessarily mean that a good coating efficiency should be expected. *Hughes* [68] suggests that typically about 40 % of the powder in a normal coating operation emerges from the gun uncharged, which can be considered as one of the factors reducing the efficiency of the coating process. In their excellent article *Mazumder* et al. [69] observe three major needs in the research for corona charging processes:

- Corona chargers that can cope with different charge to mass ratios, allowing a surface charge almost up to the *Pauthenier* limit.
- A single device to switch from positive to negative charging of the powder particles in a minimum change over time.
- An ion trap that collects ions, but not charged or uncharged powder to keep the amount of free ions on the lowest possible level.

6.2.1.2 Tribo charging guns

Tribo charging guns make use of the **principle of frictional charging** known for several thousand years as one of the earliest phenomena associated with the electrical properties of solid materials. It is well known that when two insulating materials are rubbed and subsequently separated they become charged by opposite charges. Many examples in everyday life give evidence of frictional charging, such as combing the hair, walking on nylon-made carpets or changing clothes made out of synthetic fibers.

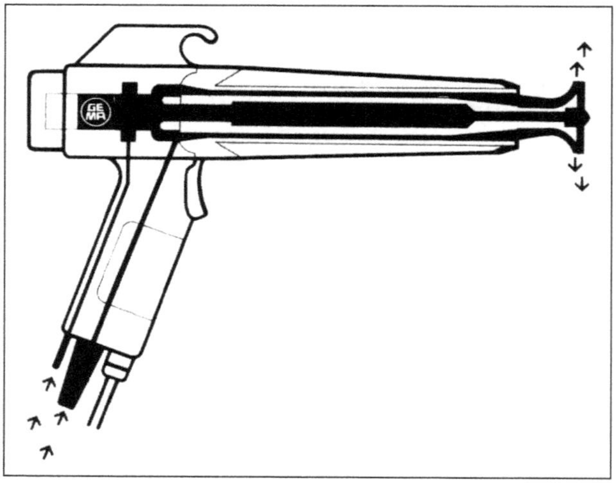

Figure 6.8: Tribo charging gun *source: Ransburg Gema AG*

The mechanism of tribo charging has been studied by *Davies* [70] and *Henry* [71] who explain the phenomenon by the contact potential difference model in which the insulator is treated as a metal. It is well known that when two metals are brought in contact then an electron transfer will take place due to the difference in the mean energy level between the electrons in the different materials. However, since the metals are conductive materials, the charge transfer will occur simultaneously with neutralization, so that the net result of charge reorganization at the interface will not be obvious. If two insulators are brought in contact, the charge transfer will create on longer term unipolar charge on the surfaces in contact, which will remain after their separation. It is believed that this mechanism itself cannot be good enough to explain the relatively high degree of charging obtained by sliding two insulators. An additional question that naturally arises is why it is necessary for **sufficient charge transfer** to rub the materials. The explanation that in this way the contact surface is increased is not very convincing. Therefore, another mechanism of charge transfer has been suggested which involves ionic transfer or bulk material transfer. *Hughes* suggests that charging of the coating powders in the tribo gun charging installation is a result of combining these two mechanisms [46, p.10].

The construction of the tribo gun is very simple. It is in most cases a Teflon-made tube through which the powder coating is propelled by means of compressed air, exhibiting a turbulent flow inside the gun. The tube walls are earthed, so that the charge generated on the walls is permanently drained to the earth (*Figure 6.8*). The high voltage supply in the tribo gun compared to corona is completely eliminated. However, the **cost savings** because of this are not the most important gain. The **Faraday Cage effect is almost completely eliminated** since there is only a very weak electric field created between the workpiece and the charged powder cloud. This makes it possible to coat objects having cavities and convex shape, since the powder particle trajectories are directed by the air stream rather than the weak electric field.

Long-term charging efficiency of the tribo guns is not always constant. In many cases, depending on the type of the coating, there is a tendency for powder particles to adhere on the gun walls. This changes the gun performance with time and it is believed to be one of the reasons why results have not been consistent on long uninterrupted runs. Therefore, it is suggested that for new unknown types of coatings, the tribo gun evaluation should be extended to at least five-hour continuous runs so that a realistic picture can be obtained about the suitability of this application technique.

Depending on the binder and crosslinker used in the powder coating formulation, quite different results can be achieved with respect to the charging efficiency of the same gun. The charging efficiency depends very much on the dielectric constant of the binder/crosslinker system. The so-called **dielectric series** tabulated in *Table 6.3* in order of their relative dielectric constants can give an indication of the expected behavior of different binders [42]. The charge achieved by rubbing two materials, given in *Table 6.3*, will depend on the distance between these materials in this list. Since all commercial tribo guns use Teflon as material for the gun walls, it should be expected that epoxy powder coatings will be more suitable for tribo application than powder coatings based on polyester resins. However, powder coating has a rather complex composition, and the chargeability will also be affected by the other constituents of the coating. Considerable work has been done to improve the charging efficiency in the case of polyester powder coatings. According to the patent literature [72-77] the problem of insufficient chargeability of the polyester powder coating has been successfully overcome.

Table 6.3: Dielectric series of different materials [42] in order of their relative dielectric constants *source: DSM Resins BV*

Order of relative dielectric constants	Material	Play a role in tribo charging
1	"Teflon"	yes
2	Polyethylene	
3	Polypropylene	
4	PVC	
5	Polyacrylonitrile	
6	Nitrocellulose	
7	Silicone	
8	Anthracene	
9	Polystyrene	
10	Polyester resins	yes
11	Cellulose acetate	
12	Rubber	
13	Metals	
14	Cellulose	
15	Polyvinyl alcohol	
16	Polyamide	
17	Wool	
18	Urea formaldehyde resins	
19	Glass	
20	Polyvinyl acetate	
21	Epoxy resins	yes
22	Polymethyl methacrylate	
23	Polyurethanes	
24	Polyethylene oxide	

Reports referring to the experience of using tribo guns in practice indicate that the initial doubts about the potentials of this technique have disappeared [78, 79] and that **the tribo equipment has become well established** [80]. Apart from elimination of the Faraday cage effect and the unique spray patterns, the tribo application process offers additional advantages. The absence of high voltages increases the **safety of the coating operation**. When in use, all electrically conductive components of the gun are earthed, eliminating the possibility of sparks igniting. Since the tribo gun is not an electrical appliance it is not subject to a compulsory check before installation. More compact powder coating films are easily obtained as a result of the absence of free ions and lower back spray tendency.

Tribo guns have usually a **lower capacity** than the corona charging guns. This can be considered as a serious disadvantage especially in applications where large quantities of powder coatings have to be applied in a short time. Certain improvements in this respect have been achieved by the new construction of the tribo guns where the friction surface is interrupted by a large number of trucks. This allows the surface charge to flow easily to earth without having an adverse effect on the powder charge [42]. Recent models now match the capacity and efficiency of corona systems [80].

Tribo guns are **more sensitive to particle size** than corona guns. Generally spoken, larger particles will be more effectively charged, due to the greater strength of their collisions with the "Teflon" charging surfaces of the gun. Moreover, the larger surface area of these particles contributes to their better charging.

Therefore, the particle size distribution of powders asks close attention when spraying with tribo guns. A too wide distribution causes deposition of the larger particles at a higher transfer efficiency than the smaller ones. By consequence, small particles will accumulate in the recycling system. To avoid this, powders with a rather **narrow particle size distribution are recommended**, bearing in mind that tribo systems are also more sensitive to powder humidity and temperature. In the already mentioned article *Mazumda* et al. [69] identified the following needs for research for tribocharging:

- reliable unipolar charging of powder coating particles
- application of fine particles (1 to 25 μm)
- a self-cleaning tribocharger with minimum contamination

6.2.1.3 Alternative guns

Several new types of electrostatic spray guns have been developed as an **answer to the problems associated with the corona and tribo guns**. The slow but steady decrease in the charging efficiency of the tribo guns is claimed to be overcome by adoption of internal corona augmentation [81].

The **corona augmented tribo gun** has been tested at Southampton University [52] leading to the conclusion that the gun gives uniform particle charging and high charging efficiency. High q/m values are obtained for epoxy/polyester powder coatings which are rather problematic for tribo application, together with low free-ion emission, enabling good powder penetration in cavities.

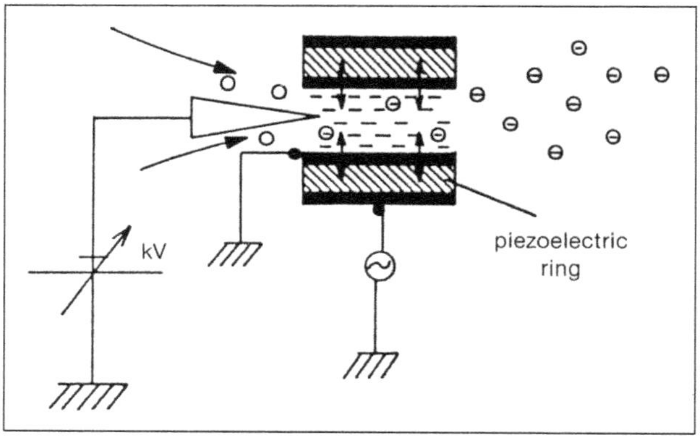

Figure 6.9: Piezoelectric attractor ring electrode [52]
source: Elsevier Science Publishers BV

Considerable efforts have been made to improve the long term efficiency of the internal low voltage corona gun. Two different approaches involving a **piezoelectric ceramic ring electrode** which undergoes an oscillatory deformation (*Figure 6.9*) and a **curtain electrode with a double helix configuration** (*Figure 6.10*) have been considered as a solution of the problem [52, 82].

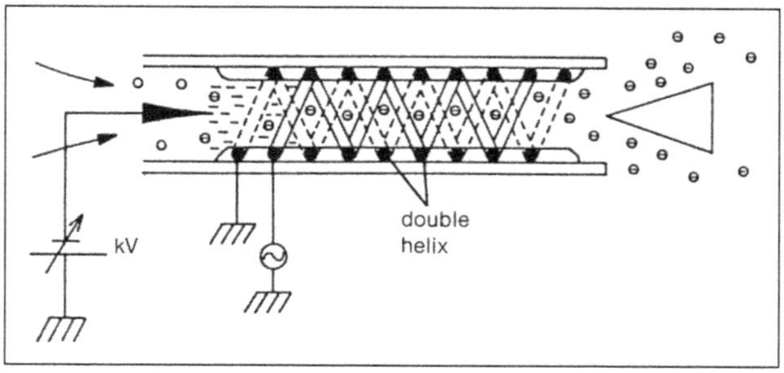

Figure 6.10: Electric curtain low-voltage gun
source: Elsevier Science Publishers BV

Despite the promising results, neither attempt resulted in a commercial gun system. It seems that the conventional porous metal ring electrode which is cleaned by air purge is the simplest and most efficient construction used in the previously described low voltage internal corona gun.

In 1994 two spraying equipment manufacturers claimed to have developed - independently from each other - new spraying devices for corona guns, by which it should be possible to spray powders without orange peel [83, 84]. The principle of their inventions is based on a **considerable reduction of the quantity of free ions** that normally reach the workpiece. As the free ions have a very small mass and are very mobile, they always move to the nearest earthed object (mostly the workpiece). By bringing an earthed counter-electrode (e.g. a ring) in the neighborhood of (behind) the high-voltage electrode(s), it is possible to "catch" the surplus of free ions and to direct the lines of force away from the workpiece. Practical experiences with these "additional sets" are very positive, for not only **orange peel is (nearly) eliminated**, but benefits are also provided by a better penetration into Faraday cages, less tendency to back-ionisation, improved transfer efficiency and a more uniform film thickness without "picture frames" (thick edges).

Another alternative gun is the **powder bell** (or powder electrostatic rotation gun) [85]. Here the powder is electrostatically charged and atomized by means of a rotating bell-shaped atomizer. At speeds of up to 10,000 rpm an extremely fine and homogeneous powder cloud is achieved, leading to an extraordinary uniform distribution of the film thickness.

As the powder bell distributes the ejected powder very evenly over a large area, it may operate at higher powder output rates. Powder bells are ideal for coating of workpieces with large flat surfaces and where high throughputs are required. Due to the fact that powders, ejected by rotating heads, have a reduced forward velocity, powder bells are not suitable for coating objects with recesses (Faraday cages).

6.2.1.4 Factors affecting the spraying process

The coating process with electrostatically charged powder particles is controlled by both electrostatic and aerodynamic factors. In several excellent articles *Hardy* [86] and *Golovoy* [50, 87, 88] analyze the influence of these factors on the deposition efficiency of the coating process and the growth of the film thickness during the powder deposition. *Hardy* has measured the potential at the surface of a layer of the powder after it has been deposited on a flat surface and compared the obtained results with those calculated from the following equation, which he derived for such a case:

Equation 6.3 $$V_s = 3E_{ps} \cdot m^2/k \cdot B \cdot d \cdot D$$

where:

V_s = potential at the surface
m = mass of the deposited powder per unit surface
B = bulk density of the powder layer
d = density of powder particles
D = powder particle diameter
k = dielectric constant of the powder layer
E_{ps} = electric field which the charge produces at the surface of the particle

Figure 6.11 confirms the excellent agreement between the experimental and calculated values. *Hardy* has found that the E_{ps} values are independent of the nature of the powder coatings and the type of gun used for spraying. However, the E_{ps} values are dependent on the voltage applied to the gun as can be concluded from Equation 6.1. Equation 6.3 can be written in the following form:

Equation 6.4 $$V_s/m = 3E_{ps} \cdot m/k \cdot B \cdot d \cdot D$$

The quantity V_s/m which is equivalent to the potential gradient across the insulating layer of deposited powder is obviously proportional to m, i.e. to the mass of deposited charged

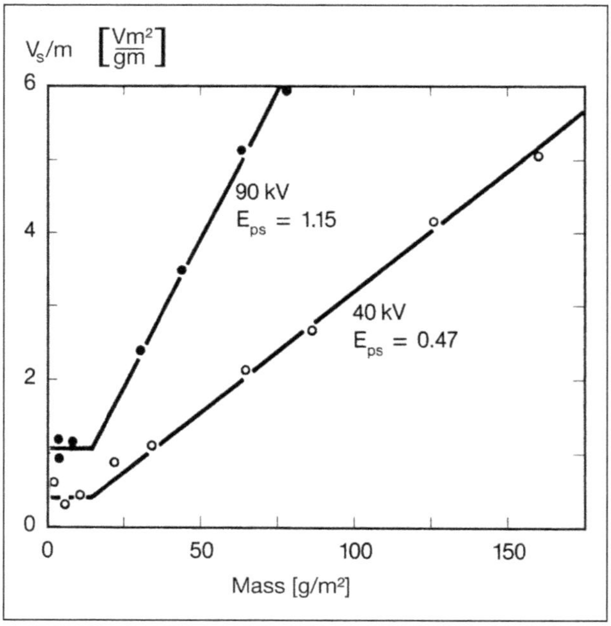

Figure 6.11: Plot of the surface potentials observed at various coating thicknesses, for two different gun voltages. The curves have been drawn according to equation (6.3) [86]

Source: Societies for Coatings Technology

particles per unit coated surface. In other words, by increasing the coating thickness the potential gradient increases linearly as long as it reaches the dielectric strength of the powder layer. At this moment an electrical breakdown will occur and the excess of the voltage will be drained to the earth. Powder particles arriving to the surface after this **critical layer thickness** has been reached will increase the surface potential again above the dielectric strength of the layer causing a localized discharge and will lose their charge. Since the charge of the particles is responsible for keeping the powder adhered to the substrate, they will held weakly on the surface and will be blown away by the air flow or will detach from the surface due to the gravitational forces. This is the explanation offered by *Hardy* for the well-known self-limiting phenomenon which can be useful for obtaining a uniform film build, but creates problems when thick coating layers have to be applied.

Hughes [46] associates the **self-limiting effect** with the back-ionization phenomenon. According to him, the positive ions created by the electrical discharge, when the dielectric strength of the powder layer has been exceeded, drift away from the substrate towards the gun. In their way to the gun they collide with the oncoming negative ions or negatively charged particles. The resulting neutralization will decrease the amount of negatively charged particles reaching the workpiece. Depending on the thickness of the deposited layer and the voltage supplied to the electrode, the coating ceases at a certain moment due to this back-ionization. He uses the same arguments to explain the differences in the critical layer thickness between the tribo and corona gun application techniques. The back-ionization which commences almost immediately after trigging the corona gun, appears in the case of tribo charging systems after 10 to 20 s. Since no ions contribute to the potential build up in the deposited layer, it takes much longer for the breakdown potential to be reached. Therefore, it takes 10 to 20 s before the onset of back-ionization when the tribo charging system is used, compared to only one second in the case of corona guns. This makes it possible to apply **much thicker films by tribo guns** because the self-limiting effect does not play an important role. While this can be an advantage in the case of functional coatings when the protection efficiency is proportional to the coating thickness, in many other industrial applications it is not an acceptable situation. However, the absence of the self-limiting effect can be compensated by careful choice of the conveyor speed and the output of the gun.

Once adhered to the substrate, powder particles may lose their charge slowly by electric conduction through the layer to the earthed workpiece. *Hardy* has found that the powder layers applied by corona guns don't lose more than 10 % of the initial charge during a period of 10 min after the application [86]. Even considerable delays in the transfer of the coated objects to the curing oven will not affect the coating thickness very much. Moreover, some charge losses can be favorable if they cause the loosely held particles to drop off, decreasing in this way the roughness of the coated surface.

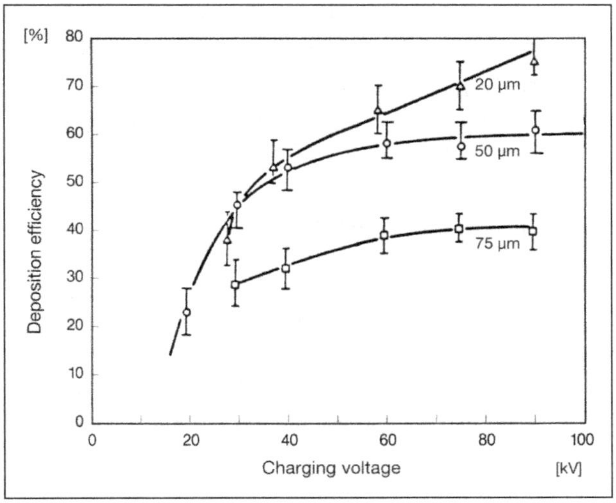

Figure 6.12: Plots of deposition efficiency versus charging voltage at film thicknesses of 20, 50 and 75 µm [50]

source: Federation of Societies for Coatings Technology

While the **critical film thickness** is dependent on the strength of the electrostatic field created between the gun and the workpiece, the deposition (or transfer) efficiency of the spraying process depends on a combination of electrostatic and aerodynamic factors and workpiece geometry. The **deposition efficiency** of the electrostatic corona spray process **increases with an increase of the charging voltage**, although as already seen the critical layer thickness decreases. This is due to a higher average charge per particle which can be achieved by a higher charging voltage, and a stronger electric field between the gun and the workpiece. Since the field is responsible for delivering the charged particles to the object, this will result in an increase of the deposition efficiency. However, as can be seen in *Figure 6.12*, in the case of thicker films the percentage of powder particles adhering to the substrate decreases.

With the following empirical equation *Golovoy* [50] describes the relation between transfer (or deposition) efficiency μ (in %) and the desired film thickness δ:

Equation 6.5 $$\mu = 93\,(1 - \delta/\delta_{max})$$

where

μ = transfer (or deposition) efficiency in [%]
δ = desired film thickness in [µm]
δ_{max} = saturation (or critical) thickness, at which back-ionisation starts

Thus, this equation says: the higher δ_{max} and/or the smaller δ, the higher is the deposition efficiency μ.

With the following empirical equation *de Lange* [89] describes a linear logarithmic relation between deposition efficiency μ and charge-to-mass ratio q/m, using a 4-electrode corona gun with integrated high-voltage (see *Figure 6.12*):

Equation 6.6 $$\mu = (0.43 \log q/m + 3.33) \cdot 100$$

where

μ = deposition efficiency in [%]
q/m = charge-to-mass ratio of tested powder in [C/g]

The electric field between the gun and the workpiece weakens when the distance between them is increased. At the same time the powder particles are exposed for a longer time to the air flow and self-repulsion effect on their way to the workpiece. This results in a **decrease**

of the deposition efficiency by increasing the spray distance. This is much more pronounced when a lower voltage is supplied to the charging electrodes.

A Canadian research group [90] studied the effect of the particle size on the coating efficiency and uniformity by means of a numerical simulation. They observed a higher deposition efficiency for coarse particles whereas the ultrafine powder showed a more uniform charging and deposition behavior [91].

Once the powder particle is deposited on the surface it has to remain there until it enters the stoving zone. Particles with high electrical surface resistivity retain their original charge for a longer time. If the surface resistivity is below 10^8 Ωm the adhesion of the powder particles to the earthed substrate will be very poor, decreasing the overall deposition efficiency of the coating process [92]. On the other hand, too high a surface resistivity produces an adverse effect by the deposited particle field on powder particle transportation and deposition efficiency. For optimum adhesion and deposition efficiency *Golovoy* recommends a region of **particle surface resistivity between 10^8 and 10^{14} Ωm** [50].

It is normal to expect that the booth air velocity should have an adverse effect on the deposition efficiency. This effect is however negligible when the booth air flow is parallel to the spraying direction. It is very pronounced when the booth air flow is normal to the spraying direction since it increases the velocity component of the particles parallel to the surface. The adverse effect of the booth air velocity can be drastically minimized by decreasing the spraying distance. The experiments performed by *Golovoy* showed that at spraying distance of 13 cm and booth air velocity of 0.6 m/sec parallel to the spraying direction the deposition efficiency is ca. 58 % compared to 35 % in the case of the same air velocity normal to the spraying direction. The decrease of the spraying distance to 5 cm compensates almost completely the negative effect of the latter bringing the deposition efficiency to the same level as in the case of parallel booth air flow.

The booth air flow and the flow of the compressed air used for transportation of the powder particles through the gun are responsible for the low deposition efficiency of powder coatings with a fine particle size. This is to be expected since the motion of the small particles is greatly affected by the air flow. The **maximum deposition efficiency** is observed by *Golovoy* **at a particle size of 35 μm**, after which the deposition efficiency decreases again. This is because above 35 μm fluidization and transportation of the particles from the powder reservoir to the gun becomes difficult, requiring an increase of the velocity of the carrying air, thus reducing the deposition efficiency.

The **size of the substrate affects the deposition of the powder coating**. This is especially emphasized for objects with small dimensions. *Golovoy* has found that for long objects with width W, the ratio W/D where D is the spray diameter at the gun exit, plays an important role with respect to the deposition efficiency independently of the operating conditions. When $W/D < 3$, the deposition efficiency rapidly decreases. For values above 3, the efficiency of the coating process increases slowly, leveling off at a limiting value that depends on the other operating conditions [50].

Tests by *de Lange* [89] show, that chemical composition of the binder, mean particle size and dry flow properties of the powder influence deposition efficiency. His results indicate that the **chemical composition has by far the greatest influence on deposition efficiency**.

6.2.2 Fluidized bed process

Powder coatings which appeared in 1950's were first applied by the fluidized bed technique. This rather **simple application process** is still in use although the electrostatic spraying plays by far the most important role among the different application techniques for powder coatings. In this process the preheated workpiece enters the space where by means of the air stream the powder particles are kept in a fluidized bed. Powder particles coming in contact with the preheated surface melt and adhere to the substrate. Depending on the temperature, heat capacity of the workpiece and its residence time in the fluidized bed, coating layers with various thicknesses can be applied. Post heating is usually not necessary in the case of thermoplastic powder coatings, especially when the workpiece is a massive object with large heat capacity. When thermosetting powder coatings are used, additional post heating follows the application step in order to complete the curing of the coating.

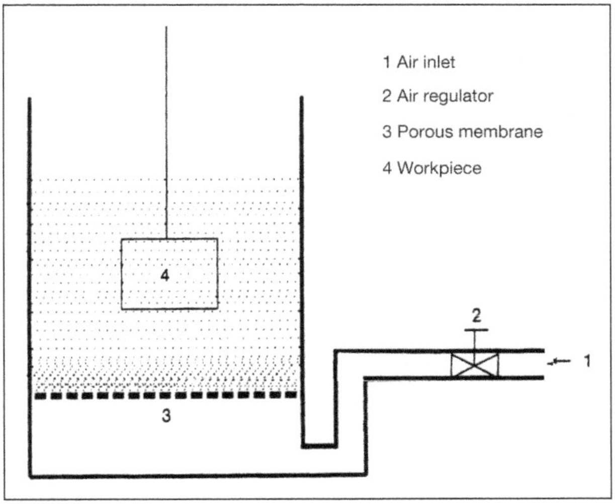

Figure 6.13: Scheme of fluidized bed installation

The basic construction of the fluidized bed chamber is presented in *Figure 6.13*. In its simplest embodiment it comprises a container within which the coating powder is confined. The upper compartment of the container is open for dipping the part to be coated and for allowing the fluidizing air to escape. This upper chamber where the coating process takes place is separated from the bottom compartment by a porous membrane. Compressed air is blown into the bottom compartment to be dispersed uniformly through the porous membrane before entering the fluidizing chamber. Fluidization of the powder in the upper compartment is effected by the lifting effect of the air stream. The amount of the air required per m^3 of the bed depends on the powder characteristics affecting the air dynamics of the fluidized bed process such as the density of the powder, the particle size and the particle size distribution. It is understandable that even distribution of the air through the membrane is a prerequisite for proper operation of the whole installation. Complete aeration of the bed from the bottom to the top is necessary in order to obtain a uniform layer thickness all over the workpiece. It is very important to avoid the formation of a dead compact layer at the bottom of the fluidizing chamber immediately above the membrane which leads to progressive stratification affecting the coating results in the dipping section [16, p.62].

Although not essential it is preferable to have a continuous feed with fresh powder to compensate for the consumed powder, instead of discontinuous addition. However, very consistent results can be obtained by supplying more frequent discontinuous additions of new powder to the dipping section.

The rate of **coating deposition depends very much on the particle size**. The fine particles are consumed in principle faster than the coarse particles. Therefore, the composition of the bed with respect to the particle size distribution changes continuously with time. However, this is a problem that exists only for a short period after the start of the coating operation. Very quickly, by subsequent addition of fresh powder the fluidized bed reaches steady-state operation conditions. By the proper maintenance of a constant temperature of the preheated object, careful control of the air supply, frequent feed with fresh powder, constant residence time of the workpiece and constant time sequence between two dipping operations, good high-quality coatings can be obtained.

Figure 6.14: Photo of a fluidized bed coating process source: Evonik Industries

The fluidized bed process is a **very attractive application technique** in many respects. The powder is confined to an enclosure and therefore is much easier to handle without the need to recycle and re-blend the overspray. Due to the absence of recycling and re-blending equipment, fluidized bed installations are in principle very compact compared to the spray gun/booth combinations. *Figure 6.14* shows a photo of a fluidized bed coating process.

6.2.3 Electrostatic fluidized bed

A combination of fluidized bed and electrostatic charge of the powder particles is effected in the electrostatic fluidized bed process. The design of the electrostatic fluidized bed installation is almost the same as that of the normal fluidized bed. By a combination of an air flow (very often accompanied by mechanical vibration) and charging electrodes in the upper chamber or in the air plenum under the porous membrane, the powder is fluidized and at the same time electrically charged. The depth of the bed is only 5 to 10 cm above the porous membrane through which the air is blown. The charging electrodes are usually located, because of safety reasons, under the membrane so as to charge the air used to fluidize the bed. The electrodes may be sharp points or more usually small diameter wires. By applying high voltage to the electrodes, a high charge density ion cloud is created in the same way as in the case of corona charging guns. The ionized air charges the powder particles as it moves upward through the powder bed. The electrostatically charged powder particles with an identical charge repel each other and move upward. In this way a cloud of charged particles is formed above the upper surface of the powder bed. The earthed workpiece entering the cloud is covered by the powder particles due to the difference in the electrical potential creating an electrical field of attraction. In fact only this field is responsible for the deposition of the powder on the workpiece. Therefore, preheating of the parts to be coated is not necessary. In contrast to the conventional fluidized bed process little air is used to control the volume of the powder cloud and to prevent stratification of the bed.

The **cloud density depends more on the voltage applied rather than on the air flow rate**. The higher the voltage the more dense the cloud. The deposition rate becomes less as the workpiece is moved away from the upper surface of the charged bed. This technique is **especially suitable for coating small objects**. Due to the Faraday Cage effect the deposition of the powder particles into holes or other convex surfaces is rather poor. This is typical for any electrostatic spray coating technique characterized by the existence of a strong electric field between the charging electrodes and the workpiece. However, the electrostatic fluidized bed process maintains all the advantages of the fluidized bed with respect to the absence of recycling and re-blending equipment. In fact the overspray is continuously recycled and recharged within the confines of the bed chamber.

The voltage employed with this method is in the range of 30 to 100 kV. This requires special **safety precautions**. The possibility of electrical sparking between the earthed workpiece and the high voltage charging electrodes may lead to ignition of the powder. This can happen especially if the workpiece comes close to the electrodes. Because of this reason, contemporary electrostatic fluidized bed installations have a unit design that incorporates a so-called inherently safe charging system with the electrodes located below the porous membrane. In this way sparking between the workpiece and the electrodes is practically eliminated. Moreover the particle trajectories are influenced much more by the air flow pattern than the electrostatic field, which is not the case when the charging electrodes are located in the upper compartment of the container. It is believed that superior powder penetration can be expected with such a design of the application unit, due to the less pronounced Faraday Cage effect.

Recently a Roman university research group [93] investigated the development of matting effects in electrostatic and conventional fluidized bed applications. They found a relationship between film morphology and the degree of curing in addition to the evolution of the complex viscosity.

The Material Science Corporation have developed a new method for applying powder coatings via an electrostatic fluidized bed method, which is called "MSC **Powder Cloud**". The method is described for **coil coating application** [94-96], but it seems feasible to apply the system in other areas too. With this new system powder is fed from an inlet hopper to a distribution brush by means of an auger feed brush, that uniformly distributes the powder to the distribution brush. The required amount of air is blown across a rotating brush and the powder is further distributed by an atomizing drum. The air source carries the powder into a cloud chamber, where a "powder cloud" is formed. The aerated powder cloud is electrostatically applied to the substrate by a series of corona wires, distributed in the cloud chamber. The charge-to-mass ratio is very high, much higher than with contemporary technology due to line source electrostatics. This means, that a single and uniform electrostatic field is generated across the entire width of the coil strip, which gives a deposition efficiency of 95 %.

The MSC Powder Cloud process generates an **electrostatic potential 1000 times higher than a conventional gun** (approx. 2500 kV/m) at approx. 20 mA. This large potential allows for effective penetration of the boundary and pilot plant line tests have shown effective coating at speeds up to 290 m/min [95, 96].

In 1999 a high speed industrial scale coil line using cloud chamber technology was installed in Middletown, Ohio [96]. This line can handle coil widths of 60 to 150 cm and has a line speed of up to 160 m/min. The powder coating is cured in 15 s by induction heating.

Xerox [97] patented in 2005 a powder cloud based technology, in which the particles are charged by triboelectrical means. This technique is claimed to be appropriate in particular for 3-dimensional substrates.

6.2.4 Flame-spray technique

The flame-spray coating technique has been developed about 25 years ago for application of thermoplastic powder coatings. Polyethylene, copolymers of ethylene and vinyl acetate, nylon and polyester powder coatings have been successfully applied by flame spraying. This technique permits powder coatings to be applied practically to any substrate, since the coated article does not undergo additional heating to ensure film formation. In this way substrates such as metal, wood, rubber, masonry can be successfully coated with powder paints if the coating itself has a proper adhesion to the substrate [98]. The technique itself is relatively simple. Powder coating is fluidized by a compressed air and fed into the flame gun. The powder is then injected at high velocity through a flame of propane. The residence time of the powder in the flame and its vicinity is short but just enough to allow complete melting of the powder particles. The molten particles in the form of high viscous droplets deposit on the substrate forming a high-build film upon solidification.

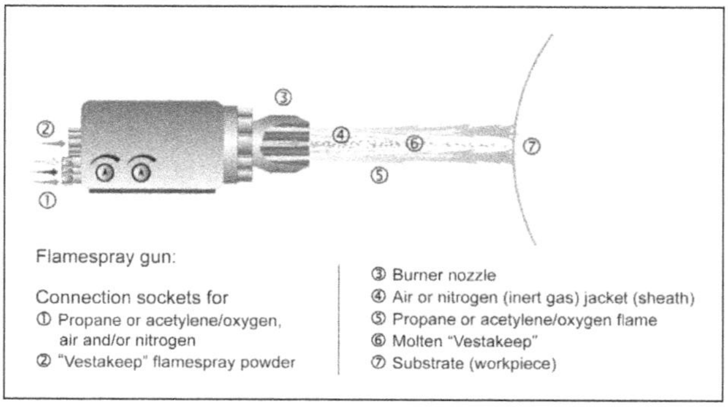

Figure 6.15: Flame-spray powder coating gun

source Evonik Industries

An example of a flame-spray gun is sketched out in *Figure 6.15*. The gun has a body with air, combustion gas and powder material supply channels. The outlet of the powder channel is axially positioned at the gun mouthpiece with the channels for the combustion gas outlet situated at equal distance on the circumference concentric to the axial powder channel. The efficiency is increased by preventing the powder from burning in the flame since the concentric circumference diameter is 2.85 to 4 times of the powder outlet channel diameter. The coating quality is increased when using liquefied gas since the combustion gas outlet channels axis is at 6 to 9° to the powder channel axis forming a diverging flame. The amounts of air and combustion gas are regulated by valves. The air passes through ejectors. The air and liquefied gas mix in chambers forming a combustible mixture which flows to the mouthpiece nozzles. The powder particles entering the flame are heated and in a molten form are supplied onto the surface being coated.

Since the flame spray process does not involve oven heating it is very **suitable for field application** on workpieces which are large or permanently fixed and thus not able to fit inside an oven. It has been reported, that objects such as bridges, pipelines, storage tanks and railcars are suitable to be coated by this technique [41].

A variant on the flame-spraying technique is the **High Velocity Thermal Spray** (HVTS) method for thermoplastic powders, described by *Blanton*, *Thames* and *Halliwell* [99] and abstracted by *Harris* [100]. HVTS also called HVOF (high velocity oxygen fuel) produces a coating, in which molten or semi-molten powder particles are applied by impact onto a workpiece at very high velocities of 900 to 1200 m/s, generated by a high-pressure flame.

More or less the reverse of flame-spraying is "flocking", for it is the application of a cold fusible powder to a preheated substrate, using an electrostatic spray gun. With flocking much higher coating thicknesses (200 to 300 μm) can be obtained and the problem of Faraday Cages can be overcome. This method is mainly used for **coating of pipes and reinforcing bars** ("rebars") and is also-called "**fusion bonding**". In pipe and rebar coating, the parts are preheated to temperatures of 230 to 270 °C, using ovens or induction coils, and then conveyed to the booth, where they are immediately electrostatically coated (usually with an epoxy powder), fused and cured by the heat of the object. Then the pipe or rebar is cooled by water quenching.

6.2.5 Comparison between different application techniques

Electrostatic spray, fluidized bed and electrostatic fluidized bed process are commercially well established, while the flame-spray process is not widely used. The performances of the first three processes have been compared by *van Buren* and *Martin* [41]. Some characteristics emanating from the nature of the flame spray technique can be compared with the other three processes (see *Table 6.4*).

The fifth application technique, the EMB-technology (see next chapter) has not been included in this comparison, due to the lack of relevant information about the practical experience.

Table 6.4: Characteristics of different powder coating application techniques *source: Decision Resources, Inc.*

Characteristics of workpiece	Electrostatic spray	Fluidized-bed and electro-static fluidized-bed	Flame-spray
Size	Larger	Smaller	Not limited
Material	Metallic, must be conductive	Any, not necessarily conductive	Any, not necessarily conductive
Temperature resistance	Relatively high	High	Not relevant
Aesthetic value	High	Low, not suitable for decorative purposes	Low, not suitable for decorative purposes
Coating thickness	Thinner films	Thick high-build films with excellent uniformity	Thick high-build films; uniformity dependent on the operator
Type of coatings	Thermoplasts and thermosets	Thermoplastic and thermosets	Thermoplasts only
Color change	Difficult	Relatively difficult	Easy
Capital investments	Moderate to high	Low	Very low
Labor	Low since highly automated	Moderate depending on the automatization	Relatively high
Energy consumption	Only post-heating	Preheating and often postheating	Low, no pre- and postheating
Coating waste	Very little	Very little	Dependent on the workpiece geometry

6.2.6 Electromagnetic brush (EMB) technology

The regulation of film thickness, low line speed and consistency of the operation on the long continuous run, are problems, associated with the above described conventional techniques, when applied to high speed **can** and **coil coating lines**. While the running speed of can and coil coating lines, using solventborne coatings, is more than 100 m/min, the installed coil coating lines that use powder coatings, operate with a line speed of 15 m/min, with a possibility for an increase to 30 m/min. The only exception is made by the MSC Powder Cloud technique (see before).

Higher line speeds and **very low film thickness** remain general prerequisites for real entrance of powder coatings in the can and coil coating industries. Both problems are exclusively related to the electrostatic spraying technique used at present for application of powder coatings.

In the case of electrostatic spraying, the transfer of the powder particles from the gun to the workpiece is done by the combination of mechanical forces (compressed air) and electrostatic forces. Much better control of the deposition process, however, can be obtained when the use of compressed air is eliminated and the transfer of the powder particles is only affected by means of electrostatic forces. A similar technique is already used in high speed laser printers, where powder coating is transferred to paper in a very well controlled way.

DSM Resins and *Huber*[43] developed a new application technology to apply powder coatings on flat substrates. The most pronounced characteristics of this application technology are the obtained high line speed and the resulting very thin films. This new technology is named EMB (**electro-magnetic brush**) technology. The principles of the EMB-technol-

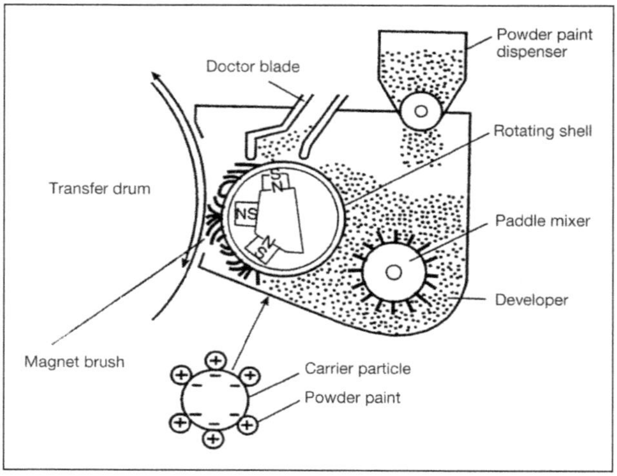

Figure 6.16: Schematic picture of the EMB device [101]

ogy have been derived from the well-known and very well studied copy- and laser printing processes, from which parallels can be drawn to apply powder coatings on flat substrates.

Figure 6.16 is a schematic picture of the EMB device, developed on laboratory scale. Powder particles are mixed with so-called carrier particles, which have a ferromagnetic iron core coated with "Teflon". By means of continuous mixing, both particle types are kept in a permanent movement. Due to this intensive mixing and friction, the powder particles are charged positively (tribo-charging), while the Teflon surfaces of the carrier particles are charged negatively. Consequently, the powder particles adhere onto the carrier particles, due to their opposite electrical charges.

The application unit contains a magnetic roll or drum which, while rotating, picks up the carrier particles, which are covered with oppositely charged powder particles. In this way an electro-magnetic "brush" (EMB) is formed on the surface of the rotating magnetic roll (see again *Figure 6.16*). This brush can be brought in the vicinity of the coil sheet. If the coil sheet is earthed or, if an electrostatic potential of right polarity is applied between the magnetic roll and the coil substrate, the powder particles are attracted by the substrate, leaving the carrier surface. The carrier particles are kept on the surface of the magnetic roll by magnetic forces, until they reach the scraper (or doctor blade remover), that cleans the surface of the magnetic roll. The carrier particles are continuously recycled back to the powder container. In this way a clean powder coating, free from carrier particles, is applied onto the coil sheet surface.

With the laboratory machine a line speed of 130 m/min has been reached, which is in the speed range typical for can and coil coating. Films of high uniformity without defects are easily applied. The film thickness can be regulated between 5 and 50 μm by changing the potential difference between the electro-magnetic brush and the coil sheet. The transfer efficiency is almost 100 %, ensuring a dustless clean working environment and making the recycling system for "overspray" powder superfluous.

A number of interesting experiments with this laboratory machine are described by *Binda* at the PCE 98 Congress [101] and in 2000 and 2002. *Binda* [102, 103] describes the first prototype 43 cm wide EMB machine that can perform continuous application. At the core of this machine is a 43 cm wide two-roller magnetic developer station, which is used in high-speed laser printers. The goal of experimenting with this prototype EMB is to further substantiate the concept of the EMB technology, namely: homogeneous application of powder coatings on high speed moving substrates, according to *Binda*.

With this machine interesting experiments have been carried out, e.g. applying a UV curable powder coating to durable luxurious wallpaper, with the objective of replacing the current solventbased clear coat (with 16 % solids) by a VOC-free powder topcoat. At speeds of 60 m/min closed, non-porous topcoats of 10 to 20 μm (and even 5 to 7 μm) have been realized, and, after off-line curing by IR and UV radiation, the coating had the right functional and decorative aspects for this type of wallpaper, while previously tested waterborne clear coats have failed in this respect. In addition, cost calculations show that, in this case, **powder coating is the more economical paint technology**.

From the results of the many experiments with the EMB, it can be concluded, that the EMB technology has potential as an alternative "dry roller" coating technology. Its main advantages are very high speed of application and very thin non-porous coating layers. However, it is also expected [104], that EMB will allow for **quick color change** (simple exchange of powder containers) and will give more even coatings in terms of surface roughness, leading to the use of less coating powder to obtain a minimum film thickness.

The EMB technology has made important steps into its development as commercially viable powder coating application technology. Several projects and co-operations are being discussed or have already been started in the field of can and coil coatings, paper coatings, and some other systems, according to *Binda*. But in order to exploit the EMB potential, co-operations should be initiated between coating applicators, powder coating producers, equipment manufacturers and carrier producers, to further test the potential of EMB as an alternative coating application.

When making a comparison between the EMB technology and the MSC Powder Cloud technique, as both methods allow for high speed application, the **EMB** will mainly find its way on **heat sensitive substrates** (using UV curable powder coatings) and there, where very thin film thicknesses (<15 μm) are required. Examples are can coatings, paper and plastic coatings, flexible packaging (aluminum) interior coatings. The **MSC Powder Cloud** will be more suitable for **coil coating**, where film thicknesses of 20 to 100 μm are required.

6.2.7 Electrostatic spraying with "Kompstat" technology

In the field of plastics, automotive add-on parts (such as bumpers, spoilers, moldings) present a large market potential for electrostatic powder coating. In the field of wood there is an expanding market in medium density fiberboards (MDF).

On the one hand, the application on these non-conductive substrates is still very limited; on the other hand, there is an increasing demand on these **environmentally friendly, cost-effective and high-quality powder coatings**.

New generations of powder coating systems (such as the highly reactive low-bake powders and of course the UV curable powders) show great advantages over the conventional powders with their high curing temperatures (≥140 °C), which are not suitable for **application on these heat sensitive materials**. The high surface resistivity of plastics, wood and wood composites would lead to costly measures (e.g. application of a conductive primer) to enable electrostatic spraying of these new powder coatings.

The "Kompstat" technology [105] (developed by the Fraunhofer Institute for Manufacturing Engineering and Automation, at Stuttgart) offers a new solution, realizing the **particle transport with compensating charges**. The technology is based on the physical property of mutual attraction between particles of opposite charge. Using positive and negative charges during the coating

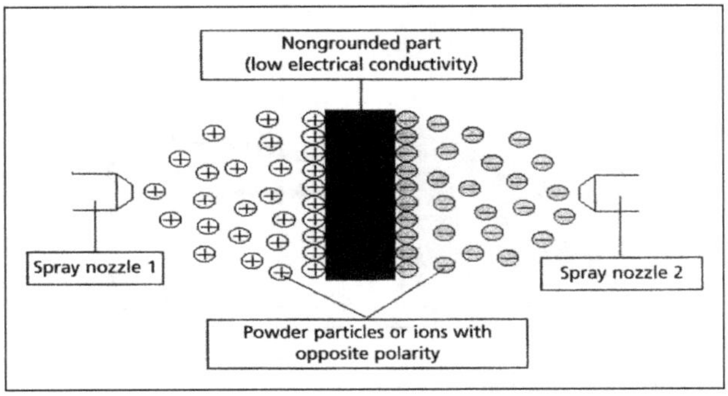

Figure 6.17: Principle of the "Kompstat" technology [105]

process, both sides of the substrate are hit simultaneously by charges with opposite polarity - either with powder or ionized air. The charge exchange is prevented by the very low electrical conductivity of the substrate. The opposite charges compensate through the substrate. *Figure 6.17*[105] illustrates the principle of the "Kompstat" technology. The "Kompstat" technology offers many benefits:

- Normal "standard" powder formulations can be used.
- Additional measures to enable the electrostatic application are not necessary.
- Due to the directed charge transport, the film thickness distribution on geometrically complex workpieces, including recessed areas, can be controlled.
- A partial coating of the substrate is possible.
- Single or double-sided coatings can be realized.

Coating experiments at Fraunhofer with powder on plastic and wooden substrates confirm the effectiveness of "Kompstat" technology and show that the **transfer efficiency on non-earthed MDF parts**, coated with "Kompstat" method, is nearly the same as the transfer efficiency on earthed metal parts, coated by conventional electrostatic powder coating technology. Fraunhofer illustrates this with the following test results:

- With single-sided coating the transfer efficiency for MDF, with "Kompstat" technology, is 60 % and for metal with conventional technology also 60 %.
- With double-sided coating the transfer efficiency for MDF with the "Kompstat" method is 48 % and for metal with conventional spraying 54 %.

An obvious disadvantage of the "Kompstat"method is the **need of two guns**, one for the front side and one for the back side of the workpiece.

6.3 Design of the spraying booths

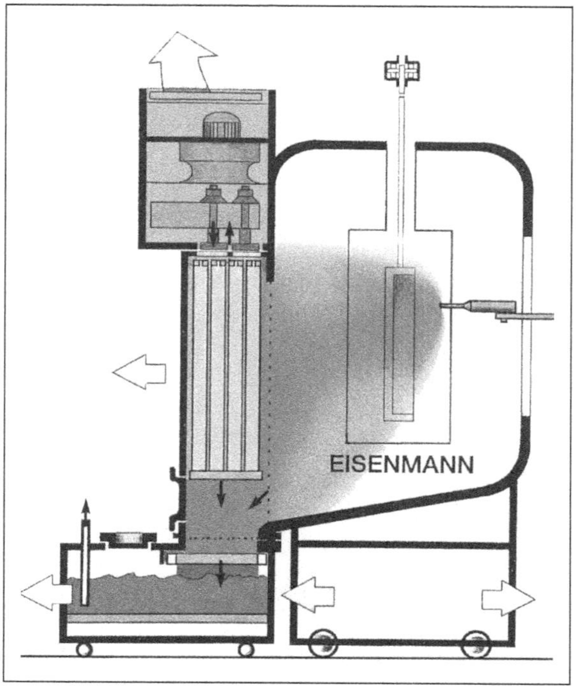

Figure 6.18: Compact recovery system

source: Eisenmann Maschinenbau KG

One of the greatest advantages of the powder coating process is almost 100 % utilization of the material achieved by **re-circulation of the oversprayed powder** which does not adhere to the workpieces. In order to combine this advantage with a high quality of the coating process, the design of the spraying booth system has to satisfy the following two requirements: recapturing the oversprayed powder as completely as possible and ensuring that the powder is continuously dressed to a **consistent quality and with nearly particle size distribution** [106]. There are three different principles used in the construction of the spraying booths to meet these requirements. A scheme of the so-called compact recovery system which makes use of air filters for recycling the powder is presented in *Figure 6.18*. The powder spray booth is provided with a replaceable filter which is part of the recovery unit. To obtain flawless surface quality, powder deposited on the filters must be continuously removed and added to the powder circuit as uniformly as possible. This is done by intermittent cleaning of the filters with a short blast of air. **Two types of filters** are used in this operation: sturdy cartridge filters with rotary wings and a relatively long service life and plate filters with exceptional separation efficiency for microfine powders and a virtually unlimited service life. The powder particles which are blown off the filters fall on a vibrating sieve or rotary sieve for automatic elimination of dirt particles. Re-blending of the powder passing through the sieve is done in the fluidized bed powder hopper where fresh powder is supplied manually or automatically from the fresh powder hopper depending on the size of the coating operation. The continual addition of the virgin powder is essential if high quality coating and efficient operation is to be maintained. The booth exhaust air is drawn through an exhaust fan, filtered and returned to the work area. The design of the powder booth provides a horizontal path of the powder to the workpiece resulting in wrap-around coverage and a uniform coating. Although in principle any material can be used to make the booth, stainless steel booths were especially suitable providing reduced cleaning expenditures, simple maintenance and extended service life.

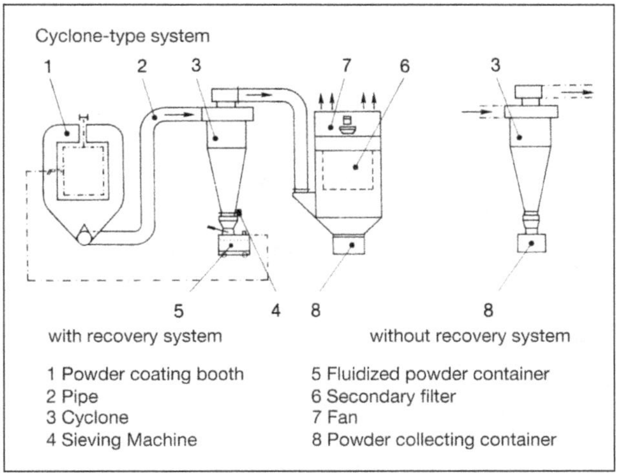

Figure 6.19: Cyclone-type system source: Ransburg Gema AG

But nowadays stainless steel is more and more replaced by non-conductive materials, such as polyethylene and PVC, which more or less repel the powder.

The **cyclone-type powder recovery system** is another option which is widely used in the coating process is depicted in *Figure 6.19*. Any powder which does not adhere to the workpiece passes through an exhaust device usually located in the lower part of the booth and via a pipe enters the cyclone where it is deposited. Recovered powder is then passed through the sieving machine and collected in the fluidized bed container for recycling. Any powder particles not extracted in the cyclone are captured by the secondary filter and collected in the container beneath the filter.

The powder coating spraying booth, that is equipped with a filter belt system, is **specially suitable for multicolor operation**. Filter belt systems have two separate air circuits. The booth exhaust air is handled by a medium pressure fan while for powder recovery a side channel fan is used. The oversprayed powder is deposited on a continuously running filter belt and carried to the suction head located outside the booth. The suction head cleans the filter bed across its entire width conveying the particles to either filter separators for single color operation or mini cyclone for multicolor operation. The powder is collected in the fluidized bed container for recirculation.

In order to maintain the powder in the fluidized bed container at a constant level, the powder feed from the fresh powder hopper is controlled by a level sensor in the fluidized powder container.

Another option in multicolor operation with recovery rates up to 120 kg/h is replacement of the sieving machine by a turbo sieve in the powder air circuit as presented in *Figure 6.17*. The delivery air filtered by the filter separators is returned to the working area via the

Figure 6.20: Control sieving and recovery of powder coatings "Cyclone Screener E240" source: *AZO GmbH*

side channel fan. The exhaust air from the booth after passing through the filter belt and an exhaust fan provided with secondary filter is returned back to the work area.

There are also booths with a combination of a cyclone- and a compact (filter cartridge) recovery system available on the market. *Figure 6.20* shows a picture of a production scale powder coating recovery equipment with an integrated control sieving device.

6.3.1 Color change in the powder coating process

Color change in the powder coating process is a rather problematic operation. Contrary to the liquid coating systems where there is no recirculation of the oversprayed material and the color change is a matter of seconds, in the powder coating operation process transfer to a different color can be done only after being sure that the whole system is free of the powder particles from the previous color. Many different solutions have been suggested and accepted by the end users for reducing the costs of this time consuming operation that affects the capacity of the coating plant very much.

Floyd [107] lists some of the possibilities such as mobile powder booths, detachable cartridge recovery modules, spray to waste and booths with moving parts with minimum fixed structure.

Another possibility is to try to create booths, which remain as clean as possible during operation. This can be done by taking plastic booths, in which either the booth exhaust is situated immediately under the gun(s), so that most of the overspray powder is continuously removed ("Magic" system [108]), or a powder extraction belt with integral cyclonic suction is used, by which the overspray is continuously fed back from the booth directly into the powder circulation ("FBC" plastic booth [109]). Another example is Nordson's cylindrically shaped "Sure-Clean" booth [110], which has a rotating floor with high velocity suction above it ("Aero-Duct" vacuum pipe system), for continuous cleaning of any powder on the floor. With a vacuum speed of 53 m/min, the air flow in the pipe guarantees that the powder particles are immediately transferred from the booth to the cyclone. In the "Sure-Clean" booth the double-walled PVC is replaced by a plastic material, called "Apogee" [111], which is as strong as steel and repels powder even better than PVC. With all three plastic booth systems powder is transported directly from the original powder container by inserted suction pipes. Glass booths have also been introduced some years ago [112].

The first option of *Floyd* with movable powder booths is quite suitable for fast color changes. However it suffers from several obvious disadvantages. There is an enormous **increase of the capital investments** due to duplication of the system. The floor space is twice as large compared to a single booth system. The savings are made with respect to the production time only. In the case of applying multiple colors, the expenses for man hours necessary to clean down the off-line system are not reduced at all.

Separate recovery units are economical in cases of extended intervals between color changes and low total output. It is still necessary, however, to clean down the interior surfaces of the booth, which is by far the most time consuming part of the color change.

Spray to waste is obviously the fastest way to switch from one color to another, with the complete loss of one of the main advantages of the powder coating process being the most economical one with respect to the material utilization. If spray to waste is chosen, a good "first pass" transfer efficiency is of utmost importance. *Bryan* [113] has calculated that, if the powder application equipment (in such a favorable case) can be cleaned within 5 to 6 min, it becomes an economic alternative to liquid coatings, concerning color change.

Floyd claims that the time necessary to complete a color change with full powder recovery, in the case of spraying booths with movable parts, is between 5 and 15 min. Since the booth parts are made from polyethylene, or PVC, they can be easily cleaned from the powder particles because the charged powder (nearly) does not adhere to non-conductive surfaces. This principle is also used in the Magic and "FBC" plastic booths, and in Nordson's Sure-Clean Apogee booth. In a recent article *Wohlgenannt* [114] reported a change time of 5 to 7 min at a recovery rate of 95 % achieved with equipment of the Wagner-group.

For several colors and not very short intervals between the color change, the booths with **compact recovery systems** making use of filter elements are a very economical solution. Spare filter units and powder containers are necessary for frequent and rapid color changes. While powder deposition continues in the booth with a second recovery unit, the filters are automatically processed in a separate cleaning system in preparation for the next color change. Economics of cost and space are the main advantages compared to the mobile powder recovery unit system.

Filter belt systems are mainly used for average to long product runs and are intended for high-quality manual and automatic work. Their flexibility allows more than one recovery system and a quick color change. Due to the small size of the recovery unit, space requirements are small. Compared with both the cyclone type and compact recovery systems, longer intervals for color changes are economical because of the better separation efficiency and powder recovery.

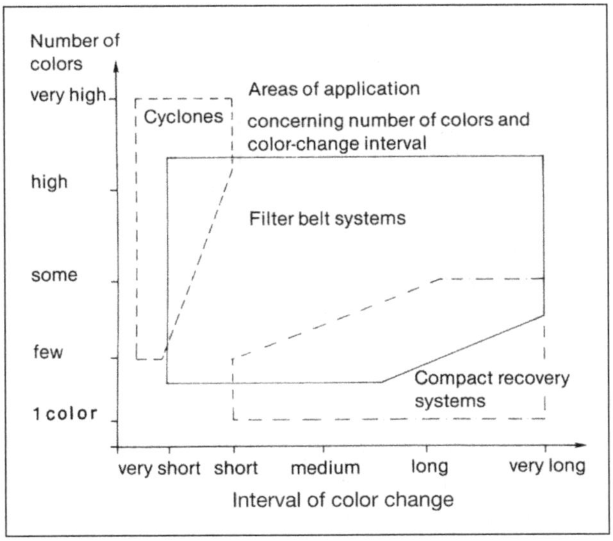

Figure 6.21: Areas of application of different spraying booth systems *source: Ransburg Gema AG*

Cyclone-type systems are the traditional solution for small batches with frequently changing colors. They are simple to operate, make few demands on the operators and are relatively inexpensive. Some limitations, however, exist due to the volume of powder to be handled and/or the volume of powder to be recovered. This system is suitable for plants with small capacities with respect to the applied powder coating per unit time characterized by very frequent color changes.

Combined cyclone/filter cartridge systems provide the operator more efficiency and flexibility in one booth. So that one can use the cyclone type for small batches of frequently changing colors. For larger volume colors the change over to the filter cartridge and utilize the higher reclaim percentage is possible. To avoid a speckled appearance Valspar [115] recommended mixing powder particles with a pigment dispersion to generate a colored free flowing powder. The diagram presented in *Figure 6.21* can be used as a guide to choosing the right system dependent on the number of different colors and the time interval between the color changes.

In addition to these criteria, the color range with respect to similarities of the shades, the plant utilization, the quality standards, the available space and the price should also be considered when choosing a spraying booth system.

6.4 Troubleshooting

The liquid nature of solvent or waterbased coatings permits **correction of the coating composition** by post addition of components which can improve certain property or remedy certain deficiencies. This is very often done by adding flow, anti-crater or defoaming agents in order to improve film formation properties, catalysts in order to increase the curing rate, crosslinker or binder to correct mechanical properties of the cured film, solvents for adjusting the rheological properties, pigment pastes to match color, etc.

Problems of this nature which can occur during the production or application of powder coatings cannot be solved in the same way. Very often every large deviation from the required specification needs a complete reworking, starting from the very first production step. Nowadays 2 to 5 % of the delivered powder coatings are rejected [116]. However, many of the problems can be overcome by simple interventions during the production, application or curing of the powder coatings. Certain guidelines for the causes of the problems with suggestions how to overcome them, collected from the published and presented data [16, 117, 118], are summarized in *Table 6.5*. *Herrmann* [119] believes that 50 to 70 % of the problems could be identified before the coating process. A thorough control of substrates is crucial. Most problems are arising at hot-dip galvanized steel due to an outgassing process in silicon rich substrates. In a follow up [120] article a reception control of powder coating formulation is recommended and different possibilities of coating repair are discussed, e.g. polishing and top coating. The same author published a review [121] regarding strain cracks due to embrittlement. The most likely reasons are insufficient homogenization during the extrusion step, inadequate through cure and inappropriate cooling down process. *Crompton* [122] recommended an automated image analysis to control particle size and shape, thus avoiding handling and flow problems. The Swedish automotive components supplier SMT uses robots for powder coating application, claiming to achieve higher quality at lower powder consumption [123].

In the case of critical film deteriorations the powder coating has to be removed. For this purpose different agents are commercially available. *Zollitsch* [124] reports on methyl pyrrolidone containing paint removers, which are able to strip a 100 μm polyester powder coating at 80 °C in 5 to 10 min. According to the author coating removal at room temperature is possible too, if special removers (e.g. "Stripal 2000 K/2K") are used.

Table 6.5: Trouble shooting guide lines for production and application of powder coatings [16, 117, 118]

source: The Society of Manufacturing Engineers and others

	Possible causes		Possible solutions
A	**Problems at production**		
A1.	Bad dispersion of the pigments and fillers		
1.	Inhomogeneous pre-blend composition	1.	Increase the residence time in the blender
2.	Viscosity of the compounding mass too low	2.	Decrease the extrusion temperature
3.	Too high pigment volume concentration	3.	Decrease the pigment and filler content
4.	Too short residence time in the extruder	4.	Decrease the extruder capacity
5.	Pigment and filler particles too coarse	5.	No help. Contact the producer
A2.	Sticking of the powder particles during grinding		
1.	Too high temperature in the grinding mill	1.	Increase the cooling efficiency
2.	Too low Tg of the coating composition	2a. 2b.	Check the Tg of the binder and crosslinker Reformulate the coating
A3.	Poor physical stability during storage		
1.	Too low Tg of the coating composition	1a. 1b.	Check the Tg of the binder and crosslinker Reformulate the coating
A4.	Poor chemical stability during storage		
1.	Too reactive system	1a. 1b.	Reduce the amount of catalyst Reformulate the coating adding resin with lower functionality
A5.	Poor impact resistance of the cured film		
1.	Under-cured film	1.	Increase the catalyst concentration
2.	Inhomogeneous powder coating	2a. 2b.	Increase the premixing time Increase the dwell time in the extruder
3.	Non-stoichiometric binder/crosslinker ratio	3.	Determine the acid, hydroxyl, isocyanate and epoxy equivalent values. Reformulate the powder with corrected ratio.
A6.	Poor flow with too strong orange peel effect		
1.	Too reactive system	1a. 1b.	Reduce the amount of catalyst Add resin with lower functionality
B	**Fluidized bed process in hopper**		
B1.	Powder blowing out of hopper		
1.	Air pressure too high	1.	Adjust the pressure
2.	Too fine powder particles	2a. 2b.	Too much reclaim added to the virgin powder Too fine originally manufactured powder Contact powder producer
B2.	No air percolation through the powder		
1.	Insufficient air pressure	1.	Check air supply line
2.	Plugged membrane	2a. 2b.	Check the membrane Check the air purity
3.	Compacted powder	3.	Try to loosen the powder manually
4.	Powder level in hopper too high	4.	Remove part of the powder
5.	Powder particles too fine	5a. 5b.	Too much reclaim added to the virgin powder Too fine virgin powder Contact powder supplier

B	Possible causes **Fluidized bed process in hopper**		Possible solutions
B3.	Air blowing canals (large holes) through the powder		
1.	Powder level too low	1.	Add more powder. Keep the hopper 2/3 full
2.	Packed or moist powder	2a. 2b.	Try to loosen the powder manually Check the humidity of the compressed air
3.	Blocked powder	3.	Check T_g of powder coating If too low contact the powder producer
4.	Membrane obstructed	4.	Check bottom of the bed
5.	Plugged or broken membrane	5.	Check the membrane
C	**Electrostatic coating operation**		
C1.	Fluctuations in the coating thickness		
1.	Irregular powder feed	1a. 1b. 1c.	Clean powder feed hoses, pump venturis and guns Check humidity of the air supply which can cause powder compaction Check powder free flowing properties
2.	Inadequate gun distance to the workpiece	2.	Readjust the distance between gun and workpiece
3.	Inadequate conveyer speed in relation to the high voltage of the gun	3.	Readjust the conveyer speed
4.	High-voltage fluctuations	4.	Systematically check electrical continuity from the voltage source to the electrode
5.	Powder blown off from the workpiece	5a. 5b.	Powder delivery air too high Turn down air setting Move gun position further away from the workpiece
6.	Pressure fluctuations in the air supply system	6.	Check the air supply system
C2.	Poor powder build-up and wrap on the workpiece		
1.	Poor earthling	1.	Check the earth connections. All contact areas must be free of powder build, grease or other insulating material
2.	Powder output too high	2.	Turn down powder feed until adequate charging is obtained
3.	Electrode(s) insulated by powder layer	3.	Clean the electrode(s)
4.	Particle size distribution not ideal	4.	Too much reclaim added to the virgin powder
5.	Powder coating composition contains resins which are not adequate for the application equipment used	5.	Contact the supplier of powder coatings
C3.	Poor penetration into recesses and gaps (Faraday cages)		
1.	Gun voltage too high	1.	Turn voltage setting down
2.	Powder delivery too low	2.	Turn up powder delivery air setting
3.	Poor earthing	3.	Check earthing connections and contact areas
4.	Inadequate gun position	4.	Adjust the gun position so that powder cloud has a direct path to the recess area

C	Possible causes **Electrostatic coating operation**		Possible solutions
	Continuation C3. Poor penetration into recesses and gaps (Faraday cages)		
5.	The spray pattern is too wide.	5.	Wrong deflector plate. Change it
6.	Delivery air speed in the gun too high	6.	Lower the air velocity
C4.	Back-ionization – powder repelled from the workpiece in spots		
1.	Voltage too high	1.	Turn down voltage setting
2.	Gun too close to the workpiece	2.	Relocate the gun
3.	Too fine powder particle size	3a 3b.	Too much reclaim added to the virgin powder Originally delivered powder is too fine. Contact powder supplier
D.	**Properties of the cured film**		
D1.	Poor impact resistance		
1.	Under-cured film	1.	Increase oven temperature or increase stoving time.
2.	Too high film thickness	2.	Readjust the application equipment.
3.	Poor cleaning or pretreatment	3.	Check pretreatment equipment and chemicals
4.	Powder coating formulation problem	4.	Contact powder manufacturer
D2.	Poor adhesion		
1.	Poor cleaning or pretreatment	1.	Check pretreatment equipment and chemicals
2.	Under-cured film	2.	Increase oven temperature or increase stoving time
D3.	Poor corrosion resistance		
1.	Poor cleaning or pretreatment	1.	Check pretreatment equipment and chemicals
2.	Under-cured film	2.	Increase oven temperature or increase stoving time
3.	Powder coating formulation problem	3.	Contact powder manufacturer
D4.	Poor chemical resistance		
1.	Under-cured film	1.	Increase oven temperature or increase stoving time
2.	Powder coating formulation problem	2.	Contact powder manufacturer
D5.	Film hardness too low		
1.	Under-cured film	1.	Increase oven temperature or increase stoving time
2.	Powder coating formulation problem	2.	Contact powder manufacturer
D6.	Poor film flow – orange peel effect		
1.	Too low film thickness	1.	Increase the film thickness
2.	Too slow heating rate	2.	Increase the heating rate
3.	Too fast curing system	3.	Contact powder manufacturer
4.	Too coarse powder particles	4.	Contact powder manufacturer

D.	Possible causes Properties of the cured film		Possible solutions
D7.	Craters in the film		
1.	Oil or grease residues on the metal surface	1.	Check the pretreatment
2.	Oil in the compressed air	2.	Check the air supply system
3.	Contamination with silicone	3.	Clean the spraying
4.	Contamination with silicone oils from the chain lubrication	4.	Clean the spraying area thoroughly
5.	Contamination with incompatible powder	5.	Clean the application installation thoroughly
D8.	Pinholes in the film		
1.	Moisture content of the powder too high	1a. 1b.	Check the powder storage conditions Check humidity of the air supply
2.	Air inclusions (due to a porous object)	2.	Preheat the workpiece (before application) to temperature just above the stoving temperature
D9.	Bubbles in the film		
1.	Water on the workpiece	1.	Check the drier and the workpiece hangers
2.	Powder film too thick (especially in the case of powders, containing caprolactam blocked isocyanates, or "HAA" or "Powderlink 1174")	2.	Decrease film thickness
D10.	Gloss and color adjustment		
1.	Gloss too low	1a. 1b. 1c.	Decrease curing temperature or time Increase film thickness Avoid contamination
2.	Gloss too high	1.	Increase curing temperature or time
3.	Inconsistent color	1a. 1b.	Control film thickness Control curing conditions

6.5 References

[1] *Treacher, S.*, Pretreatment Prior to Powder Coating, Powder Coatings 87 Conference, Manchester 1987
[2] *Lund, P.J.*, Pretreatment for Powder Coatings, Powder Coatings 87 Conference, Manchester 1987
[3] *Short, N.R., Dennis, J.K., Agbonlahor, S.O.*, R/P Trans. Inst. Metal Finish., 66, 3, 107, 1988
[4] *Bridger, R., Burden, P.*, Product Finishing, p. 6, April 1988
[5] *Marsh, C.*, Product Finishing, p. 14, April 1988; Brooks, M.E., Product Finishing, p. 8, April 1988
[6] *Kent, G.D., Petschel, M.*, Product Finishing, p. 56, September 1988
[7] *Johnson, J.H.*, Product Finishing, p. 48, September 1988
[8] *Gorecki, G.*, Metal Finishing, p. 15, December 1988
[9] *Agbonlahor, S.O., Short, N.R., Dennis, J.K.*, R/P Trans. Inst. Metal Finish, 64, 115, 1986
[10] *Linck, W.*, Oberfläche + JOT, 3, 46, 1988
[11] Pretreatments for Powder Coatings, Paint Research Association, Notes to Industry No.7, April 1989
[12] *Gruss, B.B.*, Metal Finishing, 29, May 1988
[13] *Barletta, M.*, Progress in Organic Coatings, 56(2-3), 185-198, 2006
[15] *Kuhn, A. T.*, Jahrbuch fuer Oberflaechentechnik, 63 268-279, 2007
[16] *Miller, E.*, Users Guide to Powder Coatings, p. 35, Society of Manufacturing Engineers, Dearborn Michigan, 1985
[17] *Neise, H.; Rump, K.*, JOT, Journal fuer Oberflaechentechnik, 44(4), 84-90, 2004
[18] *Alvarez, A.*, Congress Papers PCE 2000, Amsterdam, Jan. 2000, p. 61
[19] *Nunes, S.; Alves, P.; Pereira, I.*, Corrosao e Proteccao de Materiais, 27(2), 47-53, 2008
[20] *Fedel, M.; Olivier, M.; Poelman, M.; Deflorian, F.; Rossi, S.; Druart, M.-E.*, Progress in Organic Coatings, 66(2), 118-128, 2009

[21] *Yu, Z.; Liu, Y.; Yong, Z.; Liu, D.; Fan, J.*, Diandu Yu Jingshi, 27(1), 29-34, 2005

[22] Nippon Paint Co., EP 1669475, 2006

[23] Adopted from *Galjaard, D., Leeuwen van A.*, Chemische Oppervlaktebehandelingen van Metalen, Spruyt, van Mantgem & de Does N.V., Leiden, 1969, p. 53

[24] *Droniou, P.*, Congress Papers PCE 2000, Amsterdam, Jan.2000, p. 75

[25] *Tepe, B.; Gunay, B.*, Progress in Organic Coatings, 62(2), 134-144, 2008

[26] *Silvernail, N. J.; Mackay, M. L.*, US 20110177346 A1 2011

[27] *Dyllus, T.*, JOT, Journal fuer Oberflaechentechnik, 44(2), 36-38, 2004

[28] *Lenzer, S.*, JOT, Journal fuer Oberflaechentechnik, 50(6), 80-85, 2010

[29] *Jüptner, U.*, Congress Papers PCE 2002, Nuremberg, Jan.2002, p. 89

[30] *Roland, W.A.*, Congress Papers PCE 98, Amsterdam, Jan. 1998, p. 329

[31] *Heitbaum, J.*, XXVI. FATIPEC Congress "Quo Vadis - Coatings", Dresden, 2002 in Molecular Symposia 187, p. 43, Ed. Wiley-VCH, Weinheim Sept. 2002

[32] *Milczarek, R.*, Congress Papers PCE 2002, Nuremberg, Jan. 2002, p. 109

[33] *Adler, H-J. P., Jaehne, E., Henke, A., Lu Yan, Pich, A.*, XXVI. FATIPEC Congress "Quo Vadis-Coatings", Dresden 2002, in Molec. Symp. 187, p. 53, Ed. Wiley-VCH, Weinheim Sept. 2002

[34] *Harm, U., Bürgler, R., Fürbeth, W., Mangold, K.-M., Jüttner, K.*, XXVI. FATIPEC Congress "Quo Vadis-Coatings", Dresden 2002, in Molec. Symp. 187, p. 65, Ed. Wiley-VCH, Weinheim Sept. 2002

[35] *Biestek, T., Weber, J.*, in "Electrolytic and Chemical Conversion Coatings", p. 128, Portcullis Press Ltd., Redhill, 1976

[36] *Levitina, E.J.*, Zh. Prikl. Khim., 34, 984, 1960

[37] *Ropper, M.E.*, Metal Finishing, 14, 320, 1968

[38] *Farr, J.P., Kulkarni, S.V.*, Trans. Inst. Metal Finishing, 44, 21, 1966

[39] *Burns, R.M., Bradley, W.W.*, "Protective Coatings for Metals", Reinhold Publishing Corp., 3rd Ed., New York, 1967, p.553

[40] a) *Schütz, A. Triebert, J., Schubert, I.*, Farbe + Lack, 113, 2, 24-27, 2007
b) *Schütz, A., Schubert, I., Triebert, J.*, JOT, Journal fuer Oberflaechentechnik, 47(11), 48-50, 2007

[41] *Buren van, M.F., Martin, J.F.*, Spectrum, Advanced Materials and Chemical Specialties - Products and Technologies, March 1988

[42] *Lehmann, R.P.*, "Optimized Powder Application of Critical Objects", in DSM Resins Powder Coating Symposium, Noordwijk, May 1988

[43] DSM Resins/Michael Huber, WO 96/15199

[44] Rohm and Haas Company, US 20040235984, 2004

[45] *Ikezaki, K.*, Rigaku Janaru, 35(1), 7-12, 2004

[46] *Hughes, J.F.*, in Electrostatic Powder Coating, Research Studies Press Ltd., Letchworth, Hertfordshire, 1984, p.1

[47] *Corbett, R.P., Makin, B.*, Trans. Inst. Metal. Fin., 51, 160, 1973

[48] *Corbett, R.P.*, Dechema-Monographien, 72, 1, 1974

[49] *Pauthenier, M.M.*, Moreau-Hanot, M., J. Phys. Radium., 7, (3), 590, 1932

[50] *Golovoy, A.*, J. Paint Technol., 45, (580), 42, 1973

[51] *Hughes, J.F.*, J. Oil Col. Chem. Ass., 4, 145, 1989

[52] *Moyle, B.D., Hughes, J.F.*, J. Electrostatics, 16, 277, 1985

[53] *Lehman. R.*, "Electrostatic Powder Coatings, Developments and Trends" in Proc. Powder Coating Seminar, Lucerne, October 1988

[54] *Meng, X.; Zhang, H.; Zhu, J.*, Powder Technology, 195(3), 264-270, 2009

[55] *Athwall, C.S., Coventry, P.F., Hughes, J.F.*, Inst. of Physics. Conf. Ser. No. 66, Electrostatics, p. 167, 1983

[56] *Singh, S., Bright, A.W.*, 3rd Intern. Congress on Static Electricity, Grenoble 1977, Conf. Proc. 29G, p. 729

[57] *Singh, S., O'Neill, B.C.*, Bright, A.W., J. Electrostatics, 4, 325, 1978

[58] *Trigwell, S.; Biris, A. S.; Sims, R. A.; Mazumder, M. K.*, Particulate Science and Technology, 26(2), 145-157, 2008

[59] *Cross, J., Bassett, J.D.*, Trans. Inst. Metal Fin., 52, 112, 1974

[60] *Masuda, S., Mitzuno, A., Akutsu, K.*, 3rd Int. Congr. Static. Electricity Europ Fed. Chem. Eng., Grenoble 1977

[61] *Hughes, J.F., Ting, Y.C.*, IEEE/IAS Conf. Proc. 35G, p. 906, 1977

[62] *Sing, S., Bright, A.W.*, IEEE/IAS Conf. Proc. 29G, p. 729, 1977

[63] *Sing, S., Bright, A.W.*, IEEE/IAS Conf. Proc. 3F, p.105, 1978

[64] *Sing, S., Hughes, J.F., Bright, A.W.*, Inst. Phys. Conf. Ser. No. 48, p. 17, 1979

[65] *O'Neill, B.C.*, 3rd Int. Cong. Stat. Elect., Euro. Fed. Chem. Eng., Grenoble 1977

[66] *Hughes, J.F., Ting, Y.C.*, 3rd Int. Cong. Stat. Elect., Eur. Fed. Chem. Eng., Grenoble 1977

[67] *Hughes, J.F.*, J. Oil Col. Chem. Ass., 4, 145, 1989

[68] *Hughes, J.F.*, Finishing, 7, 6, 1982

[69] *Mazumder, M. K.; Sims, R. A.; Biris, A. S.; Srirama, P. K.; Saini, D.; Yurteri, C. U.; Trigwell, S.; De, S.; Sharma, R.*, Chemical Engineering Science, 61(7), 2192-2211, 2006

[70] *Davies, D.K.*, Inst. of Phys. Conf. Ser. No. 4, p.29, 1967

[71] *Henry, P.S.H.*, Inst. of Phys. Conf. Ser. No. 11, Static Electrification, p. 195, 1971

[72] *Misev, A.T., Binda, P.H.G., Hardeman, G.*, Stamicarbon B.V., NL Pat. 8.800.640, 1988
[73] *Misev, A.T., Binda, P.H.G.*, Stamicarbon B.V., NL Pat. 8.802.913, 1988
[74] *Binda, P.H.G., Misev, A.T.*, Stamicarbon B.V., NL Pat. 8.802.748, 1988
[75] Hoechst AG, EP 315 082, 1987
[76] Hoechst AG, EP 315 083, 1987
[77] Hoechst AG, EP 315 084, 1987
[78] *Stahlschmidt, K.H.*, British-German Powder Coating Conference, Stratford-Upon-Avon, 1-2 Dec. 1986
[79] *Kleber, W.*, 52nd Conference of the Coatings and Pigment Section, Soc. Germ. Chem., 1985
[80] *Guskov, S.*, Powder Coating, June 2000, 20-31, 22-26
[81] *Bauch, H., Auerbach, D., Kleber, W.*, Elektrostatisches Beschichten Vom 24 Bis 27, Hochschule f. Verkehrswesen Friedrich List, Dresden, 1978, p. 109
[82] *Masuda, S.*, IEEE/IAS Conf. Proc. 35D, 887, 1977
[83] Wagner Pulver News 1 / 94, "Corona-Star combats Orange Peel"
[84] Gema Volstatic, Technical Leaflet "Die Super-Corona", 1994
[85] Wagner: Powder News, The PER-C1 powder electrostatic rotation gun, 11/97
[86] *Hardy, G.F.*, J. Paint Technol., 46, 499, 73, 1974
[87] *Golovoy, A.*, J. Paint Technol., 45, 585, 68, 1973
[88] *Golovoy, A.*, J. Paint Technol., 45, 585, 74, 1973
[89] *Lange, P.G. de*, FATIPEC Lecture, Budapest 1978, published in Defazet 32, (6), 244, 1978
[90] *Li, Z.; Zhang, C.; Zhu, J.*, Canadian Journal of Chemical Engineering, 83(5), 882-888, 2005
[91] *Meng, X.; Zhu, J., Zhang, H.;* Journal of Electrostatics , 67(4), 663-671, 2009
[92] *Corbett, R.P.*, Conference on Electrical Methods, IEEE, London, March 1970
[93] *Barletta, M.; Bolelli, G.; Guarino, S.; Lusvarghi, L.*, Progress in Organic Coatings, 59(1), 53-67, 2007
[94] ECCA General Meeting, Barcelona, 19-22 Nov. 1996
[95] *Michaelis, R., Graziano, F. and Edwards, D.*, Congress Papers PCE 98 Amsterdam, Jan. 1998, p. 103
[96] *Schwaller, R., and Mallett, K.*, Congress Papers PCE 2000, Amsterdam, Jan. 2000, p. 419
[97] Xerox Corporation, US 20050287306, 2005
[98] *Chandler, B.W.*, Proceedings S.S.P.C. Symposium on Technology for long term protection of steel structures, Feb. 1986, Atlanta, p. 205-207
[99] *Blanton, M.D., Thames, S.F., Halliwell, M.J.*, 25th Int. Waterborne, High Solids & Powder Coatings Symp., New Orleans, Feb. 1999
[100] *Harris, S.T.*, Focus on Powder Coatings, July 1999, p. 4
[101] *Binda, P.H.*, Congress Papers PCE 98, Amsterdam, Jan. 1998, p. 265
[102] *Binda, P.H.*, Congress Papers PCE 2000, Amsterdam, Jan. 2000, p. 377
[103] *Binda, P.H.*, Congress Papers PCE 2002, Nuremberg, Jan. 2002, p. 125
[104] *Binda, P.H.*, Verfkroniek, 72, (9), 12, 1999
[105] Fraunhofer IPA, Technical Information 600/59e, 1998 "Kompstattechnology"
[106] *Kakossy, V. von*, Polymers Paint Colour J., 179, (4236), 266, 1989
[107] *Floyd, R.*, Polymers Paint Colour Jornal, 179, (4236), 270, 1989
[108] Gema Volstatic, Brochures "Magic Cylinder" and "Magic Plus" (Easy Colour Change in 15 Minutes)
[109] Wagner Powder News, The FBC plastic booth in multi-colour operations, 11/97
[110] Nordson Corp., Technical information from the PCE 2002 Exhibition, Nuremberg, and from internet "Automatic Systems Booths", 2003
[111] Nordson Corp., in Oppervlakte Technieken VOM, 47, (04), 5, 2003
[112] Eisenmann, Brochure "Pulverbeschichtung"
[113] *Bryan, B.*, PCI's Powder Coating 2000 Conf., Indianapolis, Sept. 2000
[114] *Lin, C.T.*, JCT, Journal of Coatings Technology, 5, 3, 48-55, 2008
[114] *Wohlgenannt, M.*, JOT, Journal für Oberflächentechnik, 47, 3, 44-46, 2007
[115] Valspar Sourcing, WO 2005080511, 2005
[116] *Linak, E., Kishi, A., Yang, V.*, Thermosetting Powder Coatings, SRI CONSULTING, 2008
[117] *Misev, T.*, Course on Polymer Coatings: Alternative to Solvent Based Coatings, CEI/Elsevier, Davos, September 1988
[118] *Kakossy, V. von*, The Powder Circuit and its Problem in Practical Application, Powder Coating Seminar, Lucerne, October 1988
[119] *Herrmann, T.*, JOT, Journal fuer Oberflaechentechnik, 44(5), 20-24, 2004
[120] *Herrmann, T.; Glatt-Jugel, A.*, JOT, Journal fuer Oberflaechentechnik, 44(6), 28-32, 2004
[121] a) *Herrmann, T.; Witter, M.*, JOT, Journal fuer Oberflaechentechnik, 45(2), 34-39, 2005
b) *Herrmann, T.; Ringling, C. A.*, JOT, Journal fuer Oberflaechentechnik, 51(5), 32-35, 2011
[122] *Crompton, C.*, European Coatings Journal, (4), 56-57, 2006
[123] *Horber, M.*, JOT, Journal für Oberflächentechnik, 47, 2, 14-18, 2007
[124] *Zollitsch, C.*, JOT, Journal fuer Oberflaechentechnik, 50(8), 42-43, 2010

7 Future developments

7.1 General trends of the powder coating market

"It is difficult to make predictions, especially about the future" said *Mark Twain*. As is usually done in such cases, some of the trends of today will be extrapolated to the future in order to forecast the opportunities of tomorrow. Opportunities always arise where technical developments (tech push) meet market requirements (market pull). Environmental compliances and governmental regulations play a catalytic role in these developments. This will be taken as a guideline in this short look to the possible future of powder coatings.

The worldwide annual growth rate for powder coating has been around 9 % from 2000 to 2010. Although this value was somewhat lower in the year of the financial crisis 2009 it has recovered rapidly in the last two years. In particular the demand in the Asia Pacific (China, India) region is responsible for this development. This region is forecasted to continue to show growth rates close to double-digit numbers in the coming years, whereas growth in North America and Europe will be more on the order of 2 to 3 %.

We will start with some general factors which will have a substantial influence on the powder coating market:

Penetration

The market share of powder coatings in industrial applications has reached about 18 % in Europe and 15 % in North America and Japan, which seems to be an asymptotic value. Nevertheless in emerging regions (e.g. East Europe, South America, Middle East) this number is much lower (5 to 8 %), so an above average growth can be expected in these areas.

Consolidation and relocation

The overcapacity in Europe and North America leads to closing down of production sites in the western part of the world and to a shift in the Asia Pacific section. Just recently (in 2011) the world largest powder coating producer Akzo Nobel announced the opening of the sixth plant in Wuhan (China), in line with the plan to double revenue in this region in the next five years.

Detoxification

Hydroxyalkylamides (HAA) have already substituted TGIC in Europe in the last couple of years due to toxicological reasons. Although TGIC is still market leader worldwide for exterior powder coatings, the same trend starts to be observed now even in China. Governmental regulations like REACH will lead in the long run to a decline of the use of compounds that are regarded as harmful. Up to now, polymers are exempted from these regulations. Since most of the components in powder coatings are polymers, powders have a natural advantage over other environmental compliant coating systems like liquid UV formulations containing monomeric reactive diluents.

Customization

Not only smaller regional powder coating producers but also the largest global players offer small amounts (down to 20 kg) of tailor-made powder coating formulations to meet the

Emmanouil Spyrou: Powder Coatings: Chemistry and Technology

ISBN: 978-3-86630-884-8

requirements of smaller job coaters. A huge variety of colors, textures and special effects are available nowadays, often for next day delivery. This flexibility will help to reach the maximum possible penetration of powder coatings in industrial applications.

Commodification

Due to the overcapacity in western countries with matured markets, little differentiation and "squeezed between two formidable opposing forces - petrochemicals and mass retail" (*Frauman*, Rohm and Haas, 2006), powder coating raw materials, formulations and, to a minor extent, job coaters are under considerable price pressure. One way to overcome the dependence on the oil industry is to use biobased raw materials. Although this is definitely a promising long run option, the influence on the powder coatings market will be insignificant in short term.

Automatization

Fully automated powder coating lines find a remedy for one of the major cost factors in the powder industry: down time. This automatization, which has been observed for the first time in Italy, will play a model role for other companies, in particular for larger producers.

Optimization

Lowering the film thickness of powder coatings might offer the highest cost saving potential that can be tapped short time. For this the particle size of the powder formulation must be decreased substantially (at least $<40\ \mu m$). The required grinding and classifying techniques for this task are already available. A narrow particle size distribution and a spherical shape (not easily achieved by conventional techniques) of particles lead to an even better flow and decrease the time for cleaning. Additives and an appropriate particle surface treatment can take care of the necessary fluidization of the fine powder.

7.2 Special effects

Besides the general trends there are many influencing factors on the market, which concern only individual powder coating systems, processes, applications or regions.

Driving forces

No major changes are expected in general industrial, domestic appliances, functional coatings, metal furniture and architectural applications, which are the major fields that consume almost 85 % of the powder coatings production. The **replacement of solvents** has been one of the main driving forces for increasing powder coating consumption in many of these application areas and this has already been completed.

LTC and UV powder coatings

Low temperature and radiation curable powder coatings, which are used predominantly for MDF and plastics, account for 0.5 to 1 % of the total powder market. **Low temperature curable** (LTC) powder coatings (e.g. catalyzed blocking agent free polyurethane systems or accelerated TGIC based formulations) have some restrictions concerning storage stability and leveling; though still they show considerable growth rates. In addition these highly reactive LTC powder coatings can be used at conventional temperatures just minimizing the curing time and thus increasing the productivity. Although **UV powder** coatings have distinctive advantages - short processing times (melting, coalescence and flow in 2 min by IR and curing in some seconds by UV), superior film flow (due to the separation of melting and curing), minimal thermal stress to the MDF substrate, and a reduced space requirement for the process – they have been up to now accepted only sluggishly in market. 25 % higher raw material costs and the required change of curing equipment have been some of the reasons for their lack of adoption. Since the usage of both technologies is quite low at the moment, no significant impact on the total powder coating market is to be expected in the near future. In the long run, of course, these techniques might lead to fundamental transformations in the powder coating industry.

High line speed application

Speed application techniques like the "Cloud Chamber" or the "Electromagnetic Brush" (EMB) allow line speeds of over 100 m/min and thus are predestined for **powder coil coating** applications. Up to date these novel technologies have been used only in few production sites. Still the penetration of powder in coil coatings is roughly one percent, and most of those plants are running in a line speed of 10 to 20 m/min. History has shown that the conservative coil coating industry implements changes in a remarkably slow pace, in contrary to the high line speeds.

Pre-coating techniques

The technique of **pre-coating metal** (PCM) sheets with powder coatings in a continuous way, which was developed in the Far East, will certainly spread around the world. However, it is very difficult to predict to what degree the PCM powder coatings will penetrate the market.

Architectural applications

The use of powder coatings in architectural applications has been well established since the 1970's. In Europe about 85 % of all aluminum claddings and panels are now powder coated. A similar trend is developing in Asia, particularly in China, where it is the preferred finish for architectural aluminum. In the USA this application has lagged, but the need to replace aging plants and the pressures on VOCs will cause the industry to use more powder coatings. Moreover, with the introduction of super-durable powder coatings, designed for higher UV and humidity regions, growth of the use of architectural powder coatings is another reason. **Super-durable powder coatings** now offer the building industry a cost-effective alternative to PVDF powders which have an extremely long service life. Therefore, super-durable polyester powder coatings are now being preferred for protection and decoration of precious buildings.

Coatings for PC

Powder coatings are now being widely used in the **IT industry** and most computer server cabinets are now powder coated. Many IT OEMs already specify powders as decorative coatings on the metal parts of their equipment.

Rebar coating plants

The premature corrosion of reinforcing steel initiated a wide acceptance of the so-called FBE (**fusion bonded epoxy**) powder coatings for concrete reinforcement bars ("rebars") in the USA. Excellent experience from North America for the last forty years will certainly support the tendency in Europe and the Far East to install high-speed fully automated rebar coating plants.

Protection of pipelines

FBE powder coatings have also played their role in the protection of pipelines for more than forty years, especially in the United States. They will continue to do so, as many of the existing pipelines are reaching the limit of their life expectancy and should be replaced. In addition there exists an increasing demand in the western countries to be directly fed by oil and gas via new pipelines from the Middle East and Russia. That means there is a great potential for pipe coating powders in the near future.

Primer and clear topcoats in OEMs

Automotive primer-surfacer and topcoats account for a substantial share of powder coatings in this field. The combination of the high utilization rate and the elimination of paint sludge (collected in the spraying booths with wet paints) make powder primer-surfacers very interesting paints from an economical point of view and can be seen as one of the major driving forces for a breakthrough in the automotive industry. The automotive acrylic powder clear topcoat, another subject that does not need color change, is also adopted by the car industry and is already commercially established in Europe (e.g. BMW, Volvo, Peugeot, Ferrari) but is not yet fully accepted in the United States. Though progress has been made, further improvements of the application equipment are required in order to solve the problem of the still relatively slow color change. In the short term it is therefore not likely that automotive (metallic) base-coat powders will enter the car industry.

Epoxy coatings for functional coatings

The use of bisphenol-A epoxy resins in functional pure epoxy powder coatings for pipeline protection will follow the normal growth without spectacular changes. **Epoxy novolac** resins could be expected to find broader application in functional coatings since they offer increased hardness, chemical resistance and toughness to powder coatings. Epoxy novolacs with higher glass transition temperature could be a potential solution for formulating powders for electrical insulation, where resistance of the coating towards high service temperature is a required property.

Metallic effect powder coatings

Metallic powder coatings once lacked the brilliance of liquid paints. However, recent developments in specialized metallic pigments seem to have solved this problem. These pigments are based on coated aluminum flakes (e.g. with SiO_2 and an organic layer), which are grafted with suitable prepolymers, enabling them to be easily embedded in the binder system. Thus, they participate fully in the overall film forming process during stoving, providing the desired brilliance. Metallic effect coatings have shown remarkable above average annual growth values in the last couple of years.

Tribo application technique

The use of the tribo application technique will certainly grow in the near future, due to the developing markets for the tribo-friendly HAA powders and the powder coating of wood (MDF) by UV or thermal cure.

Mechanical resistance

Scratch and mar resistance of powder coatings for both interior and exterior applications generally need to be improved. Promising results have been reported, using nano-particles in powder coatings (or nano-particle fillers). The next generation of OEM powder topcoats will have to compete with the state of the art scratch resistant liquid counterparts, like “CeramiClear” (PPG) or “iGloss” (BASF).

Functional powder coatings

First examples of powder based easy-to-clean, soft feel, FDA-compliant or antimicrobial coatings and substitutes for electroplating or anodizing have been recently observed in the literature. Higher market penetration of these **specialized powder coatings** is soon to be expected.

Matt powder coatings

The appearance of (semi-) **matt powder coatings** (which comprise 40 % of all powder applications) often suffers from manufacturing inconsistencies, problems during application, curing temperature fluctuations and inconsistent film performance, when using the conventional dryblending of two different powders. Chemical matting with a "one-shot" extrusion process led to a great improvement of the smoothness of (semi-) matte films. In particular, a one-shot hardener for the booming HAA-based powder coatings will find a substantial interest in the market.

Natural wood and powder coil coating

The biggest mistakes in forecasting the future come mostly from breakthroughs which are rather unpredictable, so called "**disruptive technologies**". Two of the biggest market potentials for powders are coatings of natural wood and powder coil coating. A major penetration in either of these technologies would open up a huge market, blowing every conservative prediction. Although appropriate prototype powder coating formulations and application processes for both of these promising applications have already been available for years, the implementation is more likely to proceed slowly than suddenly.

7.3 Conclusions

Powder coatings have been widely accepted by the market and recognized as one of the few systems which still enjoy growth in the steady-state market pattern of the coating industry. Even in 2008 and 2009 during the serious financial crisis, the booming tendency of powder coatings was unhindered in its growth. The developments in the 1970s and 1980s erased the question *to be or not to be.* The task is now - *to boldly go where no powder coating has gone before,* to find new applications and new markets.

This Chapter 7 has tried to indicate some possible directions. No matter how accurate these predictions are, it is obvious that powder coatings have a bright future. Complicated technical problems which have to be solved before being able to meet the sophisticated requirements of new market segments should be regarded as a challenge rather than as an obstacle.

8 Appendix

8.1 List of abbreviations

ABS	acrylonitrile butadiene styrene polymer
AHK	α-hydroxy ketone
APS	aqueous powder slurries
ASTM	American society for testing and materials
ATR	attenuated total reflection
ATRP	atomic transfer polymerization radical polymerization
AV	acid value
BAPO	bis-acylphosphine oxides
BMC	bulk molding compounds
B-stage	intermediate stage in the curing process
CAB	cellulose acetobutyrate
CHDA	cyclohexyl dicarboxylic acid
CPCSP	continuous powder coating process
CPVC	critical pigment volume concentration
C-stage	final stage in the curing process
DABCO	1,4-diazabicyclo [2.2.2] octane
DBTDL	dibutyltin dilaurate
DBTDLM	dibutyltin dilauryl mercaptide
DBTM	dibutyltin maleate
DBTO	dibutyltin oxide
DDDA	dodecanedioic acid
DEA	dielectric analysis
DGEBA	diglycidylether bisphenol A
DGT	diglycidyl terephthalate
DICY	dicyandiamide
DMA	dynamic mechanical analysis
DMTDC	dimethyltin dichloride
DOI	distinctness of image
DOP	dioctyl phthalate
DPC	differential photo calorimetry
DSC	differential scanning calorimetry
DTA	differential thermal analysis
EBC	electron beam curing (EB)
ECHA	European Chemicals Agency
ECS	European Coatings Show
ECTFE	ethylene chlorotrifluorothylene
EEW	epoxy equivalent weight
EGBM	ethylene glycol bistrimellitate
EMB	electro magnetic brush

Emmanouil Spyrou: Powder Coatings: Chemistry and Technology

ISBN: 978-3-86630-884-8

EMM	epoxy molar mass
ENCAT	epoxy nucleophile catalyzed transesterification
ESTA	high speed rotary bell atomizers
FDA	food and drug administration
FTIR	Fourier transform infra red spectrometry
GMA	glycidyl methacrylate (resins)
GPC	gel permeation chromatography
$H_{12}MDI$	dicyclohexylmethylene diisocyanate
HAA	β-hydroxyalkyl amide
HALS	hindered amine light stabilizer
HAP	α-hydroxy acetophenone
HBl	blocking agent
HBPA	hydrogenated bisphenol A
HDI	hexamethylene diisocyanate
HDT	hexamethylene diisocyanate trimer
HMF	heavy metal free
HMMM	hexamethoxymethyl melamine
HPHP	hydroxypivalyl hydroxypivalate
HPLC	high performance liquid chromatography
HVOF	high velocity oxygen fuel
HVTS	high velocity thermal spray
Hybrid	epoxy polyester powder coating
ICTA	International confederation of thermal analysis
IPA	isophthalic acid
IPDI	isophorone diisocyanate
IR	infra red
LC	liquid chromatography
LD50	lethal dose, 50%
LEPC	low emission paint consortium
LTC	low temperature cure
MA	maleic acid containing resin
MB	masterbatch
MDF	medium density fiber board
MDI	diphenylmethane diisocyanate
MDSC	modulated differential scanning calorimetry
MEK	methyl ethyl ketone
NCA	notched coating adhesion
NCO	isocyanate
NIR	near infra red
NMR	nuclear magnetic resonance
NPG	neopentyl glycol
OTB	ortho tolyl bisguanidine
P2VP-β-PCL	poly (2-vinylpyridine)-β-poly (ε-caprolactone) copolymers
PA12	polylauryllactame
PA6	polycaprolactame
PA66	polyamide-66

PAS	photo acoustic spectroscopy
PBT	polybutylene terephthalate
PBTS	persistent, bioaccumulative and toxic substances
PC	polycarbonate
PDO	1,3-propane diol
PE	polyethylene
PEEK	polyether ether ketones
PEI	polyether imide
PEO	polyethylene oxide
PES	polyester
PET	polyethylene terephthalate
PGSS	particles from gas saturated solutions
PMDA	pyromellitic dianhydride
PMMA	polymethyl methacrylate
POM	polyoxymethylene
PPORD	product and process orientated research and development
PS	polystyrene
PSD	particle size distribution
PTFE	polytetrafluoroethylene
PUR	polyurethane
pvc	pigment volume concentration
PVC	polyvinylchloride
PVDF	polyvinyl difluoride
QUV	accelerated weathering by means of UV-light irradiation
REACH	registration, evaluation, authorization and restriction of chemicals
SAN	styrene acrylonitrile polymer
SAM	self assembling molecules
SBP	sodium benzene phosphinate
SMA	styrene maleic anhydride
SMC	sheet molding compounds
SnOct	tin dioctoate
SVHC	substances of very high concern
TACT	trisalkoxycarbonyl amino triazine
TBA	torsional braid analysis
TBAC	tetrabutyl ammonium chloride
tBIA	5-tert.-butyl isophthalic acid
TBPA	tetrabutylphosphonium acetate
TBPBr	tetrabromo phosphonium bromide
TDI	toluene diisocyanate
T_g	glass transition temperature
TGA	thermogravimetric analysis
TGIC	triglycidyl isocyanurate
TMA	thermo mechanical analysis
TMA	trimellitic anhydride
TML	triglycidyl trimellitate
TMMGU	tetramethyl methoxyglycouril
TMP	trimethylol propane
TMXDI	tetramethylxylene diisocyanate
ToF-SIMS	time-of-flight secondary ion mass spectrometry

TPA	terephthalic acid
Tribo	charging of powder particles by friction
UV	ultra violet
VAMP	VEDOC advanced manufacturing process
vPvB	very persistent, very bioaccumulative substances
VE	vinyl ether containing resin
VOC	volatile organic compounds
WPE	weight per epoxide
Z-blade	mixing blades formed like a “Z”

8.2 Weblinks

American coatings association (NPCA), U.S.	www.paint.org
American society for Testing and Materials, USA	www.astm.org
Association for UV-/EB-Curing Technology, NL	www.radtech-europe.com
British Coatings Federation (BCF), UK	www.coatings.org.uk
Coatings Research Institute, B	www.cori-coatings.be
Coatings World, U.S.	www.coatingsworld.com
Deutsche Forschungsges. für Oberflächenbehandlung e.V., D	www.dfo-online.de
Deutsches Lackinstitut (Vdl), D	www.lacke-und-farben.de
ECCA, Brussels, B	www.eccacoil.com
ECHA homepage (REACH)	www.echa.europa.eu
Elsevier Science Publisher, U.S., Progress in Organic Coatings	www.elsevier.com
Europäische Gesellschaft für Lackiertechnik e.V., D	www.egl-online.de
European Committee for Standardization	www.cen.eu
Europ. Verein. der Lack-, Druckfarb. u. Künstlerf.ind. (CEPE), B	www.cepe.org
European Coatings Conference, Vincentz Network, D	www.coatings.de
European Coil Coating Association, D	www.ecca.de
Fachhochschule für Technik, Esslingen, D	www.fht-esslingen.de
Fédération des ind. des peintures, et adhésifs (FIPEC), F	www.fipec.org
Federation of Societies for Coatings Techn., PA, U.S.	www.coatingstech.org
Finishing industry, U.S.	www.finishing.com
Forschungsinstitut für Pigmente und Lacke, Uni Stuttgart, D	www.fpl.uni-stuttgart.de
Fraunhofer Inst. für Schicht- und Oberfl.technik, D	www.ist.fhg.de
GDCh-Tagungen, D	www.gdch.de/tagungen
Informationen zu Lackprodukten und Lieferanten, D	www.chemcompass.de
Information platform of JPCL, PCE, Paintsquare Inc, U.S.	www.protectivecoatings.com
Institut für Lacke und Farben, D	www.lackinstitut.de
Institut für Polymerforschung, D www.ipfdd.de	
Journal of Coatings Technology, FSCT, PA, U.S	www.coatingstech.org
Paint and Coatings Resource Center (EPA), U.S.	www.paintcenter.org
Paint Research Association, UK	www.pra-world.com
Phänomen Farbe, Verlag Albert, D	www.colour-europe.de, www.phaenomen-farbe.de

Powder Coating Institute (PCI), U.S.	www.powdercoating.org
Qualitätsgemeinschaft f. die Stückbeschichtung von Bauteilen, D	www.gsb-international.de
Qualicoat, quality control in the coatings industry	www.qualicoat.net/
Rodman Publishing, Ramsey, NJ, U.S.	www.specialchem4coatings.com
Verband der deutschen Lackindustrie, D	www.lackverband.de
Verband der Ingenieure des Lack- und Farbfaches e.V., D	www.vilf.de
Vereinigung der Farben- und Druckerfarbenherst. (VVVF), NL	www.vvvf.nl
Vieweg & Sohn Verlagsges. mbH, D	www.springer-vieweg.de, www.jot-oberflaeche.de
Vincentz-Network, D, European Coatings Journal, Farbe und Lack, besser lackieren!	www.european-coatings.com, www.farbeundlack.de, www.lackiernetz.de
Wiley-VCH Verlagsgesellschaft, D	www.wiley-vch.de

www.youtube.com (a short selection of catchword)

Audi 4 B8 Black Powder Coat Wheels
Brooker`s Powder Coating
Cube Bikes Powder Coatings
DIY Powder Coating
How rims are being coatet
King coat powder coating gun MP4
Long Life of E-Z Pack Powder Coat Paint System
Powder Coating Aluminum
Powder Coating Engine Covers/Part 1 and 2
Powder Coating Plant Equipment For Aluminum Profile
Powder Coating with MDB
Splatter Powder Coat by Prismatic Powders
UV-Cured Coating Process DVUV and Keyl and Polymer
UV Floor Coating RapidShield
UV Rollercoater

Author

Emmanouil Spyrou received his PhD in organic chemistry in 1993 at the WWU Münster (Germany). After a post-doctoral stint at the University of Los Angeles he joined Evonik Industries (former Hüls AG) in 1995 in the department of research and development of coating raw materials. Since 2003 he is responsible for the innovation management of powder coatings and has worked on various aspects of polyurethanes, ß-hydroxyalkyl amides, epoxies and radiation curing. He is author of more than 100 publications, lectures and patents in the field of coatings.

Index

Symbols

A

B

E

F

J

K

L

M

N

O

P

Q

R

S

T

www.ingramcontent.com/pod-product-compliance
Lightning Source LLC
Chambersburg PA
CBHW080933240926
51849CB00003BA/450
* 9 7 8 3 8 6 6 3 0 8 8 4 8 *